Boundary Value Problems for Operator Differential Equations

Mathematics and Its Applications (*Soviet Series*)

Volume 48

Boundary Value Problems for Operator Differential Equations

by

V. I. Gorbachuk

and

M. L. Gorbachuk

Institute of Mathematics,
Academy of Sciences of the Ukrainian SSR,
Kiev, U.S.S.R.

KLUWER ACADEMIC PUBLISHERS

DORDRECHT / BOSTON / LONDON

Library of Congress Cataloging-in-Publication Data

CIP

Gorbachuk, V. I. (Valentina Ivanovna)
[Granichnye zadachi dli͡a different͡sial'no-operatornykh uravneniĭ. English]
Boundary value problems for operator differential equations / by V.I. Gorbachuk and M.L. Gorbachuk.
p. cm. -- (Mathematics and its applications (Soviet series))
Translation of: Granichnye zadachi dli͡a different͡sial'no -operatornykh uravneniĭ.
Includes bibliography and index.
ISBN 0-7923-0381-4
1. Boundary value problems. 2. Differential equations, Partial. 3. Operator equations. I. Gorbachuk, M. L. (Miroslav L'vovich) II. Title. III. Series: Mathematics and its applications (Kluwer Academic Publishers). Soviet series.
QA379.G6713 1990
515'.35--dc20 89-15471

ISBN 0-7923-0381-4

Published by Kluwer Academic Publishers,
P.O. Box 17, 3300 AA Dordrecht, The Netherlands

Kluwer Academic Publishers incorporates
the publishing programmes of
D. Reidel, Martinus Nijhoff, Dr W. Junk and MTP Press.

Sold and distributed in the U.S.A. and Canada
by Kluwer Academic Publishers,
101 Philip Drive, Norwell, MA 02061, U.S.A.

In all other countries, sold and distributed
by Kluwer Academic Publishers Group,
P.O. Box 322, 3300 AH Dordrecht, The Netherlands

Printed on acid-free paper

This is the expanded and revised translation of the original work
Granichnye zadachi dlya differentsial'no-operatornykh uravneniĭ
Published by Naukova Dumka, Kiev, ©1984

Printed in the Netherlands

SERIES EDITOR'S PREFACE

'Et moi, ..., si j'avait su comment en revenir,
je n'y serais point allé.'

Jules Verne

The series is divergent; therefore we may be
able to do something with it.

O. Heaviside

One service mathematics has rendered the
human race. It has put common sense back
where it belongs, on the topmost shelf next
to the dusty canister labelled 'discarded non-
sense'.

Eric T. Bell

Mathematics is a tool for thought. A highly necessary tool in a world where both feedback and non-linearities abound. Similarly, all kinds of parts of mathematics serve as tools for other parts and for other sciences.

Applying a simple rewriting rule to the quote on the right above one finds such statements as: 'One service topology has rendered mathematical physics ...'; 'One service logic has rendered computer science ...'; 'One service category theory has rendered mathematics ...'. All arguably true. And all statements obtainable this way form part of the raison d'être of this series.

This series, *Mathematics and Its Applications*, started in 1977. Now that over one hundred volumes have appeared it seems opportune to reexamine its scope. At the time I wrote

> "Growing specialization and diversification have brought a host of monographs and textbooks on increasingly specialized topics. However, the 'tree' of knowledge of mathematics and related fields does not grow only by putting forth new branches. It also happens, quite often in fact, that branches which were thought to be completely disparate are suddenly seen to be related. Further, the kind and level of sophistication of mathematics applied in various sciences has changed drastically in recent years: measure theory is used (non-trivially) in regional and theoretical economics; algebraic geometry interacts with physics; the Minkowsky lemma, coding theory and the structure of water meet one another in packing and covering theory; quantum fields, crystal defects and mathematical programming profit from homotopy theory; Lie algebras are relevant to filtering; and prediction and electrical engineering can use Stein spaces. And in addition to this there are such new emerging subdisciplines as 'experimental mathematics', 'CFD', 'completely integrable systems', 'chaos, synergetics and large-scale order', which are almost impossible to fit into the existing classification schemes. They draw upon widely different sections of mathematics."

By and large, all this still applies today. It is still true that at first sight mathematics seems rather fragmented and that to find, see, and exploit the deeper underlying interrelations more effort is needed and so are books that can help mathematicians and scientists do so. Accordingly MIA will continue to try to make such books available.

If anything, the description I gave in 1977 is now an understatement. To the examples of interaction areas one should add string theory where Riemann surfaces, algebraic geometry, modular functions, knots, quantum field theory, Kac-Moody algebras, monstrous moonshine (and more) all come together. And to the examples of things which can be usefully applied let me add the topic 'finite geometry'; a combination of words which sounds like it might not even exist, let alone be applicable. And yet it is being applied: to statistics via designs, to radar/sonar detection arrays (via finite projective planes), and to bus connections of VLSI chips (via difference sets). There seems to be no part of (so-called pure) mathematics that is not in immediate danger of being applied. And, accordingly, the applied mathematician needs to be aware of much more. Besides analysis and numerics, the traditional workhorses, he may need all kinds of combinatorics, algebra, probability, and so on.

In addition, the applied scientist needs to cope increasingly with the nonlinear world and the extra mathematical sophistication that this requires. For that is where the rewards are. Linear models are honest and a bit sad and depressing: proportional efforts and results. It is in the non-linear world that infinitesimal inputs may result in macroscopic outputs (or vice versa). To appreciate what I am hinting at: if electronics were linear we would have no fun with transistors and computers; we would have no TV; in fact you would not be reading these lines.

There is also no safety in ignoring such outlandish things as nonstandard analysis, superspace and anticommuting integration, p-adic and ultrametric space. All three have applications in both electrical engineering and physics. Once, complex numbers were equally outlandish, but they frequently proved the shortest path between 'real' results. Similarly, the first two topics named have already provided a number of 'wormhole' paths. There is no telling where all this is leading - fortunately.

Thus the original scope of the series, which for various (sound) reasons now comprises five subseries: white (Japan), yellow (China), red (USSR), blue (Eastern Europe), and green (everything else), still applies. It has been enlarged a bit to include books treating of the tools from one subdiscipline which are used in others. Thus the series still aims at books dealing with:

- a central concept which plays an important role in several different mathematical and/or scientific specialization areas;
- new applications of the results and ideas from one area of scientific endeavour into another;
- influences which the results, problems and concepts of one field of enquiry have, and have had, on the development of another.

A differential equation of the form $y''(t)+A(t)y(t)=0$ looks very familiar and certainly a great many volumes have been written about the corresponding boundary-value problems. In this book, however, the equation above is an **operator** equation, and that makes it unique. The spectral analysis of Sturm-Liouville differential operator equations in the case of an infinite-dimensional space, began very recently (in spite of the many potential applications) and this book, by two well known researchers in the area, aims to present the subject systematically together with its natural links to the important area of extensions of symmetric operators.

The shortest path between two truths in the real domain passes through the complex domain.

J. Hadamard

La physique ne nous donne pas seulement l'occasion de résoudre des problèmes ... elle nous fait pressentir la solution.

H. Poincaré

Never lend books, for no one ever returns them; the only books I have in my library are books that other folk have lent me.

Anatole France

The function of an expert is not to be more right than other people, but to be wrong for more sophisticated reasons.

David Butler

Bussum, January 1990

Michiel Hazewinkel

Contents

Preface

The book deals with the theory of boundary value problems for second-order operator differential equations of the form

$$y''(t) + A(t)y(t) = 0 \qquad (t \in [a,b], \quad -\infty < a < b < \infty),$$

where the $A(t)$ are semi-bounded self-adjoint operators on a separable Hilbert space $\mathfrak{H}$. The study of differential equations whose coefficients are unbounded operators on a Hilbert or Banach space is useful not only because these include many partial differential equations but also because it offers the possibility of looking at ordinary as well as partial differential operators from a single viewpoint.

The studies of the last 30 years have enriched the theory of operator differential equations with significant results. The presentation of the Gauchy problem and the stability theory of solutions can be found both in textbooks on the theory of operators (Hille–Phillips [1], for example) and in special monographs (see, for example, Lions [1], S. Krein [1], Daletsky–M. Krein [1]). The spectral analysis of the Sturm–Liouville operator differential equation, which was given a lot of attention in the scalar case and in the case of a finite-dimensional $\mathfrak{H}$, began its development quite recently in the case of an infinite-dimensional space and an unbounded operator potential $A(t)$. Naturally, then, there are no books which reflect on this trend. In this book we would like to fill this gap, if not completely, then at least partially.

For the scalar Sturm–Liouville equations one usually considers two cases, that of a bounded and that of an unbounded interval, i.e. the regular and the singular case. They are known to differ as regards formulation of problems, methods of investigation, and fields of applications.

When studying operator equations one must take into account not only boundedness or unboundedness of the interval, but also the character of unboundedness of the potential. The fact whether the operators $A(t)$ are lower or upper semi-bounded proved to be fundamentally important. In this connection, the equations are divided into elliptic ($A(t) \leq 0$) and hyperbolic ($A(t) \geq 0$). The Laplace equation and the D'Alembert equation serve as respective models for these. In view of the limited volume of the book and the unlimited stream of results we mainly consider the case of a bounded interval and present the theory of dissipative (in particular, self-adjoint) boundary value problems. It is quite natural that while selecting material the authors' personal interests somewhat prevailed.

In the first chapter we give basic definitions and (almost without proofs) classical theorems from the theory of Banach, Hilbert, and locally convex topological spaces, and from the theory of linear operators on them. This information is necessary for understanding the subsequent chapters. Since the principal object of study of this book is a vector-function with values in infinite-dimensional spaces, while the major instrument of investigation is the operational calculus of self-adjoint operators, these are given greater attention.

The second chapter deals with the theory of boundary values of solutions of second-order elliptic operator differential equations which are smooth inside the interval. On the one hand, this theory plays an important role in the formulation and investigation of boundary value problems for such equations; on the other hand, it gives a uniform approach to the theory of boundary values of analytic functions, allowing one to obtain, in particular, well-known theorems concerning existence of boundary values of harmonic (analytic) functions $u(x,t)$ in the upper half-plane, in the Schwartz space of distributions if $u(x,t)$ has power growth as t approaches the real axis and in the spaces of ultradistributions if $u(x,t)$ has exponential growth as t approaches the real axis. This theory also makes it possible to establish analogous results for solutions of homogeneous partial differential equations different from the Laplace equation which are smooth inside the domain.

The proofs of the principal results are based on the spectral representation theorem for a self-adjoint operator on a Hilbert space. Also, chains of spaces with positive and negative norms and their inductive and projective limits are essentially used. Their theory is set forth in sufficient detail.

The third chapter consists in fact of two parts. The first part is devoted to the theory of extensions of abstract symmetric operators. Its presentation somewhat differs from the traditional one and is adapted to the theory of boundary value problems. The description of various classes of extensions (maximal dissipative, self-adjoint, solvable and others), as well as the structure of the spectrum of extensions from these classes, is given in terms of so-called boundary value spaces. The latter are convenient and natural because they turn into the usual boundary condition in certain concrete situations. Here, an important place is occupied by theorems about various representations of binary relations in a Hilbert space. These are the starting point in constructing the theory of extensions.

In the second part this theory is applied to investigating boundary value problems for the formally self-adjoint Sturm–Liouville expression with operator potential of hyperbolic type given on a bounded interval. The minimal operator generated by it is symmetric and has infinite deficiency numbers when $\dim \mathfrak{H} = \infty$. Each extension of it is associated with some boundary value problem in the sense that vector-functions in the domain of the extension satisfy a definite boundary condition at the ends of the interval. Therefore, a lot of properties of extensions (self-adjointness, maximal dissipativeness, structure of the spectrum, etc.) can be completely described in terms of the coefficients of the equation and the boundary conditions corresponding to these extensions.

The fourth chapter contains results concerning the spectral theory of boundary value problems in the elliptic case. The various classes of dissipative problems are described in it, and the asymptotic distribution of their eigenvalues is studied. Particular attention is paid to self-adjoint boundary value problems with discrete spectrum. The behaviour of the distribution function of the eigenvalues of such a problem depends essentially on the boundary conditions. Classes of self-adjoint boundary value problems for which the dominant term in the asymptotics of the distribution function has given order of growth are singled out. For some of them the second-order terms of the asymptotics are studied, and an estimate of the remainder is given. We also establish a connection between the asymptotic behaviour of the distribution function of the eigenvalues and the smoothness up to the boundary of elements in the domain of the self-adjoint extension corresponding to the boundary value problem considered.

The fifth, and last, chapter deals with the theory of boundary values at zero of solutions of a first-order differential equation of the form $y'(t) + Ay(t) = 0$ $(t \in (0, \infty))$ in a Banach space. One of the reasons to construct such a theory is the hope to find a general approach to the well-known Riesz theorems concerning boundary values in L_p spaces $(p \neq 2)$ of analytic functions, from the viewpoint of evolution equations.

Since infinity as well as zero is a singular point for such an equation, we also discuss results related to the behaviour of solutions at ∞, which is related to stability theory.

The book provides a number of examples which prove that the operator approach makes it possible not only to extend the class of already studied partial differential equations and their boundary value problems, but also to look from another point of view to the spectral theory of self-adjoint boundary value problems for such classical expressions as those of Laplace and D'Alembert.

We will not always formulate the results in the most general form. We have preferred to select a somewhat average level of generality (a "golden mean"). The rest is added by way of comments and references.

To make reading easy, the principal statements are distinguished in the form of theorems, lemmas, corollaries, and remarks as well as formulas.

It gives us pleasure to thank M.G. Krein and Ju.M. Berezansky, whose great influence we felt throughout our scientific activities. Their work on the theory of boundary value problems and our continual contact with them determined the subject of our investigations and the subject matter of the book.

In writing the manuscript we were helped by our pupils and colleagues. Sections 1–3 (Chapter 3) were written together with A. N. Kochubei, Section 6 (Chapter 1) and Section 3 (Chapter 4) - with V. A. Mikhailets, Section 6 (Chapter 4) - with L. I. Vainerman. V. V. Gorodetsky, A. I. Kashpirovski, A. V. Knyaziuk, V. V. Levchuk, L. B. Fedorova participated in the discussion of some sections of the book. We sincerely thank all of them.

In preparing the present version we were helped very much by A. N. Kochubei and we would like to express our particular gratitude to him.

CHAPTER 1

Some Information from the Theory of Linear Operators

1. Banach Spaces and Continuous Linear Operators on Them

1.1. NORMED AND BANACH SPACES

A set $\mathfrak{X}$ is called a complex normed space if

(1) $\mathfrak{X}$ is a vector space over the field $\mathbb{C}^1$ of complex numbers;

(2) for each element $x \in \mathfrak{X}$ there is defined a non-negative number $\|x\|$ (the norm of x) possessing the following properties:

(i) $\|\alpha x\| = |\alpha|\,\|x\| \quad (\forall x \in \mathfrak{X}, \forall \alpha \in \mathbb{C}^1)$;

(ii) $\|x + y\| \leq \|x\| + \|y\| \quad (\forall x, y \in \mathfrak{X})$;

(iii) $\|x\| = 0$ if and only if $x = 0$.

A sequence $x_n \in \mathfrak{X}$ $(n \in \mathbb{N} = \{1, 2, \dots\})$ is said to converge in $\mathfrak{X}$ to an element x if $\lim_{n\to\infty} \|x_n - x\| = 0$. A sequence x_n $(n \in \mathbb{N})$ from $\mathfrak{X}$ is called fundamental if $\lim_{m,n\to\infty} \|x_n - x_m\| = 0$.

A normed space $\mathfrak{B}$ is called a Banach space if it is complete, i.e. each fundamental sequence in it has a limit. Any incomplete normed space can be completed, i.e. imbedded into a certain Banach space as a dense linear subset.

A Banach space is called separable if it contains a countable dense set. We will consider only separable Banach spaces.

If another norm $\|x\|_2$ is given in the normed space $\mathfrak{X}$ with norm $\|x\|_1$, then for convergence with respect to $\|\cdot\|_2$ of each sequence that is convergent with respect to $\|\cdot\|_1$ it is necessary and sufficient that there exists a positive constant c such that

$$\|x\|_2 \leq c\|x\|_1.$$

Two norms $\|\cdot\|_1$ and $\|\cdot\|_2$ are called topologically equivalent if convergence with respect to one of them implies convergence with respect to the other. It is clear that in a normed space $\mathfrak{X}$, two norms $\|\cdot\|_1$ and $\|\cdot\|_2$ are topologically equivalent if and only if there exist positive constants c_1, c_2 such that

$$c_1\|x\|_1 \leq \|x\|_2 \leq c_2\|x\|_1. \tag{1.1}$$

1.2. ALGEBRAS, BANACH ALGEBRAS, IDEALS

A complex linear space $\mathfrak{A}$ over the field $\mathbb{C}^1$ is called an algebra if a multiplication of its elements is introduced in such a way that the multiplication operation is distributive relative to addition, and commutes with multiplication by complex numbers. If multiplication is commutative, i.e. $xy = yx$ $(\forall x, y \in \mathfrak{A})$, then the algebra is called commutative.

If the algebra $\mathfrak{A}$ is a Banach space and multiplication is continuous with respect to each of the factors, then $\mathfrak{A}$ is called a Banach algebra. A unit element in $\mathfrak{A}$ is a vector $e \in \mathfrak{A}$ such that $xe = ex = x$ for each $x \in \mathfrak{A}$. An algebra containing a unit element is called an algebra with a unit. If an algebra has no unit it can always be complemented so that the extension contains a unit. In any Banach algebra with a unit one can replace the norm by an equivalent one for which the following properties are satisfied:

$$\|xy\| \leq \|x\|\|y\| \quad (\forall x, y \in \mathfrak{A}); \quad \|e\| = 1.$$

Therefore, it is usually assumed that a Banach algebra has a unit, and the above-mentioned properties for a norm hold.

A set I of elements from the algebra $\mathfrak{A}$ is called a (two-sided) ideal if:

(i) $x + y \in I$ for any $x, y \in I$;

(ii) $xz \in I$, $zx \in I$ for any $x \in I$, any $z \in \mathfrak{A}$.

An ideal I of $\mathfrak{A}$ is called proper if $I \neq \mathfrak{A}$. A maximal ideal is a proper ideal which is not contained in other proper ideals.

1.3. CONCRETE BANACH SPACES

The space $\mathbb{C}^n$ consists of all ordered n-tuples $\{\alpha_1, \cdots, \alpha_n\}$ of complex numbers. It is an n-dimensional Banach space with respect to the norm

$$\|\alpha\|_{\mathbb{C}^n} = \left(\sum_{i=1}^{n} |\alpha_i|^2\right)^{\frac{1}{2}}.$$

We denote by $\mathbb{R}^n$ its subset of all n-tuples of real numbers.

The space $C^k[a, b]$ is the set of all k times continuously-differentiable complex-valued functions on the closed interval $[a, b]$ of the real line $\mathbb{R}^1$. The norm of a function $f \in C^k[a, b]$ is defined by

$$\|f\|_{C^k[a,b]} = \max_{t \in [a,b]} \{|f(t)|, \ldots, |f^{(k)}(t)|\}.$$

With this norm, $C^k[a, b]$ forms a Banach space as well as a Banach algebra. Let $C^0[a, b] = C[a, b]$. It is the space of all continuous functions $f(t)$ $(a \leq t \leq b)$, and

$$\|f\|_{C[a,b]} = \max_{t \in [a,b]} |f(t)|.$$

The space $L_p(a, b)$. The elements of this space are the complex-valued functions

f on (a,b) for which

$$\|f\|_{L_p(a,b)} = \left(\int_a^b |f(t)|^p \, dt \right)^{\frac{1}{p}} < \infty \tag{1.2}$$

(functions coinciding almost everywhere are identified). $L_p(a,b)$ with the norm (1.2) is a Banach space. The space $L_p(\Omega)$, where Ω is a domain in $\mathbb{R}^n$, is similarly defined.

1.4. BOUNDED LINEAR OPERATORS ON A BANACH SPACE

Let $\mathfrak{B}_1$ and $\mathfrak{B}_2$ be Banach spaces. A mapping $A : \mathfrak{B}_1 \to \mathfrak{B}_2$ is called a linear operator if

$$A(\alpha x + \beta y) = \alpha Ax + \beta Ay \qquad (\forall \alpha, \beta \in \mathbb{C}^1).$$

The linear operator A is continuous if $x_n \to x$ in $\mathfrak{B}_1$ implies $Ax_n \to Ax$ $(n \to \infty)$ in $\mathfrak{B}_2$, and it is called bounded if

$$\|Ax\|_{\mathfrak{B}_2} \le c\|x\|_{\mathfrak{B}_1} \tag{1.3}$$

for some positive constant c. The least of the constants c in the inequality (1.3) is the norm of the operator A:

$$\|A\| = \sup_{x \in \mathfrak{B}_1} \frac{\|Ax\|_{\mathfrak{B}_2}}{\|x\|_{\mathfrak{B}_1}} = \sup_{x : \|x\|_{\mathfrak{B}_1} = 1} \|Ax\|_{\mathfrak{B}_2}. \tag{1.4}$$

It is not difficult to show that a linear operator A is continuous if and only if it is bounded.

The set of all bounded linear operators $A : \mathfrak{B}_1 \to \mathfrak{B}_2$ is denoted by $[\mathfrak{B}_1, \mathfrak{B}_2]$. This set is a Banach space with the norm (1.4) and the natural definition of addition of operators and multiplication of an operator by a number:

$$(A + B)x = Ax + Bx, \qquad (\alpha A)x = A(\alpha x) = \alpha Ax.$$

If $B : \mathfrak{B}_2 \to \mathfrak{B}_3$, then the formula

$$Cx = B(Ax) \qquad (x \in \mathfrak{B}_1)$$

defines a linear operator $C : \mathfrak{B}_1 \to \mathfrak{B}_3$, called the product of A and B. For bounded operators A, B the operator C is bounded too, and

$$\|C\| \le \|A\|\|B\|.$$

The inverse of A is the operator $A^{-1} : \mathfrak{B}_2 \to \mathfrak{B}_1$ for which

$$A^{-1}A = E_1, \qquad AA^{-1} = E_2,$$

where $E_1 : \mathfrak{B}_1 \to \mathfrak{B}_1$ and $E_2 : \mathfrak{B}_2 \to \mathfrak{B}_2$ are the identity operators on $\mathfrak{B}_1$ and $\mathfrak{B}_2$, respectively.

The set of bounded linear operators from $\mathfrak{B}$ into $\mathfrak{B}$ is denoted by $[\mathfrak{B}]$, i.e. $[\mathfrak{B}] = [\mathfrak{B}, \mathfrak{B}]$. With the above-defined multiplication of operators, $[\mathfrak{B}]$ forms a Banach algebra.

Two Banach spaces are called isomorphic if there exists a one-to-one continuous linear mapping of one space onto another which has a continuous inverse. Such a mapping is called an isomorphism. A norm-preserving isomorphism is called an isometric isomorphism.

Let us formulate some important theorems.

THEOREM 1.1. (Banach). *Assume that an operator $A \in [\mathfrak{B}_1, \mathfrak{B}_2]$ maps a Banach space $\mathfrak{B}_1$ onto a Banach space $\mathfrak{B}_2$ one-to-one. Then $A^{-1} \in [\mathfrak{B}_2, \mathfrak{B}_1]$, hence the spaces $\mathfrak{B}_1$ and $\mathfrak{B}_2$ are isomorphic.*

It follows from this theorem that if a space $\mathfrak{B}$ is Banach with respect to $\|\cdot\|_1$ and $\|\cdot\|_2$, then for the topological equivalence of these norms the realization of any one-side bound in (1.1) is sufficient.

THEOREM 1.2. (Banach–Steinhaus). *Let $\mathfrak{M}$ be the set of operators from $[\mathfrak{B}_1, \mathfrak{B}_2]$ such that*

$$\sup_{A \in \mathfrak{M}} \|Ax\| < \infty$$

for each $x \in \mathfrak{B}_1$. Then the set $\mathfrak{M}$ is bounded, that is

$$\sup_{A \in \mathfrak{M}} \|A\| < \infty.$$

This theorem is also called the uniform boundedness principle. It follows from this theorem that if a sequence of operators $A_n \in [\mathfrak{B}_1, \mathfrak{B}_2]$ converges on each element $x \in \mathfrak{B}_1$, then the inequality

$$Ax = \lim_{n \to \infty} A_n x$$

defines a continuous linear operator from $\mathfrak{B}_1$ into $\mathfrak{B}_2$.

A linear operator $A : \mathfrak{B}_1 \to \mathfrak{B}_2$ is called compact, or completely continuous, if it maps any bounded set of $\mathfrak{B}_1$ to a compact set of $\mathfrak{B}_2$. A compact operator is continuous. A linear operator is compact if it maps the unit sphere of $\mathfrak{B}_1$ into a compact set of $\mathfrak{B}_2$.

1.5. DIRECT SUMS OF SUBSPACES AND PROJECTORS

A closed linear subset of a Banach space $\mathfrak{B}$ is called a subspace of it. A Banach space is said to be the direct sum of its subspaces $\mathfrak{B}_1$ and $\mathfrak{B}_2$, which is written as

$$\mathfrak{B} = \mathfrak{B}_1 \dot{+} \mathfrak{B}_2, \tag{1.5}$$

if an arbitrary element $x \in \mathfrak{B}$ can be uniquely represented in the form

$$x = x_1 + x_2 \quad (x_i \in \mathfrak{B}_i). \tag{1.6}$$

The direct decomposition (1.5) generates two linear operators $P_1 : \mathfrak{B} \to \mathfrak{B}_1$ and $P_2 : \mathfrak{B} \to \mathfrak{B}_2$, defined as follows: if x is represented in the form (1.6), then

$$P_1 x = x_1, \quad P_2 x = x_2.$$

The following properties of the operators $P_i \quad (i = 1, 2)$ are trivial:

$$P_i^2 = P_i, \qquad P_1 + P_2 = E, \qquad P_1 P_2 = P_2 P_1 = 0$$

(E is the identity operator). Moreover, the P_i are continuous. In fact, introduce on $\mathfrak{B}$ a new norm

$$\|x\|_1 = \|x_1\| + \|x_2\|.$$

Since $x = x_1 + x_2$, $\|x\| \leq \|x_1\| + \|x_2\| = \|x\|_1$. If a sequence $x^n = x_1^n + x_2^n$ is fundamental with respect to the norm $\|\cdot\|_1$, then the sequences x_1^n and x_2^n are fundamental. So, $x_1^n \to x_1 \in \mathfrak{B}_1$, $x_2^n \to x_2 \in \mathfrak{B}_2$. Then $x^n \to x_1 + x_2 \in \mathfrak{B}$ in both norms $\|\cdot\|$ and $\|\cdot\|_1$. Therefore, $\mathfrak{B}$ is also complete under the norm $\|\cdot\|$. Hence there exists a constant $c > 0$ such that

$$\|x\|_1 \leq c\|x\|,$$

whence

$$\|P_i x\| = \|x_i\| \leq \|x\|_1 \leq c\|x\| \qquad (i = 1, 2),$$

i.e. the operators P_i are continuous.

An operator $P \in [\mathfrak{B}]$ is called a projection operator or projector if $P^2 = P$. Thus, the operators P_1, P_2, constructed above, are projectors. The converse assertion is also valid. If P_1, P_2 are projectors on $\mathfrak{B}$ and $P_1 + P_2 = E$, then they generate a direct decomposition (1.5), where $\mathfrak{B}_1 = P_1\mathfrak{B}$, $\mathfrak{B}_2 = P_2\mathfrak{B}$.

A decomposition of $\mathfrak{B}$ into a direct sum of several subspaces is defined in the natural way.

1.6. THE DIRECT SUM OF BANACH SPACES

Let $\mathfrak{B}_1, \cdots, \mathfrak{B}_n$ be Banach spaces with norms $\|\cdot\|_1, \cdots, \|\cdot\|_n$, respectively. By the direct sum

$$\mathfrak{B} = \mathfrak{B}_1 \dotplus \mathfrak{B}_2 \dotplus \cdots \dotplus \mathfrak{B}_n$$

we mean the Banach space of ordered n-tuples $x = \{x_1, x_2, \cdots, x_n\}$ $(x_i \in \mathfrak{B}_i)$ with the natural definition of addition of n-tuples and multiplication of an n-tuple by a number. The norm in $\mathfrak{B}$ is given by

$$\|x\| = \left(\sum_{i=1}^{n} \|x_i\|_i^2 \right)^{\frac{1}{2}}.$$

Note that each space $\mathfrak{B}_i$ is isometrically imbedded in $\mathfrak{B}$ by identifying the element $x_i \in \mathfrak{B}_i$ with the n-tuple $\{0, \dots, 0, x_i, 0, \dots, 0\}$. So we obtain the uniquely determined expansion

$$x = \sum_{i=1}^{n} x_i \qquad (x_i \in \mathfrak{B}_i).$$

1.7. CONTINUOUS LINEAR FUNCTIONALS. DUAL SPACES. ADJOINT OPERATORS

An arbitrary operator $f \in [\mathfrak{B}, \mathbf{C}^1]$ is called a continuous linear functional on the Banach space $\mathfrak{B}$.

Let f be a linear functional defined on a subspace $\mathfrak{M} \subset \mathfrak{B}$, i.e. a linear mapping from $\mathfrak{M}$ into $\mathbf{C}^1$. A linear functional $\tilde{f}$ defined on $\mathfrak{B}$ is called an extension of f if $\tilde{f}(x) = f(x)$ for all $x \in \mathfrak{M}$. The following theorem holds.

THEOREM 1.3 (Hahn–Banach). *Let $\mathfrak{M}$ be a subspace of $\mathfrak{B}$ and f a bounded linear functional on $\mathfrak{M}$ (i.e. $|f(x)| \leq c\|x\|$, $\forall x \in \mathfrak{M}$, c=const). The functional f can be extended to a continuous linear functional $\tilde{f}$ defined on the whole space $\mathfrak{B}$ so that*

$$\sup_{x \in \mathfrak{B}} \frac{|\tilde{f}(x)|}{\|x\|} = \|\tilde{f}\| = \|f\| = \sup_{x \in \mathfrak{M}} \frac{|f(x)|}{\|x\|}.$$

It follows from this theorem that for every element $x_0 \in \mathfrak{B}$ distinct from zero there exists a continuous linear functional $f(x)$ on $\mathfrak{B}$ with the properties

$$\|f\| = 1, \quad f(x_0) = \|x_0\|.$$

The set $\mathfrak{B}'$ of all continuous linear functionals on a Banach space $\mathfrak{B}$ is a vector space under addition of functionals and multiplication of a functional by a number, defined as the corresponding operations for bounded operators. Taking for the norm of an element $f \in \mathfrak{B}'$ the number $\|f\|$ (the norm of the functional f) the space $\mathfrak{B}'$ becomes a normed space. This space is called the dual space of the Banach space $\mathfrak{B}$. It is not difficult to check that $\mathfrak{B}'$ is complete. Thus, the dual space $\mathfrak{B}'$ of a Banach space $\mathfrak{B}$ is always a Banach space.

A sequence $\{x_n\}_{n=1}^{\infty}$ of elements of a Banach space $\mathfrak{B}$ is said to weakly converge to an element $x_0 \in \mathfrak{B}$ if $\lim_{n\to\infty} f(x_n) = f(x_0)$ for every continuous linear functional f on $\mathfrak{B}$. To contrast with weak convergence, convergence in norm is often called strong convergence. Strong convergence evidently implies weak convergence. It follows from the uniform boundedness principle that for the weak convergence of a sequence $\{x_n\}_{n=1}^{\infty}$ to an element x_0 it is necessary and sufficient that this sequence is bounded and that the relation $\lim_{n\to\infty} f(x_n) = f(x_0)$ holds on a set of linear functionals f which is dense in $\mathfrak{B}'$. For example, in the space $C[a, b]$ weak convergence of a sequence $x_n(t)$ is equivalent to uniform boundedness of this sequence and convergence at any point of the interval $[a, b]$.

A sequence of functionals $f_n \in \mathfrak{B}'$ is said to weakly converge to a functional $f_0 \in \mathfrak{B}'$ if $\lim_{n\to\infty} f_n(x) = f_0(x)$ for all $x \in \mathfrak{B}$. Weak convergence always follows from strong convergence (from convergence in norm). The converse is known to be wrong. A sequence $\{f_n\}_{n=1}^{\infty}$ converges weakly to f_0 if and only if the sequence $\|f_n\|$ is bounded and the equality $\lim_{n\to\infty} f_n(x) = f_0(x)$ holds on a set of elements x which is dense in $\mathfrak{B}$.

Suppose $A \in [\mathfrak{B}_1, \mathfrak{B}_2]$. If $g \in \mathfrak{B}_2'$, then the functional

$$f(x) = g(Ax),$$

considered on $\mathfrak{B}_1$, is linear and continuous, i.e. $f \in \mathfrak{B}_1'$; in addition

$$\|f\|_{\mathfrak{B}_1'} \leq \|g\|_{\mathfrak{B}_2'} \|A\|. \tag{1.7}$$

In this way a unique functional $f \in \mathfrak{B}_1'$ associated with $g \subset \mathfrak{B}'$ is determined. Define the operator $A^* : \mathfrak{B}_2' \to \mathfrak{B}_1'$ by $A^*g = f$. This operator is called the adjoint of A.

It can be seen from formula (1.7) that the adjoint of A is a bounded linear operator, and

$$\|A^*\| = \|A\|.$$

If $A, B \in [\mathfrak{B}_1, \mathfrak{B}_2]$, $\lambda \in \mathbb{C}^1$, then $(\lambda A)^* = \lambda A^*$, $(A + B)^* = A^* + B^*$.

2. Hilbert Spaces and Bounded Operators on Them

2.1. THE GEOMETRY OF HILBERT SPACES

Let $\mathfrak{H}$ be a linear space over the field $\mathbb{C}^1$ with elements $f, g, \ldots$. An inner product on $\mathfrak{H}$ is a scalar-valued function $(.,.)$ on $\mathfrak{H} \times \mathfrak{H}$ ($\times$ is the cartesian product sign) satisfying the following conditions (axioms):

(i) $(f, g) = \overline{(g, f)}$;
(ii) $(f_1 + f_2, g) = (f_1, g) + (f_2, g)$;
(iii) $(\lambda f, g) = \lambda(f, g) \qquad (\forall \lambda \in \mathbb{C}^1)$;
(iv) $(f, f) > 0 \quad \text{if} \quad f \neq 0$.

A linear space $\mathfrak{H}$ with an inner product is called an inner product space (or pre-Hilbert space). It is obvious that the function $(.,.)$ is antilinear with respect to the second argument, i.e. $(f, \lambda_1 g_1 + \lambda_2 g_2) = \overline{\lambda}_1 (f, g_1) + \overline{\lambda}_2 (f, g_2)$.

The Cauchy–Buniakowski inequality follows from the axioms of an inner product:

$$|(f, g)|^2 \leq (f, f)(g, g).$$

Now it is not difficult to show that $\|f\| = (f, f)^{\frac{1}{2}}$ $(f \in \mathfrak{H})$ possesses all the properties of a norm. Therefore, $\mathfrak{H}$ is also a normed space. If $\mathfrak{H}$ is complete in the sense of this norm, it is called a Hilbert space. In the sequel all Hilbert spaces are assumed to be separable.

It is sometimes necessary to consider subspaces $\mathfrak{N}$ on which a function $(.,.)$ satisfies all axioms of an inner product except for (iv) (we denote such a function by $<.,.>$). In place of axiom (iv) the following condition should hold:

(iv)' $< f, f > \geq 0$, without the stipulation that the equality sign is not possible when $f \neq 0$.

The number $< f, g >$ is called a quasi-inner product of the vectors f, g. The Cauchy–Buniakowski inequality holds also for a quasi-inner product. Therefore, the totality of elements f for which $< f, f >= 0$ is a subspace. Indeed, if $< f, f >= 0$, then for any $h \in \mathfrak{N}$,

$$|< f, h >|^2 \leq < f, f >< h, h >= 0.$$

For arbitrary f, g such that $< f, f >= 0$, $< g, g >= 0$,

$$< \alpha f + \beta g, \alpha f + \beta g >= |\alpha|^2 < f, f > + |\beta|^2 < g, g > + \alpha\overline{\beta} < f, g > + \\ + \overline{\alpha}\beta < g, f >= 0.$$

Let us denote this subspace by $\mathfrak{M}$. Identifying elements f and g when $f - g \in \mathfrak{M}$ we obtain the factor space $\hat{\mathfrak{H}} = \mathfrak{N}/\mathfrak{M}$, which is a Hilbert space if we define an inner product in it by

$$(\hat{f}, \hat{g}) = < f, g >,$$

where $\hat{f}, \hat{g} \in \mathfrak{N}/\mathfrak{M}$, $f \in \hat{f}$, $g \in \hat{g}$. As usual, this product does not depend on the choice of $f \in \hat{f}$ and $g \in \hat{g}$.

Two elements $f, g \in \mathfrak{H}$ are called orthogonal (written as $f \perp g$) if $(f, g) = 0$. Two sets $\mathfrak{M}, \mathfrak{N}$ are said to be orthogonal in $\mathfrak{H}$ (this is denoted by $\mathfrak{M} \perp \mathfrak{N}$) if every vector $f \in \mathfrak{M}$ is orthogonal to all elements $g \in \mathfrak{N}$. It is easy to verify that the set of all vectors orthogonal to a certain set $\mathfrak{M} \subset \mathfrak{H}$ forms a subspace in $\mathfrak{H}$. This subspace is called the orthogonal complement of $\mathfrak{M}$ in $\mathfrak{H}$ and is denoted by $\mathfrak{M}^\perp$. The next statement is very important.

THEOREM 2.1. *Let $\mathfrak{M}$ be a subspace of $\mathfrak{H}$. Then*

$$\mathfrak{H} = \mathfrak{M} \oplus \mathfrak{M}^\perp.$$

Here, the sign $\oplus$ denotes the direct sum of mutually orthogonal subspaces. The projectors P_1 and P_2 onto $\mathfrak{M}$ and $\mathfrak{H} \ominus \mathfrak{M} = \mathfrak{M}^\perp$, respectively, are called ortoprojectors.

A set $\mathcal{E}$ of vectors from $\mathfrak{H}$ forms a complete system if there is no non-zero vector orthogonal to all elements from $\mathcal{E}$. Thus, completeness of a system in a Hilbert space $\mathfrak{H}$ means that the closed linear span of its vectors coincides with $\mathfrak{H}$.

A system is called orthogonal if its vectors are pairwise orthogonal. A system of vectors is said to be orthonormal if it is orthogonal and if $\|e\| = 1$ for each element e of this system. For non-orthogonal systems there exists a procedure for orthonormalizing them. If e is an element of an orthonormal system, then the number (f, e) is called the Fourier coefficient of the vector f with respect to e. Any orthonormal system in a Hilbert space is finite or countable (recall that we consider only separable spaces). We will call a complete orthonormal system an orthonormal basis. In a Hilbert space there always exists an orthonormal basis.

If $\{e_i\}_{i=1}^{\infty}$ is an orthonormal basis in $\mathfrak{H}$, then any vector $f \in \mathfrak{H}$ can be represented by the series

$$f = \sum_{k=1}^{\infty} \alpha_k e_k, \quad \alpha_k = (f, e_k), \tag{2.1}$$

which is convergent to f. In addition,

$$(f, g) = \sum_{k=1}^{\infty} \alpha_k \overline{\beta}_k,$$

whence

$$\|f\|^2 = \sum_{k=1}^{\infty} |\alpha_k|^2.$$

The last equality is often called the Parseval equality. It makes it possible to conclude that Hilbert spaces are isometrically isomorphic if and only if they have the same dimension.

2.2. EXAMPLES OF HILBERT SPACES

The space $\mathbb{C}^n$ is a Hilbert space if for the inner product of two n-tuples $\{\alpha_1, \ldots, \alpha_n\}$ and $\{\beta_1, \ldots, \beta_n\}$ we take

$$(\alpha, \beta)_{\mathbb{C}^n} = \sum_{k=1}^{n} \alpha_k \overline{\beta}_k.$$

The space l_2. Its elements are infinite sequences of complex numbers $x = \{x_k\}_{k=1}^{\infty}$ for which $\sum_{k=1}^{\infty} |x_k|^2 < \infty$. With the inner product

$$(x, y)_{l_2} = \sum_{k=1}^{\infty} x_k \overline{y}_k,$$

l_2 is a Hilbert space.

The space $L_2((a,b), \mathrm{d}\sigma)$. Let $\sigma(t)$ be a non-decreasing function on the interval (a, b), which determines a countably-additive measure σ on (a, b). The elements of $L_2((a,b), \mathrm{d}\sigma)$ are the complex-valued functions f on (a, b) for which

$$\int_a^b |f(t)|^2 \, \mathrm{d}\sigma(t) < \infty.$$

Equality of two functions in this space means equality of their σ – almost everywhere. The inner product of $f, g \in L_2((a,b), \mathrm{d}\sigma)$ is defined as

$$(f, g)_{L_2((a,b),\, \mathrm{d}\sigma)} = \int_a^b f(t) \overline{g(t)} \, \mathrm{d}\sigma(t).$$

The space $L_2((a,b), d\sigma)$ is a Hilbert space. If σ is Lebesgue measure, then $L_2((a,b), d\sigma)$ is denoted simply by $L_2(a,b)$.

In the case of a bounded interval (a,b) the set of functions

$$\left\{ \frac{1}{\sqrt{b-a}} e^{i\pi kt/(b-a)} \right\}_{k=-\infty}^{\infty}$$

is an orthonormal basis in $L_2(a,b)$, while the expansion (2.1) in this basis coincides with the trigonometric Fourier series.

Let us consider the space $L_2(-\infty,\infty)$. The sequence of Hermite functions

$$h_n(t) = (-1)^n \pi^{-\frac{1}{4}} (2^n n!)^{-\frac{1}{2}} e^{\frac{t^2}{2}} \frac{d^n e^{-t^2}}{dt^n} \qquad (n = 0, 1, \ldots)$$

is an orthonormal basis in this space.

The direct orthogonal sum of Hilbert spaces. Let $\mathfrak{H}_k$ $(k = 1, \ldots, n)$ be Hilbert spaces. Into their direct sum $\mathfrak{H}$ we introduce an inner product by

$$(f,g)_{\mathfrak{H}} = \sum_{k=1}^{n} (f_k, g_k)_{\mathfrak{H}_k} \quad (f = \{f_1, \ldots, f_n\}, g = \{g_1, \ldots, g_n\}).$$

Relative to it, $\mathfrak{H}$ is a Hilbert space in which all terms in the direct sum $\mathfrak{H} = \mathfrak{H}_1 \dot{+} \cdots \dot{+} \mathfrak{H}_n$ are pairwise orthogonal. Therefore this direct sum is called orthogonal; it is written as

$$\mathfrak{H} = \mathfrak{H}_1 \oplus \cdots \oplus \mathfrak{H}_n = \oplus_{k=1}^{n} \mathfrak{H}_k.$$

We will also use direct sums of a countable number of Hilbert spaces. By definition, the elements of $\mathfrak{H} = \oplus_{k=1}^{\infty} \mathfrak{H}_k$ are the sequences $\{f_k\}_{k=1}^{\infty}$ $(f_k \in \mathfrak{H}_k)$ for which

$$\sum_{k=1}^{\infty} \|f_k\|_{\mathfrak{H}_k}^2 < \infty.$$

The inner product of two sequences $\{f_k\}_{k=1}^{\infty}$ and $\{g_k\}_{k=1}^{\infty}$ is given by

$$\sum_{k=1}^{\infty} (f_k, g_k)_{\mathfrak{H}_k}.$$

It is not difficult to prove that $\mathfrak{H}$ is a Hilbert space.

2.3. LINEAR AND BILINEAR FUNCTIONALS ON HILBERT SPACES

For continuous linear functionals on a Hilbert space $\mathfrak{H}$ there is the F. Riesz theorem, which describes their general representation.

THEOREM 2.2. (F. Riesz). *Any continuous linear functional Φ on a Hilbert space $\mathfrak{H}$ can be represented in the form*

$$\Phi(f) = (f, h),$$

where h is a certain element from $\mathfrak{H}$ uniquely determined by the functional Φ and

$$\|\Phi\| = \|h\|.$$

In view of this Riesz representation theorem the space $\mathfrak{H}'$, the dual of $\mathfrak{H}$, can be identified with $\mathfrak{H}$ up to isomorphism.

Let f, g be elements of $\mathfrak{H}$. Define a number $\mathcal{B}(f,g)$ so that

(i) $\mathcal{B}(\alpha_1 f_1 + \alpha_2 f_2, g) = \alpha_1 \mathcal{B}(f_1, g) + \alpha_2 \mathcal{B}(f_2, g)$;

(ii) $\mathcal{B}(f, \beta_1, g_1 + \beta_2 g_2) = \overline{\beta}_1 \mathcal{B}(f, g_1) + \overline{\beta}_2 \mathcal{B}(f, g_2)$;

(iii) $\|\mathcal{B}\| = \sup_{\|f\|\le 1, \|g\|\le 1} |\mathcal{B}(f,g)| < \infty$.

The function $\mathcal{B}(f,g)$ is called a continuous bilinear functional (a continuous bilinear form) on $\mathfrak{H}$. If we set $g = f$ in a continuous bilinear form $\mathcal{B}(f,g)$, then we obtain a continuous quadratic functional (a quadratic form), $\mathcal{B}(f,f)$ $(\forall f \in \mathfrak{H})$. The number $\|\mathcal{B}\|$ is called the norm of the bilinear functional $\mathcal{B}(f,g)$. Since

$$|\mathcal{B}(f,g)| \le \|\mathcal{B}\| \|f\| \|g\| \qquad (\forall f, g \in \mathfrak{H}),$$

$\mathcal{B}(f,g)$ is a continuous function of its arguments. An inner product is the simplest example of a continuous bilinear functional.

THEOREM 2.3. *Any continuous bilinear functional* $\mathcal{B}(f,g)$ *on* $\mathfrak{H}$ can be represented in the form

$$\mathcal{B}(f,g) = (Bf, g) \quad (f, g \in \mathfrak{H}), \tag{2.2}$$

where B is a bounded operator on $\mathfrak{H}$. The operator B determines the form $\mathcal{B}(f,g)$ uniquely and $\|B\| = \|\mathcal{B}\|$. A bilinear continuous form $\mathcal{B}(f,g)$ is called Hermitian if

$$\mathcal{B}(f,g) = \overline{\mathcal{B}(g,f)}.$$

Every Hermitian bilinear functional on $\mathfrak{H}$ can be expressed in terms of a quadratic functional as follows:

$$\mathcal{B}(f,g) = \tfrac{1}{4}(\mathcal{B}(f+g,f+g) - \mathcal{B}(f-g,f-g) + i\mathcal{B}(f+ig,f+ig) - i\mathcal{B}(f-ig,f-ig)).$$

2.4. ADJOINT OPERATORS

Let A be a bounded linear operator on $\mathfrak{H}$. It is obvious that the expression (f, Ag) defines a continuous bilinear functional on $\mathfrak{H}$ with norm $\|A\|$. It follows from Theorem 2.3 that there exists only one bounded linear operator A^* for which

$$(f, Ag) = (A^* f, g).$$

Its norm is equal to $\|A\|$. The operator A^* is called the adjoint of A.

One can easily see that $A^{**} = (A^*)^* = A$ and

$$(AB)^* = B^* A^*, \quad (A+B)^* = A^* + B^*, \quad (\lambda A)^* = \overline{\lambda} A^*$$

for any $A, B \in [\mathfrak{H}]$, any $\lambda \in \mathbb{C}^1$.

It should be noted that the definition of an adjoint operator given here agrees with the definition in §1.7 up to involution, i.e. up to a mapping I from $\mathfrak{H}$ onto $\mathfrak{H}$ such that

$$I^2 = E, \quad I(\alpha f + \beta g) = \overline{\alpha} If + \overline{\beta} Ig, \quad (If, Ig) = \overline{(f, g)}.$$

If $A = A^*$, then the operator A is called self-adjoint. An orthoprojector may be used as an example of a bounded self-adjoint operator.

If $\mathcal{B}(f, g)$ is a Hermitian continuous bilinear functional, then the operator B in (2.2) is self-adjoint.

2.5. ISOMETRIC AND UNITARY OPERATORS

An operator U given on the whole Hilbert space $\mathfrak{H}$ and mapping it onto $\mathfrak{H}$ is called unitary if

$$(Uf, Ug) = (f, g) \qquad (\forall f, g \in \mathfrak{H}).$$

It follows from this definition that $U^*U = UU^* = E$, i.e. $U^* = U^{-1}$, and the operator U^{-1} is unitary too. Further, if U_1, U_2 are unitary operators, then the operator U_1U_2 is also unitary. Consequently, the set of all unitary operators on $\mathfrak{H}$ forms a group.

As an example of a unitary operator one can take the operator of multiplication by the function $e^{i\alpha t}$ ($\operatorname{Im} \alpha = 0$) on $L_2(a, b)$. The shift $(U_s f)(t) = f(t + s)$ is a unitary operator on the space $L_2(-\infty, \infty)$. The Fourier transformation

$$(Ff)(\lambda) = \tilde{f}(\lambda) = \frac{1}{\sqrt{2\pi}} \int_{-\infty}^{\infty} f(t)\, e^{-i\lambda t}\, dt,$$

where the integral is understood as the mean square limit (in the sense of $L_2(-\infty, \infty)$) is unitary on the space $L_2(-\infty, \infty)$. Its inverse is given by the formula

$$(F^{-1}f)(t) = \frac{1}{\sqrt{2\pi}} \int_{-\infty}^{\infty} f(\lambda)\, e^{i\lambda t}\, d\lambda.$$

Let two Hilbert spaces $\mathfrak{H}_1$ and $\mathfrak{H}_2$ with inner products $(\cdot, \cdot)_{\mathfrak{H}_1}$ and $(\cdot, \cdot)_{\mathfrak{H}_2}$, respectively, be given. A linear operator V defined on the whole space $\mathfrak{H}_1$ and mapping it onto $\mathfrak{H}_2$ is called isometric if for any $f, g \in \mathfrak{H}_1$

$$(Vf, Vg)_{\mathfrak{H}_2} = (f, g)_{\mathfrak{H}_1}. \tag{2.3}$$

If the range $\mathfrak{R}(V)$ of the operator V does not coincide with $\mathfrak{H}_2$ and the equality (2.3) holds, then V is called semi-isometric. In particular, $\mathfrak{H}_1$ and $\mathfrak{H}_2$ can be subspaces of the same space $\mathfrak{H}$. In this case the subscripts in the inner products may be removed. The term "isometric operator" is often used just in this particular case, in the general case the term "isometric mapping" is commonly used.

A unitary operator is always isometric ($\mathfrak{H}_1 = \mathfrak{H}_2 = \mathfrak{H}$). However, an isometric operator is not necessarily unitary. For example, the shift $Vf = \sum_{k=1}^{\infty} f_k e_{k+1}$, where $\{e_k\}_{k=1}^{\infty}$ is an orthonormal basis in $\mathfrak{H}$ and $f_k = (f, e_k)$ are the Fourier coefficients of the vector f, is isometric but not unitary, because it maps $\mathfrak{H}$ onto the orthogonal complement in $\mathfrak{H}$ of the one-dimensional subspace spanned by the vector e_1.

3. Vector-Valued Functions

3.1. CONTINUITY. DIFFERENTIABILITY

We consider abstract functions (vector-functions) $x(t) : t \to \mathfrak{B}$ ($t \in I$, I is an interval of the real line $\mathbb{R}^1$) with as values elements of a Banach space $\mathfrak{B}$.

A vector-function $x(t)$ is called continuous at the point $t_0 \in I$ if $\|x(t) - x(t_0)\| \to 0$ as $t \to t_0$. We say that $x(t)$ is continuous on I if it is continuous at each point of this interval. It is obvious that if $x(t)$ is continuous on I, then $\|x(t)\|$ is a continuous scalar function on this interval.

By $C(\mathfrak{B}, [a, b])$ we denote the space of all continuous vector-functions $x(t)$ on $[a, b]$ $(-\infty < a < b < \infty)$ with values in $\mathfrak{B}$. It is a Banach space with respect to the norm

$$\|x\|_{C(\mathfrak{B},[a,b])} = \max_{a \le t \le b} \|x(t)\|.$$

In particular, if $\mathfrak{B} = \mathbb{C}^1$, then $C(\mathbb{C}^1, [a, b]) = C[a, b]$.

A vector-function $x(t) : I \to \mathfrak{B}$ is called differentiable at the point t_0 if there exists an element $y \in \mathfrak{B}$ such that

$$\left\| \frac{x(t_0 + \Delta t) - x(t_0)}{\Delta t} - y \right\| \to 0, \quad \Delta t \to 0.$$

The element y is called the derivative of the vector-function $x(t)$ at the point t_0. It is denoted by $x'(t_0)$. The function $x(t)$ is said to be differentiable on the interval I if it is differentiable at every point of this interval. If the vector-function $x'(t)$ is continuous, then $x(t)$ is called continuously differentiable. Higher-order derivatives are defined in the usual way.

We denote by $C^k(\mathfrak{B}, [a, b])$ the space of k times continuously-differentiable vector-functions $x(t)$ on $[a, b]$. It is a Banach space with respect to the norm

$$\|x\|_{C^k(\mathfrak{B},[a,b])} = \max\{\|x(t)\|, \ldots, \|x^{(k)}(t)\|\}$$

($x^{(j)}(t)$ is the j-th derivative of $x(t)$). When $\mathfrak{B} = \mathbb{C}^1$ we obtain the space $C^k[a, b]$.

As a rule, a continuous (differentiable) vector-function is called strongly continuous (strongly differentiable), in contrast to a weakly continuous (weakly differentiable) function. A vector-function $x(t)$ is called weakly continuous at the point t_0 or on the interval I if for any continuous linear functional f on $\mathfrak{B}$ the scalar function $f(x(t))$ is continuous at t_0 or on I.

A vector-function $x(t)$ is called weakly differentiable at the point t_0 if there exists a limit of $\frac{1}{\Delta t} f(x(t_0 + \Delta t) - x(t_0))$ $(\forall f \in \mathfrak{B}')$ as $\Delta t \to 0$. Weak differentiability follows from strong differentiability; the converse is not true.

If a vector-function $x(t)$ is weakly continuous on $[a, b]$ $(-\infty < a < b < \infty)$, then it follows from the uniform boundedness principle that $\|x(t)\| \leq M$ $(t \in [a, b]$, $M = \text{const})$. For a weakly differentiable vector-function $x(t)$ on $[a, b]$ we have the inequality

$$\|x(b) - x(a)\| \leq (b - a) \sup_{t \in [a,b]} \|x'(t)\|.$$

It shows that if a vector-function $x(t)$ is weakly differentiable on $[a, b]$ and its weak derivative vanishes, then $x(t) \equiv x$ (i.e., x is independent of t).

3.2. ANALYTIC VECTOR-FUNCTIONS

Let Ω be a domain in the complex plane. Let $x(z)$ $(z \in \Omega)$ be a function with values in a Banach space $\mathfrak{B}$. The element $x'(z_0) \in \mathfrak{B}$ is called the (strong) derivative of the vector-function $x(z)$ at the point z_0 if

$$\left\| \frac{x(z_0 + \Delta z) - x(z_0)}{\Delta z} - x'(z_0) \right\| \to 0, \quad \Delta z \to 0.$$

The vector-function $x(z)$ is called (strongly) analytic in the domain Ω if it has a derivative at every point of Ω. It is called weakly analytic in Ω if the scalar function $f(x(z))$ is analytic in Ω for every continuous linear functional f on $\mathfrak{B}$.

It is clear that strong analyticity of $x(z)$ implies weak analyticity of it. It turns out that the converse statement is also true. Thus, for a vector-function strong analyticity coincides with weak analyticity. In fact, it follows from the analyticity of the scalar function $f(x(z))$ $(\forall f \in \mathfrak{B}')$ in Ω that

$$\left| f\left(\frac{1}{\alpha - \beta} \left(\frac{1}{\alpha}(x(\varsigma + \alpha) - x(\varsigma)) - \frac{1}{\beta}(x(\varsigma + \beta) - x(\varsigma)) \right) \right) \right|$$
$$= \left| \frac{1}{2\pi} \int_\gamma \frac{f(x(t))dt}{(t - \varsigma)(t - \varsigma - \alpha)(t - \varsigma - \beta)} \right| \leq c_f \quad (c_f = \text{const}),$$

where $\varsigma \in \Omega$; α, β and γ are chosen in such a way that $\varsigma + \alpha$ and $\varsigma + \beta$ are contained in a sufficiently small disc with centre at ς, while γ encircles this disc at a positive distance from it. In this case the uniform boundedness principle permits us to conclude that

$$\left\| \frac{1}{\alpha - \beta} \left(\frac{1}{\alpha}(x(\varsigma + \alpha) - x(\varsigma)) - \frac{1}{\beta}(x(\varsigma + \beta) - x(\varsigma)) \right) \right\| \leq \text{const},$$

implying the existence of the strong derivative of $x(z)$ at the point ς.

Let Γ be a rectifiable Jordan curve. For a vector-function $x(z)$ we form the integral sums $\sum_{k=1}^n x(z_k)\Delta z_k$, where $\{z_k\}_{k=1}^n$ is a partition of the curve Γ, $\Delta z_k = z_{k+1} - z_k$. If there exists a limit of the integral sums as $\max |\Delta z_k| \to 0$, then this limit is called the integral of the function $x(z)$ along the curve Γ; it is denoted by $\int_\Gamma x(z)\,dz$:

$$\int_\Gamma x(z)\,dz = \lim_{\max|\Delta z_k|\to 0} \sum_{k=1}^{n} x(z_k)\Delta z_k.$$

A vector-function $x(z)$ which is continuous on Γ is integrable, and if f is a linear continuous functional on $\mathfrak{B}$, then the equality

$$f\Big(\int_\Gamma x(z)\,dz\Big) = \int_\Gamma f(x(z))\,dz$$

is valid. Therefore, the Cauchy integral formula

$$x(z) = \frac{1}{2\pi i}\int_\Gamma \frac{x(\varsigma)}{\varsigma - z}d\varsigma$$

holds for an analytic function $x(z)$, where Γ is any rectifiable Jordan curve around the point z and lying inside the domain of analyticity. The derivatives of an anlytic vector-function $x(z)$ are expressed by the formulas

$$x^{(n)}(z) = \frac{n!}{2\pi i}\int_\Gamma \frac{x(\varsigma)\,d\varsigma}{(\varsigma - z)^{n+1}} \qquad (n = 0, 1, \ldots).$$

In a neighbourhood of any point z_0 in the domain of analyticity, the vector-function $x(z)$ has a Taylor series expansion:

$$x(z) = \sum_{n=0}^{\infty} a_n(z - z_0)^n, \quad a_n = \frac{1}{n!}x^{(n)}(z_0).$$

Conversely, any series of this kind defines an analytic vector-function in the disc (of convergence) with radius $r = \big(\overline{\lim}\sqrt[n]{\|a_n\|}\big)^{-1}$.

A vector-function which is analytic in the whole complex plane is called an entire vector-function. If an entire vector-function is bounded, then it is a constant.

3.3. THE BOCHNER INTEGRAL AND INTEGRABLE VECTOR-FUNCTIONS

Let $\sigma(t)$ be a left-continuous non-decreasing function on the interval $[a, b]$. Suppose that σ determines a countably-additive measure on this interval. A vector-function $x(t)$ $(t \in [a, b])$ with values in a Banach space $\mathfrak{B}$ is called simple if it takes only a finite number of values on measurable sets $\Delta_j : x(t) = x_j$ $(t \in \Delta_j)$, $\bigcup \Delta_j = [a, b]$. If the measure of the set Δ_j equals ∞, we assume $x(t) = 0$ $(t \in \Delta_j)$.

A vector-function $x(t)$ is called strongly measurable if there exists a sequence $x_n(t)$ of simple vector-functions which converges to $x(t)$ almost everywhere. It is called weakly measurable if for every continuous linear functional f on $\mathfrak{B}$ the scalar function $f(x(t))$ is measurable on $[a, b]$. In a separable Banach space both concepts of measurability for a vector-function coincide, so we will speak just about measurable vector-functions. If $x(t)$ is measurable, then $\|x(t)\|$ is a measurable scalar function. For a simple vector-function $x(t)$ we define its integral by

$$\int_a^b x(t)\, d\sigma(t) = \sum_j x_j \operatorname{mes}\Delta_j.$$

A vector-function $x(t)$ is said to be Bochner integrable on $[a,b]$ if there is a sequence $x_n(t)$ of simple vector-functions which is convergent to $x(t)$ almost everywhere and is such that

$$\lim_{n\to\infty} \int_a^b \|x(t) - x_n(t)\|\, d\sigma(t) = 0.$$

By definition,

$$\int_a^b x(t)\, d\sigma(t) = \lim_{n\to\infty} \int_a^b x_n(t)\, d\sigma(t).$$

A necessary and sufficient condition for a vector-function $x(t)$ to be Bochner integrable is that it is strongly measurable and that $\int_a^b \|x(t)\|\, d\sigma(t) < \infty$. The integral of a vector-function over a measurable subset M of $[a,b]$ is defined as $\int_a^b \chi_M(t)x(t)\, d\sigma(t)$, where $\chi_M(t)$ is the characteristic function of M.

The Bochner integral posesses many of the properties of the Lebesgue integral. In particular,

$$\left\| \int_a^b x(t)\, d\sigma(t) \right\| \le \int_a^b \|x(t)\|\, d\sigma(t).$$

If $\sigma(t) = t$, i.e. σ is Lebesgue measure, then any vector-function $y(t)$ which is representable in the form

$$y(t) = \int_a^t x(\tau)\, d\tau, \tag{3.1}$$

where $x(t)$ is a Bochner integrable vector-function on $[a,b]$, has a derivative at almost all points of $[a,b]$; moreover, $y'(t) = x(t)$. A vector-function $y(t)$ of the form (3.1) with a Bochner integrable function $x(t)$ is called absolutely continuous. It should be noted that in the case of an infinite-dimensional $\mathfrak{B}$ this definition is not equivalent to the generally accepted concept of an absolutely continuous function.

We would like to note one more property of the Bochner integral which will be often used later. If A is a bounded linear operator from a Banach space $\mathfrak{B}_1$ to a Banach space $\mathfrak{B}_2$ and if $x(t)$ is a Bochner integrable vector-function with values in $\mathfrak{B}_1$, then the vector-function $(Ax)(t)$ with values in $\mathfrak{B}_2$ is Bochner integrable, and

$$\int_a^b (Ax)(t)\, dt = A \int_a^b x(t)\, dt. \tag{3.2}$$

3.4. SPACES OF INTEGRABLE VECTOR-FUNCTIONS. THE SPACE $L_2(\mathfrak{H}, (a,b), d\sigma)$

The totality of all σ-integrable vector-functions on $[a,b]$ with values in $\mathfrak{B}$ forms the vector space $L_1(\mathfrak{B}, (a,b), d\sigma)$. It is a Banach space with respect to the norm

$$\|x\|_{L_1(\mathfrak{B},(a,b),\,d\sigma)} = \int_a^b \|x(t)\|\, d\sigma(t).$$

Similarly to the scalar case we introduce the Banach spaces $L_p(\mathfrak{B}, (a,b), d\sigma)$ of vector-functions $x(t)$ with finite norm

$$\|x\|_{L_p(\mathfrak{B},(a,b),\,d\sigma)} = \left(\int_a^b \|x(t)\|^p\, d\sigma(t)\right)^{\frac{1}{p}}.$$

In particular, $L_p(\mathbb{C}^1, (a,b), d\sigma) = L_p((a,b), d\sigma)$.

In the sequel we will often need the case when $\mathfrak{B} = \mathfrak{H}$ is a Hilbert space, while $p = 2$. So, $L_2(\mathfrak{H}, (a,b), d\sigma)$ is the set of σ-measurable vector-functions $x(t)$ with values in $\mathfrak{H}$ for which

$$\int_a^b \|x(t)\|^2\, d\sigma(t) < \infty.$$

We introduce an inner product in this space by

$$(x,y)_{L_2(\mathfrak{H},(a,b),\,d\sigma)} = \int_a^b (x(t), y(t))\, d\sigma(t) \quad (x, y \in L_2(\mathfrak{H}, (a,b), d\sigma)).$$

This definition is correct. Indeed, let $\{e_k\}_{k=1}^{\infty}$ be an orthonormal basis in $\mathfrak{H}$ and put

$$x_k(t) = (x(t), e_k).$$

It follows from the Cauchy–Buniakowski inequality that $x_k(t) \in L_2((a,b), d\sigma)$. Moreover, since $\sum_{k=1}^{\infty} |x_k(t)|^2 = \|x(t)\|^2$ almost everywhere,

$$\sum_{k=1}^{\infty} \int_a^b |x_k(t)|^2\, d\sigma(t) = \int_a^b \sum_{k=1}^{\infty} |x_k(t)|^2\, d\sigma(t) = \int_a^b \|x(t)\|^2\, d\sigma(t).$$

Similarly, for a vector-function $y(t)$ put

$$y_k(t) = (y(t), e_k).$$

Then the function

$$(x(t), y(t)) = \sum_{k=1}^{\infty} x_k(t)\overline{y_k(t)}$$

is σ-measurable. Since $|(x(t), y(t))| \le \|x(t)\|\ \|y(t)\|$,

$$\left(\int_a^b |(x(t), y(t))|\, d\sigma(t)\right)^2 \le \int_a^b \|x(t)\|^2\, d\sigma(t) \int_a^b \|y(t)\|^2\, d\sigma(t)$$
$$= \|x\|^2_{L_2(\mathfrak{H},(a,b),\,d\sigma)} \|y\|^2_{L_2(\mathfrak{H},(a,b),\,d\sigma)},$$

whence

$$\int_a^b (x(t), y(t))\, d\sigma(t) = \sum_{k=1}^{\infty} \int_a^b x_k(t)\overline{y_k(t)}\, d\sigma(t).$$

The space $L_2(\mathfrak{H}, (a,b), d\sigma)$ is complete. In fact, if a sequence $x^n(t)$ is fundamental, then for any k the sequence of scalar functions

$$x_k^n(t) = (x^n(t), e_k)$$

is fundamental in $L_2((a,b), d\sigma)$. By virtue of the completeness of $L_2((a,b), d\sigma)$, $x_k^n(t) \to x_k(t)$ in $L_2((a,b), d\sigma)$. It is not difficult to see that the sequence of vector-functions $x^n(t)$ converges to $x(t) = \sum_{k=1}^{\infty} x_k(t) e_k$ in $L_2(\mathfrak{H}, (a,b), d\sigma)$.

The space $L_2(\mathfrak{H}, (a,b), d\sigma)$ is separable, since the set $\{\varphi_j(t)e_i\}_{i,j=1}^{\infty}$, where $\varphi_j(t)$ is an orthonormal basis in $L_2((a,b), d\sigma)$, is an orthonormal basis in it. Because the set $C_0^\infty(a,b)$ of all infinitely-differentiable functions with compact support in (a,b) is dense in $L_2((a,b), d\sigma)$, the set of vector-functions $\sum_{k=1}^{m} \varphi_k(t) f_k$ $(\forall m \in \mathbb{N})$, where $\varphi_k(t) \in C_0^\infty(a,b)$ while f_k runs through a set which is dense in $\mathfrak{H}$, is dense in $L_2(\mathfrak{H}, (a,b), d\sigma)$.

Let $\mathfrak{H} = L_2(\Omega)$, where Ω is an open subset of the space $\mathbb{R}^n$. Then

$$L_2(\mathfrak{H}, (a,b)) = L_2((a,b) \times \Omega).$$

This follows from the fact that the set of all elements of the form $\varphi(t)f(s)$, where $\varphi(t) \in C_0^\infty(a,b)$ while $f(s) \in C_0^\infty(\Omega)$ (i.e. $f(s)$ is infinitely differentiable and of compact support in Ω) forms a complete system in $L_2(\mathfrak{H}, (a,b))$. According to Fubini's theorem the same set is also a complete system in the space $L_2((a,b) \times \Omega)$.

3.5. OPERATOR-FUNCTIONS

Let $\mathfrak{B}$ be a Banach space. By an operator-valued function, or simply an operator-function, on $\mathfrak{B}$ we mean a vector-function given in a domain $\Omega \subset \mathbb{C}^1$ and taking values in the Banach space $[\mathfrak{B}]$ of bounded linear operators from $\mathfrak{B}$ to $\mathfrak{B}$. In contrast to the case of general vector-functions, for these functions we have the concepts of uniform, strong, and weak continuity (differentiability, etc.).

Uniform continuity (differentiability, etc.) of an operator-function $A(t)$ is its continuity (differentiability, etc.) as a vector-function with values in $[\mathfrak{B}]$. Strong continuity (differentiability, etc.) of $A(t)$ means continuity (differentiability, etc.) of the vector-functions $A(t)x$ $(\forall x \in \mathfrak{B})$ with values in $\mathfrak{B}$. Weak continuity (differentiability, etc.) means the weak continuity (differentiability, etc.) of the family of vector-functions $A(t)x$ $(\forall x \in \mathfrak{B})$. Uniform continuity (differentiability) implies strong continuity and strong continuity implies weak continuity. The converses are, in general, not true. As to analyticity, all three kinds of it are equivalent.

If $A(t)$ is a strongly continuous (differentiable) operator-function on $\mathfrak{B}$ and $x(t)$ is a strongly continuous (differentiable) vector-function on the same space, then $A(t)x(t)$ is a continuous (differentiable) vector-function on $\mathfrak{B}$. Moreover, in the second case

$$\frac{d}{dt}(A(t)x(t)) = \frac{dA(t)}{dt}x(t) + A(t)\,\frac{dx(t)}{dt}.$$

We would like to note that it is possible to consider vector-functions and operator-functions of several variables. The extension of all principal assertions from the case of one variable to that of several variables does not present any difficulties: it is done in the same way as for scalar functions.

3.6. INTEGRAL OPERATORS WITH OPERATOR-KERNELS

Let K be an operator-valued function on $\mathfrak{H}$ defined on the square $[a,b] \times [a,b]$ $(-\infty < a < b < \infty)$, i.e. at all points of this square $K(t,s)$ is a continuous linear operator on the Hilbert space $\mathfrak{H}$. For simplicity of exposition we assume that K is a continuous function of two variables on $[a,b] \times [a,b]$ relative to the uniform operator topology.

We define an integral operator $\mathcal{K}$ by

$$(\mathcal{K}u)(t) = \int_a^b K(t,s)u(s)\,ds, \quad u \in L_2(\mathfrak{H},(a,b)).$$

It is a continuous operator on the space $L_2(\mathfrak{H},(a,b))$, and

$$\|\mathcal{K}\| \le \Big(\int_a^b \int_a^b \|K(t,s)\|^2\,dt\,ds\Big)^{\frac{1}{2}}.$$

If $\mathfrak{H}$ is finite-dimensional, then $\mathcal{K}$ is completely continuous. In the infinite-dimensional case this is, generally speaking, not so. But if we assume that the operator $K(t,s)$ on $\mathfrak{H}$ is completely continuous for all $\{t,s\} \in [a,b] \times [a,b]$, then the integral operator $\mathcal{K}$ is also completely continuous. Indeed, by virtue of the continuity of the kernel K, for an arbitrary $\epsilon > 0$ we can choose a partition of the interval $[a,b]$ by the points $t_0 = a < t_1 < \ldots < t_n = b$ in such a way that $\|K(t,s) - K_\epsilon(t,s)\| < \epsilon$, where

$$K_\epsilon(t,s) = K(t_i,t_j) \quad \text{for} \quad \{t,s\} \in (t_i,t_{i+1}] \times (t_j,t_{j+1}].$$

It is obvious that $\|\mathcal{K} - \mathcal{K}_\epsilon\| \to 0$ $(\epsilon \to 0)$, where

$$(\mathcal{K}_\epsilon u)(t) = \int_a^b K_\epsilon(t,s)u(s)\,ds.$$

Therefore, the integral operator $\mathcal{K}$ is completely continuous on $L_2(\mathfrak{H},(a,b))$ if the integral operator $\mathcal{K}_\epsilon$ is such. This is reduced in turn to the proof of the fact that an integral operator $\mathcal{L}$ with kernel

$$L(t,s) = \begin{cases} A & \text{for } \{t,s\} \in [\alpha,\beta] \times [\gamma,\delta], \\ 0 & \text{for all other } \{t,s\} \end{cases}$$

is completely continuous. Here A is a completely continuous operator on $\mathfrak{H}$; $\alpha,\beta,\gamma,\delta \in [a,b]$. Since the integral operator $\mathcal{L} : L_2(\mathfrak{H},(a,b)) \to L_2(\mathfrak{H},(a,b))$ can be represented as a product $\mathcal{L}_1\mathcal{L}_2\mathcal{L}_3$, where $\mathcal{L}_1 : \mathcal{L}_1 u = \int_0^\delta u(s)\,ds$ acts continuously from $L_2(\mathfrak{H},(a,b))$ into $\mathfrak{H}$, $\mathcal{L}_2 : \mathcal{L}_2 f = Af$ is completely continuous on $\mathfrak{H}$, $\mathcal{L}_3 : (L_3 f)(t) = f$ is continuous from $\mathfrak{H}$ into $L_2(\mathfrak{H},(a,b))$, the integral operator $\mathcal{L}$ is completely continuous.

4. Unbounded Operators. Spectral Expansion of Self-Adjoint Operators

This section deals with general unbounded operators on a Banach space $\mathfrak{B}$. Such operators need not be defined on the whole space, but only on a certain linear subset of $\mathfrak{B}$, which we call the domain of the respective operator and denote by $\mathfrak{D}$.

4.1. CLOSED OPERATORS

A linear operator A on $\mathfrak{B}$ is called closed if the existence of the limits $\lim_{n\to\infty} f_n = f$, $\lim_{n\to\infty} Af_n = g$ implies that $f \in \mathfrak{D}(A)$ and $Af = g$. Any continuous operator is closed. The converse is, in general, not true, for the existence of $\lim_{n\to\infty} f_n$ does not need imply that the sequence Af_n converges.

An operator $\widetilde{A}$ is called an extension of A (we write $\widetilde{A} \supset A$ or $A \subset \widetilde{A}$) if $\mathfrak{D}(\widetilde{A}) \supset \mathfrak{D}(A)$ and $\widetilde{A}f = Af$ for any $f \in \mathfrak{D}(A)$. If the operator A itself is not closed, but if it follows from $f_n \to 0$ and $Af_n \to g$ that $g = 0$, then the operator A is closeable, i.e. it has closed extensions. Denote by $\overline{A}$ the minimal such extension. It is the so-called closure of A. To obtain $\overline{A}$ the set $\mathfrak{D}(A)$ should be supplemented with all vectors $f \in \mathfrak{B}$ for which there exists a sequence of elements $f_n \in \mathfrak{D}(A)$ such that $f_n \to f$ and $Af_n \to g$ $(n \to \infty)$; then we set $\overline{A}f = g$.

The totality of all pairs $\{f, Af\}$ $(f \in \mathfrak{D}(A))$ in the direct sum $\mathfrak{B} \dotplus \mathfrak{B}$ forms the graph of A. Obviously, A is closed if and only if its graph is a closed set in $\mathfrak{B} \dotplus \mathfrak{B}$. Also it is not difficult to prove the following statements:

(i) if A is closed and $B \in [\mathfrak{B}]$, then the operator $A + B$ is closed;

(ii) if A is closed and A^{-1} exists, then the operator A^{-1} is closed;

(iii) the closed graph theorem: if a closed operator is defined on the whole space, then it is bounded.

Define on $\mathfrak{D}(A)$ the norm $\|f\|_A = \left(\|Af\|^2 + \|f\|^2\right)^{\frac{1}{2}}$. If A is closed, then $\mathfrak{D}(A)$ is a Banach space with respect to this norm. We will call convergence in this norm convergence in the sense of the graph of A. If $\mathfrak{B} = \mathfrak{H}$ is a Hilbert space, then $\|\cdot\|_A$ is induced by the inner product

$$(f,g)_A = (Af, Ag) + (f,g).$$

4.2. ADJOINT OPERATORS ON A HILBERT SPACE

Let $\mathfrak{H}$ be a Hilbert space and A a linear operator with domain $\mathfrak{D}(A)$ dense in $\mathfrak{H}$. Consider the inner product (Af, g), $f \in \mathfrak{D}(A)$. There exist pairs of elements g, g^*

for which

$$(Af, g) = (f, g^*) \qquad (\forall f \in \mathfrak{D}(A)). \tag{4.1}$$

For example, the equality (4.1) holds for $g = g^* = 0$. In view of the denseness of $\mathfrak{D}(A)$ in $\mathfrak{H}$ the element g^* is uniquely determined by the vector g. Setting $A^*g = g^*$ we obtain an operator A^*, called the adjoint of A. Its domain consists of all elements g for which there exists an element g^* satisfying (4.1).

From the definition of an adjoint operator one immediately obtains the following properties:

(i) the operator A^* is linear;
(ii) if $B \subset A$, then $B^* \supset A^*$;
(iii) the operator A^* is closed;
(iv) if A is closeable, then $(\overline{A})^* = A^*$;
(v) if A is closed, then $\mathfrak{D}(A^*)$ is dense in $\mathfrak{H}$; hence there the operator $A^{**} = A$ exists; if A is not closed but is closeable, then $A^{**} = \overline{A}$.

A linear operator A with domain dense in $\mathfrak{H}$ is called symmetric (Hermitian) if $A \subseteq A^*$. If $A = A^*$, then the operator A is called self-adjoint. Clearly, a symmetric operator is always closeable. If $\widetilde{A}$ is a symmetric extension of a symmetric operator A, then $\widetilde{A} \subset A^*$.

A symmetric operator A is said to be essentially self-adjoint on a linear set $\mathfrak{D} \subset \mathfrak{D}(A)$ if its closure on $\mathfrak{D}$ is self-adjoint. A symmetric operator A is essentially self-adjoint on $\mathfrak{D}$ if and only if $\overline{(A - zE)\mathfrak{D}} = \mathfrak{H}$ for any z ($\operatorname{Im} z \neq 0$).

The operator A is called lower (upper) semi-bounded if there exists a constant μ such that $(Af, f) \geq \mu(f, f)$ ($(Af, f) \leq \mu(f, f)$) for all $f \in \mathfrak{D}(A)$. The operator A is called non-negative if $(Af, f) \geq 0$ for any $f \in \mathfrak{D}(A)$. It is called positive if the equality sign in the last relation is possible only for $f = 0$. We will write $A \geq B$ ($A > B$) if $\mathfrak{D}(A) \subseteq \mathfrak{D}(B)$ and the operator $A - B$ is nonnegative (positive). A positive operator A is called positive definite if there exists a constant $\mu > 0$ such that $A - \mu E \geq 0$.

A densely defined closed operator A is called normal if $A^*A = AA^*$. A direct consequence of this definition is that A^* is normal. It is also clear that self-adjoint and unitary operators are normal. Without difficulty one can establish that the conditions

$$\mathfrak{D}(A) = \mathfrak{D}(A^*), \quad \|Ax\| = \|A^*x\| \qquad (\forall x \in \mathfrak{D}(A))$$

are necessary and sufficient for a densely defined closed operator A to be normal.

4.3. J-UNITARY OPERATORS

A Hilbert space $\mathfrak{H}$ is said to be a space with indefinite metric if next to the scalar product $(.,.)$ in it a so-called indefinite scalar product $[.,.]$ is given which satisfies the conditions (i)-(iii) of a usual inner product and also the following properties:

(iv) the functional $[f, f]$ ($f \in \mathfrak{H}$) takes values of both signs;

(v) the functional $[f,g]$ $(f,g \in \mathfrak{H})$ is a continuous bilinear form on $\mathfrak{H}$.

In view of Theorem 2.3,

$$[f,g] = (Bf,g),$$

where B is a bounded self-adjoint operator on $\mathfrak{H}$. Later on we will consider the case when B is invertible. Then we can introduce in $\mathfrak{H}$ a new inner product $(.,.)_1$, equivalent to $(.,.)$ (in the sense that $\|\cdot\|_1$ and $\|\cdot\|$ are equivalent), in such a way that

$$[f,g] = (Jf,g)_1,$$

where $J \in [\mathfrak{H}]$, $J = J^*$, $J^2 = E$. Therefore, we may assume from the very beginning that

$$[f,g] = (Jf,g) \qquad (\forall f,g \in \mathfrak{H}).$$

A linear operator U is called J–unitary if it maps $\mathfrak{H}$ onto $\mathfrak{H}$ and if $[Uf,Ug] = [f,g]$ $(\forall f,g \in \mathfrak{H})$.

THEOREM 4.1. *Any J–unitary operator is bounded.*

The proof follows from the fact that a J–unitary operator is closed, a consequence of the definition and the fact that the operator J is invertible.

4.4. THE SPECTRUM. THE RESOLVENT

Let A be a densely defined closed linear operator on a Banach space $\mathfrak{B}$ with domain $\mathfrak{D}(A)$. A complex number λ for which $R_\lambda(A) = (A - \lambda E)^{-1}$ exists as a bounded operator with as domain the whole space $\mathfrak{B}$ is called a regular point of A. The resolvent set of A is the set of all its regular points. It is denoted by $\rho(A)$. The operator-function $R_\lambda(A)$ defined on $\rho(A)$ is called the resolvent of A. The resolvent set is always open in the complex plane and the operator-function $R_\lambda(A)$ is analytic in it. For any $\lambda, \mu \in \rho(A)$ the Hilbert resolvent identity

$$R_\lambda(A) - R_\mu(A) = (\lambda - \mu) R_\lambda(A) R_\mu(A) \tag{4.2}$$

holds. We also note that $R_\lambda R_\mu = R_\mu R_\lambda \quad (\forall \lambda, \mu \in \rho(A))$.

If A is a continuous operator defined on all of $\mathfrak{B}$, then $\rho(A)$ is not empty and contains the set $\{\lambda : |\lambda| > \|A\|\}$. For λ belonging to this set, the expansion

$$R_\lambda(A) = -\sum_{n=0}^{\infty} \lambda^{-(n+1)} A^n$$

holds. The series converges in the uniform operator topology, that is, in $[\mathfrak{B}]$. This expansion allows us to obtain for large $|\lambda|$ the estimate

$$\|R_\lambda(A)\| \le \frac{c}{|\lambda|}. \tag{4.3}$$

The complement of $\rho(A)$ (in the complex plane) is called the spectrum of the operator A. We will denote it by $\sigma(A)$. Consequently, the set $\sigma(A)$ is always closed. If A is a bounded operator, then $\sigma(A)$ is a closed, bounded and non-empty set. It is clear that

$$\sigma(A) \subseteq \{\lambda : |\lambda| \leq \|A\|\}.$$

The radius of the least disc around the origin containing $\sigma(A)$ is called the spectral radius of A. It can be found by the formula $r(A) = \lim_{n\to\infty} \sqrt[n]{\|A^n\|}$.

It is convenient to introduce the following classification of points in a spectrum.

The set $\sigma_p(A)$ of numbers $\lambda \in \sigma(A)$ such that the mapping $A - \lambda E$ is not one-to-one is called the point spectrum of the operator A. Thus, $\lambda \in \sigma_p(A)$ if and only if $Af = \lambda f$ for some non-zero element $f \in \mathfrak{B}$. A point $\lambda \in \sigma_p(A)$ is called an eigenvalue of the operator A and f is an eigenvector of A corresponding to it. The set of all eigenvectors corresponding to the eigenvalue λ forms a subspace, $\mathfrak{B}_\lambda(A)$, in $\mathfrak{B}$. Its dimension $\dim \mathfrak{B}_\lambda(A)$ is called the multiplicity of the eigenvalue λ.

A vector f is called an adjoined vector of the operator A corresponding to the eigenvalue λ if $(A - \lambda E)^{n-1} f \neq 0$ but

$$(A - \lambda E)^n f = 0.$$

for some natural number $n \neq 1$. The set of all eigenvectors and adjoined vectors of A corresponding to the eigenvalue λ forms a lineal. The dimension of this lineal is called the algebraic multiplicity of λ.

The set $\sigma_c(A)$ of all $\lambda \in \sigma(A)$ such that the mapping $A - \lambda E$ is one-to-one and the set $(A - \lambda E)\mathfrak{D}(A)$ is dense in $\mathfrak{B}$ but not equal to $\mathfrak{B}$ is called the continuous spectrum of the operator A.

The set $\sigma_r(A)$ of all $\lambda \in \sigma(A)$ for which the mapping $A - \lambda E$ is one-to-one but for which $(A - \lambda E)\mathfrak{D}(A)$ is not dense in $\mathfrak{B}$ is called the residual spectrum of A. Clearly, $\sigma_p(A)$, $\sigma_c(A)$ and $\sigma_r(A)$ are disjoint, and

$$\sigma(A) = \sigma_p(A) \cup \sigma_c(A) \cup \sigma_r(A).$$

Normal operators have no residual spectrum. The spectrum of a self-adjoint operator is real, while for a unitary operator it lies on the unit circle. In addition, eigenvectors corresponding to different eigenvalues are orthogonal. For a completely continuous operator A any point $\lambda \in \sigma(A)$ $(\lambda \neq 0)$ is an eigenvalue of finite multiplicity, and only zero may be a limit point of the set of all eigenvalues. If it is, it may belong to any of the parts σ_p, σ_c or σ_r of the spectrum.

4.5. SPECTRAL EXPANSION OF A SELF-ADJOINT OPERATOR

We define a resolution of the identity as an operator-function E_λ on a Hilbert space $\mathfrak{H}$, given on the whole λ-axis and satisfying the following properties:

(i) at any $\lambda \in \mathbb{R}^1$ the operator E_λ is an orthogonal projector;

(ii) $E_\lambda \leq E_\mu$ for $\lambda \leq \mu$;

(iii) the vector-function $E_\lambda h$ ($\forall h \in \mathfrak{H}$) is right-continuous, i.e.

$$\lim_{\epsilon\to 0_+} \|E_{\lambda+\epsilon}h - E_\lambda h\| = 0;$$

(iv) for arbitrary $h \in \mathfrak{H}$,

$$\lim_{\lambda\to-\infty} \|E_\lambda h\| = 0, \qquad \lim_{\lambda\to+\infty} \|E_\lambda h - h\| = 0.$$

It follows from property (ii) that for any vector $h \in \mathfrak{H}$ the limits

$$\lim_{\lambda\to\mu-0} E_\lambda h = E_{\mu-0}h, \qquad \lim_{\lambda\to\mu+0} E_\lambda h = E_{\mu+0}h$$

exist. Condition (iii) means that $E_\lambda = E_{\lambda+0}$. If E_λ satisfies the conditions (i), (ii), (iv) but not condition (iii), then we can replace E_λ by the function $E'_\lambda = E_{\lambda+0}$. This last operator-function satisfies all conditions (i)–(iv).

Below we will use the following notation. If Δ is one the intervals (a,b), $[a,b)$, $(a,b]$, or $[a,b]$, then E_Δ denotes, respectively, $E_{b-0} - E_{a+0}$, $E_{b-0} - E_{a-0}$, $E_{b+0} - E_{a+0}$, or $E_{b+0} - E_{a-0}$. In particular, $E_\Delta = E_b - E_a$ if $\Delta = (a,b]$. The operator E_Δ is an orthoprojector. It is also not difficult to check that $E_{\Delta_i}E_{\Delta_j} = 0$ when $\Delta_i \cap \Delta_j = \emptyset$. Property (iv) implies that the set $\mathcal{E} = \{h_\Delta : h_\Delta = E_\Delta h, \forall h \in \mathfrak{H}, \Delta$ is any finite interval$\}$ is dense in $\mathfrak{H}$.

Let now $f(\lambda) \in C[a,b]$ $(-\infty < a < b < \infty)$. Take a partition of $(a,b]$ by points $\lambda_0 = a, \lambda_1, \ldots, \lambda_{n-1}, \lambda_n = b$. Set $\Delta_j = (\lambda_{j-1}, \lambda_j]$ and form the sum

$$S = \sum_{i=1}^{n} f(\xi_i)E_{\Delta_i},$$

where ξ_i is an arbitrary point in the interval Δ_i. This S is a bounded operator on $\mathfrak{H}$. One can show that the sum S tends in $[\mathfrak{H}]$ to a certain operator as $\max|\Delta_j| = \max|\lambda_j - \lambda_{j-1}| \to 0$. This operator does not depend on the choice of a partition of $(a,b]$ and on the points ξ_i. It is called the integral of the function $f(\lambda)$ with respect to the measure $\mathrm{d}E_\lambda$, and is denoted by

$$\int_a^b f(\lambda)\,\mathrm{d}E_\lambda. \tag{4.4}$$

The existence of the integral (4.4) implies the existence of the integral

$$\int_a^b f(\lambda)\,\mathrm{d}E_\lambda h = \lim_{\max|\Delta_j|\to 0} \sum_{j=1}^{n} f(\xi_j)E_{\Delta_j}h = \Big(\int_a^b f(\lambda)\,\mathrm{d}E_\lambda\Big)h. \tag{4.5}$$

Moreover,

$$\Big\|\int_a^b f(\lambda)\,\mathrm{d}E_\lambda h\Big\|^2 = \int_a^b |f(\lambda)|^2\,\mathrm{d}(E_\lambda h, h) = \int_a^b |f(\lambda)|^2\,\mathrm{d}\|E_\lambda h\|^2. \tag{4.6}$$

On the right-hand side of this formula we have an ordinary Stieltjes integral.

If the function $f(\lambda)$ is continuous for all values $\lambda \in (-\infty, \infty)$, then we can define the integral $\int_{-\infty}^{\infty} f(\lambda)\,\mathrm{d}E_\lambda h$ as the limit of the integral (4.5) as $a \to -\infty$, $b \to \infty$.

Because of the Cauchy convergence criterion and formula (4.6), this integral exists if and only if the ordinary Stieltjes integral

$$\int_{-\infty}^{\infty} |f(\lambda)|^2 \, \mathrm{d}\|E_\lambda h\|^2$$

exists, and also

$$\left\| \int_{-\infty}^{\infty} f(\lambda) \, \mathrm{d}E_\lambda h \right\|^2 = \int_{-\infty}^{\infty} |f(\lambda)|^2 \, \mathrm{d}(E_\lambda h, h).$$

For an arbitrary self-adjoint operator A on a Hilbert space $\mathfrak{H}$ the following statement, which is called the principal spectral theorem, is valid.

THEOREM 4.2. *Let A be a self-adjoint operator on $\mathfrak{H}$. Then there exists a unique resolution of the identity E_λ such that:*

(i) *a vector h belongs to $\mathfrak{D}(A)$ if and only if*

$$\int_{-\infty}^{\infty} |\lambda|^2 \, \mathrm{d}(E_\lambda h, h) < \infty;$$

(ii) *under the condition* (i),

$$Ah = \int_{-\infty}^{\infty} \lambda \, \mathrm{d}E_\lambda h, \tag{4.7}$$

hence

$$\|Ah\|^2 = \int_{-\infty}^{\infty} |\lambda|^2 \, \mathrm{d}(E_\lambda h, h).$$

Conversely, any operator A determined by the conditions (i), (ii) using a certain resolution of the identity is self-adjoint. If $A, B \in [\mathfrak{H}]$, then $(AB = BA) \Rightarrow (BE_\lambda = E_\lambda B)$.

We also note that the resolution of the identity of a selfadjoint operator can be expressed in terms of its resolvent. Namely, if $\Delta = (a, b)$ is a bounded open interval, then

$$E_\Delta f = \lim_{\delta,\epsilon \to 0} \frac{1}{2\pi i} \int_{a+\delta}^{b-\delta} (R_{\mu - i\epsilon}(A) - R_{\mu + i\epsilon}(A)) f \, \mathrm{d}\mu$$

for every $f \in \mathfrak{H}$.

By analogy with the one-dimensional case a resolution of the identity is introduced on the complex plane $\mathbb{C}^1$. Let $\mathcal{B}$ be the Borel σ–algebra on $\mathbb{C}^1$. Consider the mapping $\Delta \to E_\Delta$ ($\forall \Delta \in \mathcal{B}$), where E_Δ is an orthogonal projector on $\mathfrak{H}$. We say that E_Δ is a resolution of the identity on $\mathbb{C}^1$ if the following conditions are satisfied:

(i) $E_\emptyset = 0$, $E_{\mathbb{C}^1} = E$;

(ii) for every sequence $\{\Delta_k\}_{k=1}^{\infty}$ ($\Delta_k \in \mathcal{B}$, $\Delta_k \cap \Delta_j = \emptyset$ $(k \neq j)$),

$$E_{\cup \Delta_k} = \sum_{k=1}^{\infty} E_{\Delta_k}$$

(the series on the right-hand side of the identity converges strongly);

(iii) $E_{\Delta_1 \cap \Delta_2} = E_{\Delta_1} E_{\Delta_2}$ for any $\Delta_1, \Delta_2 \in \mathcal{B}$.

Let $x, y \in \mathfrak{H}$. Then $(E_\Delta x, x)$ is a scalar non-negative countably-additive measure and $(E_\Delta x, y)$ is a charge (a complex-valued measure) of bounded variation on $\mathbb{C}^1$, and for its total variation we have $\operatorname{Var}(E_\Delta x, y) \leq \|x\|\|y\|$.

Now take a continuous function of the complex variable $f(\lambda)$. For $K_n = \{\lambda : |\lambda| \leq n\}$ we can define the integral

$$\int_{K_n} f(\lambda)\, \mathrm{d}E_\lambda x \qquad (\forall x \in \mathfrak{H})$$

as the limit of the integral sum $\sum_{j=1}^N f(\xi_j) E_{\Delta_j} x$ as $|\Delta_j| = \max |z_1 - z_2| \to 0$, where $\{\Delta_j\}_{j=1}^N$ is a partition of K_n and the ξ_j are points inside Δ_j. By definition,

$$\int_{\mathbb{C}^1} f(\lambda)\, \mathrm{d}E_\lambda x = \lim_{n \to \infty} \int_{K_n} f(\lambda)\, \mathrm{d}E_\lambda x.$$

The limit is taken in the strong sense and exists if and only if

$$\int_{\mathbb{C}^1} |f(\lambda)|^2\, \mathrm{d}\|E_\lambda x\|^2 = \lim_{|\Delta_j| \to 0} \sum_{j=1}^N f(\xi_j)(E_{\Delta_j} x, x)$$

exists; also

$$\left\| \int_{\mathbb{C}^1} f(\lambda)\, \mathrm{d}E_\lambda x \right\|^2 = \int_{\mathbb{C}^1} |f(\lambda)|^2\, \mathrm{d}(E_\lambda x, x) = \lim_{|\Delta_j| \to 0} \sum_{j=1}^N |f(\xi_j)|^2 (E_{\Delta_j} x, x).$$

The next statement, similar to the principal spectral theorem 4.2, is rather important: an operator A is normal if and only if there exists a resolution of the identity E_Δ on the complex plane $\mathbb{C}^1$ such that

$$\mathfrak{D}(A) = \{x \in \mathfrak{H} : \int_{\mathbb{C}^1} \lambda^2\, \mathrm{d}(E_\lambda x, x) < \infty\},$$

and for $x \in \mathfrak{D}(A)$

$$Ax = \int_{\mathbb{C}^1} \lambda\, \mathrm{d}E_\lambda x.$$

The operator A determines E_Δ uniquely.

4.6. THE SPECTRUM OF A SELF-ADJOINT OPERATOR

By means of a resolution of the identity we can give the following characterization of the spectrum of a self-adjoint operator A:

(a) a real number λ_0 is a regular point of the operator A if and only if its resolution of the identity E_λ is constant in some neighbourhood of the point λ_0;

(b) a real number λ_0 is an eigenvalue of the operator A if and only if $E_{\lambda_0} - E_{\lambda_0 - 0} \neq 0$; in this case $E_{\lambda_0} - E_{\lambda_0 - 0}$ is the orthoprojector onto the eigensubspace of A corresponding to λ_0.

If A is a completely continuous self-adjoint operator, then its resolution of the identity is a jumpfunction; formula (4.7) for such operator can be written in the form $Ah = \sum_{k=1}^{\infty} \lambda_k (h, e_k) e_k$, where the λ_k $(k = 1, 2, \ldots)$ are the eigenvalues of A and the e_k $(k \in \mathbb{N})$ form an orthonormal basis of eigenvectors corresponding to λ_k.

We will say that A is an operator with discrete spectrum if its resolvent set $\rho(A)$ is non-empty and $R_\lambda(A)$ is a completely continuous operator at any $\lambda \in \rho(A)$. A self-adjoint operator A is an operator with discrete spectrum if and only if its spectrum consists of eigenvalues of finite multiplicity, with as only possible limit point the point at infinity.

The points of the continuous spectrum, the limit points of the set of all eigenvalues and the eigenvalues of infinite multiplicity of a self-adjoint operator form its essential or so-called limit spectrum. For example, the essential spectrum of a completely continuous self-adjoint operator is either empty or consists of the point 0 only. A point λ_0 belongs to the essential spectrum of A if and only if there exists an infinite orthonormal system $e_n \in \mathfrak{D}(A)$ such that $\|(A - \lambda_0 E)e_n\| \to 0$ as $n \to \infty$.

All eigenvalues of the operator A of finite multiplicity and not belonging to the essential spectrum make up the discrete spectrum of this operator. The spectrum of the operator A is said to be discrete in a certain interval if all points of its spectrum lying inside this interval belong to the discrete spectrum of A.

Let $\Delta = (\lambda_0 - \delta,\ \lambda_0 + \delta)$. The dimension of the subspace $E_\Delta \mathfrak{H}$, which is invariant under A, coincides with the maximal dimension of the linear sets $\mathfrak{M} \subset \mathfrak{D}(A)$ whose elements satisfy the inequality

$$\|(A - \lambda_0 E)h\| \leq \delta \|h\| \quad (h \in \mathfrak{M}).$$

In the case when the spectrum of the operator A is discrete in Δ, the dimension of the subspace $E_\Delta \mathfrak{H}$ coincides with the number of eigenvalues of A (counted with multiplicity) lying in Δ. If the interval $(\lambda_0 - \delta,\ \lambda_0 + \delta)$ is the semi-axis $(-\infty, \lambda_0)$ then the number of points of the spectrum of A lying in $(-\infty, \lambda_0)$ equals the maximal dimension of the linear sets $\mathfrak{M} \subset \mathfrak{D}(A)$ on which the inequality $((A - \lambda_0 E)h, h) < 0$ holds.

The following statement is due to H. Weyl: if a completely continuous self-adjoint operator B is added to a self-adjoint operator A, then the essential spectrum of the operator A does not change.

4.7. THE TENSOR PRODUCT OF HILBERT SPACES. OPERATORS ADMITTING SEPARATION OF VARIABLES

Let $\mathfrak{H}_1$, $\mathfrak{H}_2$ be Hilbert spaces and let $\{e_j^{(i)}\}_{j=1}^{\infty}$ be an orthonormal basis in $\mathfrak{H}_i$ $(i = 1, 2)$. We form the formal products

$$e_\alpha = e_{\alpha_1 \alpha_2} = e_{\alpha_1}^{(1)} \otimes e_{\alpha_2}^{(2)} \quad (\alpha = \{\alpha_1, \alpha_2\}, \alpha_i = 1, 2, \ldots), \tag{4.8.}$$

and span a Hilbert space on the formal vectors (4.8) by assuming that these vectors form an orthonormal basis in it. This Hilbert space is called the tensor product of the given spaces $\mathfrak{H}_1$ and $\mathfrak{H}_2$. It is denoted by $\mathfrak{H}_1 \otimes \mathfrak{H}_2$. Thus, its vectors have the form

$$f = \sum_{\alpha \in \mathbf{N} \times \mathbf{N}} f_\alpha e_\alpha, \quad f_\alpha \in \mathbf{C}^1, \quad \sum_\alpha |f_\alpha|^2 = \|f\|^2_{\mathfrak{H}_1 \otimes \mathfrak{H}_2} < \infty. \tag{4.9}$$

The inner product in $\mathfrak{H}_1 \otimes \mathfrak{H}_2$ is defined by

$$(f, g)_{\mathfrak{H}_1 \otimes \mathfrak{H}_2} = \sum_{\alpha \in \mathbf{N} \times \mathbf{N}} f_\alpha \overline{g_\alpha},$$

where $g = \sum_{\alpha \in \mathbf{N} \times \mathbf{N}} g_\alpha e_\alpha$ is another vector of the kind of (4.9).

Let $f^{(i)} = \sum_{j=1}^\infty f_j^{(i)} e_j^{(i)}$ be a vector in $\mathfrak{H}_i$ $(i = 1, 2)$. We set, by definition

$$f = f^{(1)} \otimes f^{(2)} = \sum_{\alpha_1, \alpha_2 = 1}^\infty f_{\alpha_1}^{(1)} f_{\alpha_2}^{(2)} e_{\alpha_1}^{(1)} \otimes e_{\alpha_2}^{(2)}. \tag{4.10}$$

The coefficients $f_\alpha = f_{\alpha_1}^{(1)} f_{\alpha_2}^{(2)}$ satisfy the condition (4.9), therefore the vector f belongs to $\mathfrak{H}_1 \otimes \mathfrak{H}_2$. In addition, $\|f\|_{\mathfrak{H}_1 \otimes \mathfrak{H}_2} = \|f^{(1)}\|_{\mathfrak{H}_1} \|f^{(2)}\|_{\mathfrak{H}_2}$. The function $\mathfrak{H}_1 \oplus \mathfrak{H}_2 \ni \{f^{(1)}, f^{(2)}\} \to f \in \mathfrak{H}_1 \otimes \mathfrak{H}_2$ is linear with respect to each variable, while the linear span of the vectors (4.10) is dense in $\mathfrak{H}_1 \otimes \mathfrak{H}_2$.

The definition of the tensor product $\mathfrak{H}_1 \otimes \mathfrak{H}_2$ clearly depends on the choice of bases in $\mathfrak{H}_1$ and $\mathfrak{H}_2$; however, if the bases are changed, we obtain tensor products which are isomorphic to the original one.

EXAMPLE. Let $\mathfrak{H}_1 = \mathfrak{H}$, $\mathfrak{H}_2 = L_2((a, b), d\sigma)$ where $\mathfrak{H}$ is an arbitrary Hilbert space and σ is a countably-additive measure on (a, b). Then

$$\mathfrak{H} \otimes L_2((a, b), d\sigma) = L_2(\mathfrak{H}, (a, b), d\sigma). \tag{4.11}$$

This follows from the fact that the set $\{\varphi_k(t) e_j\}_{k,j=1}^\infty$, where $\varphi_k(t)$ and e_j are orthonormal bases in $L_2((a, b), d\sigma)$ and $\mathfrak{H}$, respectively, is an orthonormal basis in both spaces appearing in (4.11).

Let now A_1, A_2 be bounded linear operators acting on Hilbert spaces $\mathfrak{H}_1$ and $\mathfrak{H}_2$, respectively. Their tensor product $A_1 \otimes A_2$ is defined by the formula

$$\begin{aligned}(A_1 \otimes A_2) f &= (A_1 \otimes A_2)\Big(\sum_{\alpha \in \mathbf{N} \times \mathbf{N}} f_\alpha e_{\alpha_1}^{(1)} \otimes e_{\alpha_2}^{(2)}\Big) \\ &= \sum_{\alpha \in \mathbf{N} \otimes \mathbf{N}} f_\alpha (A_1 e_{\alpha_1}^{(1)}) \otimes (A_2 e_{\alpha_2}^{(2)}) \qquad (f \in \mathfrak{H}_1 \otimes \mathfrak{H}_2).\end{aligned} \tag{4.12}$$

The series on the right-hand side of (4.12) weakly converges in $\mathfrak{H}_1 \otimes \mathfrak{H}_2$ and defines an operator $A_1 \otimes A_2 \in [\mathfrak{H}_1 \otimes \mathfrak{H}_2]$ with

$$\|A_1 \otimes A_2\| = \|A_1\| \|A_2\|.$$

If the operators A_1, A_2 are unbounded with dense domains, then we define $\mathfrak{D}(A_1) \otimes \mathfrak{D}(A_2)$ as the set of finite linear combinations of vectors $f^{(1)} \otimes f^{(2)}$ with $f^{(i)} \in$

$\mathfrak{D}(A_i)$. It is dense in $\mathfrak{H}_1 \otimes \mathfrak{H}_2$. Define on $\mathfrak{D}(A_1) \otimes \mathfrak{D}(A_2)$ the operator $A_1 \otimes A_2$ by the formula

$$(A_1 \otimes A_2)(f^{(1)} \otimes f^{(2)}) = A_1 f^{(1)} \otimes A_2 f^{(2)}$$

and extend it in a linear way. This definition is correct. Moreover, if the A_i $(i = 1, 2)$ are closeable, then $A_1 \otimes A_2$ has the same property.

An operator A in $\mathfrak{H}_1 \otimes \mathfrak{H}_2$ is said to admit separation of variables if it can be represented as

$$A = A_1 \otimes E + E \otimes A_2,$$

where A_1, A_2 are operators on $\mathfrak{H}_1$ and $\mathfrak{H}_2$ respectively. If the A_i are closeable, then the same holds for A.

The following statements are true for self-adjoint operators A_1 and A_2:

(i) the operators $A_1 \otimes A_2$ and $A_1 \otimes E + E \otimes A_2$ are essentially self-adjoint on $\mathfrak{D}_1 \otimes \mathfrak{D}_2$, where $\mathfrak{D}_1$, $\mathfrak{D}_2$ are the domains of essential self-adjointness of the operators A_1, A_2, respectively;

(ii) $\sigma(\overline{A_1 \otimes A_2}) = \sigma(A_1) \cdot \sigma(A_2)$; $\sigma(\overline{A_1 \otimes E + E \otimes A_2}) = \sigma(A_1) + \sigma(A_2)$ (here $\sigma(A_1) \cdot \sigma(A_2)$ (respectively $\sigma(A_1) + \sigma(A_2)$) is understood to be the set of all points λ from the complex plane which can be represented as $\lambda_1 \lambda_2$ (respectively, $\lambda_1 + \lambda_2$), where $\lambda_i \in \sigma(A_i)$ $(i = 1, 2)$.

This implies that if A_i $(i = 1, 2)$ are non-negative operators with discrete spectrum, then the operator $A_1 \otimes E + E \otimes A_2$ has the same properties; its spectrum coincides with the set $\{\lambda_{ik} = \lambda_i(A_1) + \lambda_k(A_2)\}_{i,k=1}^{\infty}$, where $\lambda_j(A_i)$ are the eigenvalues of the operator A_i.

5. The Operational Calculus

5.1. FUNCTIONS OF A BOUNDED OPERATOR

Let A be a bounded operator on a Banach space $\mathfrak{B}$. If $f(z)$ is an entire function of the complex variable z and

$$f(z) = \sum_{k=0}^{\infty} a_k z^k, \tag{5.1}$$

then it is possible to define the function $f(A)$ of the operator A by

$$f(A) = \sum_{k=0}^{\infty} a_k A^k. \tag{5.2}$$

This operator also belongs to $[\mathfrak{B}]$.

The given definition can be extended to a certain wider class of functions. $\mathcal{F}(A)$ will denote the family of all complex functions $f(z)$ each of which is analytic in some neighbourhood of the spectrum $\sigma(A)$ of the operator A. The neighbourhood need not be connected and may depend on $f \in \mathcal{F}(A)$. Let us construct the operator

$$f(A) = -\frac{1}{2\pi i} \int_\Gamma f(z) R_z(A)\, dz, \tag{5.3}$$

where Γ is a contour consisting of a finite number of closed rectifiable Jordan curves lying in the domain of analyticity of f and enclosing $\sigma(A)$. The contour is oriented in the (positive) sense customary in the theory of complex variables. Note that the integral (5.3) does not depend on the choice of Γ.

This definition of a function of an operator agrees with the previous one in as far as for a function $f(z) \in \mathcal{F}(A)$ which has a series expansion (5.1) convergent in a certain neighbourhood of $\sigma(A)$ both definitions are identical; the series (5.2) converges in the uniform operator topology.

Formula (5.3) defines the operational calculus for bounded operators on the class of functions that are analytic in some neighbourhood of the spectrum in the sense that if $f, g \in \mathcal{F}(A)$ and $\alpha, \beta \in \mathbb{C}^1$, then $\alpha f + \beta g \in \mathcal{F}(A)$, $fg \in \mathcal{F}(A)$ and

(i) $(\alpha f + \beta g)(A) = \alpha f(A) + \beta g(A)$;
(ii) $(fg)(A) = f(A) g(A)$;
(iii) $f(A) = E$ for $f(z) \equiv 1$;
(iv) $f(A) = A$ for $f(z) = z$.

The spectral mapping theorem is also valid, namely, if $f \in \mathcal{F}(A)$, then $\sigma(f(A)) = f(\sigma(A))$.

However, we will often need not analytic functions of an operator but only continuous or sufficiently smooth functions. It is possible to define such functions only for special classes of operators.

5.2. THE OPERATIONAL CALCULUS OF UNBOUNDED OPERATORS

Since the spectrum of an unbounded operator may be an arbitrary set, in particular, the empty set or the whole plane, it is difficult and sometimes impossible to construct an operational calculus using formulas (5.2), (5.3). However, much of the theory has very satisfactory generalizations to certain special classes of closed operators.

Suppose a closed operator A has at least one regular point. We will call such operators regular. Let $z_0 \in \rho(A)$. Then the operator $R_{z_0}(A)$ is bounded and it is possible to construct the operational calculus of $R_{z_0}(A)$ for the functions $\varphi \in \mathcal{F}(R_{z_0}(A))$.

Denote by $\mathcal{F}(A)$ the family of all functions $f(z)$ that are analytic in some neighbourhood of the spectrum $\sigma(A)$ of A and also at infinity. On the extended complex plane $\overline{\mathbb{C}^1}$ we define the mapping $\Phi : \overline{\mathbb{C}^1} \to \overline{\mathbb{C}^1}$ by

$$\Phi(z) = \frac{1}{z - z_0}, \quad \Phi(\infty) = 0, \quad \Phi(z_0) = \infty.$$

Then $\Phi(\sigma(A) \cup \{\infty\}) = \sigma(R_{z_0}(A))$, and the relation $\varphi(\varsigma) = f(\Phi^{-1}(\varsigma))$ determines a one-to-one correspondence between the functions $f \in \mathcal{F}(A)$ and $\varphi \in \mathcal{F}(R_{z_0}(A))$.

We define the operational calculus of A in terms of the operational calculus of $R_{z_0}(A)$ by setting for $f \in \mathcal{F}(A)$,

$$f(A) = \varphi(R_{z_0}(A)). \tag{5.4}$$

From formula (5.4) we can without difficulties obtain the following statement: if $f \in \mathcal{F}(A)$, then $f(A)$ is independent of the choice of $z_0 \in \rho(A)$, and

$$f(A) = f(\infty)E - \frac{1}{2\pi i}\int_\Gamma f(z)R_z(A)\,\mathrm{d}z. \tag{5.5}$$

Here, as before, Γ is a contour consisting of a finite number of rectifiable Jordan arcs contained in the domain of analyticity of f and enclosing the spectrum of A; the contour is positively oriented. Formula (5.5) defines a homomorphism between the algebra $\mathcal{F}(A)$ and a certain algebra of bounded operators, i.e. the properties (i)–(iii) of the previous subsection are satisfied. Also, the spectrum of the operator $f(A)$ is the image of the extended spectrum of A, the set $\{\sigma(A) \cup \{\infty\}\}$, under the map $f(z)$.

However, the class $\mathcal{F}(A)$ is rather small because a function in it must be analytic in a neigbourhood of infinity. It is possible to enlarge this class at the expense of further restrictions on the operator. E.g., if A is a closed operator with dense domain, if $(-\infty, 0] \subset \rho(A)$ and if the inequality

$$\|R_\lambda(A)\| \le M(1+|\lambda|)^{-1} \qquad (\lambda \le 0)$$

holds, then the operational calculus of A can be extended to a wider class of functions (this will be discussed in some detail in Chapter 5). In particular, it is possible to define the operators

$$A^{-\alpha} = \frac{\sin \alpha\pi}{\pi}\int_0^\infty \lambda^{-\alpha}R_{-\lambda}(A)\,\mathrm{d}\lambda \quad (0 < \alpha < 1).$$

If now $0 < \alpha < 1$, then $(h \in \mathfrak{D}(A))$

$$A^\alpha h = A^{\alpha-1}Ah = \frac{\sin \alpha\pi}{\pi}\int_0^\infty \lambda^{\alpha-1}R_{-\lambda}(A)Ah\,\mathrm{d}\lambda. \tag{5.6}$$

The class of functions suitable for constructing an operational calculus of a self-adjoint operator is considerably larger.

5.3. THE CASE OF A SELF-ADJOINT OPERATOR

For any continuous function $f(\lambda)$ on the real line and any self-adjoint operator A on a Hilbert space $\mathfrak{H}$ we define the operator $f(A)$ by

$$f(A)h = \int_{-\infty}^\infty f(\lambda)\,\mathrm{d}E_\lambda h \tag{5.7}$$

on the set $\mathfrak{D}(f(A)) = \{h : \int_{-\infty}^\infty |f(\lambda)|^2\,\mathrm{d}(E_\lambda h, h) < \infty\}$, where E_λ is the resolution of the identity of A. Since the set $\mathcal{E}(A) = \{h_\Delta : h_\Delta = E_\Delta h, \forall h \in \mathfrak{H}, \Delta \text{ is a bounded interval on the real axis}\}$ is contained in $\mathfrak{D}(f(A))$, the set $\mathfrak{D}(f(A))$ is dense in $\mathfrak{H}$.

It follows from definition (5.7) that the operator $f(A)$ is closed; it coincides with the closure of its restriction to $\mathcal{E}(A)$. Since E_λ is constant on $\rho(A) \cap (-\infty, \infty)$, the integral (5.7) is in fact taken only over the spectrum $\sigma(A)$, i.e.

$$f(A)h = \int_{\sigma(A)} f(\lambda)\, \mathrm{d}E_\lambda h; \tag{5.8}$$

and also

$$\|f(A)h\|^2 = \int_{\sigma(A)} |f(\lambda)|^2\, \mathrm{d}(E_\lambda h, h).$$

The adjoint $(f(A))^*$ of the operator $f(A)$ corresponds to the function $\overline{f}(\lambda)$:

$$(f(A))^* h = \int_{\sigma(A)} \overline{f(\lambda)}\, \mathrm{d}E_\lambda h.$$

In particular, the operator $f(A)$ is self-adjoint if and only if the values of $f(\lambda)$ are real. If $0 \leq f_1(\lambda) \leq f_2(\lambda)$, then $f_1(A) \leq f_2(A)$. For a non-negative self-adjoint operator A we can define the operator $B = \sqrt{A}$ as the function $f(\lambda) = \sqrt{\lambda}$ of A. The operator B is non-negative and also $B^2 = A$. These properties define it uniquely.

The operator $f(A)$ is bounded if and only if $f(\lambda)$ is bounded on $\sigma(A)$, and

$$\|f(A)\| = \max_{\lambda \in \sigma(A)} |f(\lambda)|.$$

The set of all continuous functions which are bounded on the spectrum of the operator A forms a Banach algebra, $\mathfrak{A}(A)$, with respect to the norm

$$\|f\| = \max_{\lambda \in \sigma(A)} |f(\lambda)|.$$

The mapping (5.8), $\mathfrak{A}(A) \ni f \to f(A) \in [\mathfrak{H}]$, is an isometric isomorphism from $\mathfrak{A}(A)$ onto a certain commutative subalgebra of the Banach algebra $[\mathfrak{H}]$. It is not difficult to convince ourselves that the operator $f(A)$ is continuously invertible if and only if the function $f(\lambda)$ does not vanish on $\sigma(A)$ and $\inf_{\lambda \in \sigma(A)} |f(\lambda)| > 0$.

We note that formula (5.7), defining the operational calculus of self-adjoint operators, may be used also for the class of Borel–measurable functions which are bounded on any bounded interval on the real axis.

5.4. THE EXPONENTIAL FUNCTION OF AN OPERATOR

Let a function $f(t, \lambda)$ be given on the set $[a, b] \times \sigma(A)$, where A is a self-adjoint operator on a Hilbert space $\mathfrak{H}$. Suppose that $f(t, \lambda)$ is continuous and bounded in λ for any fixed t. Then we can construct the operator-function $f(t, A)$ with values in $[\mathfrak{H}]$. If the function $f(t, \lambda)$ is continuous in t on some subset $\Omega \subset [a, b]$ for every finite $\lambda \in \sigma(A)$ and if

$$\sup_{t \in \Omega, \lambda \in \sigma(A)} |f(t, \lambda)| < \infty, \tag{5.9}$$

then the operator-function $f(t, A)$ is strongly continuous on Ω. This follows from the equality

$$\|(f(t,A) - f(t_0,A))h\|^2 = \int_{\sigma(A)} |f(t,\lambda) - f(t_0,\lambda)|^2 \, \mathrm{d}(E_\lambda h, h) \quad (\forall h \in \mathfrak{H}).$$

If the function $f(t, \lambda)$ is differentiable with respect to t on Ω and if

$$\sup_{t\in\Omega,\lambda\in\sigma(A)} |f_t'(t,\lambda)| < \infty, \tag{5.10}$$

then the operator-function $f(t, A)$ is strongly differentiable on Ω, and

$$\frac{\mathrm{d}}{\mathrm{d}t} f(t,A) = f_t'(t,A).$$

In the theory of differential equations a very important role is played by the operator-function e^{-At}. In the case when A is an arbitrary bounded operator it is defined by (5.2) or (5.3) with the entire (for a fixed t) function $e^{-t\lambda}$ in the capacity of $f(\lambda)$:

$$e^{-At} = \sum_{k=0}^{\infty} (-1)^k \frac{t^k}{k!} A^k = -\frac{1}{2\pi i} \int_{\Gamma} e^{-zt} (A - zE)^{-1} \, \mathrm{d}z. \tag{5.11}$$

Since the mapping $f(\lambda) \to f(A)$ is a homomorphism, the family of operators e^{-At} is a one-parameter semigroup:

$$e^{-At} e^{-A\tau} = e^{-A(t+\tau)}, \quad e^{-At}\big|_{t=0} = E.$$

It follows directly from formula (5.11) that the operator-function e^{-At} is analytic in the whole complex plane. It is easy to check that

$$\frac{\mathrm{d}}{\mathrm{d}t} e^{-At} + Ae^{-At} = 0. \tag{5.12}$$

If A is an unbounded operator, then it is not clear in general how we should define the operator e^{-At}. For a self-adjoint A it may be defined on the set of vectors h for which

$$\int_{-\infty}^{\infty} e^{-2\lambda t} \, \mathrm{d}(E_\lambda h, h) < \infty$$

by formula (5.7); this set is dense in $\mathfrak{H}$. But, generally speaking, this is an unbounded operator at every t. If A is a non-negative self-adjoint operator, then the function $e^{-t\lambda}$ is bounded and continuous on the spectrum of A for every fixed $t \geq 0$. Therefore, e^{-At} is a bounded operator on $\mathfrak{H}$ at every $t \in [0, \infty)$. As $e^{-\lambda t}$ satisfies the condition (5.9), the operator-function e^{-At} is strongly continuous on $[0, \infty)$. It is uniformly continuous there if and only if A is bounded.

It is easy to show that the operator-function e^{-At} is not strongly differentiable at $t = 0$. But if it is considered on $[\delta, \infty)$, where $\delta > 0$, then condition (5.10) is satisfied. So, e^{-At} is strongly differentiable (even analytic) on this interval and satisfies the equation (5.12). Since

$$\frac{d^n}{dt^n}e^{-At}h = (-1)^n e^{-At}A^n h \qquad (t > 0)$$

when $h \in \mathfrak{D}(A^n)$ $(n \in \mathbb{N})$ and since the vector-function on the right-hand side of this equation is continuous on $[0,\infty)$, $e^{-At}h$ is an n times continuously-differentiable vector-function, satisfying (5.12) on $[0,\infty)$.

5.5. POLAR REPRESENTATION OF A BOUNDED OPERATOR

Let$A$ be a linear operator on a Hilbert space $\mathfrak{H}$. Set

$$\operatorname{Ker} A = \{f : Af = 0\}.$$

Certainly, this is a subspace. If A is a densely defined operator on $\mathfrak{H}$, then

$$\mathfrak{H} = \overline{\mathfrak{R}(A)} \oplus \operatorname{Ker} A^* = \overline{\mathfrak{R}(A^*)} \oplus \operatorname{Ker} A.$$

An operator U is called partially isometric if it isometrically maps the subspace $\mathfrak{H} \ominus \operatorname{Ker} U$ onto $\mathfrak{R}(U)$. For a partially isometric operator $\mathfrak{R}(U)$ is a subspace of $\mathfrak{H}$.

If U is partially isometric, then U^* has the same property. Since U maps $\mathfrak{R}(U^*)$ onto $\mathfrak{R}(U)$, the operator U^* acts from $\mathfrak{R}(U)$ onto $\mathfrak{R}(U^*)$, and also $U^*U = P_1$, $UU^* = P_2$, where P_1, P_2 are the orthoprojectors from $\mathfrak{H}$ onto $\mathfrak{R}(U^*)$ and $\mathfrak{R}(U)$, respectively.

Let $A \in [\mathfrak{H}]$. Then the operator A^*A is non-negative. Consequently, there exists a unique non-negative operator $|A|$ such that $|A|^2 = A^*A$. This operator is called the modulus of A. We have

$$\|Af\|^2 = (Af, Af) = (A^*Af, f) = (|A|f, |A|f) = \||A|f\|^2 \quad (\forall f \in \mathfrak{H}).$$

The operator $U : |A|f \to Af$ isometrically maps $\mathfrak{R}(|A|)$ onto $\mathfrak{R}(A)$. Let us define it by continuity on all of $\overline{\mathfrak{R}(|A|)}$, and set $Uf = 0$ for $f \in \operatorname{Ker}|A|$. The obtained operator is partially isometric.

Thus, any bounded linear operator A on $\mathfrak{H}$ has a representation

$$A = U|A|$$

where $U : \mathfrak{R}(A^*) \to \mathfrak{R}(A)$ is a partially isometric operator. This representation is called the polar representation.

The following properties of the operators U and $|A|$ follow from their definition:

(i) $U^*A = |A|$;

(ii) $|A^*| = U|A|U^*$; $|A| = U^*|A^*|U$;

(iii) $A = |A^*|U$, $|A^*| = AU^*$.

6. Singular Numbers of Completely Continuous Operators and their Properties

6.1. THE MINIMAX PRINCIPLE FOR COMPLETELY CONTINUOUS SELF-ADJOINT OPERATORS

Let A be a completely continuous self-adjoint linear operator on a Hilbert space $\mathfrak{H}$. Then all its eigenvalues are real. It follows from Theorem 4.2 that the operator A admits a uniformly convergent representation

$$A = \sum_{j=1}^{r} \lambda_j(A)(\cdot, e_j)e_j, \tag{6.1}$$

where $\{e_j\}_{j=1}^{r}$ ($r = \operatorname{rank} A = \dim \mathfrak{R}(A)$) is the orthonormal system of eigenvectors of A. Here $\lambda_j(A) \to 0$ as $j \to \infty$, i.e. $\lambda_j(A) = o(1)$.

The expansion (6.1) may be written as

$$A = \sum_{j=1}^{r^+} \lambda_j^+(A)(\cdot, e_j^+)e_j^+ - \sum_{j=1}^{r^-} \lambda_j^-(A)(\cdot, e_j^-)e_j^-, \tag{6.2}$$

where $\lambda_j^+(A)$ and $-\lambda_j^-(A)$ are, respectively, the sequences of all positive and negative eigenvalues of A, indexed in such a way that the $\lambda_j^\pm(A)$ do not increase, and repeated according to their multiplicity; by $e_j^\pm$ we have denoted the orthonormal eigenvectors corresponding to the numbers $\lambda_j^\pm(A)$. The following minimax principle for the numbers $\lambda_n^\pm(A)$ is valid.

THEOREM 6.1. *Let $A = A^*$ be a completely continuous operator on $\mathfrak{H}$. Then*

$$\lambda_n^\pm(A) = \min_L \max_{x \in L\setminus\{0\}} \pm \frac{(Ax, x)}{(x, x)},$$

where L is an arbitrary subspace of $\mathfrak{H}$ such that $\dim L^\perp \leq n - 1$.

Proof. It is sufficient to consider the numbers $\lambda_n^+(A)$, since $\lambda_n^-(A) = \lambda_n^+(-A)$. Let us denote by F_n the linear span of the vectors $\{e_j^+\}_{j=1}^n$, i.e. $F_n =$ l.s.$\{e_1^+, \ldots, e_n^+\}$. Assume that for some L there is an $\epsilon_0 > 0$ such that $(Ax, x) \leq (\lambda_n^+(A) - \epsilon_0)(x, x)$ ($\forall x \in L, x \neq 0$). If, in addition, $\dim(\mathfrak{H} \ominus L) \leq n - 1$, then there exists an $x \neq 0$, $x \in L \cap F_n$. Consequently, $x = \sum_{i=1}^n \alpha_i e_i^+$ and

$$(Ax, x) = \sum_{i=1}^{n} \lambda_i^+(A)|\alpha_i|^2 > \lambda_n^+(A) \sum_{i=1}^{n} |\alpha_i|^2 = \lambda_n^+(A)(x, x),$$

which contradicts the assumption. Thus,

$$\lambda_n^+(A) \leq \max_{x \in L\setminus\{0\}} \frac{(Ax, x)}{(x, x)} \quad (\dim L^\perp \leq n - 1). \tag{6.3}$$

It remains to note that equality holds in (6.3) if $L = \mathfrak{H} \ominus F_{n-1}$. □

COROLLARY 6.1. *Let $A = A^*$, $B = B^*$ be completely continuous operators on $\mathfrak{H}$ and $0 \leq A \leq B$. Then*

$$\lambda_j^+(A) \leq \lambda_j^+(B) \quad (j \in \mathbb{N}).$$

6.2. S-NUMBERS OF A COMPLETELY CONTINUOUS OPERATOR, AND SIMPLEST PROPERTIES

We denote by $\mathfrak{S}_\infty(\mathfrak{H})$, or simply $\mathfrak{S}_\infty$, the set of all completely continuous operators on $\mathfrak{H}$. Let $A \in \mathfrak{S}_\infty$. Then the operators A^*A and $|A| = (A^*A)^{\frac{1}{2}}$ are non-negative (hence, self-adjoint) and completely continuous. Therefore the eigenvalues $\lambda_n(|A|)$ of $|A|$ are non-negative and $\lambda_n(|A|) \to 0$ monotonically as $n \to \infty$. These numbers are usually called the s-numbers of A, and are denoted by $s_n(A)$ $(n \in \mathbb{N})$. It is clear that if rank $|A| = r < \infty$, then $s_n(A) = 0$ for $n \geq r+1$. And also, since $\|A^*A\| = \|A\|^2$, we have $s_1(A) = \|A\|$. If $A = A^*$, then $s_n(A) = |\lambda_n(A)|$, where the $\lambda_n(A)$ are indexed according to non-increasing order of their moduli.

Let $A = U|A|$ be the polar representation of A. In view of (6.1)

$$|A| = \sum_{n=1}^{\infty} s_n(A)(\cdot\,, e_n)e_n.$$

By letting the operator U act on both sides of the equality we obtain

$$A = \sum_{n=1}^{\infty} s_n(A)(\cdot\,, e_n)e_n', \tag{6.4}$$

where the vectors $e_n' = Ue_n$ form an orthonormal system. The expansion (6.4) is called the Schmidt expansion of the completely continuous operator A. It follows from this expansion that

$$A^* = \sum_{n=1}^{\infty} s_n(A)(\cdot\,, e_n')e_n. \tag{6.5}$$

The relations (6.4), (6.5) and property (i) of the polar representation of A yield

$$A^*Ae_n = s_n^2(A)e_n, \quad AA^*e_n' = s_n^2(A)e_n' \qquad (n \in \mathbb{N}),$$

whence

$$s_n(A) = s_n(A^*). \tag{6.6}$$

Let B be any continuous operator on $\mathfrak{H}$. Then the operators AB and BA are completely continuous, and it is possible to define the s-numbers for them. We prove that

$$s_n(BA) \leq \|B\|s_n(A), \tag{6.7}$$

$$s_n(AB) \leq \|B\|s_n(A). \tag{6.8}$$

From the definition we have

$$s_n^2(BA) = \lambda_n(A^*B^*BA), \quad s_n^2(A) = \lambda_n(A^*A).$$

Moreover,

$$(A^*B^*BAf, f) = \|BAf\|^2 \leq \|B\|^2\|Af\|^2 = \|B\|^2(A^*Af, f) \quad (\forall f \in \mathfrak{H}).$$

Since the vector $f \in \mathfrak{H}$ is arbitrary, $A^*B^*BA \leq \|B\|^2 A^*A$. According to Corollary 6.1,

$$s_n^2(BA) = \lambda_n(A^*B^*BA) \leq \|B\|^2\lambda_n(A^*A) = \|B\|^2 s_n^2(A),$$

which is equivalent to (6.7). Since $s_n(AB) = s_n(B^*A^*)$ and $s_n(B^*A^*) \leq \|B^*\| s_n(A^*) = \|B\| s_n(A)$, inequality (6.8) holds too.

6.3. A CHARACTERIZATION OF S-NUMBERS VIA APPROXIMATION AND RELATED INEQUALITIES

We denote by $\mathfrak{K}_n$ the set of all finite-dimensional linear operators on $\mathfrak{H}$ whose dimension is less than or equal to n, i.e. the set of operators in $\mathfrak{H}$ whose ranks do not exceed n.

THEOREM 6.2. *For an operator $A \in \mathfrak{S}_\infty$ the following equalities are valid:*

$$s_{n+1}(A) = \min_{K \in \mathfrak{K}_n} \|A - K\| \quad (n = 0, 1, \ldots). \tag{6.9}$$

Proof. Let $K \in \mathfrak{K}_n$ and rank $K = n$. From the Schmidt expansion

$$K = \sum_{k=1}^{n} s_k(K)(\cdot, g_k)g'_k$$

it follows that $\dim(\mathfrak{H} \ominus \operatorname{Ker} K) = n$. So, by virtue of the minimax principle,

$$s_{n+1}(A) \leq \max_{x \in \operatorname{Ker} K \setminus \{0\}} \frac{\|Ax\|}{\|x\|}.$$

Since $\|Ax\| = \|(A - K)x\| \leq \|A - K\|\|x\|$ for all $x \in \operatorname{Ker} K$, we have $s_{n+1}(A) \leq \|A - K\|$. It remains to observe that $\|A - K_n\| = s_{n+1}(A)$, where

$$K_n = \sum_{j=1}^{n} s_j(A)(\cdot, e_j)e'_j. \tag{6.10}$$

This concludes the proof. □

Equality (6.9) shows that $s_{n+1}(A)$ coincides with the distance of the operator A to the set $\mathfrak{K}_n$. It may be interpreted as a new definition of s-numbers, equivalent to the definition above. Sometimes it is more convenient than the original one.

COROLLARY 6.2. *Let V be an operator of rank r. Then*

$$s_{n+r}(A) \le s_n(A+V) \le s_{n-r}(A). \tag{6.11}$$

Proof. Let K_n be an operator defined by (6.10). Then, in view of Theorem 6.2,

$$s_{n+1}(A) = \|(A+V) - (V+K_n)\| \ge s_{n+r+1}(A+V). \tag{6.12}$$

Interchanging the operators A and $A+V$, we have

$$s_{n+1}(A+V) \ge s_{n+r+1}(A). \tag{6.13}$$

The inequalities (6.12) and (6.13) yield (6.11). □

COROLLARY 6.3. *For arbitrary completely continuous operators A and B,*

$$s_{m+n-1}(A+B) \le s_m(A) + s_n(B) \quad (m, n \in \mathbb{N}) \tag{6.14}$$

$$s_{m+n-1}(AB) \le s_m(A)s_n(B) \quad (m, n \in \mathbb{N}). \tag{6.15}$$

Proof. Let an $(m-1)$-dimensional operator K_1 and an $(n-1)$-dimensional operator K_2 be such that

$$s_m(A) = \|A - K_1\|, \quad s_n(B) = \|B - K_2\|.$$

Then

$$s_{m+n-1}(A+B) \le \|A+B-(K_1+K_2)\| \le \|A-K_1\| + \|B-K_2\|$$
$$= s_m(A) + s_n(B).$$

Moreover, since the dimension of the operator $AK_2 + K_1(B-K_2)$ does not exceed $m+n-2$ and $(A-K_1)(B-K_2) = AB - AK_2 - K_1(B-K_2)$,

$$s_{m+n-1}(AB) \le \|A-K_1\|\|B-K_2\|,$$

whence (6.15) follows. □

The well-known inequality of H. Weyl for non-negative completely continuous operators C_1 and C_2, i.e.

$$\lambda_{n+m-1}(C_1+C_2) \le \lambda_n(C_1) + \lambda_m(C_2),$$

is a special case of relation (6.14). This inequality can be generalized to the case of arbitrary completely continuous self-adjoint operators. Making use of Theorem 6.1 we can show that

$$\lambda^{\pm}_{n+m-1}(C_1+C_2) \le \lambda^{\pm}_n(C_1) + \lambda^{\pm}_m(C_2). \tag{6.16}$$

6.4. THE DISTRIBUTION FUNCTION OF S-NUMBERS

In certain situations it is more convenient to use not a sequence of s-numbers itself but a function of a continuous argument associated with it. Most often this is the so-called distribution function of s–numbers:

$$n(r; A) = \max\{n \ : \ s_n(A) \geq r^{-1}\} \quad (r > 0),$$

This function is non-negative, it increases monotonically together with r. From the results in subsection 6.2 it immediately follows that if $A \in \mathfrak{S}_\infty$, $B \in [\mathfrak{H}]$, then

$$n(r; A) = n(r; A^*), \quad n(r; AB) \leq n(\|B\|r; A),$$
$$n(r; BA) \leq n(\|B\|r; A).$$

LEMMA 6.1. *For all $r_1 > 0$ and $r_2 > 0$ the inequalities*

$$n\left(\frac{r_1 r_2}{r_1 + r_2}; A + B\right) \leq n(r_1; A) + n(r_2; B), \tag{6.17}$$
$$n(r_1 r_2; AB) \leq n(r_1; A) + n(r_2; B)$$

are valid.

Proof. For any $r_1 > 0$, any $r_2 > 0$ we set

$$n = \max\{k : s_k(A) \geq r_1^{-1}\},$$
$$m = \max\{k : s_k(B) \geq r_2^{-1}\},$$
$$p = \max\{k : s_k(A + B) \geq r_1^{-1} + r_2^{-1}\},$$
$$q = \max\{k : s_k(AB) \geq r_1^{-1} r_2^{-1}\}$$

(for the sake of convenience we assume that $s_0(\cdot) = +\infty$).
It is clear, that $n(r; A) = n$, $n(r_2; B) = m$, $n((r_1^{-1} + r_2^{-1})^{-1}; A + B) = p$, $n(r_1 r_2; AB) = q$. Taking into account the inequalities (6.14), (6.15) for s–numbers, namely

$$s_{m+n+1}(A + B) \leq s_{n+1}(A) + s_{m+1}(B),$$
$$s_{m+n+1}(AB) \leq s_{n+1}(A) s_{m+1}(B),$$

from the definition of n and m we obtain

$$s_{m+n+1}(A + B) < r_1^{-1} + r_2^{-1},$$
$$s_{m+n+1}(AB) < r_1^{-1} r_2^{-1}.$$

This implies, because of the extremal nature of the numbers p and q, that $m+n+1 > p$, $m + n + 1 > q$, that is, $m + n + 1 \geq p + 1$, $m + n + 1 \geq q + 1$. Hence,

$$n((r_1^{-1} + r_2^{-1})^{-1}; A + B) \leq n(r_1; A) + n(r_2; B),$$
$$n(r_1 r_2; AB) \leq n(r_1; A) + n(r_2; B)$$

and the proof is complete. □

REMARK 6.1. In a similar way, taking the inequality (6.16) instead of (6.14), it is possible to establish that for completely continuous self-adjoint operators C_1 and C_2 the relation

$$n^{\pm}((r_1^{-1} + r_2^{-1})^{-1}; C_1 + C_2) \leq n^{\pm}(r_1; C_1) + n^{\pm}(r_2; C_2), \tag{6.18}$$

where $n^{\pm}(r;C) = \max\{n : \lambda_n^{\pm}(C) \geq r^{-1}\}$, is valid.

Let θ denote one of the symbols o or O.

COROLLARY 6.4. *If* $n(r;A) = \theta_1(r^{\alpha})$, $n(r;B) = \theta_2(r^{\beta})$, *then*

$$n(r;A+B) = \theta_1(r^{\alpha}) + \theta_2(r^{\beta}),$$
$$n(r;AB) = (\theta_1+\theta_2)(r^{\alpha\beta/(\alpha+\beta)}).$$

Proof. Set in (6.17) $r_1 = r_2 = 2r$. We have

$$n(r;A+B) \leq n(2r;A) + n(2r;B) = \theta_1(r^{\alpha}) + \theta_2(r^{\beta}).$$

Analogously, taking in (6.17) $r_1 = r^{\beta/(\alpha+\beta)}$, $r_2 = r^{\alpha/(\alpha+\beta)}$ we obtain

$$n(r;AB) \leq n(r^{\beta/(\alpha+\beta)};A) + n(r^{\alpha/(\alpha+\beta)};B) = (\theta_1+\theta_2)(r^{\alpha\beta/(\alpha+\beta)}). \qquad \square$$

Let A be a completely continuous operator. There exists a close connection between the rate with which its s–numbers approach zero and the asymptotic behaviour of $n(r;A)$ in a neighbourhood of infinity. This can be seen, for example, from the equality

$$\sum_{k=1}^{\infty} s_k^p(A) = p\int_0^{\infty} \frac{n(r;A)}{r^{p+1}}\,dr \quad (p>0)$$

which results from integration by parts. A more substantial assertion of this kind is contained in the next lemma.

LEMMA 6.2. *The following asymptotic equalities are equivalent:*

(i) $s_n(A) \sim a^{-\frac{1}{\alpha}} n^{-\frac{1}{\alpha}} \quad (n \to \infty)$;
(ii) $n(r;A) \sim a^{-1} r^{\alpha} \quad (r \to \infty)$.

Proof. Let $s_n(A) \sim a^{-\frac{1}{\alpha}} n^{-\frac{1}{\alpha}}$. Then, whatever $\epsilon > 0$, for all n which are larger than a certain $n_0 = n_0(\epsilon)$,

$$(a+\epsilon)^{-\frac{1}{\alpha}} n^{-\frac{1}{\alpha}} \leq s_n(A) \leq (a-\epsilon)^{-\frac{1}{\alpha}} n^{-\frac{1}{\alpha}},$$

whence

$$[(a+\epsilon)^{-1} r^{\alpha}] \leq n(r;A) \leq [(a-\epsilon)^{-1} r^{\alpha}], \quad r > s_{n_0}^{-1}(A)$$

(here $[\cdot]$ denotes, as is customary, the integer part of a number).
Therefore,

$$(a-\epsilon)^{-1} \geq \overline{\lim_{r\to\infty}}\,(n(r;A)r^{-\alpha}) \geq \underline{\lim_{r\to\infty}}\,(n(r;A)r^{-\alpha}) \geq (a+\epsilon)^{-1}.$$

As $\epsilon > 0$ was arbitrary, $\lim_{r\to\infty}(n(r;A)r^{-\alpha})$ exists and is equal to a^{-1}.

Conversely, let $n(r;A) \sim a^{-1}r^{\alpha}$. Then for each $\epsilon > 0$ an $r_0(\epsilon)$ can be found such that

$$[(a+\epsilon)^{-1} r^{\alpha}] \leq n(r;A) \leq [(a-\epsilon)^{-1} r^{\alpha}]$$

for all $r > r_0(\epsilon)$. This implies that for sufficiently large n, $(a+\epsilon)^{-\frac{1}{\alpha}} \le s_n(A) \le (a-\epsilon)^{-\frac{1}{\alpha}} n^{-\frac{1}{\alpha}}$. So,

$$(a-\epsilon)^{-\frac{1}{\alpha}} \ge \overline{\lim_{n\to\infty}} (s_n(A) n^{\frac{1}{\alpha}}) \ge \underline{\lim}_{n\to\infty} (s_n(A) n^{\frac{1}{\alpha}}) \ge (a+\epsilon)^{-\frac{1}{\alpha}}.$$

Since $\epsilon > 0$ was arbitrary, $\lim_{n\to\infty}(s_n(A) n^{\frac{1}{\alpha}})$ exists and is equal to $a^{-\frac{1}{\alpha}}$. □

Similar arguments show that the asymptotic equalities

$$s_n(A) = \theta(n^{-\frac{1}{\alpha}}) \quad \text{and} \quad n(r; A) = \theta(r^{\alpha})$$

occur or do not occur simultaneously.

6.5. ADDITIVE PERTURBATIONS AND THE ASYMPTOTICS OF S-NUMBERS

THEOREM 6.3. *Let A and V be completely continuous operators. Suppose that*

$$n(r; A) = \sum_{k=1}^{N} a_k r^{\alpha_k} + \theta(r^{\alpha_N}) \quad (\alpha_1 > \cdots > \alpha_N > 0)$$

and

$$n(r; V) = \theta(r^{\gamma})$$

where $\gamma \le \frac{\alpha_N}{1+\alpha_1-\alpha_N}$. Then

$$n(r; A+V) = \sum_{k=1}^{N} a_k r^{\alpha_k} + \theta(r^{\alpha_N}) \quad (r \to \infty).$$

Proof. Let $\theta = o$. According to the assumption, $n(r; V) = r^{\gamma}\delta(r)$, where $\delta(r) = o(1)$ as $r \to \infty$. For any fixed $\epsilon > 0$ we set $r_1 = r(1+\epsilon r^{\alpha_N-\alpha_1})$, $r_2 = r(1+\epsilon^{-1} r^{\alpha_1-\alpha_N})$. Then $r_1^{-1} + r_2^{-1} = r^{-1}$. Let us show that

$$\left|\left(n(r; A+V) - \sum_{k=1}^{N} a_k r^{\alpha_k}\right) r^{-\alpha_N}\right| \le \epsilon. \tag{6.19}$$

Employing the results of subsection 6.4 we obtain

$$\left|\left(n(r; A+V) - \sum_{k=1}^{N} a_k r^{\alpha_k}\right) r^{-\alpha_N}\right| \le$$

$$\le \left|n(r_1; A) - \sum_{k=1}^{N} a_k r_1^{\alpha_k}\right| r_1^{-\alpha_N} (r_1^{\alpha_N} r^{-\alpha_N}) +$$

$$+ \sum_{k=1}^{N} a_k (r_1^{\alpha_k} - r^{\alpha_k}) r^{-\alpha_N} + n(r_2; V) r^{-\alpha_N}.$$

Since r_1 and r_2 tend to ∞ as $r \to \infty$, in order to prove (6.19) it is sufficient to prove that

$$\lim_{n\to\infty}(r_1 r^{-1}) = 1, \tag{6.20}$$

$$\sum_{k=1}^{N} a_k(r_1^{\alpha_k} - r^{\alpha_k})r^{-\alpha_N} \le c_1(a_k, \alpha_k)\epsilon, \tag{6.21}$$

$$n(r_2; V) = o(r^{\alpha_N}) \quad (r \to \infty). \tag{6.22}$$

The validity of (6.20) is obvious. As to (6.22) we have

$$n(r_2;V)r^{-\alpha_N} = r_2^{\gamma}\delta(r_2)r^{-\alpha_N} \le$$
$$\le r^{\frac{\alpha_N}{1+\alpha_1-\alpha_N}-\alpha_N}\left(1+\epsilon^{-1}r^{\alpha_1-\alpha_N}\right)^{\frac{\alpha_N}{1+\alpha_1-\alpha_N}}\delta(r_2) = o(1).$$

At last we verify that (6.21) is also true. Indeed,

$$\sum_{k=1}^{N} a_k(r_1^{\alpha_k} - r^{\alpha_k})r^{-\alpha_N} = \sum_{k=1}^{N} a_k r^{\alpha_k-\alpha_N}\left((1+\epsilon r^{\alpha_N-\alpha_1})^{\alpha_k} - 1\right)$$
$$= \sum_{k=1}^{N} a_k r^{\alpha_k-\alpha_N}\left(\epsilon\alpha_k r^{\alpha_N-\alpha_1}\right)(1+o(1)) \le c_1(a_k, \alpha_k)\epsilon.$$

By virtue of (6.19) and the arbitrariness of $\epsilon > 0$ we obtain the desired result.

Analogously one can consider the case $\theta = O$: it is sufficient to set $\epsilon = 1$ everywhere and to replace the relation (6.22) by $n(r_2; V) = O(r^{\alpha_N})$. □

Setting $N = 1$, $\theta = o$ and using Lemma 6.2 we arrive at the following statement.

COROLLARY 6.5. *If* $\lim_{n\to\infty}(s_n(A)n^{\frac{1}{\alpha}}) = \mathrm{d} > 0$, $\lim_{n\to\infty}(s_n(V)n^{\frac{1}{\alpha}}) = 0$, *then*

$$\lim_{n\to\infty}\left(s_n(A+V)n^{\frac{1}{\alpha}}\right) = \mathrm{d}. \tag{6.23}$$

It follows from the results of subsection 6.4 that the last assertion is also valid if $\mathrm{d} = 0$.

Using the inequality (6.18) instead of (6.17), the next theorem can be established by a similar reasoning.

THEOREM 6.4. *Let operators $A = A^*$ and $V = V^*$ be completely continuous and put*

$$n^+(\lambda; A) = \sum_{i=1}^{N} a_i\lambda^{\alpha_i} + \theta(\lambda^{\alpha_N}) \quad (\alpha_1 > \cdots > \alpha_N > 0),$$

$$n^-(\lambda; A) = \sum_{j=1}^{M} b_j\lambda^{\beta_j} + \theta(\lambda^{\beta_M}) \quad (\beta_1 > \cdots > \beta_M > 0),$$

$$n(\lambda; V) = n^+(\lambda; V) + n^-(\lambda; V) = \theta(\lambda^{\gamma}),$$

where $\gamma \leq \min\{\alpha_N(1+\alpha_1-\alpha_N)^{-1},\ \beta_M(1+\beta_1-\beta_M)^{-1}\}$. *Then*

$$n^+(\lambda; A+V) = \sum_{i=1}^{N} a_i \lambda^{\alpha_i} + \theta(\lambda^{\alpha_N}),$$

$$n^-(\lambda; A+V) = \sum_{j=1}^{M} b_j \lambda^{\beta_j} + \theta(\lambda^{\beta_M}) \quad (\lambda \to +\infty).$$

6.6. THE IDEALS $\mathfrak{S}_p$

We consider the following classes of operators:

$$\mathfrak{S}_p = \{A : A \in \mathfrak{S}_\infty,\ \sum_{j=1}^{\infty} s_j^p(A) < \infty\} \quad (p > 0).$$

It follows from property (6.6) of s-numbers that $A \in \mathfrak{S}_p$ implies $A^* \in \mathfrak{S}_p$. The properties (6.7) and (6.8) show that if $A \in \mathfrak{S}_p$, $B \in [\mathfrak{H}]$, then AB and $BA \in \mathfrak{S}_p$. From (6.14) we conclude that $A, B \in \mathfrak{S}_p$ implies $A + B \in \mathfrak{S}_p$. Thus, for $p > 0$ the class $\mathfrak{S}_p$ is a two-sided ideal in the algebra $[\mathfrak{H}]$.

It is possible to prove for $p \geq 1$ (see, for example, Schatten [1]) that $\mathfrak{S}_p$ is a Banach space with respect to the norm

$$\|A\|_p = \Big(\sum_{j=1}^{\infty} s_j^p(A)\Big)^{\frac{1}{p}}.$$

This norm possesses the following properties:

(*i*) $\|A\|_p = \|A^*\|_p \quad (A \in \mathfrak{S}_p)$;

(*ii*) $\|AB\|_p \leq \|A\|_p\|B\|, \quad \|BA\|_p \leq \|A\|_p\|B\| \quad (A \in \mathfrak{S}_p,\ B \in [\mathfrak{H}])$.

If the operator B is unitary, then

$$\|AB\|_p = \|BA\|_p = \|A\|_p.$$

Operators of the class $\mathfrak{S}_1$ (respectively $\mathfrak{S}_2$) are called nuclear (respectively Hilbert–Schmidt operators), while $\|\cdot\|_1$ (respectively ($\|\cdot\|_2$) is the trace or, which is the same, the nuclear norm (respectively the Hilbert–Schmidt norm). If $\{e_i\}_{i=1}^{\infty}$ is an arbitrary orthonormal basis in $\mathfrak{H}$, then for an operator $A \in \mathfrak{S}_2$,

$$\sum_{i=1}^{\infty} \|Ae_i\|^2 = \|A\|_2^2.$$

An operator $A \in [\mathfrak{H}]$ is called dissipative if

$$\operatorname{Im} A = \frac{1}{2i}(A - A^*) \geq 0.$$

We will need the following theorem.

THEOREM 6.5. *If a dissipative operator A is nuclear, then the system of its eigenvectors and adjoined vectors is dense in $\mathfrak{H}$.*

An operator $A \in \mathfrak{S}_\infty$ is called a Volterra operator if it has no eigenvalues different from zero. If λ is an eigenvalue of a completely continuous operator A, then $\overline{\lambda}$ is an eigenvalue of A^*. Therefore, if A is a Volterra operator, then so is A^*.

For Volterra operators the following statement is true.

THEOREM 6.6. *Let A be a Volterra operator. If its imaginary part $A_J = \mathrm{Im}\, A$ belongs to $\mathfrak{S}_p$ $(1 < p < \infty)$, then $A \in \mathfrak{S}_p$.*

7. Locally Convex Topological Spaces

7.1. LINEAR TOPOLOGICAL SPACES

If in a set $\mathfrak{F}$ a system $\{U\}$ of subsets, called neighbourhoods, is specified possessing the following properties:

(i) each point $f \in \mathfrak{F}$ belongs to a certain neighbourhood $U = U(f) \in \{U\}$;
(ii) if a point f belongs to neighbourhoods U and V from $\{U\}$, then it belongs to a neighbourhood W entirely lying in $U \cap V$;
(iii) for any pair of points $f_1 \neq f_2$ it is possible to find a neighbourhood U containing f_1 and not containing f_2,

then this set is called a topological space. All neighbourhoods and all their unions form the system of open sets in $\mathfrak{F}$. A closed set is defined as a set whose complement in all of $\mathfrak{F}$ is open. An open set is characterized by the fact that each point of it is interior, i.e. it has a neighbourhood contained within the set. This readily implies that the union of any number and the intersection of a finite number of open sets is again an open set.

A topological space $\mathfrak{F}$ is called a Hausdorff space if for two arbitrary points $f_1 \neq f_2$ of it neighbourhoods $U(f_1)$ and $U(f_2)$ can be found such that $U(f_1) \cap U(f_2) = \emptyset$.

In a topological space convergence of a sequence of elements is introduced as usual. Namely, a sequence $\{f_n\}_{n=1}^\infty$ converges to f if for any neighbourhood $U(f)$ of f we can indicate a number n_0 such that $f_n \in U(f)$ as $n > n_0$. We would like to note that in a general topological space one can not obtain the closure of a set $F \subset \mathfrak{F}$ by joining limits of convergent sequences $f_n \in F$ to this set. Moreover, a set containing all limits of convergent sequences of its elements is not closed, in general. However, all this is true if we replace "sequence" by "generalized sequence". The last notion is introduced in order to operate with limit transitions in topological spaces.

A set of indices I with an order $>$ satisfying the requirements:

(i) if $\alpha, \beta \in I$, then there exists a $\gamma \in I$ such that $\gamma > \alpha$ and $\gamma > \beta$;
(ii) $>$ is a partial order, that is,
$\alpha > \alpha$, $\forall \alpha \in I$,
if $\alpha > \beta$ and $\beta > \alpha$, then $\alpha = \beta$,
if $\alpha > \beta$ and $\beta > \gamma$, then $\alpha > \gamma$

is called a directed set. A generalized sequence (or a net) in the topological space $\mathfrak{F}$ is a mapping from I into $\mathfrak{F}$; it is denoted by the symbol $\{f_\alpha\}_{\alpha\in I}$. If we take the set of all natural numbers with its natural order as I, then every generalized sequence will be simply a sequence in $\mathfrak{F}$, so a generalized sequence is actually a generalization of a sequence.

A generalized sequence $\{f_\alpha\}_{\alpha\in I}$ in a topological space $\mathfrak{F}$ is said to converge to $f \in \mathfrak{F}$ (we write $f_\alpha \xrightarrow{\mathfrak{F}} f$) if for any neighbourhood $U(f)$ of the point f there is a $\beta \in I$ such that $f_\alpha \in U(f)$ for all $\alpha > \beta$. If $\mathfrak{F}$ is Hausdorff, then the relations $f_\alpha \xrightarrow{\mathfrak{F}} f$, $f_\alpha \xrightarrow{\mathfrak{F}} g$ imply $f = g$, i.e. a convergent generalized sequence has only one limit.

Let two topological spaces $\mathfrak{F}_1$ and $\mathfrak{F}_2$ be given. A function F mapping $\mathfrak{F}_1$ into $\mathfrak{F}_2$ is called continuous if the inverse image $F^{-1}(O)$ of any open set O in $\mathfrak{F}_2$ is an open set in $\mathfrak{F}_1$. It is not difficult to show that a function $F : \mathfrak{F}_1 \to \mathfrak{F}_2$ is continuous if and only if the generalized sequence $\{F(f_\alpha)\}_{\alpha\in I}$ converges to $F(f)$ in $\mathfrak{F}$ for any convergent generalized sequence $\{f_\alpha\}_{\alpha\in I}$ in $\mathfrak{F}$ such that $f_\alpha \xrightarrow{\mathfrak{F}} f$.

We say that two spaces $\mathfrak{F}_1$ and $\mathfrak{F}_2$ are homeomorphic (topologically isomorphic) if there exists a one-to-one continuous mapping with continuous inverse of one of the spaces onto the other.

The space $\mathfrak{F}$ is called a linear topological space if it is both a linear and a topological space, where the operations of addition and multiplication by a number are continuous relative to the topology of $\mathfrak{F}$. In a linear topological space translations of neighbourhoods of zero by any vector f_0 give a system of neighbourhoods equivalent to the original one. We would like to recall that two systems of neighbourhoods $\{U\}$ and $\{V\}$ are called equivalent if each neighbourhood from $\{U\}$ is contained in a certain neighbourhood from $\{V\}$ and vice versa. Thus, a topology in a linear topological space is completely determined by a system of neighbourhoods of zero.

A family $\{M\}$ of subsets M in the space $\mathfrak{F}$ is called a neighbourhood base of a point f if each M is a neighbourhood of f and if for a given neighbourhood $U(f)$ there exists an $M \in \{M\}$ contained in $U(f)$. If the space $\mathfrak{F}$ is such that every $f \in \mathfrak{F}$ has a countable neighbourhood base, then everywhere above we can take ordinary instead of generalized sequences.

A set M in a vector space $\mathfrak{F}$ is called convex if

$$\alpha x + (1-\alpha)y \in M$$

for any $x, y \in M$ and $0 < \alpha < 1$. A set M is balanced if $\alpha x \in M$ when $x \in M$ and $|\alpha| \le 1$. A set M is called absorbing if for any $f \in \mathfrak{F}$ we can find an $\alpha > 0$ with $\alpha f \in M$. In the most commonly used linear topological spaces there exists a convex neighbourhood base of zero. Such spaces are called locally convex. Every neighbourhood of zero in a linear topological space is absorbing and contains a balanced neigbourhood; every convex neigbourhood of zero in it contains a balanced convex neigbourhood. This implies that every locally convex space has a balanced absorbing convex neighbourhood base of zero.

A system of balanced convex neighbourhoods $\{U(0)\}$ of zero in a locally convex space $\mathfrak{F}$ such that the family of sets

$$\bigcap_{k=1}^{n} \lambda_k U_k(0) \qquad (\lambda_k \in \mathbb{C}^1)$$

is a neighbourhood base of zero is called a generating system of neighbourhoods of zero.

Local convexity is closely connected with the notion of semi-norm.

A real-valued function $p(x)$ defined on a linear space $\mathfrak{X}$ is called a semi-norm if the following conditions are satisfied:

$$\begin{aligned} p(x+y) &\leq p(x) + p(y) \qquad &&\text{(semi-additivity)} \\ p(\alpha x) &= |\alpha| p(x) &&(\forall \alpha \in \mathbb{C}^1). \end{aligned}$$

Let $P = \{p_\gamma(x) : \gamma \in \Gamma\}$ be a family of semi-norms in $\mathfrak{X}$. It is called separating if for any $x \in \mathfrak{X}$ one can find a semi-norm p_γ from P so that $p_\gamma(x) \neq 0$. Assume that P is separating. Choose a system of semi-norms from P, for example $p_{\gamma_1}(x), \dots p_{\gamma_n}(x)$, and a positive number ϵ, and consider the set

$$U = \{x \in \mathfrak{X} : p_{\gamma_j}(x) \leq \epsilon \quad (j = 1, \dots, n)\}.$$

U is a balanced convex absorbing set. The family of sets U satisfies all requirements of a neighbourhood base of zero. The space $\mathfrak{X}$ topologized in this way is locally convex, and every semi-norm $p_\gamma(x)$ is continuous on $\mathfrak{X}$ in this topology. Conversely, one can define the topology of any locally convex space using a separating family of semi-norms. For example, we can take as semi-norms the Minkowski functionals of all possible balanced convex absorbing open sets M in $\mathfrak{F}$, i.e. the functionals

$$p_M(x) = \inf_{\alpha > 0, \alpha x \in M} \alpha^{-1}.$$

A generalized sequence $\{f_\alpha\}_{\alpha \in I}$ in a locally convex space $\mathfrak{F}$ is called fundamental if for every $\epsilon > 0$ and every p_γ there exists an α_0 such that $p_\gamma(f_\alpha - f_\beta) < \epsilon$ as $\alpha, \beta > \alpha_0$. The space $\mathfrak{F}$ is complete if every fundamental generalized sequence converges in it.

A linear space Φ in which a topology is introduced as above using a countable separating family $P = \{p_\gamma\}_{\gamma \in \mathbb{N}}$ of compatible norms, is called countably normed. Recall that two norms $\|\cdot\|_1$ and $\|\cdot\|_2$ in Φ are called compatible if any sequence $f_n \in \Phi$ which is fundamental in both and which converges in one of them to zero, converges to zero in the other norm too. The given sequence of norms $p_\gamma(x)$ $(\gamma \in \mathbb{N})$ can always be assumed to be non-decreasing:

$$p_1(x) \leq p_2(x) \leq \dots \qquad (\forall x \in \Phi).$$

If this is not the case, then every norm $p_\gamma(x)$ may be replaced by the norm $p'_\gamma(x) = \max\{p_1(x), \dots, p_\gamma(x)\}$. For each γ we complete Φ with respect to the norm p_γ. We thus obtain a system of Banach spaces Φ_γ. Since all norms are compatible and the sequence $p_\gamma(x)$ is non-decreasing,

$$\Phi_1 \supset \Phi_2 \supset \cdots \supset \Phi_n \supset \cdots \supset \Phi.$$

The space Φ is complete if and only if it coincides with the intersection of all spaces Φ_n:

$$\Phi = \bigcap_{n=1}^{\infty} \Phi_n.$$

In what follows we will always assume that a countably-normed space is complete. Moreover, without loss of generality we can assume that all spaces Φ_γ are different. The following lemma is valid.

LEMMA 7.1. *For any sequence of positive numbers $\{m_n\}_{n=1}^{\infty}$ one can indicate an element $f \in \Phi$ satisfying the inequalities*

$$\|f\|_\gamma > m_\gamma \qquad (\|\cdot\|_\gamma = p_\gamma(\cdot)).$$

A set M in a topological space $\mathfrak{F}$ is dense in $\mathfrak{F}$ if its closure coincides with $\mathfrak{F}$. M is nowhere dense if its closure contains no interior point. Countable unions of nowhere dense sets are called sets of the first category. Every subset in $\mathfrak{F}$ that is not of the first category is called a set of the second category. The category theorem asserts that any countably-normed space is a set of the second category in itself. We need also the following fact, which is part of the open mapping theorem: let $\mathfrak{B}_i$ $(i = 1, 2)$ be Banach spaces and let a linear continuous operator $\Gamma : \mathfrak{B}_1 \to \mathfrak{B}_2$ be such that its image $\Gamma(\mathfrak{B}_1)$ is a set of the second category in $\mathfrak{B}_2$, then $\Gamma(\mathfrak{B}_1) = \mathfrak{B}_2$.

Denote by $\mathfrak{F}'$ the set of all continuous linear functionals on a linear topological space $\mathfrak{F}$ (continuous linear mappings from $\mathfrak{F}$ into $\mathbb{C}^1$). It is a vector space, called the dual of $\mathfrak{F}$. In the case when $\mathfrak{F}$ is locally convex this space is rather rich in elements.

Two extreme topologies in $\mathfrak{F}'$ are distinguished: the weak topology, denoted by $\sigma(\mathfrak{F}', \mathfrak{F})$, and the strong one, denoted by $\tau^*(\mathfrak{F}', \mathfrak{F})$. A weak neighbourhood of zero in $\mathfrak{F}'$ is determined by a finite set $\{f_1, \ldots, f_n\}$ $(f_i \in \mathfrak{F})$ and a number $\epsilon > 0$ as the totality of all $F \in \mathfrak{F}'$ for which

$$|F(f_k)| < \epsilon \quad (k = 1, \ldots, m).$$

A strong neighbourhood of zero in $\mathfrak{F}'$ is determined by a bounded set $M \subset \mathfrak{F}$ and a number $\epsilon > 0$ as the totality of all $F \in \mathfrak{F}'$ such that

$$|F(f)| < \epsilon \quad (\forall f \in M).$$

In other words, the strong topology is generated by the semi-norms $\rho_M(F) = \sup_{f \in M} |F(f)|$. Recall that a set M in $\mathfrak{F}$ is called bounded if it is absorbed by any neighbourhood of zero, that is, if for any neighbourhood $U(0)$ of zero a number $\lambda > 0$ exists such that $\lambda M \subset U(0)$. If $\mathfrak{F} = \mathfrak{B}$ is a Banach space, then the strong topology in $\mathfrak{B}'$ coincides with its topology as a normed space.

It is obvious that weak convergence is convergence on each element from $\mathfrak{F}$; similarly to the case of a Banach space (see section 1.7), a sequence $F_n \in \mathfrak{F}'$ converges weakly to $F \in \mathfrak{F}'$ if and only if for any $f \in \mathfrak{F}$, we have $F_n(f) \to F(f)$ $(n \to \infty)$.

Let $\mathfrak{F}$ be a locally convex space and $\mathfrak{F}'$ be its dual. Equip $\mathfrak{F}'$ with the strong topology $\tau^*(\mathfrak{F}', \mathfrak{F})$. The dual of $\mathfrak{F}'$ in the $\tau^*(\mathfrak{F}', \mathfrak{F})$–topology is called the second dual of $\mathfrak{F}$, and is denoted by $\mathfrak{F}''$: $\mathfrak{F}'' = (\mathfrak{F}')'$, if it is equipped with the topology $\tau^*(\mathfrak{F}'', \mathfrak{F}')$. The mapping $J : \mathfrak{F} \to \mathfrak{F}''$ defined by

$$(Jf)F = Ff \qquad (F \in \mathfrak{F}')$$

is called the natural imbedding of $\mathfrak{F}$ into $\mathfrak{F}''$. It is not always continuous. If it is a homeomorphism, then $\mathfrak{F}$ is called reflexive. I.e., $\mathfrak{F}$ is reflexive if

(i) $\mathfrak{F}'' = \mathfrak{F}$;

(ii) the topology $\tau^*(\mathfrak{F}, \mathfrak{F}')$ coincides on $\mathfrak{F}$ with the original topology.

A set M in a topological space $\mathfrak{F}$ is said to be compact if any covering of M by open sets contains a finite number of sets which also cover M. The following important theorem concludes this subsection.

THEOREM 7.1. (Banach–Alaoglu). *Let $\mathfrak{B}'$ be the dual of a Banach space $\mathfrak{B}$. Then the unit ball in $\mathfrak{B}'$ is compact in the weak topology $\sigma(\mathfrak{B}', \mathfrak{B})$.**

7.2. INDUCTIVE AND PROJECTIVE LIMITS OF BANACH SPACES

We will consider two constructions which allow us to form locally convex topological vector spaces starting from a given family of locally convex spaces.

Let $\{\mathfrak{F}_\tau\}_{\tau \in T}$ (τ runs through an ordered set of indices T) be a family of locally convex spaces $\mathfrak{F}_\tau$. Assume that $\mathfrak{F}_{\tau_2} \subseteq \mathfrak{F}_{\tau_1}$ if $\tau_1 < \tau_2$ and that this inclusion is continuous. Set $\mathfrak{F}_{\text{pr}} = \bigcap_{\tau \in T} \mathfrak{F}_\tau$. It is clear that $\mathfrak{F}_{\text{pr}}$ is a vector space. Let us equip it with the following topology. As a generating system of neighbourhoods of zero we take the family of all sets of the form $U_\tau(0) \bigcap \mathfrak{F}_{\text{pr}}$ where $\{U_\tau(0)\}$ is a neighbourhood base of zero in $\mathfrak{F}_\tau$. It is not difficult to check that $\mathfrak{F}_{\text{pr}}$ is, with this topology, a locally convex topological vector space. It is called the projective limit of the family $\{\mathfrak{F}_\tau\}_{\tau \in T}$: $\mathfrak{F}_{\text{pr}} = \lim_{\tau \in T} \text{pr} \mathfrak{F}_\tau$.

It follows from the definition that convergence of a sequence $f_n \in \mathfrak{F}_{\text{pr}}$ to an element $f \in \mathfrak{F}_{\text{pr}}$ is equivalent to convergence of f_n to f in every space $\mathfrak{F}_\tau$. Boundedness of a set in $\mathfrak{F}_{\text{pr}}$ is boundedness of it in every $\mathfrak{F}_\tau$.

Suppose a set $\{\mathfrak{F}_\tau\}_{\tau \in T}$ of locally convex spaces is "increasing", in the sense that $\mathfrak{F}_{\tau_1} \subseteq \mathfrak{F}_{\tau_2}$ if $\tau_1 < \tau_2$ and the inclusion is continuous. Set $\mathfrak{F}_{\text{ind}} = \bigcup_{\tau \in T} \mathfrak{F}_\tau$ and introduce in $\mathfrak{F}_{\text{ind}}$ the following topology. As a neighbourhood base of zero we take the family of sets $\bigcup U_\tau(0)$, where the $U_\tau(0)$ are such that $U_{\tau_1}(0) \subseteq U_{\tau_2}(0)$ if $\tau_1 < \tau_2$. With this topology $\mathfrak{F}_{\text{ind}}$ is a locally convex topological space, called the inductive limit of the spaces $\mathfrak{F}_\tau$: $\mathfrak{F}_{\text{ind}} = \text{limind}_{\tau \in T} \mathfrak{F}_\tau$.

* The topology $\sigma(\mathfrak{B}', \mathfrak{B})$ is usually called the weak *-topology of $\mathfrak{B}'$.

We call a set $M \subset \mathfrak{F}_{\text{ind}}$ regularly bounded if $M \subset \mathfrak{F}_\tau$ for some $\tau \in T$ and if it is bounded in this $\mathfrak{F}_\tau$. A regularly bounded set in $\mathfrak{F}_{\text{ind}}$ is bounded. The converse need not be true, even not in the case of normed $\mathfrak{F}_\tau$. If any bounded set in $\mathfrak{F}_{\text{ind}}$ is regularly bounded, then $\mathfrak{F}_{\text{ind}}$ is called a regular inductive limit. We give a few examples of regular inductive limits: (a) $\mathfrak{F}_{\text{ind}}$ when the $\mathfrak{F}_\tau$ are reflexive Banach spaces; (b) $\mathfrak{F}_{\text{ind}}$ provided that the $\mathfrak{F}_\tau$ are Banach spaces and the inclusions $\mathfrak{F}_{\tau_1} \subseteq \mathfrak{F}_{\tau_2}$ $(\tau_1 < \tau_2)$ for any τ_1, $\tau_2 \in T$ are compact; (c) the inductive limit of locally convex topological spaces $\mathfrak{F}_\tau$ such that $\mathfrak{F}_{\tau_1}$ is a subspace of $\mathfrak{F}_{\tau_2}$ as $\tau_1 < \tau_2$, i.e. $\mathfrak{F}_{\tau_1}$ is a closed subset of $\mathfrak{F}_{\tau_2}$, while the topology in $\mathfrak{F}_{\tau_1}$ coincides with the topology induced in it by the topology of $\mathfrak{F}_{\tau_2}$.

It turns out that if $\mathfrak{F}_{\text{ind}} = \text{limind}_{\tau\in T}\mathfrak{F}_\tau$ $(\mathfrak{F}_{\text{pr}} = \text{limpr}_{\tau\in T}\mathfrak{F}_\tau)$, then $\mathfrak{F}'_{\text{ind}} = \bigcap_{\tau\subset T}\mathfrak{F}'_\tau$ $(\mathfrak{F}'_{\text{pr}} = \bigcup_{\tau\in T}\mathfrak{F}'_\tau)$

In the sequel we will often use inductive and projective limits of families of reflexive Banach spaces $\mathfrak{B}_\tau$ $(\tau \in T)$ imbedded continuously and densely one into another, or of Banach spaces imbedded compactly and densely one into another. The continuity of the imbeddings in these situations means that if the family $\mathfrak{B}_\tau$ is "increasing", then $\|\cdot\|_{\mathfrak{B}_{\tau_1}} \le \|\cdot\|_{\mathfrak{B}_{\tau_2}}$ as $\tau_1 > \tau_2$; for a "decreasing" family the opposite inequalities must hold. In the "increasing" case the inductive limit $\mathfrak{B}_{\text{ind}} = \text{limind}_{\tau\in T}\mathfrak{B}_\tau$ is always regular, and if the spaces $\mathfrak{B}'_\tau$, $\mathfrak{B}'_{\text{ind}}$ and $\mathfrak{B}'_{\text{pr}} = (\text{limpr}_{\tau\in T}\mathfrak{B}_\tau)'$ are equipped with the strong topologies then the following equalities hold not only in the set-theoretic, but also in topological sense:

$$\mathfrak{B}'_{\text{ind}} = \operatorname*{limpr}_{\tau\in T}\mathfrak{B}'_\tau, \quad \mathfrak{B}'_{\text{pr}} = \operatorname*{limind}_{\tau\in T}\mathfrak{B}'_\tau. \tag{7.1}$$

7.3. THE SPACE $D(\Omega)$

Let Ω be a domain in $\mathbb{R}^n$. The least closed set in Ω outside of which a function $\varphi(t)$ $(t = \{t_1, \dots, t_n\})$ defined on Ω vanishes is called the support of φ (we denote it by $\operatorname{supp}\varphi$). Set $D(\Omega) = \{\varphi : \varphi$ is an infinitely-differentiable function with compact support in $\Omega\}$. If K is a compact set lying in Ω, then

$$D_K(\Omega) = \{\varphi : \varphi \in D(\Omega), \quad \operatorname{supp}\varphi \subseteq K\}.$$

We denote by $D_K^n(\Omega)$ the completion of $D_K(\Omega)$ in the norm $\|\cdot\|_{C^n[K]}$ and equip $D_K(\Omega)$ with the topology of the projective limit of the Banach spaces $D_K^n(\Omega)$ $(n \in \mathbb{N})$:

$$D_K(\Omega) = \lim_{n\in\mathbb{N}} \text{pr}\, D_K^n(\Omega).$$

Denoting by $\{K_n\}_{n\in\mathbb{N}}$ a sequence of extending compact sets lying in Ω and with union all of Ω, we obtain the set-theoretic equality

$$D(\Omega) = \bigcup_{n\in\mathbb{N}} D_{K_n}(\Omega).$$

Since $D_{K_n}(\Omega) \subset D_{K_{n+1}}(\Omega)$ and $D_{K_n}(\Omega)$ is a subspace of $D_{K_{n+1}}(\Omega)$, the set $D(\Omega)$ with the topology of the inductive limit of the spaces $D_{K_n}(\Omega)$ $(n \in \mathbb{N})$ is a regular inductive limit. It is possible to show that convergence of a sequence φ_n to a function φ in $\mathcal{D}(\Omega)$ is equivalent to the fact that the φ_n belong to a certain $\mathcal{D}_{K_m}(\Omega)$ and converge in this space, i.e. uniform convergence on K_m of the sequence $\{\varphi_n\}$ and the sequences $\{\varphi_n^{(k)}\}_{n\in\mathbb{N}}$ of derivatives of arbitrary orders k to the functions φ and $\varphi^{(k)}$, respectively. $\mathcal{D}(\Omega)$ is called the Schwartz space of test functions.

7.4. THE SPACE OF GENERALIZED FUNCTIONS $\mathcal{D}'(\Omega)$

The space $\mathcal{D}'(\Omega)$ equipped with the strong topology is called the Schwartz space of generalized functions (distributions). Thus, the elements of $\mathcal{D}'(\Omega)$ are the continuous linear functionals on $\mathcal{D}(\Omega)$.

If $f(t)$ is a locally integrable function in Ω, i.e. integrable on every compact set $K \subset \Omega$, then it gives rise to a generalized function F_f by the formula

$$F_f(\varphi) = \int_\Omega f(t)\varphi(t)\,dt,$$

and different distributions correspond to different locally integrable functions (equality is understood almost everywhere).

For $F \in \mathcal{D}'(\Omega)$ the derivative $\frac{\partial F}{\partial t_i}$ is determined by

$$\frac{\partial F}{\partial t_i}(\varphi) = -F\Big(\frac{\partial \varphi}{\partial t_i}\Big) \qquad (\forall \varphi \in \mathcal{D}(\Omega)),$$

which defines a continuous linear mapping from $\mathcal{D}'(\Omega)$ into $\mathcal{D}'(\Omega)$. Higher-order derivatives are defined in a natural way.

In what follows, we will often deal with the chain of spaces

$$\mathcal{D}(\Omega) \subset L_2(\Omega) \subset \mathcal{D}'(\Omega),$$

in which an element $f \in L_2(\Omega)$ is identified with the distribution $F_f \;:\; F_f(\varphi) = (f, \overline{\varphi})_{L_2(\Omega)}$ (the bar denotes the sign of complex conjugation).

7.5. SOBOLEV SPACES

Let $m \geq 0$ be an integer. Let Ω be a bounded domain in $\mathbb{R}^n$ with piecewise continuously-differentiable boundary $\partial\Omega$. The completion of the set $C^m[\Omega \cup \partial\Omega]$ of functions on $\Omega \cup \partial\Omega$ that are m-times continuously differentiable, with respect to the norm

$$\|\varphi\|_{W_2^m(\Omega)} = \Big(\sum_{|\alpha|\leq m} \|D^\alpha \varphi\|^2_{L_2(\Omega)}\Big)^{\frac{1}{2}}, \tag{7.2}$$

where $D^\alpha = \frac{\partial^{\alpha_1+\cdots+\alpha_n}}{\partial t_1^{\alpha_1}\cdots\partial t_n^{\alpha_n}}$, $\alpha = \{\alpha_1, \ldots, \alpha_n\}$, $|\alpha| = \alpha_1 + \cdots + \alpha_n$, is the so-called Sobolev space $W_2^m(\Omega)$. The completion of $\mathcal{D}(\Omega)$ in the same norm (7.2) is denoted by $\overset{\circ}{W}{}_2^m(\Omega)$. It is obvious that $\overset{\circ}{W}{}_2^m(\Omega)$ is a closed subspace of $W_2^m(\Omega) \subset L_2(\Omega)$.

It is possible to describe the space $W_2^m(\Omega)$ $(m = 0, 1, \ldots)$ in the following way. A function $\varphi(t)$ $(t \in \Omega)$ belongs to $W_2^m(\Omega)$ if and only if it belongs to $L_2(\Omega)$ together with its derivatives $D^\alpha \varphi$ $(|\alpha| \leq m)$ (in the sense of distributions). This description may also serve as a definition of the Sobolev space $W_2^m(\Omega)$ when the domain Ω is not bounded. If we introduce in $W_2^m(\Omega)$ an inner product by

$$(\varphi, \psi)_{W_2^m(\Omega)} = \sum_{|\alpha| \leq m} \left(D^\alpha \varphi,\ D^\alpha \psi\right)_{L_2(\Omega)},$$

it becomes a Hilbert space.

The use of Sobolev spaces in solving various problems of analysis is convenient, first, because they are Hilbert spaces and, secondly, because unlike for $L_2(\Omega)$ the functions in $W_2^m(\Omega)$ with sufficiently large m make sense on manifolds whose dimension is less than n. The following statements are valid: let $k = 0, 1, 2, \ldots$ be such that

$$0 \leq k < m - \frac{n}{2};$$

then $W_2^m(\Omega) \subset C^k\,[\Omega \cup \partial\Omega]$ and this imbedding is compact; if

$$k \geq m - \frac{n}{2},$$

$l = 0, 1, \ldots$ and a real number q satisfy the relations

$$l > n - 2(m-k), \quad 1 < q < \frac{2l}{n - 2(m-k)},$$

and $\mathfrak{M} \subseteq \Omega \cup \partial\Omega$ is any m–times continuously-differentiable manifold with dim $\mathfrak{M}$ $< l$, then for each function $\varphi(t) \in W_2^m(\Omega)$ we have $(D^\alpha\varphi)(t) \in L_q(\mathfrak{M})$ $(t \in \mathfrak{M},$ $|\alpha| \leq k)$, $\|D^\alpha \varphi\|_{L_q(\mathfrak{M})} \leq c\|\varphi\|_{W_2^m(\Omega)}$ (c =const) and the imbedding is compact.

These statements, known as imbedding theorems, imply that $\overset{\circ}{W}{}_2^m(\Omega)$ $(m \in \mathbb{N})$ coincides with the set of all functions $\varphi(t) \in W_2^m(\Omega)$ for which $(D^\alpha\varphi)(t) = 0$ $(t \in \partial\Omega)$ as $|\alpha| \leq m - 1$.

In the case when $\Omega = \mathbb{R}^n$ it is possible to give another definition of the space $W_2^m(\mathbb{R}^n)$, equivalent to the one above. It is based on the properties of the Fourier transformation

$$\varphi \to (F\varphi)(s) = \frac{1}{(2\pi)^{\frac{n}{2}}} \int_{\mathbb{R}^n} e^{-it\cdot s} \varphi(t)\,dt, \quad \varphi \in L_2(\mathbb{R}^n),$$

where $t \cdot s = \sum_{i=1}^n t_i s_i$. Namely,

$$W_2^m(\mathbb{R}^n) = \{\varphi\ :\ \varphi \in L_2(\mathbb{R}^n),\ (1 + |s|^2)^{\frac{m}{2}} (F\varphi)(s) \in L_2(\mathbb{R}^n)\}$$
$$(|s|^2 = s_1^2 + \cdots + s_n^2)$$

and also

$$\|\varphi\|_{W_2^m(\mathbb{R}^n)} = \|(1 + |s|^2)^{\frac{m}{2}} (F\varphi)(s)\|_{L_2(\mathbb{R}^n)}.$$

This definition can also be generalized to the case where $m > 0$ is not an integer.

We will also need the spaces $W_2^m(\mathfrak{H},(a,b))$ and $\overset{\circ}{W}{}_2^m(\mathfrak{H},(a,b))$ $(-\infty < a < b < \infty)$ of vector-functions with values in a Hilbert space $\mathfrak{H}$. They are defined as the completion with respect to the norm

$$\|y\|_{W_2^m(\mathfrak{H},(a,b))} = \left(\sum_{k=0}^{m} \|y^{(k)}\|^2_{L_2(\mathfrak{H},(a,b))}\right)^{\frac{1}{2}}$$

of the set $C^m(\mathfrak{H},[a,b])$ of vector-functions on $[a,b]$ that are m-times continuously differentiable, and, respectively, of the set $C_0^m(\mathfrak{H},[a,b])$ of functions in $C^m(\mathfrak{H},[a,b])$ that vanish at the ends a and b together with their derivatives up to order m inclusive. It is not difficult to infer from the equality

$$y^{(k)}(t) = y^{(k)}(s) + \int_s^t y^{(k+1)}(\xi)\,d\xi \qquad (k = 0,1,\ldots), \tag{7.3}$$

which is valid for any vector-function $y \in C^m(\mathfrak{H},[a,b])$, that

$$\max_{t\in[a,b]} \|y^{(k)}(t)\| \le c\|y\|_{W_2^m(\mathfrak{H},(a,b))}.$$

This inequality implies the continuity of the imbedding $W_2^m(\mathfrak{H},(a,b)) \subset C^k(\mathfrak{H},[a,b])$ $(0 \le k \le m-1)$. Equality (7.3) also shows that $W_2^m(\mathfrak{H},(a,b))$ consists of the $(m-1)$-times continuously-differentiable vector-functions whose $(m-1)$-th derivative is absolutely continuous and whose m-th derivative belongs to $L_2(\mathfrak{H},(a,b))$.

Clearly $\overset{\circ}{W}{}_2^m(\mathfrak{H},(a,b))$ is a subspace of $W_2^m(\mathfrak{H},(a,b))$ and a vector-function $y \in W_2^m(\mathfrak{H},(a,b))$ belongs to $\overset{\circ}{W}{}_2^m(\mathfrak{H},(a,b))$ if and only if all its derivatives $y^{(k)}(t)$ $(k = 0,1,\ldots)$ vanish at the points a and b. It is not difficult to convince ourselves that

$$W_2^m(\mathfrak{H},(a,b)) = W_2^m(a,b) \otimes \mathfrak{H},$$
$$\overset{\circ}{W}{}_2^m(\mathfrak{H},(a,b)) = \overset{\circ}{W}{}_2^m(a,b) \otimes \mathfrak{H}.$$

Since the space $W_2^m(\mathfrak{H},(a,b))$ is continuously imbedded into $L_2(\mathfrak{H},(a,b))$, there exists a positive self-adjoint operator Λ such that $\mathfrak{D}(\Lambda) = W_2^m(\mathfrak{H},(a,b))$ and

$$\|y\|^2_{W_2^m(\mathfrak{H},(a,b))} = \|y\|^2_{L_2(\mathfrak{H},(a,b))} + \|\Lambda y\|^2_{L_2(\mathfrak{H},(a,b))}.$$

We set, by definition,

$$W_2^s(\mathfrak{H},(a,b)) = \mathfrak{D}(\Lambda^{1-\theta}) \qquad (s \ge 0),$$
$$\|y\|^2_{W_2^s(\mathfrak{H},(a,b))} = \|y\|^2_{L_2(\mathfrak{H},(a,b))} + \|\Lambda^{1-\theta}y\|^2_{L_2(\mathfrak{H},(a,b))},$$

where $(1-\theta)m = s$, $0 < \theta < 1$. We can prove that this definition does not depend on the choice of the integer m. For integer $s = m$ it is equivalent to the definition introduced above. It is also clear that

$$W_2^m(\mathfrak{H},(a,b)) \subset W_2^s(\mathfrak{H},(a,b)) \subset L_2(\mathfrak{H},(a,b))$$

for $0 < s < m$.

7.6. THE INTERMEDIATE DERIVATIVE THEOREM

Let A be a non-negative operator on a Hilbert space $\mathfrak{H}$. We introduce in $\mathfrak{D}(A^\alpha)$ $(0 \le \alpha \le 1)$ an inner product by

$$(f,g)_{A^\alpha} = (A^\alpha f, A^\alpha g) + (f,g).$$

As we noted in subsection 4.1 , $\mathfrak{D}(A^\alpha)$ becomes a Hilbert space, which we denote by $\mathfrak{H}^\alpha$. Of course, $\mathfrak{H}^0 = \mathfrak{H}$.

Let $W_m(a,b)$ $(-\infty < a < b < \infty)$ be the completion of $C^m(\mathfrak{H}^1, [a,b])$ with respect to the norm

$$\|u\|_{W_m(a,b)} = \left(\|u\|^2_{L_2(\mathfrak{H}^1,(a,b))} + \|u^{(m)}\|^2_{L_2(\mathfrak{H},(a,b))}\right)^{\frac{1}{2}}.$$

It is a Hilbert space, contained in $W_2^m(\mathfrak{H},(a,b))$. Therefore, a vector-function $u \in W_m(a,b)$ has continuous derivatives up to order $m-1$ inclusive with values in $\mathfrak{H}$. It is found that at any point its i–th $(0 \le i \le m-1)$ derivative belongs to a space that is narrower than $\mathfrak{H}$.

THEOREM 7.2. *If a vector-function u belongs to $W_m(a,b)$, then $u^{(i)} \in L_2(\mathfrak{H}^{1-\frac{i}{m}}, (a,b))$ $(0 \le i \le m-1)$, where $u \to u^{(j)}$ is a continuous linear mapping from $W_m(a,b)$ into $L_2(\mathfrak{H}^{1-\frac{i}{m}}, (a,b))$. Moreover, the derivative $u^{(i)}(t)$ at any point from $[a,b]$ makes sense, has values in $\mathfrak{H}^{1-\frac{i+\frac{1}{2}}{m}}$, and the mapping $u \to u^{(i)}$ is continuous from $W_m(a,b)$ into $C(\mathfrak{H}^{1-\frac{i+\frac{1}{2}}{m}}, [a,b])$.*

CHAPTER 2

Boundary Values of Solutions of Homogeneous Operator Differential Equations

1. Positive and Negative Spaces

In the solution of many problems of mathematical analysis it is convenient to use triples of Hilbert spaces

$$\mathfrak{H}_+ \subseteq \mathfrak{H}_0 \subseteq \mathfrak{H}_-,$$

where $\mathfrak{H}_+$ and $\mathfrak{H}_-$ are spaces of test and generalized elements (antilinear continuous functionals on $\mathfrak{H}_+$), respectivily, instead of pairs of spaces (a test space and its dual). The significance of $\mathfrak{H}_0$ in the chain is that its inner product may be extended to a continuous bilinear form on $\mathfrak{H}_- \times \mathfrak{H}_+$, giving an action of a generalized element on a test element. The general theory of such triples is presented below.

1.1. ON POSITIVE AND NEGATIVE SPACES

Let $\mathfrak{H}_0$ be a Hilbert space over the field of complex numbers with inner product $(.,.)_{\mathfrak{H}_0}$ and norm $\|f\|_{\mathfrak{H}_0} = \sqrt{(f,f)_{\mathfrak{H}_0}}$. Let $\mathfrak{H}_+ \subset \mathfrak{H}_0$ be a dense linear set in it which itself is a Hilbert space with respect to another inner product $(.,.)_{\mathfrak{H}_+}$. Suppose also that the norms $\|\cdot\|_{\mathfrak{H}_0}$ and $\|\cdot\|_{\mathfrak{H}_+}$ satisfy the inequality

$$\|f\|_{\mathfrak{H}_0} \leq \|f\|_{\mathfrak{H}_+} \quad (f \in \mathfrak{H}_+) \tag{1.1}$$

(the more general case of an inequality $\|\cdot\|_{\mathfrak{H}_0} \leq c\|\cdot\|_{\mathfrak{H}_+}$ is reduced to (1.1) by means of renormalization). The elements of $\mathfrak{H}_+$ will be called test elements. The space $\mathfrak{H}_+$ is called positive.

Each element $f \in \mathfrak{H}_0$ generates an antilinear functional l_f on $\mathfrak{H}_+$ by the formula

$$l_f(g) = (f,g)_{\mathfrak{H}_0} \quad (g \in \mathfrak{H}_+).$$

This functional is continuous, according to the estimate

$$|l_f(g)| = |(f,g)_{\mathfrak{H}_0}| \leq \|f\|_{\mathfrak{H}_0}\|g\|_{\mathfrak{H}_0} \leq \|f\|_{\mathfrak{H}_0}\|g\|_{\mathfrak{H}_+}.$$

We introduce in $\mathfrak{H}_0$ a new norm:

$$\|f\|_{\mathfrak{H}_-} = \|l_f\| = \sup_{g \in \mathfrak{H}_+} \frac{|(f,g)_{\mathfrak{H}_0}|}{\|g\|_{\mathfrak{H}_+}}. \tag{1.2}$$

This definition is correct since $\|f\|_{\mathfrak{H}_-} = 0$ implies $(f,g)_{\mathfrak{H}_0} = 0$ for any $g \in \mathfrak{H}_+$, whence $f = 0$ in view of the denseness of $\mathfrak{H}_+$ in $\mathfrak{H}_0$. We denote by $\mathfrak{H}_-$ the completion of $\mathfrak{H}_0$ with respect to the norm $\|\cdot\|_{\mathfrak{H}_-}$. This space is termed negative. Its elements are called generalized elements.

THEOREM 1.1. *The negative space $\mathfrak{H}_-$ is a Hilbert space.*

Proof. Consider the bilinear form

$$\mathfrak{H}_0 \times \mathfrak{H}_+ \ni \{f, g\} \to (f, g)_{\mathfrak{H}_0} \in \mathbb{C}^1.$$

It is continuous: $|(f,g)_{\mathfrak{H}_0}| \leq \|f\|_{\mathfrak{H}_0}\|g\|_{\mathfrak{H}_0} \leq \|f\|_{\mathfrak{H}_0}\|g\|_{\mathfrak{H}_+}$, and therefore it has a representation

$$(f,g)_{\mathfrak{H}_0} = (f, Og)_{\mathfrak{H}_0} = (If, g)_{\mathfrak{H}_+}, \tag{1.3}$$

where $O : \mathfrak{H}_+ \to \mathfrak{H}_0$, $I = O^* : \mathfrak{H}_0 \to \mathfrak{H}_+$ are mutually adjoint continuous operators and O is the imbedding of $\mathfrak{H}_+$ into $\mathfrak{H}_0$.

Define in $\mathfrak{H}_0$ a quasi-inner product $(f,g)_{\mathfrak{H}_-}$ by

$$(f,g)_{\mathfrak{H}_-} = (If, Ig)_{\mathfrak{H}_+} = (f, Ig)_{\mathfrak{H}_0} = (If, g)_{\mathfrak{H}_0}. \tag{1.4}$$

According to (1.2)-(1.4),

$$\|f\|_{\mathfrak{H}_-} = \sup_{g\in\mathfrak{H}_+} \frac{|(f,g)_{\mathfrak{H}_0}|}{\|g\|_{\mathfrak{H}_+}} = \sup_{g\in\mathfrak{H}_+} \frac{|(If,g)_{\mathfrak{H}_+}|}{\|g\|_{\mathfrak{H}_+}} = \|If\|_{\mathfrak{H}_+} = \sqrt{(f,f)_{\mathfrak{H}_-}}.$$

Since $\|\cdot\|_{\mathfrak{H}_-}$ is a norm in $\mathfrak{H}_0$, (1.4) defines an inner product which, as a result of completing, can be extended from $\mathfrak{H}_0$ onto $\mathfrak{H}_-$. Also $\|f\|_{\mathfrak{H}_-} \leq \|f\|_{\mathfrak{H}_0}$ $(f \in \mathfrak{H}_0)$. □

Thus, we have constructed a chain of Hilbert spaces with continuous imbeddings

$$\begin{aligned} &\mathfrak{H}_+ \subseteq \mathfrak{H}_0 \subseteq \mathfrak{H}_-, \\ &\|f\|_{\mathfrak{H}_+} \geq \|f\|_{\mathfrak{H}_0} \geq \|f\|_{\mathfrak{H}_-} \quad (f \in \mathfrak{H}_+). \end{aligned} \tag{1.5}$$

The operator I in formula (1.3) can be extended from $\mathfrak{H}_0$ to an isometric operator $\mathbf{I}$ from $\mathfrak{H}_-$ into $\mathfrak{H}_+$ so that

$$\mathbf{I}\mathfrak{H}_- = \mathfrak{H}_+. \tag{1.6}$$

The latter follows because the set $\mathbf{I}\mathfrak{H}_0 = I\mathfrak{H}_0$ is dense in $\mathfrak{H}_+$. (It is dense because the existence of an element $g_0 \in \mathfrak{H}_+$ with the property $(If, g_0)_{\mathfrak{H}_+} = (f, g_0)_{\mathfrak{H}_0} = 0$ would imply $g_0 = 0$.)

Now, for arbitrary $f, g \in \mathfrak{H}_-$,

$$(f,g)_{\mathfrak{H}_-} = (\mathbf{I}f, g)_{\mathfrak{H}_0} = (f, \mathbf{I}g)_{\mathfrak{H}_0} = (\mathbf{I}f, \mathbf{I}g)_{\mathfrak{H}_+}. \tag{1.7}$$

The mapping $f \to l_f$ $(f \in \mathfrak{H}_0)$ is linear and $\|f\|_{\mathfrak{H}_-} = \|l_f\|$, so it can be extended isometrically to the whole space $\mathfrak{H}_-$. Consequently, $\mathfrak{H}_-$ may be considered as a part of the space $\mathfrak{H}'_+$, the dual of $\mathfrak{H}_+$, i.e. as a part of the space of continuous antilinear functionals[1] on $\mathfrak{H}_+ : \mathfrak{H}_- \subseteq \mathfrak{H}'_+$. Since for elements $f \in \mathfrak{H}_0$ we have

[1] Here the notation for the space of all continuous antilinear functionals is the same as for the dual space of continuous linear functionals, since they can always be identified.

$l_f(g) = (f,g)_{\mathfrak{H}_0}$ $(g \in \mathfrak{H}_+)$ and the left-hand side in this equality, as said above, makes sense for $f \in \mathfrak{H}_-$, we will denote $l_f(g)$ by $(f,g)_{\mathfrak{H}_0}$ also for $f \in \mathfrak{H}_-$, $g \in \mathfrak{H}_+$. In this case the Cauchy–Buniakowski inequality

$$|(f,g)_{\mathfrak{H}_0}| = |l_f(g)| \le \|l_f\| \|g\|_{\mathfrak{H}_+} = \|f\|_{\mathfrak{H}_-} \|g\|_{\mathfrak{H}_+}$$

holds. It turns out that $\mathfrak{H}_-$ is not only imbedded into $\mathfrak{H}'_+$, but even coincides with it.

THEOREM 1.2. *The equality*

$$\mathfrak{H}_- = \mathfrak{H}'_+$$

is valid.

Proof. It is sufficient to prove that an arbitrary functional $l \in \mathfrak{H}'_+$ coincides with some l_f, i.e. $l(g) = l_f(g) = (f,g)_{\mathfrak{H}_0}$ $(\forall g \in \mathfrak{H}_+)$, where $f \in \mathfrak{H}_-$. The Riesz theorem and the equality (1.6) imply

$$l(g) = (h,g)_{\mathfrak{H}_+} = (\mathbf{I}f,g)_{\mathfrak{H}_+} = (f,g)_{\mathfrak{H}_0} \quad (f = \mathbf{I}^{-1}h \in \mathfrak{H}_-). \qquad \square$$

EXAMPLE 1. Let $\mathfrak{H}_0 = L_2(\mathbb{R}^n)$, $\mathfrak{H}_+ = L_2(\mathbb{R}^n, p(x)\mathrm{d}x)$, $1 \le p(x) \in C(\mathbb{R}^n)$.[1] Then $\mathfrak{H}_- = L_2(\mathbb{R}^n, p^{-1}(x)\mathrm{d}x)$.

EXAMPLE 2. Let Ω be a domain in $\mathbb{R}^n$. Set $\mathfrak{H}_0 = L_2(\Omega)$ and $\mathfrak{H}_+ = W_2^m(\Omega)$ ($m \ge 0$ is an integer). Since

$$\|\varphi\|_{W_2^m(\Omega)} \ge \|\varphi\|_{L_2(\Omega)} \quad (\varphi \in W_2^m(\Omega)),$$

the space $W_2^m(\Omega)$ is positive with respect to $L_2(\Omega)$. The negative space constructed for the pair $W_2^m(\Omega)$, $L_2(\Omega)$ is denoted by $W_2^{-m}(\Omega)$; it is called a negative Sobolev space. In a similar way we construct the negative Sobolev space $W_2^{-\alpha}(\mathbb{R}^n)$ for any $\alpha > 0$.

Since the Schwartz space $\mathcal{D}(\Omega)$ of test functions is continuously imbedded in $W_2^m(\Omega)$ for any integer $m \ge 0$, the space $W_2^{-m}(\Omega)$ can be continuously imbedded in the space $\mathcal{D}'(\Omega)$ of distributions.

The chain (1.5) was constructed starting from the pair of spaces $\mathfrak{H}_0$ and $\mathfrak{H}_+$. It can also be constructed starting from the pair $\mathfrak{H}_0$ and $\mathfrak{H}_-$.

Let, in a Hilbert space $\mathfrak{H}_-$, a dense set $\mathfrak{H}_0$ be given which itself is a Hilbert space relative to another inner product $(.,.)_{\mathfrak{H}_0}$, where

$$\|f\|_{\mathfrak{H}_-} \le \|f\|_{\mathfrak{H}_0} \quad (f \in \mathfrak{H}_0).$$

Consider the bilinear form

$$\mathfrak{H}_0 \times \mathfrak{H}_0 \ni \{f,g\} \to (f,g)_{\mathfrak{H}_-} \in \mathbb{C}^1.$$

It is continuous because

$$|(f,g)_{\mathfrak{H}_-}| \le \|f\|_{\mathfrak{H}_-} \|g\|_{\mathfrak{H}_-} \le \|f\|_{\mathfrak{H}_0} \|g\|_{\mathfrak{H}_0}.$$

---

[1] $C(\mathbb{R}^n)$ is the set of all continuous functions on $\mathbb{R}^n$.

Hence it can be represented as

$$(f,g)_{\mathfrak{H}_-} = (Kf,g)_{\mathfrak{H}_0} \quad (f,g \in \mathfrak{H}_0),$$

where $K \geq 0$ is a continuous operator from $\mathfrak{H}_0$ into $\mathfrak{H}_0$ with $\|K\| \leq 1$. Its range $\mathfrak{R}(K)$ is dense in $\mathfrak{H}_0$. Indeed, if an element $h \in \mathfrak{H}_0$ is such that $(Kf,h)_{\mathfrak{H}_0} = 0$ for any $f \in \mathfrak{H}_0$, then $0 = (Kf,h)_{\mathfrak{H}_0} = (f,h)_{\mathfrak{H}_-}$, whence, since $\mathfrak{H}_0$ is dense in $\mathfrak{H}_-$, $h = 0$.

On $\mathfrak{R}(K)$ the inverse K^{-1} of K exists: if $Kf = 0$ $(f \in \mathfrak{H}_0)$, then $(f,g)_{\mathfrak{H}_-} = (Kf,g)_{\mathfrak{H}_0} = 0$ $(\forall g \in \mathfrak{H}_0)$, i.e. $f = 0$. Set

$$(f,g)_{\mathfrak{H}_+} = (K^{-1}f,g)_{\mathfrak{H}_0} \quad (f,g \in \mathfrak{R}(K)).$$

Since $\|K\| \leq 1$,

$$\|f\|^2_{\mathfrak{H}_+} = (K^{-1}f,f)_{\mathfrak{H}_0} \geq \|f\|^2_{\mathfrak{H}_0} \quad (f \in \mathfrak{R}(K)).$$

This implies that the completion $\mathfrak{H}_+$ of the set $\mathfrak{R}(K)$ with respect to the norm $\|\cdot\|_{\mathfrak{H}_+}$ satisfies all conditions of a positive space. It only remains to show that the negative space $\mathcal{K}_-$ constructed for the pair $\mathfrak{H}_+$, $\mathfrak{H}_0$ coincides with $\mathfrak{H}_-$.

Since $(If,g)_{\mathfrak{H}_+} = (f,g)_{\mathfrak{H}_0} = (Kf,g)_{\mathfrak{H}_+}$ $(f \in \mathfrak{H}_0,\ g \in \mathfrak{R}(K))$ and $\mathfrak{R}(K)$ is dense in $\mathfrak{H}_+$, $K = OI$, where the operators O and I are constructed for the pair $\mathfrak{H}_+$, $\mathfrak{H}_0$ as was done above. Therefore,

$$(f,g)_{\mathfrak{H}_-} = (Kf,g)_{\mathfrak{H}_0} = (If,g)_{\mathfrak{H}_0} = (f,g)_{\mathcal{K}_-}$$

as $f,g \in \mathfrak{H}_0$, whence, since $\mathfrak{H}_0$ is dense in $\mathfrak{H}_-$ and in $\mathcal{K}_-$, it follows that $\mathcal{K}_- = \mathfrak{H}_-$.

1.2. A RELATION BETWEEN CHAINS STARTING FROM SUBSPACES OF $\mathfrak{H}_0$

Let the chain (1.5) be given and let $\mathfrak{H}_{++}$ be a dense linear set in the space $\mathfrak{H}_+$ which is itself a Hilbert space with respect to another inner product $(.,.)_{++}$, where

$$\|f\|_{\mathfrak{H}_{++}} \geq \|f\|_{\mathfrak{H}_+} \quad (f \in \mathfrak{H}_{++}).$$

It is clear that the set $\mathfrak{H}_{++}$ is dense in $\mathfrak{H}_0$ too, and $\|\cdot\|_{\mathfrak{H}_{++}} \geq \|\cdot\|_{\mathfrak{H}_0}$. Thus, $\mathfrak{H}_{++}$ is a positive space in the pair $\mathfrak{H}_{++}$, $\mathfrak{H}_0$. Denote the corresponding negative space by $\mathfrak{H}_{--}$. The following question arises: What relation is there between the spaces $\mathfrak{H}_{--}$ and $\mathfrak{H}_-$? It is not difficult to prove the inclusion $\mathfrak{H}_- \subseteq \mathfrak{H}_{--}$ holds.

Indeed, according to Theorem 1.2, one can identify $\mathfrak{H}_-$ with the space of continuous antilinear functionals on $\mathfrak{H}_+$. But every such functional is at the same time a continuous antilinear functional on $\mathfrak{H}_{++}$. Since $\mathfrak{H}_{++}$ is dense in $\mathfrak{H}_+$, functionals different on $\mathfrak{H}_+$ will also be different on $\mathfrak{H}_{++}$, i.e. $\mathfrak{H}'_+$ is uniquely imbedded in $\mathfrak{H}'_{++} = \mathfrak{H}_{--}$.

Hence we obtain the chain

$$\mathfrak{H}_{++} \subseteq \mathfrak{H}_+ \subseteq \mathfrak{H}_0 \subseteq \mathfrak{H}_- \subseteq \mathfrak{H}_{--},$$

where

$$\|f\|_{\mathfrak{H}_{++}} \geq \|f\|_{\mathfrak{H}_+} \geq \|f\|_{\mathfrak{H}_0} \geq \|f\|_{\mathfrak{H}_-} \geq \|f\|_{\mathfrak{H}_{--}} \quad (f \in \mathfrak{H}_{++}).$$

1.3. ADJOINT OPERATORS RELATIVE TO $\mathfrak{H}_0$

Let A be an operator acting continuously from $\mathfrak{H}_+$ into $\mathfrak{H}_0$. Then A^*, its adjoint, acts from $\mathfrak{H}_0$ into $\mathfrak{H}_+$, and for $f \in \mathfrak{H}_+$, $g \in \mathfrak{H}_0$ we have

$$(Af, g)_{\mathfrak{H}_0} = (f, A^*g)_{\mathfrak{H}_+} = (f, \mathbf{I}^{-1}A^*g)_{\mathfrak{H}_0} = (f, A^+g)_{\mathfrak{H}_0},$$

where $A^+ = \mathbf{I}^{-1}A^*$ is a continuous linear operator acting from $\mathfrak{H}_0$ into $\mathfrak{H}_-$. This operator is called the adjoint of A relative to $\mathfrak{H}_0$.

1.4. THE CHAINS CONSTRUCTED FROM AN OPERATOR

Let A be a self-adjoint operator on the space $\mathfrak{H}_0$ such that

$$\|Af\|_{\mathfrak{H}_0} \geq \|f\|_{\mathfrak{H}_0} \quad (f \in \mathfrak{D}(A)). \tag{1.8}$$

The set $\mathfrak{D}(A)$ becomes a Hilbert space if we define an inner product in it by

$$(f, g)_{\mathfrak{H}_+} = (Af, Ag)_{\mathfrak{H}_0} \quad (f, g \in \mathfrak{D}(A)).$$

In this connection, inequality (1.8) means that $\|f\|_{\mathfrak{H}_+} \geq \|f\|_{\mathfrak{H}_0}$ $(f \in \mathfrak{D}(A))$. Therefore $\mathfrak{H}_+ = \mathfrak{D}(A)$ is a positive space in the pair $\mathfrak{H}_+$, $\mathfrak{H}_0$.

The operator I constructed for the chain $\mathfrak{H}_+ \subseteq \mathfrak{H}_0 \subseteq \mathfrak{H}_-$ in the situation considered coincides with A^{-2}. In fact, for $f \in \mathfrak{H}_0$, $g \in \mathfrak{H}_+$,

$$(f, g)_{\mathfrak{H}_0} = (If, g)_{\mathfrak{H}_+} = (AIf, Ag)_{\mathfrak{H}_0} = (A^2If, g)_{\mathfrak{H}_0}.$$

Since $\mathfrak{H}_+$ is dense in $\mathfrak{H}_0$, $A^2I = E$, whence $I = A^{-2}$. Consequently,

$$(f, g)_{\mathfrak{H}_-} = (If, Ig)_{\mathfrak{H}_+} = (A^{-2}f, A^{-2}g)_{\mathfrak{H}_+} = (A^{-1}f, A^{-1}g)_{\mathfrak{H}_0}$$

for any $f, g \in \mathfrak{H}_0$. So we can obtain the space $\mathfrak{H}_-$ as the completion of $\mathfrak{H}_0$ with respect to the norm

$$\|f\|_{\mathfrak{H}_-} = \|A^{-1}f\|_{\mathfrak{H}_0}.$$

1.5. INDUCTIVE AND PROJECTIVE LIMITS OF HILBERT SPACES CONNECTED BY A CHAIN

Let $\mathfrak{H}_\tau$ (τ runs through an ordered set T of indices) be a family of Hilbert spaces each of which is dense in $\mathfrak{H}_0$, and

$$\|f\|_{\mathfrak{H}_\tau} \geq \|f\|_{\mathfrak{H}_0} \quad (f \in \mathfrak{H}_\tau).$$

Suppose that also $\mathfrak{H}_{\tau_1} \supseteq \mathfrak{H}_{\tau_2}$ ($\mathfrak{H}_{\tau_1} \subseteq \mathfrak{H}_{\tau_2}$) for $\tau_1 < \tau_2$, $\|f\|_{\mathfrak{H}_{\tau_1}} \leq \|f\|_{\mathfrak{H}_{\tau_2}}$ (respectively $\|f\|_{\mathfrak{H}_{\tau_1}} \geq \|f\|_{\mathfrak{H}_{\tau_2}}$) and that $\mathfrak{H}_{\tau_2}$ is dense in $\mathfrak{H}_{\tau_1}$ (respectively $\mathfrak{H}_{\tau_1}$ is dense in $\mathfrak{H}_{\tau_2}$). Then the intersection $\bigcap_{\tau \in T} \mathfrak{H}_\tau$ is dense in each $\mathfrak{H}_\tau$. This follows from the following statement.

LEMMA 1.1. *Let $\mathfrak{B}_n$ $(n \in \mathbb{N})$ be Banach spaces, $\mathfrak{B}_{n+1} \subset \mathfrak{B}_n$ $(n = 1, 2, \ldots)$, where the inclusions are dense and continuous. Then $\bigcap_{n=1}^{\infty} \mathfrak{B}_n$ is dense in $\mathfrak{B}_1$.*

Proof. By induction on n choose for an arbitrary $x_1 \in \mathfrak{B}_1$ an element $x_n \in \mathfrak{B}_n$ so that $\|x_n - x_{n-1}\|_{n-1} < \frac{\epsilon}{2^{n-1}}$ $(n = 2, 3, \ldots)$ ($\|\cdot\|_k$ is the norm in $\mathfrak{B}_k$; without loss of generality we can assume $\|f\|_{n+1} \geq \|f\|_n$ $(f \in \mathfrak{B}_{n+1})$; $\epsilon > 0$ is arbitrarily small). Then

$$\|x_n - x_1\|_1 \leq \|x_n - x_{n-1}\|_{n-1} + \cdots + \|x_2 - x_1\|_1 < \epsilon, \tag{1.9}$$

and for $\forall k \geq 1$, $\forall m > n > k$,

$$\begin{aligned}\|x_n - x_m\|_k &\leq \|x_n - x_{n+1}\|_k + \cdots + \|x_{m-1} - x_m\|_k \\ &\leq \|x_n - x_{n+1}\|_n + \cdots + \|x_{m-1} - x_m\|_{m-1} < \epsilon\Big(\frac{1}{2^n} + \cdots + \frac{1}{2^{m-1}}\Big) < \frac{\epsilon}{2^{n-1}}.\end{aligned}$$

Consequently, for any k there exists a $y_k \in \mathfrak{B}_k$ such that $\|x_n - y_k\|_k \to 0$ $(n \to \infty)$. It is evident that $y_k \equiv y$ $(\forall k \in \mathbb{N})$. So, $y \in \bigcap_{n=1}^{\infty} \mathfrak{B}_n$. Limit transition in (1.9) as $n \to \infty$ yields $\|y - x_1\|_1 \leq \epsilon$. This concludes the proof. □

Now form $\mathfrak{H}_{\mathrm{pr}} = \operatorname{limpr}_{\tau \in T} \mathfrak{H}_\tau$, $\mathfrak{H}_{\mathrm{ind}} = \operatorname{limind}_{\tau \in T} \mathfrak{H}_\tau$.

THEOREM 1.3. *If $\mathfrak{H}_{\mathrm{pr}}$ is the projective limit of Hilbert spaces $\mathfrak{H}_\tau$ $(\tau \in T)$ (respectively, $\mathfrak{H}_{\mathrm{ind}}$ is the inductive limit of $\mathfrak{H}_\tau$) for which chains $\mathfrak{H}_\tau \subset \mathfrak{H}_0 \subset \mathfrak{H}'_\tau$ of type (1.5) have been constructed, then for the dual space $\mathfrak{H}'_{\mathrm{pr}}$ (respectively, $\mathfrak{H}'_{\mathrm{ind}}$) of continuous antilinear functionals on $\mathfrak{H}_{\mathrm{pr}}$ (respectively, $\mathfrak{H}_{\mathrm{ind}}$) the following topological equality is valid:*

$$\mathfrak{H}'_{\mathrm{pr}} = \operatorname*{limind}_{\tau \in T} \mathfrak{H}'_\tau \quad \text{(respectively, } \mathfrak{H}'_{\mathrm{ind}} = \operatorname*{limpr}_{\tau \in T} \mathfrak{H}'_\tau) \tag{1.10}$$

(here $\mathfrak{H}'_{\mathrm{pr}}$ (respectively, $\mathfrak{H}'_{\mathrm{ind}}$) is equipped with the strong topology of a dual space).

The proof follows from Theorem 1.2 and equality (7.1) of Chapter 1, taking into account that Hilbert spaces are reflexive.

REMARK 1.1. As in the case with Hilbert spaces it is possible to give a certain meaning to the expression $(f, g)_{\mathfrak{H}_0}$ for $f \in \mathfrak{H}_{\mathrm{pr}}$ and $g \in \mathfrak{H}'_{\mathrm{pr}}$ (respectively, $f \in \mathfrak{H}_{\mathrm{ind}}$, $g \in \mathfrak{H}'_{\mathrm{ind}}$) as follows: if $f \in \mathfrak{H}_{\mathrm{pr}}$, $g \in \mathfrak{H}'_{\mathrm{pr}}$ (respectively, $f \in \mathfrak{H}_{\mathrm{ind}}$, $g \in \mathfrak{H}'_{\mathrm{ind}}$), then $f \in \mathfrak{H}_\tau$ for any $\tau \in T$ and there exists a τ_0 such that $g \in \mathfrak{H}'_{\tau_0}$ (respectively, there exists a τ_0 such that $f \in \mathfrak{H}_{\tau_0}$ and $g \in \mathfrak{H}'_\tau$ for any τ); the expression $(f, g)_{\mathfrak{H}_0}$ should now be understood as the extension of the inner product from $\mathfrak{H}_0 \times \mathfrak{H}_0$ to $\mathfrak{H}_{\tau_0} \times \mathfrak{H}'_{\tau_0}$ (see subsection 1.1).

2. Some Spaces of Test and Generalized Elements

A lot of classical function spaces (for example, Sobolev spaces, the space of analytic functions, the Schwartz spaces of infinitely differentiable functions and distributions) may be interpreted as positive and negative (relative to L_2) spaces constructed from the functions of the differentiation operator and of multiplication by an independent variable, or as projective and inductive limits of such spaces. Below we consider spaces of test and generalized elements which are constructed from functions of an arbitrary self-adjoint operator. In certain concrete situations, by taking their projective and inductive limits, we obtain various spaces often used in analysis. In particular, the spaces of infinitely-differentiable vectors, analytic vectors, entire vectors, or of vectors of exponential type, of an unbounded operator are included in this scheme.

2.1. THE SPACES $\mathfrak{H}_G$ AND $\mathfrak{H}'_G$

Let A be a non-negative unbounded self-adjoint operator on a Hilbert space $\mathfrak{H}$ with resolution of the identity E_λ $(\lambda \geq 0)$. For a function $G(\lambda)$ continuous on $[0, \infty)$ and satisfying the conditions

$$G(\lambda) \geq 1, \quad \lim_{\lambda \to \infty} G(\lambda) = \infty \tag{2.1}$$

we define on the domain $\mathfrak{D}(G(A))$ of the operator $G(A) = \int_0^\infty G(\lambda)\, \mathrm{d}E_\lambda$ an inner product by

$$(f, g)_{\mathfrak{H}_G} = (G(A)f, G(A)g).$$

With this inner product $\mathfrak{D}(G(A))$ is a Hilbert space, which we will denote by $\mathfrak{H}_G = \mathfrak{H}_G(A)$.

It follows from (2.1) that

$$\|f\|_{\mathfrak{H}_G} \geq \|f\|.$$

Therefore $\mathfrak{H}_G$ may be considered as a positive space with respect to $\mathfrak{H}_0 = \mathfrak{H}$. Let $\mathfrak{H}'_G$ denote the negative space constructed from the pair $\mathfrak{H}_G$ and $\mathfrak{H}$. As was noted in subsection 1.4, we can obtain $\mathfrak{H}'_G$ by completing $\mathfrak{H}$ with respect to the norm

$$\|f\|_{\mathfrak{H}'_G} = \|G^{-1}(A)f\|.$$

If an inequality

$$G_1(\lambda) \geq \mu G_2(\lambda) \quad (0 < \mu = \text{const})$$

holds, where $G_1(\lambda)$ and $G_2(\lambda)$ are functions continuous on $[0, \infty)$ and satisfying (2.1), then

$$\mathfrak{H}_{G_1} \subseteq \mathfrak{H}_{G_2} \subseteq \mathfrak{H} \subseteq \mathfrak{H}'_{G_2} \subseteq \mathfrak{H}'_{G_1}. \tag{2.2}$$

If, in addition,

$$\lim_{\lambda \to \infty} \frac{G_1(\lambda)}{G_2(\lambda)} = \infty,$$

then all these imbeddings are strict.

EXAMPLE 1. Set $\mathfrak{H} = L_2(\mathbb{R}^1)$. Let A be multiplication by $|x|$. In this case, for a function $G(\lambda)$ continuous on $[0, \infty)$ and satisfying the conditions (2.1),

$$\mathfrak{H}_G = L_2(\mathbb{R}^1, G^2(|x|)dx), \quad \mathfrak{H}'_G = L_2(\mathbb{R}^1, G^{-2}(|x|)dx)$$

(compare with example 1 in subsection 1.1)

EXAMPLE 2. Let $\mathfrak{H} = L_2(\mathbb{R}^1)$. Take for A the operator D, the modulus of the differentiation operator, i.e. the modulus of the closure of the operator generated by the expression $\frac{d}{dx}$ on the set $C_0^\infty(\mathbb{R}^1)$ of infinitely differentiable functions with compact support. Set $G(\lambda) = \lambda^\tau + 1$ $(\tau > 0)$. Then

$$\mathfrak{H}_{\lambda^\tau+1} = W_2^\tau(\mathbb{R}^1), \quad \mathfrak{H}'_{\lambda^\tau+1} = W_2^{-\tau}(\mathbb{R}^1),$$

where $W_2^{\pm\tau}(\mathbb{R}^1)$ are the positive and negative Sobolev spaces of order τ, respectively.

In the case of an arbitrary non-negative self-adjoint operator A we set $\mathfrak{H}^\tau = \mathfrak{H}_{\lambda^\tau+1}$, $\mathfrak{H}^{-\tau} = \mathfrak{H}'_{\lambda^\tau+1}$ $(\tau > 0)$, $\mathfrak{H}^0 = \mathfrak{H}$.
The family $\{\mathfrak{H}^\tau\}_{\tau\in\mathbb{R}^1}$ is called the Hilbert scale generated by the operator A.

2.2. INFINITELY DIFFERENTIABLE VECTORS OF A CLOSED OPERATOR

Let $\mathfrak{B}$ be a Banach space. Let A be a closed linear operator on $\mathfrak{B}$ with dense domain $\mathfrak{D}(A)$. Denote

$$\mathfrak{B}^\infty(A) = \bigcap_{n=0}^{\infty} \mathfrak{D}(A^n).$$

Vectors from $\mathfrak{B}^\infty(A)$ are called infinitely differentiable vectors of the operator A.

If A is bounded, then $\mathfrak{B}^\infty(A) = \mathfrak{B}$. It is not difficult to give an example of an operator whose set of infinitely differentiable vectors consists of zero only. Therefore, the question whether $\mathfrak{B}^\infty(A)$ is dense in $\mathfrak{B}$ is natural. If $\mathfrak{B} = \mathfrak{H}$ is a Hilbert space and A is a self-adjoint operator on it, then $\overline{\mathfrak{H}^\infty(A)} = \mathfrak{H}$, because the set

$$\mathcal{E}(A) = \{E_\Delta h \ : \ \Delta \text{ is a bounded interval on the real line,}$$
$$h \text{ is any vector of } \mathfrak{H}\}$$

is dense in $\mathfrak{H}$. In a Banach space the following holds.

THEOREM 2.1. *If a densely-defined closed linear operator A on a Banach space $\mathfrak{B}$ has at least one regular point, then $\overline{\mathfrak{B}^\infty(A)} = \mathfrak{B}$.*

Proof. Set

$$\mathfrak{B}^n(A) = \mathfrak{D}(A^n) \quad (n \in \mathbb{N}),$$
$$\|f\|_{\mathfrak{B}^n} = \|f\| + \|Af\| + \cdots + \|A^n f\|.$$

It follows from the relations

$$\|(A-\lambda E)^{-1}f\|_{\mathfrak{B}^{n+1}} = \|(A-\lambda E)^{-1}f\| + \cdots + \|(A^{n+1}(A-\lambda E)^{-1}f\| \le c\|f\|_{\mathfrak{B}^n},$$

where $\lambda \in \rho(A)$ ($\rho(\cdot)$ is the resolvent set of an operator), $c = \|(A-\lambda E)^{-1}\| + \|(A-\lambda E)^{-1}A\|$, and

$$\|(A-\lambda E)f\|_{\mathfrak{B}^n} = \|(A-\lambda E)f\| + \cdots + \|(A-\lambda E)A^n f\|$$
$$\leq |\lambda|\|f\|_{\mathfrak{B}^n} + \|Af\| + \cdots + \|A^{n+1}f\| < (1+|\lambda|)\|f\|_{\mathfrak{B}^{n+1}},$$

that the operator $R_\lambda(A) = (A-\lambda E)^{-1}$ realizes an isomorphism between $\mathfrak{B}^n(A)$ and $\mathfrak{B}^{n+1}(A)$, so $\mathfrak{B}^n(A)$ is a Banach space (because $\mathfrak{B}$ is) and $\mathfrak{B}^{n+1}(A)$ is dense in $\mathfrak{B}^n(A)$ (because of $\overline{\mathfrak{D}(A)} = \mathfrak{B}$). It remains to use Lemma 1.1. □

Under the notations used in the proof of the theorem,

$$\mathfrak{B}^\infty(A) = \bigcap_{n=1}^{\infty} \mathfrak{B}^n(A).$$

We introduce in $\mathfrak{B}^\infty(A)$ the projective limit topology of the Banach spaces $\mathfrak{B}^n(A)$:

$$\mathfrak{B}^\infty(A) = \operatorname*{limpr}_{n\to\infty} \mathfrak{B}^n(A).$$

Clearly $\mathfrak{B}^\infty(A)$ is continuously imbedded in $\mathfrak{B}$. Formula (7.1) allows us to conclude that

$$\mathfrak{B}^{-\infty}(A) \overset{\text{def}}{=} (\mathfrak{B}^\infty(A))' = \operatorname*{limind}_{n\to\infty} (\mathfrak{B}^n(A))';$$

in this connection the strong topology in $(\mathfrak{B}^\infty(A))'$ coincides with the inductive limit topology of the spaces $(\mathfrak{B}^n(A))'$. If $\rho(A) \neq \emptyset$, then according to Theorem 2.1 $\overline{\mathfrak{B}^\infty(A)} = \mathfrak{B}$. In the case, when $\mathfrak{B} = \mathfrak{H}$ is a Hilbert space and $A \geq 0$ is a self-adjoint operator, the chain

$$\mathfrak{H}^\infty(A) \subset \mathfrak{H} \subset \mathfrak{H}^{-\infty}(A)$$

with continuous dense imbeddings arises, where $\mathfrak{H}^{-\infty}(A) = \operatorname{limind}_{n\to\infty} \mathfrak{H}^{-n}(A)$.

EXAMPLE 3. Set $\mathfrak{B} = C[a,b]$ ($-\infty < a < b < \infty$) and take for A the differentiation operator $Au = \frac{du}{dx}$, $\mathfrak{D}(A) = C^1[a,b]$. Then $\mathfrak{B}^\infty(A) = C^\infty[a,b]$ is the set of ordinary infinitely differentiable functions on $[a,b]$.

EXAMPLE 4. Let $\mathfrak{B} = L_p(a,b)$ ($-\infty \leq a < b \leq \infty$, $p \geq 1$) and $A : Au = \frac{du}{dx}$, where $\mathfrak{D}(A)$ is the set of absolutely continuous functions $u(x)$ on (a,b) for which $u'(x) \in L_p(a,b)$. In this case $\mathfrak{B}^\infty(A)$ coincides with the set of infinitely differentiable functions $f(x)$ on (a,b) for which $f^{(n)}(x) \in L_p(a,b)$ ($n \in \mathbb{N}$). In particular, if $\mathfrak{B} = L_2(\mathbb{R}^1)$, $A = D$, then

$$\mathfrak{B}^\infty(A) = \operatorname*{limpr}_{n\to\infty} W_2^n(\mathbb{R}^1) \overset{\text{def}}{=} W_2^\infty(\mathbb{R}^1),$$

that is, $(L_2(\mathbb{R}^1))^\infty(D)$ coincides with the space of infinitely differentiable functions on the real axis all derivatives of which are square integrable. The elements of the space $(W_2^\infty(\mathbb{R}^1))' \overset{\text{def}}{=} W_2^{-\infty}(\mathbb{R}^1)$ are derivatives of finite order (in the sense of the theory of distributions) of functions from $L_2(\mathbb{R}^1)$.

EXAMPLE 5. If $\mathfrak{H} = L_2(\mathbb{R}^1)$ and A is multiplication by $|x|$, then $\mathfrak{H}^\infty(A)$ is the set of square integrable functions $f(x)$ satisfying the condition

$$\int_{-\infty}^{\infty} |f(x)|^2 x^{2k}\, dx < \infty \quad (k \in \mathbb{N}).$$

The space $\mathfrak{H}^{-\infty}(A)$ coincides in this case with the set of all locally square integrable functions $f(x)$ for which

$$\int_{-\infty}^{\infty} |f(x)|^2 x^{-2k}\, dx < \infty \quad (k \in \mathbb{N}).$$

2.3. CLASSES OF INFINITELY DIFFERENTIABLE VECTORS

Let $\{m_n\}_{n=0}^{\omega}$ be a non-decreasing sequence of positive numbers, let A be a closed linear operator on a Banach space $\mathfrak{B}$ such that $\overline{\mathfrak{B}^\infty(A)} = \mathfrak{B}$. Set

$$\begin{aligned} C_{\{m_n\}}(A) &= \{f \in \mathfrak{B}^\infty(A) : \exists \alpha > 0, \exists c > 0 : \|A^n f\| \le c\alpha^n m_n\} \\ C_{(m_n)}(A) &= \{f \in \mathfrak{B}^\infty(A) : \forall \alpha > 0, \exists c = c(\alpha) : \|A^n f\| \le c\alpha^n m_n\}. \end{aligned} \tag{2.3}$$

In the sequel the symbol A in the notations of spaces will be omitted, exept when it is necessary to stress which operator we keep in mind.

Of course

$$C_{\{m_n\}} = \bigcup_{\alpha>0} C_\alpha < m_n >, \quad C_{(m_n)} = \bigcap_{\alpha>0} C_\alpha < m_n >,$$

where $C_\alpha < m_n >= C_\alpha < m_n > (A)$ is the Banach space of vectors $f \in \mathfrak{B}^\infty(A)$ for which an inequality as in (2.3) with fixed α is satisfied. A norm in this space is defined by

$$\|f\|_{C_\alpha < m_n >} = \sup_n \frac{\|A^n f\|}{\alpha^n m_n}.$$

For $\alpha_1 < \alpha_2$ we have $C_{\alpha_1} < m_n > \subset C_{\alpha_2} < m_n >$. Endow $C_{\{m_n\}}$ (respectively, $C_{(m_n)}$) with the inductive (projective) limit topology of the spaces $C_\alpha < m_n >$:

$$C_{\{m_n\}} = \operatorname*{limind}_{\alpha\to\infty} C_\alpha < m_n >, \quad C_{(m_n)} = \operatorname*{limpr}_{\alpha\to 0} C_\alpha < m_n > .$$

If $m_n \equiv \infty$, then the spaces $C_{\{m_n\}}$ and $C_{(m_n)}$ coincide with each other and also with the space of infinitely differentiable vectors of A. Thus

$$C_{\{\infty\}} = C_{(\infty)} = \mathfrak{B}^\infty.$$

A very important case is when $m_n = n^{n\beta}$ (equivalently, $m_n = (n!)^\beta$), $\beta > 0$. The spaces $C_{\{m_n\}}$ and $C_{(m_n)}$ corresponding to this sequence have special notations:

$$C_{\{n^{n\beta}\}}(A) \overset{\text{def}}{=} \mathfrak{G}_{\{\beta\}}(A), \quad C_{(n^{n\beta})}(A) \overset{\text{def}}{=} \mathfrak{G}_{(\beta)}(A).$$

For $\beta > 1$ they are named Gevrey classes (or classes of ultradifferentiable vectors) of Roumieu and Beurling type, respectively. These names we also preserve for

$0 < \beta < 1$. If $\beta = 1$ we obtain the spaces $\mathfrak{A}(A)$ and $\mathfrak{A}_c(A)$ of analytic and entire vectors of the operator A:

$$\mathfrak{A} \stackrel{\text{def}}{=} \mathfrak{G}_{\{1\}} = C_{\{n^n\}} = C_{\{n!\}}, \quad \mathfrak{A}_c \stackrel{\text{def}}{=} \mathfrak{G}_{(1)} = C_{(n^n)} = C_{(n!)}.$$

The space $C_{\{1\}}(A)$ constructed from the sequence $m_n \equiv 1$ has its own notation:

$$C_{\{1\}}(A) \stackrel{\text{def}}{=} \mathcal{E}xp_A \mathfrak{B}.$$

Its elements are called vectors of exponential type of A. It is obvious that

$$C_{(1)}(A) \supseteq \text{Ker } A.$$

In the concrete situation of example 3 the spaces $\mathfrak{G}_{\{\beta\}}$ and $\mathfrak{G}_{(\beta)}$ $(\beta > 1)$ are the ordinary Gevrey classes of the infinitely differentiable functions on $[a, b]$ of Roumieu and Beurling type respectively, $\mathfrak{A}$ and $\mathfrak{A}_c$ are the spaces of analytic on (a, b) and entire functions, $\mathcal{E}xp_A \mathfrak{B}$ is the set of all entire functions of exponential type. If $\mathfrak{B} = C_b(\mathbb{R}^1)$ is the space of continuous bounded functions $f(x)$ on $(-\infty, \infty)$, $A : Au = \frac{du}{dx}$, and $\mathfrak{D}(A) = C_b^1(\mathbb{R}^1)$ (i.e. the set of functions $f(x) \in C_b(\mathbb{R}^1)$ for which $f'(x) \in C_b(\mathbb{R}^1)$), then boundedness on the real axis is added to the characterization of functions from $\mathfrak{A}$, $\mathfrak{A}_c$ and $\mathcal{E}xp_A \mathfrak{B}$.

The spaces considered above are continuously imbedded in $\mathfrak{B}$. To imbed $\mathfrak{B}'$ into $C'_{\{m_n\}}$ (into $C'_{(m_n)}$) it is necessary for $C_{\{m_n\}}$ (respectively $C_{(m_n)}$) to be dense in $\mathfrak{B}$. Conditions guaranteeing this will be given in Chapter 5. Now we will sketch in detail the case when A is a normal operator on a Hilbert space. Since $C_\alpha < m_n > (A) = C_\alpha < m_n > (|A|)$ where $|A| = \sqrt{A^*A}$, it suffices to study these sets for a non-negative self-adjoint operator only.

2.4. INDUCTIVE AND PROJECTIVE LIMITS OF SPACES CONSTRUCTED FROM A FUNCTION WHICH INCREASES FASTER THAN POWERS. A THEOREM OF WIENER–PALEY TYPE

Let $G(\lambda)$ $(\lambda \in [0, \infty))$ be, in addition to the conditions (2.1), monotonically increasing. Moreover, suppose that $G(\lambda)$ is continuously differentiable on $[0, \infty)$ and that

$$\lambda G'(\lambda) G^{-1}(\lambda) \uparrow \infty \quad (\lambda \to \infty). \tag{2.4}$$

From a non-negative self-adjoint operator A and the functions $G_\alpha(\lambda) = G(\alpha\lambda)$ $(\alpha > 0)$ we construct the family of the Hilbert spaces $\mathfrak{H}_{G\alpha} = \mathfrak{H}_{G\alpha}(A)$ as in subsection 2.1. The continuous and dense imbeddings

$$\mathfrak{H}_{G\alpha} \subset \mathfrak{H} \subset \mathfrak{H}'_{G\alpha}$$

hold for them. It is clear that $\|f\|_{\mathfrak{H}_{G_s}} \geq \|f\|_{\mathfrak{H}_{G_t}}$ for $s > t$ and also that these norms are compatible, whence we have a continuous imbedding $\mathfrak{H}_{G_s} \subseteq \mathfrak{H}_{G_t}$. It is dense because the set

$$\mathcal{E}(A) = \{f = E_\Delta h : \Delta \text{ a bounded interval of } [0, \infty), \quad \forall h \in \mathfrak{H}\}$$

is dense in $\mathfrak{H}$. Set

$$\mathfrak{H}_{\{G\}}(A) = \operatorname*{limind}_{\alpha\to 0} \mathfrak{H}_{G_\alpha}(A), \quad \mathfrak{H}_{(G)}(A) = \operatorname*{limpr}_{\alpha\to\infty} \mathfrak{H}_{G_\alpha}(A).$$

As was noted in Chapter 1, subsection 7.2, convergence of a sequence f_n in $\mathfrak{H}_{(G)}$ means convergence of it in each $\mathfrak{H}_{G_\alpha}$. Convergence of f_n in $\mathfrak{H}_{\{G\}}$ is equivalent to the fact that all f_n belong to a certain $\mathfrak{H}_{G_\alpha}$ and that they converge in this space. In fact, since $\mathfrak{H}_{\{G\}}$ is a regular inductive limit, the convergence $f_n \to 0$ $(n \to \infty)$ in $\mathfrak{H}_{\{G\}}$ implies the existence of $\alpha > 0$ such that $f_n \in \mathfrak{H}_{G_\alpha}$ and $\|f_n\|^2_{\mathfrak{H}_{G_\alpha}} = \int_0^\infty G^2(\alpha\lambda)\,\mathrm{d}(E_\lambda f_n, f_n) \le c = \text{const}$. But the topology of the space $\mathfrak{H}$ is weaker than that of $\mathfrak{H}_{\{G\}}$. So, $f_n \to 0$ $(n \to \infty)$ in $\mathfrak{H}$. Consequently, for every fixed λ, $\tau_n(\lambda) = \int_0^\lambda G^2(\alpha s)\,\mathrm{d}(E_s f_n, f_n)$ tends to zero. Taking into account the boundedness of $\tau_n(\lambda)$ we obtain in view of Helly's theorem that $\|f_n\|_{\mathfrak{H}_{G_\alpha}} = \int_0^\infty \mathrm{d}\tau_n(\lambda) \to 0$ $(n \to \infty)$. With the function $G(\lambda)$ we associate the sequence

$$m_n = \sup_{\lambda\ge 1} \frac{\lambda^n}{G(\lambda)}. \tag{2.5}$$

This sequence is positive and monotonically increasing. It possesses the property

(i) for any $\alpha > 0$ there exists a constant $c = c(\alpha)$ such that

$$m_n > c\alpha^n \quad (n = 0, 1, \ldots).$$

Indeed, for $\alpha \ge 1$,

$$m_n = \sup_{\lambda\ge 1} \frac{\lambda^n}{G(\lambda)} \ge \frac{\alpha^n}{G(\alpha)},$$

whence (i) follows. If $\alpha < 1$, then

$$m_n \ge G^{-1}(1) > \frac{\alpha^n}{G(1)}.$$

Define the functions

$$T(\lambda) = \sup_k \frac{m_0\lambda^k}{m_k}, \quad \widetilde{T}(\lambda) = m_0\Big(\sum_{n=0}^\infty \frac{\lambda^{2n}}{m_n^2}\Big)^{\frac{1}{2}}, \quad \approx{T}(\lambda) = m_0 \sum_{n=0}^\infty \frac{\lambda^n}{m_n}. \tag{2.6}$$

Property (i) ensures the convergence of the series in (2.6) for each λ. The same estimate implies the relations

$$\begin{gathered} T(\lambda) \le \widetilde{T}(\lambda) \le \approx{T}(\lambda) \le c(\beta)\widetilde{T}(\beta\lambda) \le c(\beta,\beta_1)T(\beta_1\lambda) \\ \forall\beta > 1, \quad \forall\beta_1 > \beta, \quad c(\beta) = \frac{\beta}{(\beta^2-1)^{1/2}}, \quad c(\beta\beta_1) = c(\beta)\frac{\beta_1}{(\beta_1^2-\beta^2)^{1/2}} \end{gathered} \tag{2.7}$$

between these functions. The three functions (2.6) increase monotonically and satisfy (2.1).

LEMMA 2.1. *Let $G(\lambda)$ be a continuous function on $[0,\infty)$ that monotonically increases and satisfies the conditions* (2.1), (2.4). *If the function $T(\lambda)$ is constructed* *from the sequence* (2.5) *as in* (2.6), *then*

$$m_0^{-1}T(\lambda) \le G(\lambda) \le m_0^{-1}\lambda T(\lambda) \quad (\lambda \ge 1). \tag{2.8}$$

Proof. In fact,

$$T(\lambda) = \sup_k \frac{m_0\lambda^k}{m_k} = \sup_k \frac{m_0\lambda^k}{\sup_{r\ge 1}(r^k G^{-1}(r))} \le \sup_k \frac{m_0\lambda^k}{\lambda^k G^{-1}(\lambda)} = m_0 G(\lambda).$$

In order to prove the right side inequality in (2.8) we consider the function

$$H(r) = \sup_{\lambda\ge 1} \frac{\lambda^r}{G(\lambda)} \quad (r > 0).$$

Clearly, $H(k) = m_k$. It follows from the definition of $H(r)$ that we can find a number r_λ such that $H(r_\lambda) = \lambda^{r_\lambda}G^{-1}(\lambda)$, namely $r_\lambda = \lambda G'(\lambda)G^{-1}(\lambda)$. Then

$$G(\lambda) = \frac{\lambda^{r_\lambda}}{H(r_\lambda)} \le \sup_r \frac{\lambda^r}{H(r)} \stackrel{\text{def}}{=} T_H(\lambda).$$

On the other hand,

$$T_H(\lambda) = \sup_r \frac{\lambda^r}{\sup_{p\ge 1}(p^r G^{-1}(p))} \le \sup_r \frac{\lambda^r}{\lambda^r G^{-1}(\lambda)} = G(\lambda).$$

Thus, $T_H(\lambda) = G(\lambda)$. Because of the monotonicity of $H(r)$,

$$G(\lambda) = \sup_r \frac{\lambda^r}{H(r)} \le \sup_r \frac{\lambda^{[r]+1}}{H([r])} \le \frac{\lambda}{m_0}\sup_n \frac{m_0\lambda^n}{m_n} = m_0^{-1}\lambda T(\lambda)$$

(here $[\cdot]$ is the integer part of a number). □

Suppose that in addition to the conditions of Lemma 2.1 the function $G(\lambda)$ has the property:

$$\exists c_0 > 0,\ \exists \alpha_0\ (0 < \alpha_0 < 1) : G(\lambda) \ge c_0\lambda G(\alpha_0\lambda) \quad (\lambda \in [0,\infty)). \tag{2.9}$$

For a $G(\lambda)$ satisfying this inequality the sequence (2.5) satisfies the condition

(ii) there exist constants $c > 0$ and $h > 1$ such that

$$m_{n+1} \le ch^n m_n \quad (n = 0,1,\ldots).$$

Indeed,

$$\begin{aligned}
m_{n+1} &= \sup_{\lambda\ge 1}\frac{\lambda^{n+1}}{G(\lambda)} \le \sup_{\lambda\ge 1}\frac{\lambda^n}{c_0 G(\alpha_0\lambda)} = \sup_{\lambda\ge 1}\frac{(\alpha_0\lambda)^n}{c_0\alpha_0^n G(\alpha_0\lambda)}\\
&= \frac{1}{c_0\alpha_0^n}\sup_{\lambda\ge 1}\frac{(\alpha_0\lambda)^n}{G(\alpha_0\lambda)} = \frac{1}{c_0\alpha_0^n}\sup_{\mu\ge\alpha_0}\frac{\mu^n}{G(\mu)}\\
&= \frac{1}{c_0\alpha_0^n}\max\Big\{\sup_{\alpha_0\le\mu\le 1}\frac{\mu^n}{G(\mu)}, m_n\Big\} \le \frac{1}{c_0\alpha_0^n}\max\Big\{\frac{m_0}{m_0 G(\alpha_0)}, m_n\Big\}\\
&\le \frac{1}{c_0\alpha_0^n}\max\Big\{\frac{m_n}{m_0 G(\alpha_0)}, m_n\Big\} = ch^n m_n,
\end{aligned}$$

where $h = \alpha_0^{-1} > 1$, $c = \frac{1}{c_0}\max\{\frac{1}{m_0 G(\alpha_0)}, 1\}$.

From the sequence (2.5) we construct the spaces $C_{\{m_n\}}(A)$ and $C_{(m_n)}(A)$. Note that property (i) for $\{m_n\}_{n=0}^{\infty}$ implies

$$\mathcal{E}xp_A\mathfrak{H} \subset C_\alpha\!<\!m_n\!>(A) \quad (\alpha > 0).$$

PROPOSITION 2.1. *Suppose $G(\lambda)$ is as in Lemma 2.1 and also satisfies condition (2.9). If $\{m_n\}_{n=0}^{\infty}$ is the sequence corresponding to $G(\lambda)$ by formula (2.5), then the topological equalities*

$$C_{\{m_n\}} = \mathfrak{H}_{\{G\}}, \quad C_{(m_n)} = \mathfrak{H}_{(G)}$$

are valid.

Proof. The relations

$$\begin{aligned}
\|f\|^2_{C_\alpha\lt m_n\gt} &= \sup_n \frac{\|A^n f\|^2}{\alpha^{2n}m_n^2} = \sup_n \frac{\int_0^\infty \lambda^{2n}\mathrm{d}(E_\lambda f,f)}{\alpha^{2n}m_n^2} \\
&= \sup_n \frac{\int_0^\alpha \lambda^{2n}\mathrm{d}(E_\lambda f,f) + \int_\alpha^\infty \lambda^{2n}\mathrm{d}(E_\lambda f,f)}{\alpha^{2n}m_n^2} \\
&\le m_0^{-2}\|f\|^2 + \sup_n\left(m_n^{-2}\int_\alpha^\infty \frac{(\alpha^{-1}\lambda)^{2n}G^2(\alpha^{-1}\lambda)}{G^2(\alpha^{-1}\lambda)}\,\mathrm{d}(E_\lambda f,f)\right) \\
&\le m_0^{-2}\|f\|^2 + \sup_n\left(m_n^{-2}\left(\sup_{\tau\ge 1}\frac{\tau^n}{G(\tau)}\right)^2\right)\|f\|^2_{\mathfrak{H}_{G_{\alpha^{-1}}}} \\
&= (m_0^{-2}+1)\|f\|^2_{\mathfrak{H}_{G_{\alpha^{-1}}}}
\end{aligned}$$

and

$$\begin{aligned}
\|f\|^2_{\mathfrak{H}_{G_\alpha}} &= \int_0^\infty G^2(\alpha\lambda)\,\mathrm{d}(E_\lambda f,f) = \int_0^{\alpha^{-1}} G^2(\alpha\lambda)\,\mathrm{d}(E_\lambda f,f) + \\
&\qquad + \int_{\alpha^{-1}}^\infty G^2(\alpha\lambda)\,\mathrm{d}(E_\lambda f,f) \\
&\le G^2(1)\|f\|^2 + c_0^{-2}\int_{\alpha^{-1}}^\infty \frac{G^2(\alpha_0^{-1}\alpha\lambda)}{(\alpha_0^{-1}\alpha\lambda)^2}\,\mathrm{d}(E_\lambda f,f) \\
&\le G^2(1)\|f\|^2 + (c_0m_0)^{-2}\int_{\alpha^{-1}}^\infty T^2(\alpha_0^{-1}\alpha\lambda)\,\mathrm{d}(E_\lambda f,f) \\
&\le G^2(1)\|f\|^2 + (c_0m_0)^{-2}\int_0^\infty \tilde{T}^2(\alpha_0^{-1}\alpha\lambda)\,\mathrm{d}(E_\lambda f,f) \\
&= G^2(1)\|f\|^2 + c_0^{-2}\int_0^\infty \sum_{n=0}^\infty \frac{(\alpha_0^{-1}\alpha\lambda)^{2n}}{m_n^2}\,\mathrm{d}(E_\lambda f,f) \le G^2(1)\|f\|^2 + \\
&\quad + c_0^{-2}\sum_{n=0}^\infty \frac{\|A^n f\|^2\beta^{2n}}{m_n^2(\alpha_0\alpha^{-1}\beta)^{2n}} \le \left(G^2(1)m_0^2 + \frac{c_0^{-2}}{1-\beta^2}\right)\|f\|^2_{c_s\lt m_n\gt} \\
&\quad (s = \alpha_0\alpha^{-1}\beta,\ \forall\beta < 1),
\end{aligned}$$

arising from (2.7)-(2.9), give the topological equalities required. □

Conversely, let a non-decreasing sequence of positive numbers m_n satisfying property (i) be given. Associate with it functions as in (2.6). It is easily seen that these functions satisfy (2.1). From the function $T(\lambda)$ we construct the family of the Hilbert spaces $\mathfrak{H}_{T_\alpha}$, and their projective and inductive limits $\mathfrak{H}_{(T)}$ and $\mathfrak{H}_{\{T\}}$ as $\alpha \to \infty$ and $\alpha \to 0$, respectively. We can do the same relative to the functions $\widetilde{T}(\lambda)$ and $\overset{\approx}{T}(\lambda)$. The inequalities (2.8) show that $\mathfrak{H}_{(T)} = \mathfrak{H}_{(\widetilde{T})} = \mathfrak{H}_{(\overset{\approx}{T})}$; analogously, $\mathfrak{H}_{\{T\}} = \mathfrak{H}_{\{\widetilde{T}\}} = \mathfrak{H}_{\{\overset{\approx}{T}\}}$. Since

$$\begin{aligned}\|f\|^2_{\mathfrak{H}_{T_\alpha}} &= \int_0^\infty T^2(\alpha\lambda)\,\mathrm{d}(E_\lambda f, f) \le \int_0^\infty \widetilde{T}(\alpha\lambda)\,\mathrm{d}(E_\lambda f, f)\\ &= m_0^2 \int_0^\infty \sum_{n=0}^\infty \Big(\frac{\alpha^n\lambda^n}{m_n}\Big)^2 \mathrm{d}(E_\lambda f, f) = m_0^2 \sum_{n=0}^\infty \frac{\beta^{-2n}\int_0^\infty \lambda^{2n}\,\mathrm{d}(E_\lambda f, f)}{(\alpha^{-1}\beta^{-1})^{2n} m_n^2}\\ &\le m_0^2 c^2(\beta)\Big(\sup_n \frac{\|A^n f\|}{(\alpha^{-1}\beta^{-1})^n m_n}\Big)^2 = m_0^2 c^2(\beta)\|f\|^2_{C_{\alpha^{-1}\beta^{-1}\langle m_n\rangle}}\end{aligned}$$

(any $\alpha > 0$, any $\beta > 1$) and, vice versa,

$$\begin{aligned}\|f\|^2_{C_\alpha\langle m_n\rangle} &= \sup_n \frac{\|A^n f\|^2}{\alpha^{2n} m_n^2} = \sup_n \int_0^\infty \Big(\frac{\lambda^n}{\alpha^n m_n}\Big)^2 \mathrm{d}(E_\lambda f, f)\\ &\le \int_0^\infty \Big(\sup_n \frac{\lambda^n \alpha^{-n}}{m_n}\Big)^2 \mathrm{d}(E_\lambda f, f) = m_0^{-2}\int_0^\infty T^2(\alpha^{-1}\lambda)\,\mathrm{d}(E_\lambda f, f)\\ &= m_0^{-2}\|f\|^2_{\mathfrak{H}_{T_{\alpha^{-1}}}},\end{aligned}$$

the following assertion is true.

PROPOSITION 2.2. *If a non-decreasing sequence of positive numbers m_n ($n = 0, 1, \ldots$) satisfies condition* (i), *then the topological equalities*

$$C_{\{m_n\}} = \mathfrak{H}_{\{T\}}, \quad C_{(m_n)} = \mathfrak{H}_{(T)}$$

hold, where $T(\lambda)$ is the function corresponding to $\{m_n\}_{n=0}^\infty$ as in (2.6).

Now we take a non-decreasing positive sequence $\{m_n\}_{n=0}^\infty$ with the properties (i), (ii) and show that the function $\overset{\approx}{T}(\lambda)$ in (2.6) constructed from this sequence has the properties (2.1), (2.4), (2.9). In fact,

$$\begin{aligned}\overset{\approx}{T}(\lambda) &= m_0 \sum_{n=0}^\infty \frac{\lambda^n}{m_n} \ge m_0\lambda \sum_{n=1}^\infty \frac{\lambda^{n-1}}{m_n} \ge m_0\lambda \sum_{n=1}^\infty \frac{\lambda^{n-1}}{ch^{n-1}m_{n-1}}\\ &= m_0 c^{-1}\lambda \sum_{n=1}^\infty \frac{(\lambda h^{-1})^{n-1}}{m_{n-1}} = c_0\lambda \overset{\approx}{T}(\alpha_0\lambda)\end{aligned}$$

where $c_0 = m_0 c^{-1}$, $\alpha_0 = h^{-1}$. The function $\overset{\approx}{T}(\lambda)$ is entire. The Cauchy–

Buniakowski inequality yields for $\lambda > 0$,

$$(\lambda \overset{\approx}{T}'(\lambda)\overset{\approx}{T}^{-1}(\lambda))' = \Big(\sum_{n=1}^{\infty} \frac{n^2\lambda^{n-1}}{m_n}\sum_{n=0}^{\infty} \frac{\lambda^n}{m_n} - \sum_{n=1}^{\infty} \frac{n\lambda^n}{m_n}\sum_{n=1}^{\infty}\frac{n\lambda^{n-1}}{m_n}\Big)\Big(\sum_{n=0}^{\infty} \frac{\lambda^n}{m_n}\Big)^{-2}$$

$$= \Big(m_0^{-1}\sum_{n=1}^{\infty} \frac{n^2\lambda^{n-1}}{m_n} + \lambda\Big(\sum_{n=1}^{\infty} \frac{n^2\lambda^{n-1}}{m_n}\sum_{n=1}^{\infty} \frac{\lambda^{n-1}}{m_n} -$$

$$- \Big(\sum_{n=1}^{\infty} \frac{n\lambda^{n-1}}{m_n}\Big)^2\Big)\Big)\Big(\sum_{n=0}^{\infty} \frac{\lambda^n}{m_n}\Big)^{-2} > 0.$$

Therefore, $\lambda \overset{\approx}{T}'(\lambda)\overset{\approx}{T}^{-1}(\lambda) \uparrow \infty$ as $\lambda \to +\infty$. The rest is obvious.

Propositions 2.1 and 2.2 together with what was proved just now allow us to formulate as a consequence the following theorem.

THEOREM 2.2. *Let $A \geq 0$ be a self-adjoint operator on a Hilbert space $\mathfrak{H}$. If a non-decreasing positive sequence $\{m_n\}_{n=0}^{\infty}$ satisfies the conditions* (i) *and* (ii), *$T(\lambda) = \sup_n \frac{\lambda^n}{m_n}$ is the function corresponding to it by* (2.6), *$m'_n = \sup_{\lambda \geq 1}(\lambda^n T^{-1}(\lambda))$, then*

$$C_{\{m_n\}}(A) = \mathfrak{H}_{\{T\}}(A) = C_{\{m'_n\}}(A); \quad C_{(m_n)}(A) = \mathfrak{H}_{(T)}(A) = C_{(m'_n)}(A).$$

Conversely, let $G(\lambda)$ be a continuously-differentiable function on $[0, \infty)$ with the properties (2.1), (2.4), (2.9). *Set $M_n = \sup_{\lambda \geq 1} \frac{\lambda^n}{G(\lambda)}$, $G^1(\lambda) = M_0 \sup_n \frac{\lambda^n}{M_n}$. Then*

$$\mathfrak{H}_{\{G\}}(A) = C_{\{M_n\}}(A) = \mathfrak{H}_{\{G^1\}}(A); \quad \mathfrak{H}_{(G)}(A) = C_{(M_n)}(A) = \mathfrak{H}_{(G^1)}(A).$$

REMARK 2.1. If $\{m_n\}_{n=0}^{\infty}$ satisfies the condition opposite to (i), that is, there exists $\alpha \geq 1$ such that $m_n < c\alpha^n$, then the function $\tilde{T}(\lambda)$ corresponding to this sequence becomes infinite for $\lambda \geq \alpha$. In this case $\mathfrak{H}_{\{T\}}(A) = \mathfrak{H}_{\{\tilde{T}\}}(A) = \mathcal{E}(A)$, $\mathfrak{H}_{(T)}(A) = \mathrm{Ker}(A)$. On the other hand, it is not difficult to observe that $C_{\{m_n\}}(A) = \mathcal{E}xp_A\mathfrak{H}$, $C_{(m_n)}(A) = \mathrm{Ker} A$. Taking into account the, easily verifiable, equality

$$\mathcal{E}(A) = \mathcal{E}xp_A\mathfrak{H},$$

we can immediately establish the validity of Theorem 2.2 also under the condition opposite to (i) on $\{m_n\}_{n=0}^{\infty}$.

Above, it was required that the m_n take only finite values. If $m_N = \infty$ for some N, hence for any $n > N$, then $\overset{\approx}{T}(\lambda)$ is a polynomial of degree $(N-1)$ and, as is easily seen,

$$\mathfrak{H}_{\{T\}}(A) = \mathfrak{H}_{(T)}(A) = C_{\{m_n\}}(A) = C_{(m_n)}(A) = \mathfrak{D}(A^{n-1}).$$

So, the theorem is valid in this case too.

Theorem 2.2 may be considered as a certain generalization of Wiener–Paley theorems for the case when A is the differentiation operator in function spaces. For example, if $\mathfrak{H} = L_2(\mathbb{R}^1)$, A is the modulus of the differentiation operator, then the fact that $f \in \mathfrak{H}_{G_\alpha}(A)$ is, in terms of spectral theory, equivalent to the integral $\int_0^\infty |(Ff)(\lambda)|^2 G^2(\alpha\lambda)\, d\lambda$ being finite, that is, to a certain restriction on the decrease of the Fourier transform of f at infinity. At the same time the fact that f belongs to the space $C_\alpha < m_n > (A)$ imposes in the situation considered certain restrictions on the growth of its derivatives. Thus, Theorem 2.2 connects the decrease of the Fourier transform of the function f at ∞ with its analyticity properties.

2.5. ANALYTIC AND ENTIRE VECTORS

In subsection 2.3 analytic (entire) vectors of a closed linear operator A on a Banach space $\mathfrak{B}$ were defined as vectors $f \in \mathfrak{B}^\infty(A)$ satisfying the inequality

$$\|A^n f\| \le c\alpha^n n! \quad (n = 0, 1, \ldots) \tag{2.10}$$

for some $c, \alpha > 0$ (for any α; $c = c(\alpha)$). As can be easily seen this definition is equivalent to the following one: a vector $f \in \mathfrak{B}^\infty(A)$ is called analytic (entire) for the operator A if

$$\sum_{n=0}^{\infty} \frac{\|A^n f\|}{n!} t^n < \infty \tag{2.11}$$

for some (any) $t > 0$. For example, let A be the operator generated on the space $L_2(a, b)$ $(-\infty \le a < b \le \infty)$ by the expression $i\frac{d}{dx}$ on absolutely continuous functions with square integrable derivative. A function $f \in L_2(a, b)$ is an analytic (entire) vector of A if and only if it is analytic on $[a, b]$ (entire); in the case of an infinite interval all its derivatives must, in addition, be square integrable.

The set of analytic (entire) vectors of A forms a vector space. We denote it by $\mathfrak{A}(A)$ (respectively, $\mathfrak{A}_c(A)$) or simply by $\mathfrak{A}$ (respectively, $\mathfrak{A}_c$):

$$\mathfrak{A}(A) = C_{\{n!\}}(A) = C_{\{n^n\}}(A); \quad \mathfrak{A}_c(A) = C_{(n!)}(A) = C_{(n^n)}(A).$$

If A is a self-adjoint operator on a Hilbert space $\mathfrak{H}$, then the set $\mathcal{E}(A) = \{E_\Delta f : f \in \mathfrak{H},\ \Delta \subset \mathbb{R}^1 \text{ a bounded interval}\} = \mathcal{E}xp_A\mathfrak{H}$ consists of entire vectors and is dense in $\mathfrak{H}$.

On an analytic vector f of a self-adjoint operator A we have

$$e^{iAt} f = \sum_{n=0}^{\infty} \frac{(iA)^n}{n!} t^n f \tag{2.12}$$

for all t, where the series (2.11) converges.

In fact, if $f_\Delta \in \mathcal{E}xp_A\mathfrak{H}$, then because of the boundedness of the operator AE_Δ,

$$e^{iAt} f_\Delta = e^{iAE_\Delta t} f_\Delta = \sum_{n=0}^{\infty} \frac{(iAE_\Delta)^n}{n!} t^n f_\Delta = \sum_{n=0}^{\infty} \frac{(iA)^n}{n!} t^n f_\Delta.$$

Since $\sum_{n=0}^\infty \frac{\|A^n f\|}{n!} t^n < \infty$, for any $\epsilon > 0$ we can find a number N such that

$$\sum_{n=N}^{\infty} \frac{\|A^n f_\Delta\|}{n!} t^n \le \sum_{n=N}^{\infty} \frac{\|A^n f\|}{n!} t^n < \frac{\epsilon}{2}$$

independently of Δ. Then

$$\Big\|\sum_{n=0}^{\infty}\frac{t^n}{n!}(iA)^n f_\Delta - \sum_{n=0}^{\infty}\frac{t^n}{n!}(iA)^n f\Big\| \le \Big\|\sum_{n=0}^{\infty}\frac{t^n}{n!}(iA)^n (f - f_\Delta)\Big\| + \epsilon.$$

Now let the intervals $(-m, m)$ $(m \in \mathbb{N})$ play the role of Δ. Taking into account that $\|\sum_{n=0}^{N}\frac{t^n}{n!}(iA)^n(f - f_\Delta)\| \to 0$ and also that $e^{iAt}f_\Delta \to e^{iAt}f$ as $m \to \infty$, we obtain the inequality

$$\Big\|e^{iAt}f - \sum_{n=0}^{\infty}\frac{t^n}{n!}(iA)^n f\Big\| \le \epsilon$$

which leads to (2.12), since ϵ is arbitrary.

Notice that analyticity of a vector f implies that of $A^k f$ for any $k \in \mathbb{N}$. This follows from the estimate

$$\|A^n A^k f\| = \|A^{n+k} f\| \le c\alpha^{n+k}(n+k)! \le c(2\alpha)^k \alpha^n n! n^k \le c(2\alpha k)^n \alpha^n n!,$$

valid for fixed $k \in \mathbb{N}$ and sufficiently large natural n.

According to Theorem 2.2,

$$\mathfrak{A} = \operatorname*{limind}_{\alpha \to 0} \mathfrak{H}_{e,\alpha};$$

here $\mathfrak{H}_{e,\alpha}$ is, by definition, the space $\mathfrak{H}_{G_\alpha}$ constructed from the function $G(\lambda) = e^\lambda$, i.e.

$$\mathfrak{H}_{e,\alpha} = \mathfrak{H}_{e^{\alpha\lambda}}.$$

Thus,

$$\mathfrak{A}(A) = \bigcup_{\alpha>0} \mathfrak{H}_{e,\alpha}(A)$$

and its topology is the inductive limit topology of the Hilbert spaces $\mathfrak{H}_{e,\alpha}(A)$. As was shown in subsection 2.4 convergence in $\mathfrak{A}$ means the convergence in a certain $\mathfrak{H}_{e,\alpha}$.

The following simple fact will be used frequently.

THEOREM 2.3. *Let $A \ge 0$ be a self-adjoint operator on a Hilbert space $\mathfrak{H}$. Let $G(\lambda) \ge 1$ be a continuous function on $[0, \infty)$ such that*

$$G(\lambda) \le c \exp(\alpha\lambda) \tag{2.13}$$

for any $\alpha > 0$ ($c =$ const depends on α). Then

$$\mathfrak{A}(A) \subseteq \mathfrak{H}_G(A) \subseteq \mathfrak{H},$$

and these imbeddings are continuous.

Proof. The fact $f \in \mathfrak{A}$ is equivalent to $f \in \mathfrak{H}_{e,\alpha}$ for some $\alpha > 0$, i.e.

$$\int_0^\infty e^{2\alpha\lambda}\, d(E_\lambda f, f) < \infty.$$

From the estimate (2.13) we have

$$\|f\|^2_{\mathfrak{H}_G} = \int_0^\infty G^2(\lambda)\,\mathrm{d}(E_\lambda f, f) \le c^2 \int_0^\infty e^{2\alpha\lambda}\,\mathrm{d}(E_\lambda f, f) = c^2 \|f\|^2_{\mathfrak{H}_{e,\alpha}}.$$

This shows that $f \in \mathfrak{H}_G$, and the imbedding $\mathfrak{A} \subseteq \mathfrak{H}_G$ is continuous. □

In view of Theorem 1.3,

$$\mathfrak{A}'(A) = \operatorname*{limpr}_{\alpha\to 0} \mathfrak{H}'_{e,\alpha}(A).$$

Vectors from $\mathfrak{A}'(A)$ are called analytic functionals or co-analytic vectors of A. The convergence in $\mathfrak{A}'$ of a sequence $f_n \in \mathfrak{A}'$ to an element f means its convergence to f in every space $\mathfrak{H}'_{e,\alpha}$. The expression (f,g) $(f \in \mathfrak{A}',\ g \in \mathfrak{A})$ can now be understood as an extension of the inner product $(.,.)$ to a continuous bilinear form on $\mathfrak{A}' \times \mathfrak{A}$ (see Remark 1.1).

DEFINITION. Let A be a symmetric operator on $\mathfrak{H}$. For $f \in \mathfrak{H}^\infty(A)$ denote by $\mathcal{D}(f)$ the linear span of the vectors $A^k f$ $(k = 0,1,\ldots)$. Let $L(f) = \overline{\mathcal{D}(f)}$. The restriction $A \restriction \mathcal{D}(f)$ of the operator A to $\mathcal{D}(f)$ is a symmetric operator on $L(f)$. Since the continuous extension of the operator $\sum_{k=0}^m c_k A^k f \to \sum_{k=0}^m \bar{c}_k A^k f$ (c_k is a complex number) from $\mathcal{D}(f)$ defines an involution in $L(f)$ commuting with A, the deficiency numbers of $A \restriction \mathcal{D}(f)$ are equal; hence it admits a self-adjoint extension (possibly not unique) (see Chapter 3, subsection 1.1). If the extension is unique, the vector f is called a uniqueness vector of A.

LEMMA 2.2. *Every analytic vector of a symmetric operator A is a uniqueness vector of A.*

Proof. Let f be an analytic vector of A. Suppose that $A \restriction \mathcal{D}(f)$ has two different self-adjoint extensions A_1 and A_2 to $L(f)$. Since the vector $y = A^k f$ $(k = 0,1,\ldots)$ is analytic for A we have from (2.12),

$$e^{iA_1 t} y = e^{iA_2 t} y = \sum_{n=0}^\infty \frac{t^n}{n!}(iA)^n y.$$

when t is sufficiently small. The group property of the operator e^{tB} (B is a self-adjoint operator) allows us to conclude that $e^{iA_1 t} y = e^{iA_2 t} y$ for all $t \in \mathbb{R}^1$. Coincidence of A_1 and A_2 on $L(f)$ follows from this. Hence f is a uniqueness vector of A. □

DEFINITION. A symmetric operator A is called essentially self-adjoint if its closure is self-adjoint.

THEOREM 2.4. *Let A be a symmetric operator on $\mathfrak{H}$. Assume that the set $\mathcal{D} = \bigcup \mathcal{D}(f)$, where f runs through the set of all uniqueness vectors of A, is dense in $\mathfrak{H}$. Then A is essentially self-adjoint.*

Proof. Let h be an arbitrary element from $\mathfrak{H}$. Since $\mathcal{D}$ is dense in $\mathfrak{H}$, there exists a vector $y \in \mathcal{D}(f)$ (f is a certain uniqueness vector) approximating h. In turn, y

can be approximated by vectors of the form $(A \pm iE)g$ $g \in \mathcal{D}(f))$. This shows that $\overline{\mathfrak{R}(A \pm iE)} = \mathfrak{H}$, which is equivalent to the essential self-adjointness of A. □

As we could see, any self-adjoint operator has a set of analytic vectors that is dense in $\mathfrak{H}$. This property is characteristic for this class of operators. Namely, the following assertion is true.

THEOREM 2.5. *A closed symmetric operator A on $\mathfrak{H}$ is self-adjoint if and only if it has a set of analytic vectors that is dense in $\mathfrak{H}$.*

Proof. It is sufficient to note that if A has such a set of analytic vectors, then by virtue of Lemma 2.2 it has a dense set of uniqueness vectors. By Theorem 2.4 it is self-adjoint. □

2.6. GEVREY CLASSES

Let $A \geq 0$ be a self-adjoint operator on a Hilbert space $\mathfrak{H}$. According to the definition given in subsection 2.3, a vector $f \in \mathfrak{H}^\infty(A)$ belongs to the Gevrey class of order β $(0 \leq \beta < \infty)$ of Roumieu (Beurling) type if and only if there exist positive constants c, α (for any $\alpha > 0$ there exists $c = c(\alpha) > 0$) such that

$$\|A^n f\| \leq c\alpha^n n^{n\beta} \tag{2.14}$$

(clearly one can replace $n^{n\beta}$ in (2.14) by $(n!)^\beta$). The Gevrey classes constructed from the operator A are vector spaces. They are denoted by $\mathfrak{G}_{\{\beta\}}(A)$ (the Roumieu class of order β) and $\mathfrak{G}_{(\beta)}(A)$ (the Beurling class of the same order β) and also $\mathfrak{G}_{\{1\}}(A) = \mathfrak{A}(A)$, $\mathfrak{G}_{(1)}(A) = \mathfrak{A}_c(A)$. If $\mathfrak{G}_\beta^\alpha(A)$ denotes the Banach space of all vectors in $\mathfrak{H}^\infty(A)$ for which inequality (2.14) holds, with norm

$$\|f\|_{\mathfrak{G}_\beta^\alpha} = \sup_n \frac{\|A^n f\|}{\alpha^n n^{n\beta}},$$

then

$$\mathfrak{G}_{\{\beta\}} = \operatorname*{limind}_{\alpha \to \infty} \mathfrak{G}_\beta^\alpha, \quad \mathfrak{G}_{(\beta)} = \operatorname*{limpr}_{\alpha \to 0} \mathfrak{G}_\beta^\alpha.$$

Clearly $\mathfrak{G}_{\{\beta_1\}} \subset \mathfrak{G}_{\{\beta_2\}}$, $\mathfrak{G}_{(\beta_1)} \subset \mathfrak{G}_{(\beta_2)}$ for $\beta_1 < \beta_2$.

It is not difficult to compute that the function $T(\lambda) = \exp(\lambda^{\frac{1}{\beta}})$ $(\beta > 0)$ corresponds to the sequence $m_n = n^{n\beta}$ in (2.6). By Theorem 2.2,

$$\mathfrak{G}_{\{\beta\}} = \mathfrak{H}_{\{\exp \lambda^{\frac{1}{\beta}}\}}, \quad \mathfrak{G}_{(\beta)} = \mathfrak{H}_{(\exp \lambda^{\frac{1}{\beta}})}.$$

Recall once more that $\mathfrak{G}_{\{0\}}(A) = \mathcal{E}xp_A \mathfrak{H}$ is the space of vectors of exponential type of A, $\mathfrak{G}_{(0)}(A) = \operatorname{Ker} A$.

Denoting by $\mathfrak{G}'_{\{\beta\}}$ and $\mathfrak{G}'_{(\beta)}$ the spaces of continuous antilinear functionals on $\mathfrak{G}_{\{\beta\}}$ and $\mathfrak{G}_{(\beta)}$, respectively, we obtain for $0 < \beta_1 < \beta_2$ the chain

$$\begin{aligned} &\mathcal{E}xp_A \mathfrak{H} \subset \mathfrak{G}_{(\beta_1)} \subset \mathfrak{G}_{\{\beta_1\}} \subset \mathfrak{G}_{(\beta_2)} \subset \mathfrak{G}_{\{\beta_2\}} \subset \mathfrak{H}^\infty \subset \mathfrak{H} \subset \\ &\subset \mathfrak{H}^{-\infty} \subset \mathfrak{G}'_{\{\beta_2\}} \subset \mathfrak{G}'_{(\beta_2)} \subset \mathfrak{G}'_{\{\beta_1\}} \subset \mathfrak{G}'_{(\beta_1)} \subset (\mathcal{E}xp_A \mathfrak{H})', \end{aligned}$$

where all imbeddings are dense and continuous.

EXAMPLE 6. Let $\mathfrak{H} = L_2(\mathbb{R}^1)$ and let A be multiplication by $|x|$. The space $\mathfrak{G}_{\{\beta\}}$ (respectively, $\mathfrak{G}_{(\beta)}$) consists of all $f(x) \in L_2(\mathbb{R}^1)$ satisfying the condition

$$\int_{-\infty}^{\infty} \exp(\alpha |x|^{\frac{1}{\beta}}) |f(x)|^2 \, dx < \infty$$

for a certain (any) $\alpha > 0$. The space $\mathfrak{G}'_{\{\beta\}}$ (respectively, $\mathfrak{G}'_{(\beta)}$) coincides in this case with the set of all functions $f(x)$ that are locally square integrable and have the property

$$\int_{-\infty}^{\infty} \exp(-\alpha |x|^{\frac{1}{\beta}}) |f(x)|^2 \, dx < \infty$$

for any (a certain) $\alpha > 0$. In particular, $\mathfrak{A}$ is the space of square integrable functions for which

$$\int_{-\infty}^{\infty} \exp(\alpha |x|) |f(x)|^2 \, dx < \infty$$

with some α, and $\mathfrak{A}'$ is identified with the set of all locally square integrable functions $f(x)$ for which

$$\int_{-\infty}^{\infty} \exp(-\alpha |x|) |f(x)|^2 \, dx < \infty$$

for all $\alpha > 0$.

EXAMPLE 7. Let $\mathfrak{H} = L_2(\mathbb{R}^1)$ and let $A = D$ be the modulus of the differentiation operator. Then

$$\mathfrak{G}_{\{\beta\}}(D) = \{f \in L_2(\mathbb{R}^1) \,|\exists \alpha > 0, \ \exists c > 0 : \int_{-\infty}^{\infty} |f^{(n)}(x)|^2 \, dx \le c\alpha^{2n} n^{2n\beta}; \quad n = 0, 1, \ldots\}$$

$$\mathfrak{G}_{(\beta)}(D) = \{f \in L_2(\mathbb{R}^1) \,|\forall \alpha > 0 \ \exists c = c(\alpha) > 0 : \int_{-\infty}^{\infty} |f^{(n)}(x)|^2 \, dx \le c\alpha^{2n} n^{2n\beta}; \quad n = 0, 1, \ldots\}$$

If $\beta > 1$, then $\mathfrak{G}_{\{\beta\}}(D)$ (respectively, $\mathfrak{G}_{(\beta)}(D)$) is the usual Gevrey class of order β of Roumieu (Beurling) type, while $\mathfrak{G}'_{\{\beta\}}(D)$ and $\mathfrak{G}'_{(\beta)}(D)$ coincide with the Gevrey classes of ultradistributions of the corresponding type. For details about the structure of the space $\mathfrak{A}(D)$ see §7.

2.7. COMPARISON OF SPACES CONSTRUCTED FROM DIFFERENT OPERATORS

Let A and B be positive self-adjoint operators on $\mathfrak{H}$. Recall that

$$A \ge B \tag{2.15}$$

means:

(i) $\mathfrak{D}(A) \subseteq \mathfrak{D}(B)$;
(ii) $(Af, f) \geq (Bf, f) \quad (\forall f \in \mathfrak{D}(A))$.

LEMMA 2.3. *Suppose that self-adjoint operators* $A, B > 0$ *satisfy* (2.15) *and that* $0 \in \rho(A) \cap \rho(B)$. *Then* $A^{-1} \leq B^{-1}$.

Proof. The relation $A \geq B$ implies that $\mathfrak{D}(A^{\frac{1}{2}}) \subseteq \mathfrak{D}(B^{\frac{1}{2}})$. Since the operator $B^{\frac{1}{2}} A^{-\frac{1}{2}}$ is closed and defined on the whole space $\mathfrak{H}$, it is bounded.

Set $C = A^{-\frac{1}{2}} B A^{-\frac{1}{2}} = (B^{\frac{1}{2}} A^{-\frac{1}{2}})^* (B^{\frac{1}{2}} A^{-\frac{1}{2}})$. The operator $C \geq 0$ is bounded. We have for $y \in \mathfrak{D}(A^{\frac{1}{2}})$,

$$(Cy, y) = (A^{-\frac{1}{2}} B A^{-\frac{1}{2}} y, y) = (BA^{-\frac{1}{2}} y, A^{-\frac{1}{2}} y) \leq (AA^{-\frac{1}{2}} y, A^{-\frac{1}{2}} y) = (y, y),$$

whence

$$(Cy, y) \leq (y, y)$$

for all $y \in \mathfrak{H}$. So, $C \leq E$. This implies that $C^{-1} \geq E$, that is

$$(C^{-1} y, y) \geq (y, y) \quad (y \in \mathfrak{D}(C^{-1})).$$

Substituting $A^{-\frac{1}{2}} x$ for y in this inequality, where $x = BA^{-\frac{1}{2}} z$, $z \in \mathfrak{D}(A^{\frac{1}{2}})$ (it is not difficult to see that such vectors y belong to $\mathfrak{D}(C^{-1})$), we obtain

$$\begin{aligned}(C^{-1} y, y) &= (A^{\frac{1}{2}} B^{-1} x, A^{-\frac{1}{2}} x) = (B^{-1} x, x) \geq (y, y) \\ &= (A^{-\frac{1}{2}} x, A^{-\frac{1}{2}} x) = (A^{-1} x, x).\end{aligned}$$

Consequently, the inequality

$$(B^{-1} x, x) \geq (A^{-1} x, x)$$

is valid for $x \in \mathfrak{R}(BA^{-1})$. Taking into account that $\mathfrak{R}(BA^{-1})$ is dense in $\mathfrak{H}$, which can be easily verified, we obtain the result desired. □

The following assertions are direct consequences of this lemma.

PROPOSITION 2.3. *Let* $A, B \geq 0$ *be self-adjoint operators such that* $A \geq B$. *Then the relations*

$$A(A + \omega E)^{-1} \geq B(B + \omega E)^{-1} \quad (\omega > 0), \tag{2.16}$$

$$A^{\tau} \geq B^{\tau} \quad (0 \leq \tau \leq 1) \tag{2.17}$$

hold.

Inequality (2.16) is obtained because $A(A+\omega E)^{-1} = E - \omega(A+\omega E)^{-1}$. Relation (2.17) is a consequence of (2.16) and formula (5.6) from Chapter 1.

For $\tau > 1$ the inequality (2.17) can not, in general, be obtained from (2.15). However, if A and B satisfy a certain condition stronger than (2.15), then inequalities close to (2.17) are valid.

THEOREM 2.6. *Assume that for positive self-adjoint operators A, B we have the inequality*

$$e^A \geq e^B. \tag{2.18}$$

Then

$$A^n + cn^n E \geq B^n \quad (\forall n \in \mathbb{N}), \tag{2.19}$$

where $0 < c =$ const does not depend on A, B and n.

Proof. Let f be an arbitrary element from $\mathfrak{D}(e^A)$. For $0 < t < 1$, $n \in \mathbb{N}$,

$$\frac{d^n}{dt^n}(e^{At} f) = A^n e^{At} f,$$
$$e^{At} f = \frac{\sin \pi t}{\pi} \int_0^\infty r^{t-1} e^A (e^A + rE)^{-1} f \, dr \tag{2.20}$$

(see formula (5.6), Chapter 1). Passing to the limit as $t \to 0$ and taking into account the equality $r^t \sin \pi t = \text{Im}\,(e^{t(\ln r + i\pi)})$, we find

$$A^n f = \lim_{t\to 0} \frac{d^n}{dt^n}\Big(\frac{\sin \pi t}{\pi} \int_0^1 r^{t-1} e^A (e^A + rE)^{-1} f \, dr\Big) +$$
$$+ \frac{1}{\pi} \int_1^\infty \text{Im}(\ln r + i\pi)^n e^A (e^A + rE)^{-1} r^{-1} f \, dr. \tag{2.21}$$

Denote by $I_1(n, A)f$ the first term at the right-hand side of (2.21); we decompose the second summand into a sum of two integrals, $I_2(n, A)f$ and $I_3(n, A)f$, with integrationbounds from 1 to e^n and from e^n to ∞. An analogous decomposition is done for $B^n f$.

On account of (2.16) and the relations

$$0 < \arg(\ln r + i\pi) = \text{arctg}\frac{\pi}{\ln r} \leq \text{arctg}\frac{\pi}{n} < \frac{\pi}{n},$$

where $r > e^n$, i.e. $\text{Im}(\ln r + i\pi)^n \geq 0$ if $r > e^n$, the inequality $I_3(n, A) \geq I_3(n, B)$ holds. This implies that

$$\begin{aligned}(B^n f, f) \leq (A^n f, f) - (I_1(n, A)f, f) + (I_1(n, B)f, f) - \\ - (I_2(n, A)f, f) + (I_2(n, B)f, f).\end{aligned} \tag{2.22}$$

We show that the operators $I_1(n, A)$, $I_2(n, A)$ are bounded and estimate their norms. Set

$$F_1(z, A) = \frac{\sin \pi z}{\pi} \int_0^1 r^{z-1} e^A (e^A + rE)^{-1} \, dr \quad (\text{Re } z > 0),$$
$$F_2(z, A) = e^{Az} - \frac{\sin \pi z}{\pi} \int_1^\infty r^{z-1} e^A (e^A + rE)^{-1} \, dr \quad (\text{Re } z < 0).$$

It is not difficult to see that $F_1(z,A)$ and $F_2(z,A)$ are analytic operator-functions in the right and left half-plane, respectively, whose values are bounded operators on $\mathfrak{H}$. Also, the vector-function $F_2(z,A)$ is analytically continued from the half-plane $\text{Re } z < 0$ to the half-plane $\text{Re } z < 1$. By virtue of (2.20),

$$F_1(z,A)f = F_2(z,A)f \quad (0 < \text{Re } z < 1).$$

Thus, $F_1(z,A)f$ and $F_2(z,A)f$ make up an entire vector-function $F(z,A)f$. Taking into account the estimates

$$\begin{gathered} \|e^A(e^A + rE)^{-1}\| \le 1 \quad (r > 0), \\ \left|\int_0^1 r^{z-1}\,\mathrm{d}r\right| \le 1 \quad (\text{Re } z = 1), \quad \left|\int_1^\infty r^{-(z+1)}\,\mathrm{d}r\right| \le 1 \quad (\text{Re } z = -1), \\ \|e^{-Az}\| \le 1 \quad (\text{Re } z > 0), \\ \sup_{|\text{Re } z| \le 1} |e^{z^2}\sin(\pi z)| \le \exp\Big(1 + \frac{\pi^2}{4}\Big), \end{gathered} \tag{2.23}$$

and using the maximum modulus principle, we obtain for $f \in \mathfrak{D}(e^A)$, $g \in \mathfrak{H}$,

$$\begin{aligned} \sup_{|\text{Re } z| \le 1} |e^{z^2}(F(z,A)f, g)| &= \sup_{|\text{Re } z| = 1} |e^{z^2}(F(z,A)f, g)| \\ &\le \sup_{|\text{Re } z| = 1} \|e^{z^2}F(z,A)f\|\|g\| \le \big(e + e^{1+\frac{\pi^2}{4}}\big)\|f\|\|g\| = c_1\|f\|\|g\|, \end{aligned}$$

where $c_1 = e + \exp(1 + \frac{\pi^2}{4})$. Since $\mathfrak{D}(e^A)$ is dense in $\mathfrak{H}$, it follows that $e^{z^2}F(z,A)$, and therefore $F(z,A)$, is an entire operator-function whose values are bounded operators on $\mathfrak{H}$ and

$$\sup_{|z|=1} \|F(z,A)\| \le \sup_{|z|=1} |e^{-z^2}| \sup_{|\text{Re } z| = 1} \|e^{z^2}F(z,A)\| \le ec_1.$$

Further, it is easily observed that

$$I_1(n,A) = \frac{\mathrm{d}^n}{\mathrm{d}z^n}F(z,A)\Big|_{z=0}.$$

Then the well-known Cauchy inequalities imply

$$\|I_1(n,A)\| \le c_1 e n!. \tag{2.24}$$

Using (2.24), we estimate $\|I_2(n,A)\|$:

$$\begin{aligned} \|I_2(n,A)\| &\le \frac{1}{\pi}\int_1^{e^n} (\ln r + \pi)^n r^{-1}\,\mathrm{d}r = \frac{1}{\pi}\int_0^n (\rho + \pi)^n\,\mathrm{d}\rho \\ &= \frac{(\rho + \pi)^{n+1}}{\pi(n+1)}\bigg|_0^n \le (n+\pi)^n \le e^\pi n^n. \end{aligned} \tag{2.25}$$

As is seen, the estimates of the norms $\|I_1(n,A)\|$ and $\|I_2(n,A)\|$ do not depend on A. Hence, the estimates (2.24) and (2.25) are true for $\|I_1(n,B)\|$ and $\|I_2(n,B)\|$ too. As a consequence of (2.22), for $f \in \mathfrak{D}(e^A)$,

$$(A^n f, f) + cn^n(f,f) \ge (B^n f, f) \quad (\forall n \in \mathbb{N}), \tag{2.26}$$

where $c = 2(e^{\pi} + e^{2+\frac{\pi^2}{4}} + e^2)$.

If now f is an arbitrary element from $\mathfrak{D}(A^p)$, then there exists a sequence $f_m \in \mathfrak{D}(e^A)$ convergent to f in the sense of the graph of A^p. Inequality (2.26) for $n = 2p$ shows that this sequence is fundamental in the sense of the graph of B^p. Since the operator B^p is closed, $f \in \mathfrak{D}(B^p)$. Passing to the limit in the inequality

$$(A^p f_m, f_m) + cp^p(f_m, f_m) \geq (B^p f_m, f_m)$$

as $m \to \infty$, we obtain (2.19). □

COROLLARY 2.1. *Let $G(\lambda)$ be an entire function, having power series expansion*

$$G(\lambda) = \sum_{n=0}^{\infty} a_n \lambda^n \tag{2.27}$$

with $a_n \geq 0$, and suppose

$$L = \sum_{n=0}^{\infty} a_n n^n < \infty. \tag{2.28}$$

Then for operators A and B satisfying (2.18) the membership $f \in \mathfrak{D}(G^{\frac{1}{2}}(A))$ implies that $f \in \mathfrak{D}(G^{\frac{1}{2}}(B))$, and

$$\|G^{\frac{1}{2}}(A)f\|^2 + cL\|f\|^2 \geq \|G^{\frac{1}{2}}(B)f\|^2. \tag{2.29}$$

Proof. The estimate (2.29) for $f \in \mathfrak{D}(e^A)$ is obtained by means of summing the inequalities (2.19) written in the form of inequalities for the corresponding quadratic forms and multiplied by the numbers a_n. We show that

$$\mathfrak{D}(G^{\frac{1}{2}}(A)) \subseteq \mathfrak{D}(G^{\frac{1}{2}}(B)).$$

In fact, if $f \in \mathfrak{D}(G^{\frac{1}{2}}(A))$, then there exists a sequence $f_m \in \mathfrak{D}(e^A)$ convergent to f in the sense of the graph of $G^{\frac{1}{2}}(A)$. It follows from inequality (2.29), with $f_m - f_n$ instead of f, that f_m is fundamental also in the sense of the graph of $G^{\frac{1}{2}}(B)$. Therefore $f \in \mathfrak{D}(G^{\frac{1}{2}}(B))$. After limit transition we find (2.29) also for $f \in \mathfrak{D}(G^{\frac{1}{2}}(B))$. □

DEFINITION. An operator A is said to analytically dominate an operator B if any analytic vector of A is an analytic vector of B.

COROLLARY 2.2. *If for positive self-adjoint operators A and B inequality (2.18) holds, then A analytically dominates B.*

Proof. Taking in the inequality (2.17) operators e^A and e^B in the role of A and B, respectively, we obtain for $0 \leq \alpha \leq 1$,

$$e^{A\alpha} \geq e^{B\alpha}.$$

Since $\mathfrak{A}(A) = \bigcup_{0\le\alpha\le1}\mathfrak{D}(e^{A\alpha}) \subseteq \bigcup_{0\le\alpha\le1}\mathfrak{D}(e^{B\alpha}) = \mathfrak{A}(B)$, the operator A analytically dominates B. □

COROLLARY 2.3. *If two operators A and B satisfy the inequality* (2.18), *then*

$$\mathfrak{G}_{\{\beta\}}(A) \subset \mathfrak{G}_{\{\beta\}}(B), \quad \mathfrak{G}_{(\beta)}(A) \subset \mathfrak{G}_{(\beta)}(B). \tag{2.30}$$

Proof. Take the entire function

$$G(\lambda) = \sum_{n=0}^{\infty} \frac{\lambda^n}{(n!)^\beta}.$$

It clearly satifies the conditions (2.27), (2.28). Since $\mathfrak{H}_{\{G\}} = \mathfrak{H}_{\{\exp\lambda^{\frac{1}{\beta}}\}} = \mathfrak{G}_{\{\beta\}}$, $\mathfrak{H}_{(G)} = \mathfrak{H}_{(\exp\lambda^{\frac{1}{\beta}})} = \mathfrak{G}_{(\beta)}$ (see subsection 2.6), the inclusions (2.30) are immediate consequences of Corollary 2.1. □

REMARK 2.2. Corollaries 2.2 and 2.3 remain valid if the inequality

$$\mu e^{A\tau_1} \ge e^{B\tau_2},$$

with certain positive constants μ, τ_1, τ_2, holds instead of assumption (2.18). In this case it follows from Theorem 2.6 that

$$(\tau_2 B)^n \le (\tau_1 A + \ln\mu E)^n + cn^n E.$$

Therefore,

$$\mathfrak{D}(G^{\frac{1}{2}}(B)) \supseteq \mathfrak{D}(G^{\frac{1}{2}}(A))$$

and also

$$\|G^{\frac{1}{2}}(\tau_1 A + \ln\mu E)\|^2 + cL\|f\|^2 \ge G^{\frac{1}{2}}(\tau_2 B)f\|^2 \quad (f \in \mathfrak{D}(G^{\frac{1}{2}}(A)).$$

3. The Exponential Function of a Non-Negative Self-Adjoint Operator

As has been noted before (see Chapter 1, subsection 5.4) the vector-function $e^{-At}f$ ($f \in \mathfrak{H}$, $t \ge 0$), where A is a non-negative self-adjoint operator on a Hilbert space $\mathfrak{H}$, is infinitely differentiable on $(0,\infty)$. Its boundary value at $t = 0$ belongs to $\mathfrak{H}$. However, solving equations often leads to the necessity of considering the vector-function $e^{-\hat{A}t}f$, where $\hat{A}$ is an extension of the operator A onto a space of generalized elements wider than $\mathfrak{H}$, and f is a generalized element. In many cases it happens that $e^{\hat{A}t}f$ is also infinitely differentiable on $(0,\infty)$ into $\mathfrak{H}$, while its limit as $t \to 0$ exists in the generalized sense. It is quite natural to raise the question of describing all generalized elements f which give infinitely differentiable vector-functions $e^{\hat{A}t}f$ on $(0,\infty)$ into $\mathfrak{H}$ and of determining the degree of generalization of f in dependence on the behaviour of $\|e^{\hat{A}t}f\|$ as t approaches the point 0. This problem is discussed below.

3.1. EXTENDING THE OPERATOR A TO THE SPACE $\mathfrak{H}'_G$

Let A be a non-negative self-adjoint operator on $\mathfrak{H}$ with resolution of the identity E_λ. Let $G(\lambda)$ be a continuous function on $[0,\infty)$ satisfying the conditions (2.1). From A and $G(\lambda)$ we construct the chain of spaces $\mathfrak{H}_G \subset \mathfrak{H} \subset \mathfrak{H}'_G$.

LEMMA 3.1. *The restriction* $A \restriction \mathcal{E}xp_A\mathfrak{H}$ *of the operator* A *to the set* $\mathcal{E}xp_A\mathfrak{H}$ *is closeable on the space* $\mathfrak{H}'_G$. *The closure* $\hat{A}_G$ *is a non-negative self-adjoint operator, and also* $A \subset \hat{A}_G$.

Proof. It follows from the equality $(Af,g) = (f,Ag)$ for $f,g \in \mathcal{E}xp_A\mathfrak{H}$ that the operator $A \restriction \mathcal{E}xp_A\mathfrak{H}$ is closeable on the space $\mathfrak{H}'_G$. The closure, denoted by $\hat{A}_G$, is non-negative because

$$(\hat{A}_G f, f)_{\mathfrak{H}'_G} = (Af,f)_{\mathfrak{H}'_G} = (G^{-1}(A)Af, G^{-1}(A)f) \\ = (AG^{-1}(A)f, G^{-1}(A)f) \geq 0$$

for any $f \in \mathcal{E}xp_A\mathfrak{H}$.

The self-adjointness of $\hat{A}_G$ arises from the fact that the set $(A - zE)\mathcal{E}xp_A\mathfrak{H}$ (Im $z \neq 0$) is dense in $\mathfrak{H}$, hence in $\mathfrak{H}'_G$. Indeed, if an element $g \in \mathfrak{H}$ would exist for which $((A-zE)f, g) = 0$ $(\forall f \in \mathcal{E}xp_A\mathfrak{H})$, then g would belong to $\mathfrak{D}(A)$ and $(A-zE)g = 0$. But this is not possible.

Since for any $f \in \mathfrak{D}(A)$ the sequence $f_n = E_{\Delta_n} f$ $(\Delta_n = (-n,n),\ n \in \mathbb{N})$ from $\mathcal{E}xp_A\mathfrak{H}$ converges to f in the sense of the graph of A in the space $\mathfrak{H}$, we have $A \subset \hat{A}_G$. □

It follows from the inclusion $A \subset \hat{A}_G$ that a similar relation is valid for the resolvents too, i.e. $R_z(A) \subset R_z)\hat{A}_G)$ (Im $z \neq 0$). Hence, for the resolutions of the identity E_λ and $\hat{E}_{G,\lambda}$ of these operators we have $E_\lambda \subset \hat{E}_{G,\lambda}$. Therefore $\varphi(A) \subset \varphi(\hat{A}_G)$ if $\varphi(\lambda)$ is a continuous function on $[0,\infty)$. Notice also that if another function $G_1(\lambda)$ of the same type as $G(\lambda)$ satisfies the inequality $G_1(\lambda) \geq \mu G(\lambda)$ $(0 < \mu = \text{const})$, then $\varphi(\hat{A}_{G_1}) \supseteq \varphi(\hat{A}_G)$.

For the sake of convenience we denote in the sequel by $\hat{A}$ the operator $\hat{A}_G$ corresponding to the function $G(\lambda) = e^\lambda$. Its resolution of the identity in the space $\mathfrak{H}'_e \stackrel{\text{def}}{=} \mathfrak{H}'_{\exp\lambda}$, the dual of $\mathfrak{H}_e \stackrel{\text{def}}{=} \mathfrak{H}_{\exp\lambda}$, is denoted by $\hat{E}_\lambda$. Then, for arbitrary $f,g \in \mathfrak{H}$,

$$(\hat{E}_\lambda f, g)_{\mathfrak{H}'_e} = (E_\lambda f, g)_{\mathfrak{H}'_e} = (e^{-A}E_\lambda f, e^{-A}g) = \int_0^\lambda e^{-2t}\, \mathrm{d}(E_t f, g).$$

This implies that if $f,\ g \in \mathfrak{H}$, then

$$(E_\lambda f, g) = \int_0^\lambda e^{2t}\, \mathrm{d}(\hat{E}_t f, g)_{\mathfrak{H}'_e}, \tag{3.1}$$

and, hence, $\mathfrak{H} \subseteq \mathfrak{D}(e^{\hat{A}})$. It turns out that

$$\mathfrak{H} = \mathfrak{D}(e^{\hat{A}}). \tag{3.2}$$

In fact, the operator $B = e^A$ is an isometry from $\mathfrak{H}_e$ onto $\mathfrak{H}$. Therefore, the operator $B^+ = \mathbf{I}^{-1}B^*$, the adjoint of B relative to $\mathfrak{H}$, acts isometrically from $\mathfrak{H}$ onto $\mathfrak{H}'_e$. Since the operator $(B^+)^{-1}$ defined on the whole of $\mathfrak{H}'_e$ is continuous from $\mathfrak{H}'_e$ into $\mathfrak{H}$ and $(B^+)^{-1}g = e^{-A}g = e^{-\hat{A}}g$ $(g \in \mathfrak{H})$, it follows that $(B^+)^{-1}g = e^{-\hat{A}}$, since $\mathfrak{H}_e$ is dense in $\mathfrak{H}'_e$. Taking into account that $\mathfrak{D}(B^+) = \mathfrak{H}$ we find (3.2).

LEMMA 3.2. *If a continuous function* $G(\lambda)$ *on* $[0, \infty)$ *satisfies condition* (2.1) *and also the inequality*

$$G(\lambda) \leq ce^{\lambda} \quad (0 < c = \text{ const}), \tag{3.3}$$

then the space $\mathfrak{H}'_G$ *consists of those elements in* $\mathfrak{H}'_e$ *for which*

$$\int_0^\infty e^{2\lambda}G^{-2}(\lambda)\,\mathrm{d}(\hat{E}_\lambda f, f)_{\mathfrak{H}'_e} < \infty.$$

Proof. Set $F(\lambda) = e^\lambda G^{-1}(\lambda)$. In view of equality (3.1) we have for any $f \in \mathfrak{H}$,

$$\int_0^\infty e^{2\lambda}G^{-2}(\lambda)\,\mathrm{d}(\hat{E}_\lambda f, f)_{\mathfrak{H}'_e} = \int_0^\infty G^{-2}(\lambda)\,\mathrm{d}(E_\lambda f, f) < \infty,$$

that is, $\mathfrak{H} \subset \mathfrak{D}(F(\hat{A}))$. From (3.2) we obtain the inclusion $\mathcal{E}xp_{\hat{A}}\mathfrak{H}'_G \subset \mathfrak{H}$. Therefore $\mathfrak{H}$ is dense in $\mathfrak{D}(F(\hat{A}))$ in the sense of the graph of $F(\hat{A})$. Taking into account that

$$\|f\|^2_{\mathfrak{H}'_G} = \int_0^\infty G^{-2}(\lambda)\,\mathrm{d}(E_\lambda f, f) = \int_0^\infty G^{-2}(\lambda)e^{2\lambda}\,\mathrm{d}(\hat{E}_\lambda f, f)_{\mathfrak{H}'_e} \tag{3.4}$$

for an element $f \in \mathfrak{H}$ and that $\mathfrak{H}$ is dense in $\mathfrak{H}'_G$, we conclude that

$$\mathfrak{H}'_G = \mathfrak{D}(F(\hat{A})) = F^{-1}(\hat{A})_{\mathfrak{H}'_e}. \tag{3.5}$$

□

The corollary follows from (3.5) and (3.2).

COROLLARY 3.1. *If a function* $G(\lambda)$ *is as in Lemma* 3.2, *then the equality*

$$\mathfrak{H}'_G = G(\hat{A})\mathfrak{H}$$

is valid.

In the special case when $G(\lambda) = e^{\alpha\lambda}$ $(0 < \alpha \leq 1)$ we have

$$\mathfrak{H}'_{e,\alpha} = e^{\hat{A}\alpha}\mathfrak{H}, \tag{3.6}$$

where the operator $e^{-\hat{A}\alpha}$ isometrically maps $\mathfrak{H}'_{e,\alpha}$ onto $\mathfrak{H}$.

3.2. THE VECTOR-FUNCTION $e^{-\hat{A}t}f$

We consider the vector-function $e^{-\hat{A}t}f$ $(t \geq 0)$. If $f \in \mathfrak{H}'_e$, then it is infinitely differentiable on $(0, \infty)$ and continuous on $[0, \infty)$ into the space $\mathfrak{H}'_e$. If f belongs to the more narrow space $\mathfrak{H}'_G \subset \mathfrak{H}'_e$ ($G(\lambda)$ satisfies (3.3)), then the considered vector-function is continuous on $[0,\infty)$ into $\mathfrak{H}'_G$. Indeed, for $f \in \mathfrak{H}'_G$ we have $e^{-\hat{A}t}f = e^{-\hat{A}_G t}f$, and the vector-function at the right-hand side of this identity is continuous on $[0,\infty)$ into $\mathfrak{H}'_G$.

Let us turn to the following problem: what space of generalized elements f should we choose in order that the vector-function $e^{-\hat{A}t}f$ be infinitely differentiable on $(0,\infty)$ into the space $\mathfrak{H}$? The following theorem is connected with solving this problem.

THEOREM 3.1. *In order that the vector-function $e^{-\hat{A}t}f$ $(f \in \mathfrak{H}'_e)$ be infinitely differentiable on $(0,\infty)$ into the space $\mathfrak{H}$, it is necessary and sufficient for f to belong to the space $\mathfrak{A}'(A) = \operatorname{limpr}_{\alpha\to 0} \mathfrak{H}'_{e,\alpha}(A)$. Under this condition $e^{-\hat{A}t}f$ is continuous on $[0,\infty)$ into $\mathfrak{A}'(A)$, and satisfies the equation*

$$\frac{dy(t)}{dt} + Ay(t) = 0 \tag{3.7}$$

if $t \in (0,\infty)$.

Proof. Let $f \in \mathfrak{A}'$, that is, $f \in \mathfrak{H}'_{e,\alpha}$ for any $\alpha > 0$. According to formula (3.6), $f = e^{\hat{A}\alpha}h_\alpha$ where $h_\alpha \in \mathfrak{H}$, α is arbitrarily small. Therefore, values of the vector-function $e^{-\hat{A}t}f = e^{-\hat{A}(t-\alpha)}e^{-\hat{A}\alpha}f = e^{-A(t-\alpha)}h_\alpha$ are elements of $\mathfrak{H}$ if $t \in (\alpha,\infty)$. It is infinitely differentiable on this interval and satisfies equation (3.7). Since α is arbitrary, the considered vector-function is infinitely differentiable on $(0,\infty)$ and satisfies (3.7) there. The continuity of $e^{-\hat{A}t}f$ on $[0,\infty)$ into the space $\mathfrak{A}'$ follows from its continuity into each of the spaces $\mathfrak{H}'_{e,\alpha}$.

Conversely, let the vector-function $e^{-\hat{A}t}f$ $(f \in \mathfrak{A}')$ take its values in the space $\mathfrak{H} : e^{-\hat{A}t}f = h_t \in \mathfrak{H}$. Then in view of (3.6), $f = e^{\hat{A}t}h_t \in \mathfrak{H}'_{e,t}$ $(t > 0)$. Since t is arbitrary, $f \in \mathfrak{A}'$. □

LEMMA 3.3. *Let $\{f_n\}_{n=1}^\infty$ be a sequence of elements from $\mathfrak{A}'(A)$. If $f_n \xrightarrow{\mathfrak{A}'} f$ $(n \to \infty)$, then $e^{-\hat{A}t}f_n \to e^{-\hat{A}t}f$ uniformly on any interval $[\alpha,\infty)$ $(\alpha > 0)$.*

Proof. Let $f_n \xrightarrow{\mathfrak{A}'} f$ $(n \to \infty)$. Then $f_n \to f$ in the space $\mathfrak{H}'_{e,\alpha}$ for any $\alpha > 0$. Since the operator $e^{-\hat{A}\alpha}$ $(0 < \alpha \le 1)$ isometrically maps $\mathfrak{H}'_{e,\alpha}$ onto $\mathfrak{H}$, the sequence $g_n = e^{-\hat{A}\alpha}f_n$ converges to $g = e^{-\hat{A}\alpha}f$ in $\mathfrak{H}$. Therefore, if $t \ge \alpha$ then

$$\begin{aligned}\|e^{-\hat{A}t}f_n - e^{-\hat{A}t}f\| &= \|e^{-\hat{A}t}(f_n - f)\| = \|e^{-\hat{A}(t-\alpha)}e^{-\hat{A}\alpha}(f_n - f)\| \\ &= \|e^{-A(t-\alpha)}(g_n - g)\| \le \|g_n - g\|.\end{aligned}$$

□

3.3. ON A BOUNDARY VALUE OF THE VECTOR-FUNCTION $e^{-\hat{A}t}f$

Theorem 3.1 shows that if $f \in \mathfrak{A}'$, then for $t \in (0,\infty)$ the vector-function $e^{-\hat{A}t}f$ is infinitely differentiable into the space $\mathfrak{H}$. But in general it has no limit in this space as $t \to 0$. The existence of the limit is guaranteed only in $\mathfrak{A}'$. It will be called the boundary value of $e^{-\hat{A}t}f$ at the point 0. Whether the boundary value at 0 of $e^{-\hat{A}t}f$ ($f \in \mathfrak{A}'$) belongs to a space more narrow than $\mathfrak{A}'$ depends on the behaviour of this vector-function near zero. The next theorems confirm this.

THEOREM 3.2. *The vector-function $e^{-\hat{A}t}f$ ($f \in \mathfrak{A}'$) satisfies the inequality*

$$\|e^{-\hat{A}t}f\| \leq c \quad (t \in (0, b]) \tag{3.8}$$

($c =$ const, $b > 0$ is finite and arbitrary) if and only if $f \in \mathfrak{H}$.

Proof. If $f \in \mathfrak{H}$, then the inequality (3.8) is obvious. Suppose that the vector-function $e^{-\hat{A}t}f$ ($f \in \mathfrak{A}'$) satisfies (3.8). Then in view of (3.1)

$$c^2 \geq \|e^{-\hat{A}t}f\|^2 = \int_0^\infty e^{2s}\, \mathrm{d}(\hat{E}_s e^{-\hat{A}t}f, e^{-\hat{A}t}f)_{\mathfrak{H}'_e}$$

$$= \int_0^\infty e^{2s(1-t)}\, \mathrm{d}(\hat{E}_s f, f)_{\mathfrak{H}'_e}.$$

By passing to the limit in this equality as $t \to 0$ we find

$$\int_0^\infty e^{2s}\, \mathrm{d}(\hat{E}_s f, f)_{\mathfrak{H}'_e} \leq c^2.$$

According to (3.2), $f \in \mathfrak{D}(e^{\hat{A}}) = \mathfrak{H}$. □

Thus, if we do not require boundedness of $\|e^{-\hat{A}t}f\|$ ($f \in \mathfrak{A}'$) in a neighbourhood of zero, then the element f will certainly not belong to $\mathfrak{H}$. By imposing various restrictions on the growth of the function $\|e^{-\hat{A}t}f\|$ as t approaches zero it is possible to find the space, intermediate between $\mathfrak{H}$ and $\mathfrak{A}'$, in which the corresponding boundary value of the vector-function $e^{-\hat{A}t}f$ lies. The growth will be determined using a function $\gamma(t)$ tending, as a rule, to zero as $t \to 0$.

Let $\gamma(t)$ be a continuous, positive and integrable function on $(0, b]$ ($b > 0$ is finite). Set

$$G(b, \lambda) = \left(\int_0^b \gamma(t) e^{-2\lambda t}\, \mathrm{d}t\right)^{-\frac{1}{2}}.$$

If $b_1 < b$, then there exist positive constants c_1 and c_2 such that

$$c_1 G(b_1, \lambda) \leq G(b, \lambda) \leq c_2 G(b_1, \lambda).$$

Therefore, the spaces $\mathfrak{H}_{G(b,\lambda)}$ and $\mathfrak{H}_{G(b_1,\lambda)}$ coincide. Thus, the behaviour of the function $\gamma(t)$ in a neighbourhood of zero only plays an essential role in the construction of the space $\mathfrak{H}_{G(b,\lambda)}$. From now on the symbol b in the notation $G(b, \lambda)$ will be omitted because one can always substitute instead of b any b_1 smaller than b. So, we assume

$$G(\lambda) = \left(\int_0^b \gamma(t) e^{-2\lambda t}\, \mathrm{d}t\right)^{-\frac{1}{2}}. \tag{3.9}$$

It is clear that $G(\lambda) \geq c > 0$ ($c =$ const), $G(\lambda) \to \infty$ as $\lambda \to \infty$ and $G(\lambda) \leq c_\alpha e^{\alpha\lambda}$ for any $\alpha > 0$ (c_α is a constant depending on α), whence it follows that $\mathfrak{H}'_G \subset \mathfrak{A}'$.

THEOREM 3.3. *The boundary value at the point* 0 *of the vector-function* $e^{-\hat{A}t}f$ ($f \in \mathfrak{A}'$) *belongs to the space* $\mathfrak{H}'_G$ *if and only if*

$$\int_0^b \gamma(t)\|e^{-\hat{A}t}f\|^2\,\mathrm{d}t < \infty, \tag{3.10}$$

where the functions $G(\lambda)$ *and* $\gamma(t)$ *are connected by* (3.9).

Proof. In view of (3.1)

$$\begin{aligned}\int_0^b \gamma(t)\|e^{-\hat{A}t}f\|^2\,\mathrm{d}t &= \int_0^b \gamma(t)\int_0^\infty e^{2\lambda}\,\mathrm{d}(\hat{E}_\lambda e^{-\hat{A}t}f, e^{-\hat{A}t}f)_{\mathfrak{H}'_e}\\ &= \int_0^b \gamma(t)\int_0^\infty e^{2\lambda(1-t)}\,\mathrm{d}(\hat{E}_\lambda f, f)_{\mathfrak{H}'_e}\\ &= \int_0^\infty e^{2\lambda}G^{-2}(\lambda)\,\mathrm{d}(\hat{E}_\lambda f, f)_{\mathfrak{H}'_e}\\ &= \|f\|^2_{\mathfrak{H}'_G}.\end{aligned} \tag{3.11}$$

So, according to Lemma 3.2 f belongs to $\mathfrak{H}'_G$ if and only if the integral (3.10) is convergent. □

We discuss Theorem 3.3 in some concrete and rather important situations. Set, for example, $\gamma(t) = t^{2\alpha-1}$ ($\alpha > 0$). Then

$$G(\lambda) = \left(\int_0^b t^{2\alpha-1}e^{-2\lambda t}\,\mathrm{d}t\right)^{-\frac{1}{2}} = (2\lambda)^\alpha\left(\int_0^{2b\lambda} t^{2\alpha-1}e^{-t}\,\mathrm{d}t\right)^{-\frac{1}{2}}.$$

Since

$$\lim_{\lambda\to\infty}\int_0^{2b\lambda} t^{2\alpha-1}e^{-t}\,\mathrm{d}t = \int_0^\infty t^{2\alpha-1}e^{-t}\,\mathrm{d}t = c > 0$$

and the function $G(\lambda)$ is continuous and positive on any bounded interval, there exist constants $c_1, c_2 > 0$ such that

$$c_1(1+\lambda^\alpha) \le G(\lambda) \le c_2(1+\lambda^\alpha).$$

This implies that in this situation $\mathfrak{H}_G$ coincides with the space $\mathfrak{H}^\alpha \overset{\mathrm{def}}{=} \mathfrak{H}_{\lambda^\alpha+1}$. Hence the following statement holds.

THEOREM 3.4. *The boundary value at the point* 0 *of the vector-function* $e^{-\hat{A}t}f$ ($f \in \mathfrak{A}'$) *belongs to the space* $\mathfrak{H}^{-\infty} = (\mathfrak{H}^\infty)'$ *if and only if for sufficiently small* $t > 0$,

$$\|e^{-\hat{A}t}f\| \le ct^{-\alpha} \quad (c = \mathrm{const}) \tag{3.12}$$

with a certain $\alpha > 0$. *More precisely,* $f \in \mathfrak{H}^{-\alpha}$ ($\alpha > 0$) *if and only if*

$$\int_0^b t^{2\alpha-1}\|e^{-\hat{A}t}f\|^2\,\mathrm{d}t < \infty. \tag{3.13}$$

Proof. The second assertion is a direct consequence of Theorem 3.3. Let now $e^{-\hat{A}t}f$ satisfy inequality (3.12). Then for any $\epsilon > 0$,

$$\int_0^b t^{2(\alpha+\epsilon)-1}\|e^{-\hat{A}t}f\|^2\,dt \le c^2\int_0^b t^{2\epsilon-1}\,dt < \infty,$$

i.e. $f \in \mathfrak{H}^{-(\alpha+\epsilon)}$, hence $f \in \mathfrak{H}^{-\infty} = \operatorname*{limind}_{n\to\infty} \mathfrak{H}^{-n}$. Conversely, if $f \in \mathfrak{H}^{-\infty}$, then $f \in \mathfrak{H}^{-n}$ for some n and

$$e^{-\hat{A}t}f = e^{-\hat{A}t}(E+\hat{A}^n)h = (E+A^n)e^{-At}h \quad (h \in \mathfrak{H},\ t > 0).$$

Therefore we have for $0 < t < 1$,

$$\begin{aligned} t^{2n}\|e^{-\hat{A}t}f\|^2 &= t^{2n}\int_0^\infty (1+\lambda^n)^2 e^{-2\lambda t}\,d(E_\lambda h, h) \\ &< \int_0^\infty (1+(\lambda t)^n)^2 e^{-2\lambda t}\,d(E_\lambda h, h) \le c_1\|h\|^2, \end{aligned}$$

where $c_1 = \sup_x((1+x^n)^2e^{-2x})$. Thus, (3.12) holds for the vector-function $e^{-\hat{A}t}f$. □

Suppose that in addition to the above-mentioned properties the function $\gamma(t)$ satisfies:

1°) $\gamma(t_1) > \gamma(t_2)$ if $t_1 > t_2$;
2°) there exist constants $c_0 > 0$, $\beta_0 < 1$ such that

$$t\gamma(t) > c_0\gamma(\beta_0 t).$$

The function $G(\lambda)$ corresponding to $\gamma(t)$ by formula (3.9) is also continuous; it increases monotonically on $[0,\infty)$. Moreover, $G(\lambda)$ satisfies property (2.9), that is, $G(\lambda) > c\lambda G(\alpha_0\lambda)$ with some $c > 0$, $0 < \alpha_0 < 1$.

In fact, for sufficiently small $\epsilon > 0$,

$$\begin{aligned} \lambda^{-1}G(\lambda) &\ge c_1\Big(\int_0^{\beta_0^2 b} \lambda^2\gamma(t)e^{-2\lambda t}\,dt\Big)^{-\frac12} = c_1\Big(\int_0^{\beta_0^2 b} \lambda^2 t^2 e^{-2\lambda t}\frac{\gamma(t)}{t^2}\,dt\Big)^{-\frac12} \\ &\ge c_2\Big(\int_0^{\beta_0^2 b} e^{-2\lambda t(1-\epsilon)}\gamma(t\beta_0^{-2})\,dt\Big)^{-\frac12} = c\Big(\int_0^b e^{-2\alpha_0\lambda s}\gamma(s)\,ds\Big)^{-\frac12} \\ &= cG(\alpha_0\lambda), \end{aligned}$$

where c_1, c_2, c are certain constants and $\alpha_0 = (1-\epsilon)\beta_0^2$.

THEOREM 3.5. *In order that the boundary value at zero of the vector-function $e^{-\hat{A}t}f$ $(f \in \mathfrak{A}')$ belongs to the space*

$$\mathfrak{H}'_{\{G\}} = \operatorname*{limpr}_{\alpha\to 0} \mathfrak{H}'_{G_\alpha} \quad (\mathfrak{H}'_{(G)} = \operatorname*{limind}_{\alpha\to\infty} \mathfrak{H}'_{G_\alpha}),$$

where $\mathfrak{H}_{G_\alpha}$ is the space corresponding to the function

$$G_\alpha(\lambda) = G(\alpha\lambda) = \Big(\frac{1}{\alpha}\int_0^{b\alpha} \gamma\Big(\frac{t}{\alpha}\Big)e^{-2\lambda t}\,dt\Big)^{-\frac{1}{2}},$$

it is necessary and sufficient that this vector-function satisfies the inequality

$$\|e^{-\hat{A}t}f\|^2 \le c\gamma^{-1}\Big(\frac{t}{\alpha}\Big) \tag{3.14}$$

with any $\alpha > 0$, $c = c(\alpha)$ (with a certain $\alpha > 0$, $c > 0$).

Proof. Let $f \in \mathfrak{H}'_{\{G\}}$. This means that $f \in \mathfrak{H}'_{G_\alpha}$ for any α. According to Theorem 3.3

$$\int_0^{\alpha b} \gamma\Big(\frac{s}{\alpha}\Big)\|e^{-\hat{A}s}f\|^2\,ds = c < \infty \tag{3.15}$$

(any $\alpha > 0$). Since $\gamma(t)$ monotonically increases while $\|e^{-\hat{A}t}f\|$ monotonically decreases on $(0, b]$ we have

$$\frac{t}{2}\gamma\Big(\frac{t}{2\alpha}\Big)\|e^{-\hat{A}t}f\|^2 \le \int_{\frac{t}{2}}^{t} \gamma\Big(\frac{s}{\alpha}\Big)\|e^{-\hat{A}s}f\|^2\,ds \le \int_0^{\alpha b} \gamma\Big(\frac{s}{\alpha}\Big)\|e^{-\hat{A}s}f\|^2\,ds = c,$$

whence, because of property 2° of the function $\gamma(t)$,

$$c_0\alpha\gamma\Big(\frac{\beta_0 t}{2\alpha}\Big)\|e^{-\hat{A}t}f\|^2 \le c,$$

which implies

$$\|e^{-\hat{A}t}f\|^2 \le c'\gamma^{-1}\Big(\frac{t}{\alpha'}\Big),$$

where $c' = c(c_0\alpha)^{-1}$, $\alpha' = 2\alpha\beta_0^{-1}$.

Thus, the estimate (3.14) is valid for any α.

Conversely, let the vector-function $e^{-\hat{A}t}f$ ($f \in \mathfrak{A}'$) satisfy the estimate (3.14). Multiplication of this inequality by $\gamma(\frac{t}{\alpha})$ and subsequent integration over the interval $(0, \alpha b)$ leads to relation (3.15) and, in view of Theorem 3.3, to the second assertion of the theorem. The case between parentheses is proved in the same way. □

Set now $\gamma(t) = \exp(-t^{-q})$ $(q > 0)$. Then

$$G(\alpha\lambda) = \Big(\int_0^b \exp(-t^{-q} - 2\alpha\lambda t)\,dt\Big)^{-\frac{1}{2}} \quad (\alpha > 0).$$

If λ is large, then the estimate

$$c\exp(\mu\lambda^{\frac{q}{q+1}}) \le G(\alpha\lambda) \le c_\epsilon\exp((\mu+\epsilon)\lambda^{\frac{q}{q+1}})$$

holds, where $\mu = \frac{q+1}{2}\big(\frac{2\alpha}{q}\big)^{\frac{q}{q+1}}$, $\epsilon > 0$ is arbitrary small, c and c_ϵ are certain positive constants.

In fact, the function $\exp(-t^{-q} - 2\alpha\lambda t)$ has its maximum at the point $t_0 = (2\alpha\lambda q^{-1})^{-\frac{1}{q+1}}$. It is not difficult to calculate this maximum; it is equal to $\exp\bigl(-\lambda^{\frac{q}{q+1}}\bigl(\frac{2\alpha}{q}\bigr)^{\frac{q}{q+1}}(1+q)\bigr) = \exp\bigl(-2\mu\lambda^{\frac{q}{q+1}}\bigr)$. So,

$$G(\alpha\lambda) \geq b^{-\frac{1}{2}} \exp\bigl(\mu\lambda^{\frac{q}{q+1}}\bigr) = c\exp\bigl(\mu\lambda^{\frac{q}{q+1}}\bigr).$$

Conversely, if λ is sufficiently large, then $t_0 \in (0, b)$ and for $0 < \delta < 1$,

$$\int_0^b \exp(-t^{-q} - 2\alpha\lambda t)\,dt \geq \int_{t_0(1-\delta)}^{t_0} \exp(-t^{-q} - 2\alpha\lambda t)\,dt$$
$$\geq t_0\delta\exp\bigl(-t_0^{-q}(1-\delta)^{-q} - 2\alpha\lambda t_0(1-\delta)\bigr).$$

Take $\epsilon_1 = \bigl(\frac{2\alpha}{q}\bigr)^{-\frac{q}{q+1}}\epsilon$, where ϵ is arbitrarily small. There exists a δ_0, $0 < \delta_0 < 1$, such that $(1-\delta_0)^{-q} < (1+\epsilon_1)(1-\delta_0)$. Then

$$\int_0^b \exp(-t^{-q} - 2\alpha\lambda t)\,dt \geq t_0\delta_0\exp\bigl(-(1-\delta_0)\bigl((1+\epsilon_1)t_0^{-q} + 2\alpha\lambda t_0\bigr)\bigr)$$
$$= \delta_0\Bigl(\frac{2\alpha\lambda}{q}\Bigr)^{-\frac{1}{q+1}}\exp\Bigl(-(1-\delta_0)\Bigl((1+\epsilon_1)\Bigl(\frac{2\alpha\lambda}{q}\Bigr)^{\frac{q}{q+1}} + 2\alpha\lambda\Bigl(\frac{2\alpha\lambda}{q}\Bigr)^{-\frac{1}{q+1}}\Bigr)\Bigr)$$
$$\geq \delta_0\Bigl(\frac{q}{2\alpha}\Bigr)^{\frac{1}{q+1}}\lambda^{-\frac{1}{q+1}}\exp\Bigl(-\lambda^{\frac{q}{q+1}}\Bigl(\frac{2\alpha}{q}\Bigr)^{\frac{q}{q+1}}(1+\epsilon_1+q)\Bigr)$$
$$= c_1\lambda^{-\frac{1}{q+1}}\exp\bigl(-\lambda^{\frac{q}{q+1}}(2\mu+\epsilon)\bigr) \geq c_\epsilon^{-2}\exp\bigl(-\lambda^{\frac{q}{q+1}}(2\mu+2\epsilon)\bigr),$$

whence

$$G(\alpha\lambda) \leq c_\epsilon\exp\bigl(\lambda^{\frac{q}{q+1}}(\mu+\epsilon)\bigr).$$

This two-sided estimate for the function $G(\alpha\lambda)$, and the identities $\mathfrak{G}_{\{\beta\}} = \mathfrak{H}_{\{\exp\lambda^{\frac{1}{\beta}}\}}$, $\mathfrak{G}_{(\beta)} = \mathfrak{H}_{(\exp\lambda^{\frac{1}{\beta}})}$ (see subsection 2.6) allow us to conclude that in the case considered,

$$\mathfrak{H}_{\{G\}} = \mathfrak{G}_{\{\frac{q+1}{q}\}}, \quad \mathfrak{H}_{(G)} = \mathfrak{G}_{(\frac{q+1}{q})}.$$

Thus, we arrive at the following statement.

THEOREM 3.6. *In order that the boundary value of the vector-function* $e^{-\hat{A}t}f$ $(f \in \mathfrak{A}')$ *belongs to the space* $\mathfrak{G}'_{\{\beta\}}$ *(respectively,* $\mathfrak{G}'_{(\beta)}$*), where* $\mathfrak{G}_{\{\beta\}}$ *(respectively,* $\mathfrak{G}_{(\beta)}$*) is the Gevrey class of order* $\beta > 1$ *of Roumieu (respectively, Beurling) type, it is necessary and sufficient that for any* $\alpha > 0$ *there exists a constant* $c = c(\alpha) > 0$ *(respectively, there exist constants* $\alpha, c > 0$*) such that*

$$\|e^{-\hat{A}t}f\| \leq c\ \exp(\alpha t^{-q}) \quad \Bigl(q = \frac{1}{\beta - 1}\Bigr).$$

As can be easily seen, in all cases above $e^{-\hat{A}t}f \to f$ $(t \to 0)$ in the space in which f lies.

4. Operator Differential Equations of the First Order

This section deals with the equation

$$\frac{\mathrm{d}y(t)}{\mathrm{d}t} + Ay(t) = 0, \quad t \in (0,b), \tag{4.1}$$

where A is a non-negative self-adjoint operator on a Hilbert space $\mathfrak{H}$. We establish the general form of its solutions inside the interval $(0,b)$, and we also study the boundary values of solutions at zero.

4.1. SOLUTIONS INSIDE AN INTERVAL

DEFINITION. A vector-function $y(t)$ taking values in $\mathfrak{D}(A)$ is called a solution inside the interval $(0,b)$ of equation (4.1) if it is continuously differentiable on $(0,b)$ and satisfies this equation there.

THEOREM 4.1. *If two solutions inside $(0,b)$, $y_1(t)$ and $y_2(t)$, of equation (4.1) coincide at a point $t_0 \in (0,b)$, then $y_1(t) \equiv y_2(t)$ on $[t_0,b)$.*

Proof. By virtue of the linearity of equation (4.1), it is sufficient to show that if a solution $y(t)$ inside $(0,b)$ vanishes at a point t_0, then $y(t) = 0$ for $t \geq t_0$. Let $y(t_0) = 0$. Since $(y'(s), y(s)) = -(Ay(s), y(s)) \leq 0$, we have $(y'(s), y(s)) = \operatorname{Re}(y'(s), y(s)) = \frac{1}{2}\frac{\mathrm{d}}{\mathrm{d}s}(y(s), y(s))$ $(s \in (0,b))$. Hence, for any $t \in (t_0, b)$,

$$\|y(t)\|^2 - \|y(t_0)\|^2 = 2\int_{t_0}^{t} (y'(s), y(s))\,\mathrm{d}s = -2\int_{t_0}^{t} (Ay(s), y(s))\,\mathrm{d}s \tag{4.2}$$

which implies $\|y(t)\|^2 \leq 0$. This implies that $y(t) = 0$ for $t \geq t_0$. □

COROLLARY 4.1. *If $y(t)$ is a continuous vector-function on $[0,b)$ that is a solution inside the interval $(0,b)$ of equation (4.1), and if $y(0) = 0$, then $y(t) \equiv 0$.*

Proof. From (4.2) it follows that $\|y(t)\|^2 - \|y(t_0)\|^2 \leq 0$ for any t_0, $0 < t_0 < t < b$. By letting t_0 tend to zero we obtain $y(t) \equiv 0$. □

THEOREM 4.2. *A continuous vector-function $y(t)$ on $[0,b)$ is a solution inside the interval $(0,b)$ of equation (4.1) if and only if it can be represented as*

$$y(t) = \mathrm{e}^{-At} f \quad (f \in \mathfrak{H}). \tag{4.3}$$

Proof. As noted above (Chapter 1, subsection 5.4), a vector-function of the form of (4.3) is continuous on $[0,b)$ and is also a solution inside $(0,b)$ of (4.1). Conversely, if $y(t)$ is a solution inside $(0,b)$ of (4.1) that is continuous at the point 0, then the vector-function $z(t) = y(t) - \mathrm{e}^{-At} f$ $(f = y(0))$ satisfies the conditions of Corollary 4.1. Therefore $z(t) \equiv 0$, i.e. $y(t) = \mathrm{e}^{-At} f$ $(f = y(0))$. □

It turns out that all solutions inside $(0,b)$ of (4.1) are also described by formula (4.3), where f runs through the space, larger than $\mathfrak{H}$, $\mathfrak{A}'(A)$ of co-analytic vectors of the operator A, the dual of the space $\mathfrak{A}(A)$ of analytic vectors of A, while the operator A is replaced by its extension $\hat{A}$ as discussed in subsection 3.1. The main result of the present section is the following.

THEOREM 4.3. *A vector-function $y(t)$ is a solution inside the interval $(0, b)$ of equation (4.1) if and only if*

$$y(t) = e^{-\hat{A}t} f \quad (f \in \mathfrak{A}'(A)). \tag{4.4}$$

Proof. The fact that a vector-function (4.4) is a solution inside $(0, b)$ of (4.1) follows from Theorem 3.1. Conversely, let $y(t)$ be a solution inside $(0, b)$ of (4.1). Then it is continuous on $[\alpha, b)$ $(0 < \alpha < b)$. Because of Theorem 4.2 it has on $[\alpha, b)$ a representation

$$y(t) = e^{-A(t-\alpha)} g \quad (g = y(\alpha) \in \mathfrak{H}). \tag{4.5}$$

Take any β, $0 < \beta < \alpha$. The equality

$$y(t) = e^{-A(t-\beta)} g_\beta \quad (g_\beta \in \mathfrak{H}) \tag{4.6}$$

holds for $t \in [\beta, b)$. Since the representations (4.5), (4.6) coincide at $t = \alpha$, we have

$$g = e^{-A(\alpha-\beta)} g_\beta = e^{-A\delta} g_\beta \quad (0 < \delta < \alpha),$$

that is $g \in \bigcap_{0<\delta<\alpha} \mathfrak{H}_{e,\delta}$. Further,

$$e^{\hat{A}\alpha} g = e^{\hat{A}\alpha} e^{-A\delta} g_\beta = e^{\hat{A}(\alpha-\delta)} g_\beta \in \mathfrak{H}'_{e,\alpha-\delta}$$

for any δ, $0 < \delta < \alpha$. Hence, $f = e^{\hat{A}\alpha} g \in \mathfrak{A}'(A)$. Returning to the representation (4.5) we obtain

$$y(t) = e^{-A(t-\alpha)} g = e^{-\hat{A}t} f \quad (f \in \mathfrak{A}',\ t \in [\alpha, b)).$$

Since the vector-function $e^{-\hat{A}t} f$ $(f \in \mathfrak{A}')$ is a solution inside $(0, b)$ of (4.1) and coincides with $y(t)$ on $[\alpha, \beta)$, since α is arbitrary we have $y(t) = e^{-\hat{A}t} f$ on the whole interval $(0, b)$. □

This theorem, together with Theorem 3.1, shows that any solution inside $(0, b)$ of (4.1) is infinitely differentiable on $(0, b)$. Its boundary value at the point 0 belongs to $\mathfrak{A}'(A)$ and the solution can be uniquely recovered from it. Thus, the Cauchy problem for equation (4.1), that is, finding for a given $f \in \mathfrak{A}'(A)$ a solution $y(t)$ inside $(0, b)$ such that $y(t) \xrightarrow{\mathfrak{A}'} f$ $(t \to 0)$, is always solvable and its solution is unique. Therefore one can consider $\mathfrak{A}'(A)$ as the space of initial data of the Cauchy problem just formulated.

4.2. A CHARACTERIZATION OF THE BOUNDARY VALUES OF SOLUTIONS INSIDE AN INTERVAL

Having additional information about the behaviour near zero of a solution inside $(0, b)$ of equation (4.1), it is possible to establish the existence of its boundary value in a certain space of generalized elements which is a proper subspace of $\mathfrak{A}'$.

Let $\gamma(t)$ be a continuous, positive and integrable function on $(0,b)$ $(0<b<\infty)$. Denote by K_γ the set of all solutions $y(t)$ inside $(0,b)$ of equation (4.1) which belong to the space $L_2(\mathfrak{H},(0,b),\gamma(t)\,\mathrm{d}t)$, i.e. those $y(t)$ that satisfy the estimate

$$\|y\|^2_{L_2(\mathfrak{H},(0,b),\gamma(t)\,\mathrm{d}t)}=\int_0^b \gamma(t)\|y\|^2\,\mathrm{d}t<\infty. \tag{4.7}$$

Since $b<\infty$ and $\gamma(t)>0$ on any interval $[\delta,b]$, this estimate characterizes the behaviour of the solution $y(t)$ in a neighbourhood of zero.

LEMMA 4.1. *The set K_γ is a subspace of $L_2(\mathfrak{H},(0,b),\gamma(t)\,\mathrm{d}t)$.*

Proof. Let a sequence $y_n(t)\in K_\gamma$ converge in the norm $\|\cdot\|_{L_2(\mathfrak{H},(0,b),\gamma(t)\,\mathrm{d}t)}$ to a vector-function $y(t)$. According to Theorem 4.2, for an arbitrary δ, $0<\delta<b$, $y_n(t)=e^{-A(t-\delta)}f_n$ $(f_n=y_n(\delta)\in\mathfrak{H})$ on the interval $[\delta,b]$. Therefore,

$$\begin{aligned}\|y_n-y_m\|^2_{L_2(\mathfrak{H},(0,b),\gamma(t)\,\mathrm{d}t)}&\geq\int_\delta^b\gamma(t)\|y_n(t)-y_m(t)\|^2\,\mathrm{d}t\\&\geq c\int_\delta^b\|e^{-(t-\delta)A}(f_n-f_m)\|^2\,\mathrm{d}t\\&=c\int_\delta^b\mathrm{d}t\int_0^\infty e^{-2\lambda(t-\delta)}\,\mathrm{d}(E_\lambda(f_n-f_m),f_n-f_m)\\&=c\int_0^\infty\frac{1-e^{-2\lambda(b-\delta)}}{2\lambda}\,\mathrm{d}(E_\lambda(f_n-f_m),f_n-f_m)\\&\geq c_1\int_0^\infty(1+\lambda)^{-1}\,\mathrm{d}(E_\lambda(f_n-f_m),f_n-f_m)\\&=c_2\|f_n-f_m\|^2_{\mathfrak{H}^{-\frac12}}\qquad(0<c,c_1,c_2=\text{const}),\end{aligned}$$

that is, the sequence $\{f_n\}_{n=1}^\infty$ is fundamental in $\mathfrak{H}^{-\frac12}$. Hence it converges in this space to a certain element f. In view of Lemma 3.3 $e^{-\hat{A}(t-\delta)}f_n\to e^{-\hat{A}(t-\delta)}f$ in $\mathfrak{H}$ uniformly in $t\in[2\delta,b]$, hence in $L_2(\mathfrak{H},(2\delta,b),\gamma(t)\,\mathrm{d}t)$. Consequently, $y(t)=e^{-\hat{A}(t-\delta)}f$ $(t\in[2\delta,b]$, $f\in\mathfrak{H}^{-\frac12}\subset\mathfrak{A}')$. This implies that $y(t)$ is a solution inside $(2\delta,b)$. Since δ is arbitrary, $y(t)$ is a solution inside $(0,b)$ of (4.1). □

As in the previous section we construct using $\gamma(t)$ the function

$$G(\lambda)=\left(\int_0^b\gamma(t)e^{-2\lambda t}\,\mathrm{d}t\right)^{-\frac12}. \tag{4.8}$$

THEOREM 4.4. *The boundary value at 0 of a solution $y(t)$ inside the interval $(0,b)$ of equation (4.1) belongs to the space $\mathfrak{H}'_G(A)$ corresponding to the function $G(\lambda)$ determined by (4.8) if and only if $y(t)$ belongs to the space K_γ. Moreover, formula (4.4) defines an isometric isomorphism between K_γ and $\mathfrak{H}'_G$. Under the assumptions 1°, 2° from subsection 3.3 on the function $\gamma(t)$, the solution*

$y(t)$ has a boundary value in $\mathfrak{H}'_{\{G\}}(A) = \operatorname*{limpr}_{\alpha\to 0} \mathfrak{H}'_{G_\alpha}(A)$ *(respectively,* $\mathfrak{H}'_{(G)}(A) = \operatorname*{limind}_{\alpha\to\infty} \mathfrak{H}'_{G_\alpha}(A)$*) if and only if for any* $\alpha > 0$ *there exists a constant* $c = c(\alpha) > 0$ *(there exist* $\alpha, c > 0$*) such that*

$$\|y(t)\|^2 \le c\gamma^{-1}\left(\frac{t}{\alpha}\right).$$

The proof follows from Theorems 4.3, 3.3 and 3.5 taking into account (3.11). □

The following result is an immediate consequence of Theorems 4.3, 3.2 and 3.4.

COROLLARY 4.2. *The boundary value $y(0)$ at the point 0 of a solution $y(t)$ inside $(0,b)$ of equation (4.1) belongs to the space $\mathfrak{H}$ if and only if*

$$\|y(t)\| \le c \quad (c = \text{const}).$$

It exists in the space $\mathfrak{H}^{-\infty}(A)$ if and only if for sufficiently small t,

$$\|y(t)\| \le ct^{-\alpha} \quad (c = \text{const})$$

for any $\alpha > 0$. More precisely, $y(0)$ exists in the space $\mathfrak{H}^{-\alpha}(A)$ $(\alpha > 0)$ if and only if

$$\int_0^b t^{2\alpha-1}\|y(t)\|^2\,\mathrm{d}t < \infty.$$

Theorem 3.6 leads to the following conclusion.

COROLLARY 4.3. *For the boundary value $y(0)$ of a solution $y(t)$ inside $(0,b)$ of equation (4.1) to belong to the space $\mathfrak{G}'_{\{\beta\}}(A)$ (respectively, $\mathfrak{G}'_{(\beta)}(A)$), where $\mathfrak{G}_{\{\beta\}}$ (respectively, $\mathfrak{G}_{(\beta)}$) is the Gevrey class of order $\beta > 1$ of Roumieu (respectively, Beurling) type it is necessary and sufficient that for any $\alpha > 0$ there exists a constant $c = c(\alpha) > 0$ (respectively, there exist $\alpha, c > 0$) such that*

$$\|y(t)\| \le c\,\exp(\alpha t^{-q}), \quad q = \frac{1}{\beta - 1}.$$

5. Operator Differential Equations of the Second Order

In this section we describe for the differential equation

$$\frac{\mathrm{d}^2y(t)}{\mathrm{d}t^2} = A^2y(t), \quad t \in (a,b), \tag{5.1}$$

where A is, as before, a non-negative self-adjoint operator on a Hilbert space $\mathfrak{H}$, all smooth solutions inside (a,b) and we establish the existence of their boundary values at the ends of the interval in various spaces of generalized elements. The character of the space depends on the rate of growth of a solution as t approaches the boundary. At the same time we consider generalized solutions of the given equation and prove their smoothness inside the interval (a,b).

5.1. SOLUTIONS INSIDE AN INTERVAL. UNIQUENESS

DEFINITION. We will call a vector-function $y(t)$ a solution inside the interval (a,b) $(-\infty \le a < b \le \infty)$ of equation (5.1) if this vector-function is twice continuously differentiable on (a,b) into $\mathfrak{H}$, if at every $t \in (a,b)$ we have $y(t) \in \mathfrak{D}(A^2)$, and if it satisfies this equation.

THEOREM 5.1. *If two solutions $y_1(t)$, $y_2(t)$ inside the interval (a,b) of equation (5.1) coincide at points t_1, t_2 from (a,b) $(t_1 < t_2)$, then $y_1(t) \equiv y_2(t)$ on $[t_1,t_2]$.*

Proof. By virtue of the linearity of equation (5.1) it suffices to prove that if a solution $y(t)$ inside (a,b) is such that $y(t_1) = y(t_2) = 0$ $(t_1 < t_2;\ t_1,t_2 \in (a,b))$, then $y(t) \equiv 0$ on $[t_1,t_2]$.

Since $y(t)$ satisfies (5.1),

$$\int_{t_1}^{t_2} (y''(t), y(t))\,dt = \int_{t_1}^{t_2} \|Ay(t)\|^2\,dt.$$

Integrating by parts we obtain

$$(y'(t_2), y(t_2)) - (y'(t_1), y(t_1)) - \int_{t_1}^{t_2} \|y'(t)\|^2\,dt = \int_{t_1}^{t_2} \|Ay(t)\|^2\,dt, \tag{5.2}$$

whence

$$\int_{t_1}^{t_2} \|y'(t)\|^2\,dt + \int_{t_1}^{t_2} \|Ay(t)\|^2\,dt = 0.$$

This implies that $y(t) = \text{const}$ on $[t_1,t_2]$, so $y(t) \equiv 0$ on this interval. □

COROLLARY 5.1. *If $y(t)$ is a continuous vector-function on $[a,b]$ that is a solution inside (a,b) of equation (5.1) with $\|y'(t)\| \le \text{const}$ and $y(a) = y(b) = 0$, then $y(t) \equiv 0$.*

Proof. Limit transition in (5.2) as $t_1 \to a$, $t_2 \to b$ yields the proof. □

5.2. REPRESENTATION OF A SOLUTION INSIDE AN INTERVAL IN THE CASE OF A POSITIVE DEFINITE A

In the present subsection the operator A is supposed to be self-adjoint and positive definite: $(Af,f) \ge \mu(f,f)$ $(\mu > 0,\ f \in \mathfrak{D}(A))$.

LEMMA 5.1. *A continuously differentiable vector-function $y(t)$ on $[a,b]$ $(-\infty < a < b < \infty)$ into $\mathfrak{H}$ such that $y(t) \in \mathfrak{D}(A)$ at every $t \in [a,b]$ is a solution inside the interval (a,b) of equation (5.1) if and only if this vector-function can be represented in the form*

$$y(t) = e^{-A(t-a)} f_1 + e^{-A(b-t)} f_2 \quad (f_1, f_2 \in \mathfrak{D}(A)). \tag{5.3}$$

Proof. Direct substitution shows that a vector-function $y(t)$ of the same type as (5.3) is continuously differentiable on $[a,b]$ into $\mathfrak{H}$ and is a solution inside (a,b) of equation (5.1) taking values in $\mathfrak{H}^1(A) = \mathfrak{D}(A)$.

Conversely, let $z(t)$ be a solution inside (a, b) of equation (5.1) satisfying the conditions of the lemma. Select vectors f_1, f_2 so that the vector-function (5.3) coincides with $z(t)$ at the points a and b. To this end we have to solve the system of equations

$$\left.\begin{aligned} z(a) &= f_1 + e^{-A(b-a)} f_2, \\ z(b) &= e^{-A(b-a)} f_1 + f_2. \end{aligned}\right\} \tag{5.4}$$

Since the operator $E - e^{-2A(b-a)}$ is continuously invertible, the system (5.4) is solvable. The solution $\{f_1, f_2\}$ has the form

$$\begin{aligned} f_1 &= \left(E - e^{-2A(b-a)}\right)^{-1} \left(z(a) - e^{-A(b-a)} z(b)\right), \\ f_2 &= \left(E - e^{-2A(b-a)}\right)^{-1} \left(z(b) - e^{-A(b-a)} z(a)\right), \end{aligned} \tag{5.5}$$

and as we see, the vectors f_1, f_2 belong to the space $\mathfrak{H}^1(A)$. Thus, $z(a) = y(a)$, $z(b) = y(b)$, where $y(t)$ is a vector-function of the type (5.3) with f_1, f_2 being expressed by the formulas (5.5). Since $y(t)$ is a solution of equation (5.1) which also satisfies the conditions of the lemma, we conclude that, because of Corollary 5.1, $y(t) \equiv z(t)$ $(t \in [a, b])$. Hence $z(t)$ has a representation (5.3). □

THEOREM 5.2. *Let the operator A in equation (5.1) be positive definite. A vector-function $y(t)$ is a solution inside the interval (a, b) of this equation if and only if $y(t)$ has a representation*

$$y(t) = e^{-\hat{A}(t-a)} f_1 + e^{-\hat{A}(b-t)} f_2 \quad (f_1, f_2 \in \mathfrak{A}'(A)). \tag{5.6}$$

Proof. According to Theorem 3.1 the vector-function $e^{-\hat{A}(t-a)} f_1$ $(e^{-\hat{A}(b-t)} f_2)$ $(f_1, f_2 \in \mathfrak{A}'(A))$ is infinitely differentiable on $(a, b]$ (on$[a, b)$) into the space $\mathfrak{H}$ and continuous on $[a, b]$ into $\mathfrak{A}'(A)$. It is not difficult to verify that each of these vector-functions is a solution inside (a, b) of (5.1).

Conversely, let $y(t)$ be a solution inside the interval (a, b) of equation (5.1). Then it satisfies the conditions of Lemma 5.1 on the closed interval $[a + \delta_0, b - \delta_0]$, where $0 < \delta_0 < \frac{b-a}{4}$. Hence, $y(t)$ can be represented on this interval as

$$y(t) = e^{-A(t-(a+\delta_0))} g_1 + e^{-A((b-\delta_0)-t)} g_2 \quad (g_1, g_2 \in \mathfrak{H}^1(A)). \tag{5.7}$$

A similar representation for $y(t)$ can also be given on the interval $[a+\delta, b-\delta]$ where $0 < \delta < \delta_0$, namely:

$$y(t) = e^{-A(t-(a+\delta))} g_{1,\delta} + e^{-A((b-\delta)-t)} g_{2,\delta} \quad (g_{i,\delta} \in \mathfrak{H}^1(A);\ i = 1, 2).$$

By equating both representations at the points $a + \delta_0$ and $b - \delta_0$ we obtain

$$g_1 = e^{-A(\delta_0 - \delta)} g_{1,\delta}, \quad g_2 = e^{-A(\delta_0 - \delta)} g_{2,\delta}.$$

These equalities show that $g_i \in \bigcap_{0<\delta<\delta_0} \mathfrak{H}_{e,\delta}(A)$ $(i = 1, 2)$, hence the vectors

$$f_1 = e^{\hat{A}\delta_0} g_1, \quad f_2 = e^{\hat{A}\delta_0} g_2$$

belong to the space $\mathfrak{A}'(A)$. The representation (5.7) yields (5.6) on $[a+\delta_0, b-\delta_0]$. Since the right-hand side of (5.6) is a solution inside (a,b) of equation (5.1) and coincides with $y(t)$ on $[a+\delta_0, b-\delta_0]$ and since δ_0 is arbitrary we obtain the representation (5.6) for $y(t)$ on the interval (a,b). □

5.3. THE DIRICHLET PROBLEM

Theorems 5.2 and 3.1 show that any solution $y(t)$ inside (a,b) of equation (5.1) with a positive definite operator A is infinitely differentiable on (a,b) into the space $\mathfrak{H}$; its boundary values $y(a) = \lim_{t\to a} y(t)$ and $y(b) = \lim_{t\to b} y(t)$ exist in the space $\mathfrak{A}'(A)$. The solution can be recovered from its boundary values by the formula

$$y(t) = 2\left(E - e^{-2\hat{A}(b-a)}\right)^{-1}\left(\mathrm{sh}(\hat{A}(b-t))e^{-\hat{A}(b-a)}y(a) + \right.$$
$$\left. + \mathrm{sh}(\hat{A}(t-a))e^{-\hat{A}(b-a)}y(b)\right), \tag{5.8}$$

which is obtained from (5.6) using (5.5), where the operator $\hat{A}$ plays the role of A.

Since any solution inside (a,b) can be written in the form (5.6) and since f_1, f_2, as follows from (5.5), are uniquely expressed in terms of $y(a)$ and $y(b)$, a solution is uniquely recovered from its boundary values.

Thus, for arbitrary elements $g_1, g_2 \in \mathfrak{A}'(A)$ there exists a unique solution $y(t)$ inside (a,b) of equation (5.1) which coincides at the ends of the interval with g_1 and g_2. The latter should be understood as $y(t) \xrightarrow{\mathfrak{A}'} g_1$ when $t \to a$, $y(t) \xrightarrow{\mathfrak{A}'} g_2$ when $t \to b$. The solution is determined by the formula (5.8), where $y(a) = g_1$, $y(b) = g_2$.

However, it is possible to give a meaning to formula (5.8) also when A is a non-negative operator. Indeed, the functions

$$\omega_1(t,\lambda) = \frac{2\mathrm{sh}(\lambda(b-t))e^{-\lambda(b-a)}}{1-e^{-2\lambda(b-a)}}, \quad \omega_2(t,\lambda) = \frac{2\mathrm{sh}(\lambda(t-a))e^{-\lambda(b-a)}}{1-e^{-2\lambda(b-a)}}$$

have the following properties:

(1) $\omega_i(t,\lambda)$ $(i = 1,2)$ are meromorphic in λ for every $t \in [a,b]$ and uniformly bounded on $[a,b] \times [0,\infty)$;

(2) $\omega_i(t,\lambda)$ $(i = 1,2)$ are infinitely differentiable in t on $[a,b]$ and the $\frac{\partial^k \omega_i(t,\lambda)}{\partial t^k}$ $(k = 0,1,\ldots)$ are bounded on $[0,\infty)$ in λ for any fixed $t \in (a,b)$;

(3) $\frac{\partial^2 \omega_i(t,\lambda)}{\partial t^2} = \lambda^2 \omega_i(t,\lambda)$ $(i = 1,2)$;

(4) $\omega_1(a,\lambda) \equiv 1$, $\omega_1(b,\lambda) \equiv 0$, $\omega_2(a,\lambda) \equiv 0$, $\omega_2(b,\lambda) \equiv 1$;

(5) $\sup_{t,\lambda} \left|\lambda^{-k} e^{\lambda(t-a)} \frac{\partial^k \omega_1(t,\lambda)}{\partial t^k}\right| < \infty$,
$\sup_{t,\lambda} \left|\lambda^{-k} e^{\lambda(b-t)} \frac{\partial^k \omega_2(t,\lambda)}{\partial t^k}\right| < \infty$ $(k = 0,1,2,\ldots)$.

Therefore, if $\hat{A}$ is a non-negative operator, because of property (5) of $\omega_i(t,\lambda)$, the operators $\omega_1(t,\hat{A})$ and $\omega_2(t,\hat{A})$ (any $t \in (a,b)$) map the space $\mathfrak{A}'(A)$ into $\mathfrak{H}_{e,t-a-\delta}$

and $\mathfrak{H}_{e,b-t-\delta}$, respectively, for any sufficiently small $\delta > 0$. Consequently, the vector-functions $\omega_1(t, \hat{A})f$ and $\omega_2(t, \hat{A})f$ $(f \in \mathfrak{A}'(A))$ take values in $\mathfrak{H}$ if $t \in (a, b]$ and, respectively, if $t \in [a, b)$. By virtue of the properties (3) and (5), $\omega_i(t, \hat{A})f$ $(f \in \mathfrak{A}'(A);\ i = 1, 2)$ is infinitely differentiable on (a, b) into the space $\mathfrak{H}$, and satisfies equation (5.1). By property (4), $\omega_1(t, \hat{A})f \xrightarrow{\mathfrak{A}'} f$ as $t \to a$, $\omega_2(t, \hat{A})f \xrightarrow{\mathfrak{A}'} f$ as $t \to b$.

Thus, a vector-function of the form

$$y(t) = \omega_1(t, \hat{A})g_1 + \omega_2(t, \hat{A})g_2 \quad (g_1, g_2 \in \mathfrak{A}'(A)) \tag{5.9}$$

is a solution inside (a, b) of equation (5.1) taking the values g_1 and g_2 at the ends a and b of the interval, that is, $y(t)$ is a solution of the Dirichlet problem. Moreover, if $g_1, g_2 \in \mathfrak{D}(A^2)$, then a function $y(t)$ of the type (5.9) is a solution on $[a, b]$ that is twice continuously differentiable into $\mathfrak{H}$. We show that any solution inside (a, b) of equation (5.1) has the form (5.9).

THEOREM 5.3. *A vector-function $y(t)$ is a solution inside the interval (a, b) of equation* (5.1) *if and only if it is representable in the form* (5.9).

Proof. It is sufficient to prove that if $y(t)$ is a solution inside (a, b) of equation (5.1), then it can be written in the form (5.9).

Since $y(t)$ is a solution inside (a, b), the vectors $y(a+\delta), y(b-\delta)$ belong to $\mathfrak{D}(A^2)$ for arbitrary δ, $0 < \delta < b - a$. Because of this the vector-function

$$z(t) = \omega_1^\delta(t, A)y(a + \delta) + \omega_2^\delta(t, A)y(b - \delta),$$

where

$$\omega_1^\delta(t, \lambda) = \frac{2\mathrm{sh}(\lambda(b - \delta - t))e^{-\lambda(b-a-2\delta)}}{1 - e^{-2\lambda(b-a-2\delta)}},$$

$$\omega_2^\delta(t, \lambda) = \frac{2\mathrm{sh}(\lambda(t - a - \delta))e^{-\lambda(b-a-2\delta)}}{1 - e^{-2\lambda(b-a-2\delta)}},$$

is a twice continuously differentiable solution of (5.1) on $[a + \delta, b - \delta]$ which coincides at the points $a + \delta, b - \delta$ with the solution $y(t)$, which is twice continuously differentiable on the same interval. According to Corollary 5.1, $y(t) = z(t)$ for $t \in [a + \delta, b - \delta]$.

Let us fix δ_0, $0 < \delta_0 < \frac{b-a}{4}$, and take δ, $0 < \delta < \delta_0$. Then

$$y(a + \delta_0) = \omega_1^\delta(a + \delta_0, A)y(a + \delta) + \omega_2^\delta(a + \delta_0, A)y(b - \delta),$$
$$y(b - \delta_0) = \omega_1^\delta(b - \delta_0, A)y(a + \delta) + \omega_2^\delta(b - \delta_0, A)y(b - \delta),$$

whence

$$y(a + \delta) = \psi_1(\delta, A)y(a + \delta_0) + \psi_2(\delta, A)y(b - \delta_0),$$
$$y(b - \delta) = \psi_2(\delta, A)y(a + \delta_0) + \psi_1(\delta, A)y(b - \delta_0),$$

where

$$\psi_1(\delta, \lambda) = \frac{\mathrm{sh}(\lambda(b - a - \delta - \delta_0))(1 - e^{-2\lambda(b-a-2\delta)})e^{\lambda(b-a-2\delta)}}{\mathrm{sh}^2(\lambda(b - a - \delta - \delta_0)) - \mathrm{sh}^2(\lambda(\delta_0 - \delta))},$$

$$\psi_2(\delta,\lambda) = -\frac{\operatorname{sh}(\lambda(\delta_0-\delta))(1-e^{-2\lambda(b-a-2\delta)})e^{\lambda(b-a-2\delta)}}{\operatorname{sh}^2(\lambda(b-a-\delta-\delta_0))-\operatorname{sh}^2(\lambda(\delta_0-\delta))}.$$

Since

$$\begin{aligned}\omega_1^\delta(a+\delta_0,\lambda) &< ce^{-\lambda(\delta_0-\delta)}, & \omega_2^\delta(a+\delta_0,\lambda) &< ce^{-\lambda\frac{b-a}{2}},\\ \omega_1^\delta(b-\delta_0,\lambda) &< ce^{-\lambda\frac{b-a}{2}}, & \omega_2^\delta(b-\delta_0,\lambda) &< ce^{-\lambda(\delta_0-\delta)}\end{aligned}\qquad (0<c=\text{const})$$

at all $\lambda \geq 0$ and $y(a+\delta), y(b-\delta) \in \mathfrak{H}$, and since δ, $0<\delta<\delta_0$, is arbitrary, we conclude that

$$y(a+\delta_0) \in \bigcap_{0<\delta<\delta_0} \mathfrak{H}_{e,\delta}, \quad y(b-\delta_0) \in \bigcap_{0<\delta<\delta_0} \mathfrak{H}_{e,\delta}.$$

But the functions $\psi_1(\delta,\lambda)e^{-\lambda\delta_0}$ and $\psi_2(\delta,\lambda)e^{-\lambda\delta_0}$ are uniformly bounded on $[0,\delta_0]\times[0,\infty)$. For every $\lambda\in[0,\infty)$ they tend to $\psi_1(0,\lambda)e^{-\lambda\delta_0}$ and $\psi_2(0,\lambda)e^{-\lambda\delta_0}$, respectively, as $\delta\to 0$. Therefore $y(a+\delta), y(b-\delta)$ converge in the space $\mathfrak{A}'(A)$ to certain elements g_1, g_2.

Let now t be any point in (a,b). Take $\delta<\min\{t-a, b-t\}$. Then

$$y(t) = \omega_1^\delta(t,A)y(a+\delta)+\omega_2^\delta(t,A)y(b-\delta).$$

Since the $\omega_i^\delta(t,\lambda)$ are uniformly bounded in δ and λ and because $\omega_i^\delta(t,\lambda)\to\omega_i(t,\lambda)$ as $\delta\to 0$, by limit transition in the last equality we deduce the representation (5.9). □

Thus, formula (5.9) determines a one-to-one correspondence between $\mathfrak{A}'(A)\times\mathfrak{A}'(A)$ and the set of all solutions inside (a,b) of equation (5.1). This is why we can regard $\mathfrak{A}'(A)$ as the natural space for formulating the Dirichlet problem for equation (5.1).

REMARK 5.1. There are two terms in the representation (5.9). The first one, $\omega_1(t,\hat{A})g_1$ ($g_1\in\mathfrak{A}'(A)$), is, in view of the properties of the function $\omega_1(t,\lambda)$, an infinitely differentiable vector-function on $(a,b]$ into $\mathfrak{H}$ such that $\omega_1(t,\hat{A})g_1\to 0$ as $t\to b$. If $g_1\in\mathfrak{H}'_G$, then $\omega_1(t,\hat{A})g_1\to g_1$ as $t\to a$ in the same space. The vector-function $\omega_2(t,\hat{A})g_2$ ($g_2\in\mathfrak{A}'(A)$) is, because of the properties of $\omega_2(t,\lambda)$, infinitely differentiable on $[a,b)$ into $\mathfrak{H}$, and $\omega_2(t,\hat{A})g_2\to 0$ as $t\to a$. If $g_2\in\mathfrak{H}'_G$, then $\omega_2(t,\hat{A})g_2\to g_2$ in $\mathfrak{H}'_G$ as $t\to b$.

REMARK 5.2. If the operator A is non-negative, but not positive definite, then the representation (5.6) for a solution $y(t)$ inside (a,b) of equation (5.1) does not hold, in general. Indeed, in this situation $\mathfrak{R}(E-e^{-2A(b-a)})\neq\mathfrak{H}$. Hence, there exists an element $h\in\mathfrak{H}$, $h\notin\mathfrak{R}(E-e^{-2A(b-a)})$. Since the Dirichlet problem $y(a)=0$, $y(b)=h$ for the equation considered is solvable, there exists a solution inside (a,b) that is continuous on the closed interval $[a,b]$ and whose values at the ends are 0 and h. If this solution would have a representation (5.6), then the elements f_1, f_2 would belong to $\mathfrak{H}$, and $h=(E-e^{-2A(b-a)})f_2$, which leads to a contradiction.

5.4. VECTOR-VALUED DISTRIBUTIONS

Let $\mathcal{D}[a,b]$ (respectively, $\mathcal{D}(a,b)$) ($-\infty < a < b < \infty$) be the space of infinitely differentiable functions on $[a,b]$ vanishing together with their derivatives at the ends a and b (the space of infinitely differentiable functions on (a,b) with compact support). It is easy to see that

$$\mathcal{D}[a,b] = \bigcap_{m\in\mathbb{N}} \overset{\circ}{W}{}_2^m(a,b)$$

where $\overset{\circ}{W}{}_2^{m+1}(a,b) \subset \overset{\circ}{W}{}_2^m(a,b)$ and the imbedding is nuclear (i.e. the imbedding operator is nuclear). The set $\mathcal{D}[a,b]$ is endowed with the topology of the projective limit of the Hilbert spaces $\overset{\circ}{W}{}_2^m(a,b)$:

$$\mathcal{D}[a,b] = \operatorname*{limpr}_{m\to\infty} \overset{\circ}{W}{}_2^m(a,b).$$

It is clear that convergence of a sequence f_n in $\mathcal{D}[a,b]$ means uniform convergence on $[a,b]$ of the sequences $\{f_n^{(k)}\}_{n=1}^{\infty}$ ($k = 0,1,2,\ldots$). It follows from the definition of $\mathcal{D}(a,b)$ (see also Chapter 1, subsection 7.3) that

$$\mathcal{D}(a,b) = \operatorname*{limind}_{n\to\infty} \mathcal{D}[\alpha_n,\beta_n],$$

where $a < \alpha_{n+1} < \alpha_n < \beta_n < \beta_{n+1} < b$, $\alpha_n \to a, \beta_n \to b$ as $n \to \infty$.

DEFINITION. A continuous linear mapping from $\mathcal{D}[a,b]$ (respectively, $\mathcal{D}(a,b)$) into $\mathfrak{H}$ is called a vector-valued distribution. These form the class $\mathcal{D}'(\mathfrak{H},[a,b])$ (respectively, $\mathcal{D}'(\mathfrak{H},(a,b))$).

If $\mathfrak{H} = \mathbb{C}^1$, we obtain the ordinary distributions.

The space $L_2(\mathfrak{H},(a,b))$ is naturally imbedded in the class $\mathcal{D}'(\mathfrak{H},[a,b])$: the action of $u \in L_2(\mathfrak{H},(a,b))$ on a test function $\varphi \in \mathcal{D}[a,b]$ is defined by $u(\varphi) = \int_a^b u(t)\varphi(t)\,dt$.

On a vector-valued distribution $\omega \in \mathcal{D}'(\mathfrak{H},[a,b])$ ($\omega \in \mathcal{D}'(\mathfrak{H},(a,b))$) there is defined the differentiation operation: the $n-$th derivative of ω is the unique element $\omega^{(n)}$ from $\mathcal{D}'(\mathfrak{H},[a,b])$ ($\mathcal{D}'(\mathfrak{H},(a,b))$) for which

$$\omega^{(n)}(\varphi) = (-1)^n\omega(\varphi^{(n)}), \quad \forall\varphi \in \mathcal{D}[a,b] \quad (\forall\varphi \in \mathcal{D}(a,b)). \tag{5.10}$$

If a vector-valued distribution coincides with a continuously differentiable vector-function, then its derivative in the sense of (5.10) is its ordinary derivative.

If $\mathfrak{H}_1 \subset \mathfrak{H}$ and this imbedding is continuous, then the imbedding $\mathcal{D}'(\mathfrak{H}_1,[a,b]) \subset \mathcal{D}'(\mathfrak{H},[a,b])$ ($\mathcal{D}'(\mathfrak{H}_1,(a,b)) \subset \mathcal{D}'(\mathfrak{H},(a,b))$) is continuous, and the differentiation operation on $\mathcal{D}'(\mathfrak{H}_1,[a,b])$ (on $\mathcal{D}'(\mathfrak{H}_1,(a,b))$) is the restriction of the same operation on $\mathcal{D}'(\mathfrak{H},[a,b])$ (on $\mathcal{D}'(\mathfrak{H},(a,b))$) .

We denote by $\overset{\circ}{W}{}_2^m(\mathfrak{H},(a,b))$ the completion of the linear set $C_0^\infty(\mathfrak{H},[a,b])$ of vector-functions $u(t) = \sum_{k=1}^p \varphi_k(t)h_k$ ($\varphi_k \in \mathcal{D}[a,b]$, $h_k \in \mathfrak{H}$, $p \in N$) with respect to the norm

$$\|u\|_{\overset{\circ}{W}{}_2^m(\mathfrak{H},(a,b))} = \Big(\sum_{k=0}^{m} \int_a^b \|u^{(k)}(t)\|^2\,dt\Big)^{\frac{1}{2}}.$$

Let $W_{-m}(\mathfrak{H},(a,b))$ be the negative space constructed from the positive space $\overset{\circ}{W}{}_2^m(\mathfrak{H},(a,b))$ relative to $L_2(\mathfrak{H},(a,b))$.

LEMMA 5.2. *The equality* $\mathcal{D}'(\mathfrak{H},[a,b]) = \bigcup_{m=1}^{\infty} W_{-m}(\mathfrak{H},(a,b))$ *is valid.*

Proof. We show that $W_{-m}(\mathfrak{H},(a,b))$ consists of those vector-valued distributions that are m-th derivatives in the sense of (5.10) of vector-functions from $L_2(\mathfrak{H},(a,b))$. For this purpose we identify the space $\overset{\circ}{W}{}_2^m(\mathfrak{H},(a,b))$ with a certain closed subspace of $\oplus_{(m+1)\text{times}} L_2(\mathfrak{H},(a,b))$ using the mapping

$$u \to \{u^{(j)}\}_{j=0}^{m}$$

According to the Hahn–Banach and Riesz theorems any antilinear functional $y \in W_{-m}(\mathfrak{H},(a,b))$ can be represented in the form

$$\begin{aligned}(y,u)_{L_2(\mathfrak{H},(a,b))} &= \sum_{0\le j\le m} \int_a^b (g_j(t), u^{(j)}(t))\,dt \\ &= (-1)^m \int_a^b (y_0(t), u^{(m)}(t))\,dt \qquad (5.11)\\ &(\forall u \in C_0^\infty(\mathfrak{H},[a,b]);\ g_j, y_0 \in L_2(\mathfrak{H},(a,b)))\end{aligned}$$

(here $(y,u)_{L_2(\mathfrak{H},(a,b))}$ denotes the action of the functional y on a test element u), where the element $y_0(t)$ in the representation (5.11) is, in general, not unique.

Now we associate with an element y a vector-valued distribution $\hat{y} \in \mathcal{D}'(\mathfrak{H},[a,b])$ as follows:

$$\hat{y}(\varphi) = (-1)^m \int_a^b y_0(t)\varphi^{(m)}(t)\,dt \quad (\forall \varphi \in \mathcal{D}[a,b]).$$

This correspondence does not depend on the choice of y_0 in (5.11). In fact, if $y_1(t)$ is another vector-function from $L_2(\mathfrak{H},(a,b))$ which yields the representation (5.11), then for an arbitrary $h \in \mathfrak{H}$,

$$\begin{aligned}(-1)^m\Big(\int_a^b y_0(t)\varphi^{(m)}(t)\,dt, h\Big) &= (-1)^m \int_a^b (y_0(t), \overline{\varphi^{(m)}(t)h})\,dt \\ &= (y, \overline{\varphi^{(m)}}h)_{L_2(\mathfrak{H},(a,b))} \\ &= (-1)^m \int_a^b (y_1(t), \overline{\varphi^{(m)}(t)h})\,dt \\ &= (-1)^m\Big(\int_a^b y_1(t)\varphi^{(m)}(t)\,dt, h\Big),\end{aligned}$$

whence

$$(-1)^m \int_a^b y_0(t)\varphi^{(m)}(t)\,dt = (-1)^m \int_a^b y_1(t)\varphi^{(m)}(t)\,dt.$$

Thus, each $y \in W_{-m}(\mathfrak{H}, (a, b))$ is imbedded into $\mathcal{D}'(\mathfrak{H}, [a, b])$, i.e. $\bigcup_{m \in \mathbb{N}} W_{-m}(\mathfrak{H}, (a, b)) \subset \mathcal{D}'(\mathfrak{H}, [a, b])$; moreover, $y = y_0^{(m)}$ in the sense of (5.10).

We prove the converse inclusion. Suppose $\omega \in \mathcal{D}'(\mathfrak{H}, [a, b])$. Since ω is a continuous linear mapping from $\mathcal{D}[a, b]$ into $\mathfrak{H}$ and $\mathcal{D}[a, b] = \operatorname{limpr}_{m \to \infty} \overset{\circ}{W}{}_2^m(a, b)$, we can find $m \in \mathbb{N}$ and $c = \text{const}$ such that

$$\|\omega(\varphi)\| \le c\|\varphi\|_{\overset{\circ}{W}{}_2^m(a,b)} \quad (\forall \varphi \in \mathcal{D}[a, b]).$$

Let $\{\psi_k\}_{k=1}^{\infty}$ be an orthonormal basis in $\overset{\circ}{W}{}_2^{m+1}(a, b)$ consisting of real-valued functions from $\mathcal{D}[a, b]$. On the linear set $\Phi \subset C_0^{\infty}(\mathfrak{H}, [a, b])$ of vector-functions $u(t)$ of the form

$$u(t) = \sum_{k=1}^{p} \psi_k(t) h_k \quad (h_k \in \mathfrak{H}, p \in \mathbb{N}),$$

the vector-valued distribution ω generates in a natural way an antilinear functional

$$\hat{\omega}(u) = \sum_{k=1}^{p} (\omega(\psi_k), h_k).$$

(Since the above-mentioned expression for $u(t)$ is unique, this definition is correct on Φ.) Since the imbedding of $\overset{\circ}{W}{}_2^{m+1}(a, b)$ into $\overset{\circ}{W}{}_2^m(a, b)$ is of Hilbert–Schmidt class, we have

$$\sum_{k=1}^{\infty} \|\psi_k\|^2_{\overset{\circ}{W}{}_2^m(a,b)} = c_1^2 < \infty.$$

The estimates now follow:

$$\begin{aligned}
|\hat{\omega}(u)| &\le \sum_{k=1}^{p} |(\omega(\psi_k), h_k)| \le \sum_{k=1}^{p} \|(\omega(\psi_k)\| \|h_k\| \\
&\le c \sum_{k=1}^{p} \|\psi_k\|_{\overset{\circ}{W}{}_2^m(a,b)} \|h_k\| \le c \Big(\sum_{k=1}^{p} \|\psi_k\|^2_{\overset{\circ}{W}{}_2^m(a,b)} \Big)^{\frac{1}{2}} \Big(\sum_{k=1}^{p} \|h_k\|^2 \Big)^{\frac{1}{2}} \\
&\le cc_1 \Big(\sum_{k=1}^{p} \|\psi_k\|^2_{\overset{\circ}{W}{}_2^{m+1}(a,b)} \|h_k\|^2 \Big)^{\frac{1}{2}} = cc_1 \|u\|_{\overset{\circ}{W}{}_2^{m+1}(\mathfrak{H},(a,b))}.
\end{aligned}$$

Hence, the norm of the functional $\hat{\omega}$, defined on Φ, calculated with respect to the norm in the space $\overset{\circ}{W}{}_2^{m+1}(\mathfrak{H}, (a, b)) \supset \Phi$, does not exceed cc_1. Since Φ is dense in $\overset{\circ}{W}{}_2^{m+1}(\mathfrak{H}, (a, b))$, the functional $\hat{\omega}$ can be extended to the whole space $\overset{\circ}{W}{}_2^{m+1}(\mathfrak{H}, (a, b))$. □

5.5. GENERALIZED SOLUTIONS

DEFINITION. We will call a vector-valued distribution $\omega \in \mathcal{D}'(\mathfrak{H}, (a,b))$ a generalized solution of equation (5.1) if

$$(\omega(\varphi''), h) - (\omega(\varphi), A^2 h) = 0 \tag{5.12}$$

for any $\varphi \in \mathcal{D}(a,b)$, $h \in \mathfrak{H}^{\infty}(A)$.

It is not difficult to observe that any solution inside (a,b) of equation (5.1) is a generalized solution of it. Conversely, if a generalized solution ω coincides with a twice continuously differentiable vector-function $y(t)$ on (a,b) into $\mathfrak{H}$ whose values belong to $\mathfrak{D}(A^2)$, then $\omega = y(t)$ is a solution inside (a,b) of the same equation.

THEOREM 5.4. *A vector-valued distribution* $\omega \in \mathcal{D}'(\mathfrak{H}, [a,b])$ *is a generalized solution of equation* (5.1) *if and only if it coincides with an ordinary vector-function of the form*

$$y(t) = \omega_1(t, \hat{A}) f_1 + \omega_2(t, \hat{A}) f_2 \quad (f_1, f_2 \in \mathfrak{H}^{-\infty}(A)). \tag{5.13}$$

Proof. We assume that $\omega \in \mathcal{D}'(\mathfrak{H}, [a,b])$ is a generalized solution of (5.1). According to Lemma 5.2, $\omega \in W_{-(m-1)}(\mathfrak{H}, (a,b))$ for some m and $\omega = y_0^{(m)}$, where $y_0(t)$ is a vector-function that is continuous on $[a,b]$ and has values in $\mathfrak{H}$ (the derivative is understood in the sense of the theory of distributions). Since ω satifies (5.12), we arrive on account of (5.10) at the equation (in y_0)

$$\int_a^b (y_0(t), h)\varphi^{(m+2)}(t)\,dt - \int_a^b (y_0(t), A^2 h)\varphi^{(m)}(t)\,dt = 0 \tag{5.14}$$
$$(\forall h \in \mathfrak{H}^{\infty}(A), \forall \varphi \in \mathcal{D}(a,b)).$$

Let

$$w(t) = \int_a^t (t-\xi)\hat{A}^2 y_0(\xi)\,d\xi.$$

The vector-function $w(t)$, which takes values in $\mathfrak{H}^{-2}$, is twice continuously differentiable into this space. Integrating by parts twice the second term in (5.14), we obtain for any function $\varphi(t) \in \mathcal{D}(a,b)$,

$$\int_a^b \varphi^{(m+2)}(t)(y_1(t), h)\,dt = 0, \tag{5.15}$$

where $y_1(t) = y_0(t) - w(t) - \sum_{i=0}^{m+1} g_i t^i$ $(g_i \in \mathfrak{H}^{-2})$. Limit transition yields (5.15) for an arbitrary function $\varphi(t) \in \overset{\circ}{W}{}_2^{m+2}(a,b)$. In particular, (5.15) holds for the function

$$\varphi(t) = \frac{1}{(m+1)!}\int_a^t (t-\xi)^{m+1}(h, y_1(\xi))\,d\xi,$$

where the vectors g_i $(i = 0, 1, \ldots, m+1)$ in $y_1(t)$ are selected so that $y_1^{(k)}(b) = 0$ $(k = 0, 1, \ldots, m+1)$. These conditions determine g_i uniquely because

$$\det \left\| \int_a^b (b-\xi)^j \xi^k \, d\xi \right\|_{j,k=0}^{m+1} \neq 0.$$

Substituting this function into (5.15) we deduce $\int_a^b |(y_1(t), h)|^2 \, dt = 0$, hence $(y_1(t), h) = 0$ (any $h \in \mathfrak{H}^\infty$) almost everywhere. By virtue of the continuity of $y_1(t)$ into the space $\mathfrak{H}^{-2}$, $y_1(t) \equiv 0$, i.e.

$$y_0(t) = w(t) + \sum_{i=0}^{m+1} g_i t^i.$$

It can thus be seen that the vector-function $y_0(t)$ is twice continuously differentiable on $[a, b]$ into $\mathfrak{H}^{-2}$. It is a generalized solution from $\mathcal{D}'(\mathfrak{H}^{-2}, [a, b])$ of the equation

$$y''(t) - \hat{A}^2 y(t) = r(t), \quad r(t) = \frac{d^2}{dt^2}\Big(\sum_{i=0}^{m+1} g_i t^i\Big),$$

that is,

$$\int_a^b (y_0(t), h)\varphi''(t) \, dt - \int_a^b (y_0(t), \hat{A}^2 h)\varphi(t) \, dt = \int_a^b (r(t), h)\varphi(t) \, dt$$
$$(\forall h \in \mathfrak{H}^\infty(A), \forall \varphi \in \mathcal{D}(a, b))$$

(here we mean by $\hat{A}$ the restriction of the operator $\hat{A}$ to $\mathfrak{H}^{-2}$; its set of infinitely differentiable vectors coincides with that of the operator A). The twice continuously differentiable vector-function (on $[a, b]$ into $\mathfrak{H}^{-3}$)

$$z(t) = \tfrac{1}{2} \int_a^b e^{-\hat{A}|t-s|} \hat{A}^{-1} r(s) \, ds \tag{5.16}$$

is a particular solution of this equation. Then the twice continuously differentiable vector-function $y_0 - z$ (on $[a, b]$ into $\mathfrak{H}^{-3}$) takes values in $\mathfrak{H}^{-1}$. Since the domain of the restriction of $\hat{A}^2$ to $\mathfrak{H}^{-3}$ coincides with $\mathfrak{H}^{-1}$, the vector-function $y_0 - z$ is a solution of the equation $y'' - \hat{A}^2 y = 0$, considered in the Hilbert space $\mathfrak{H}^{-3}$, inside the interval (a, b). By Theorem 5.3 it can be written in the form

$$y_0(t) - z(t) = \omega_1(t, \hat{A}) f_1' + \omega_2(t, \hat{A}) f_2' \tag{5.17}$$
$$(f_1' = (y_0 - z)(a), \; f_2' = (y_0 - z)(b), \; f_i' \in \mathfrak{H}^{-1} (i = 1, 2)).$$

But $z(t), \omega_1(t, \hat{A}) f_1', \omega_2(t, \hat{A}) f_2'$ are $(m+2)$ times continuously differentiable on $[a, b]$ into $\mathfrak{H}^{-(m+3)}$, therefore so is $y_0(t)$. This implies that the vector-valued distribution $y = y_0^{(m)} \in \mathcal{D}'(\mathfrak{H}, [a, b]) \subset \mathcal{D}'(\mathfrak{H}^{-(m+3)}, [a, b])$ coincides with a twice continuously differentiable vector-function on $[a, b]$ into $\mathfrak{H}^{-(m+3)}$. Since $y_0^{(m)}$, as can be seen from (5.16) and (5.17), takes values in $\mathfrak{H}^{-(m+1)}$, the vector-function $y(t)$ is a twice continuously differentiable solution on $[a, b]$ into $\mathfrak{H}^{-(m+3)}$ of the equation

$y'' - \hat{A}^2 y = 0$ (notice that the domain of $\hat{A}^2$ as an operator on $\mathfrak{H}^{-(m+3)}$ coincides with $\mathfrak{H}^{-(m+1)}$). According to Theorem 5.3, $y(t)$ has a representation

$$y(t) = \omega_1(t, \hat{A}) f_1 + \omega_2(t, \hat{A}) f_2 \quad (f_1, f_2 \in \mathfrak{H}^{-(m+1)} \subset \mathfrak{H}^{-\infty}).$$

Conversely, let a vector-function $y(t)$ be represented as in (5.13) with $f_1, f_2 \in \mathfrak{H}^{-m}$ for some m. Since $\omega_i(t, \hat{A}) f_i$ ($f_i \in \mathfrak{H}^{-m}$ $(i = 1, 2)$) are infinitely differentiable inside (a, b) (see Remark 5.1), $f_i = (\hat{A}^2 + E)^m g_i$ ($g_i \in \mathfrak{H}$) and

$$\frac{d^{2k} \omega_i(t, \hat{A}) f_i}{dt^{2k}} = \omega_i(t, \hat{A}) \hat{A}^{2k} f_i \quad (k = 0, 1, \ldots),$$

we have that

$$y(t) = \left(\frac{d^2}{dt^2} + E\right)^m (\omega_1(t, \hat{A}) g_1 + \omega_2(t, \hat{A}) g_2),$$

where $\omega_1(t, \hat{A}) g_1 + \omega_2(t, \hat{A}) g_2$ is continuous on $[a, b]$ into $\mathfrak{H}$.

The vector-function $y(t)$ defines a linear mapping from $\mathcal{D}[a, b]$ into $\mathfrak{H}$ as follows:

$$\omega(\varphi) = \int_a^b (\omega_1(t, \hat{A}) g_1 + \omega_2(t, \hat{A}) g_2) \left(\frac{d^2}{dt^2} + 1\right)^m \varphi(t)\, dt.$$

Since

$$\left\| \int_a^b (\omega_1(t, \hat{A}) g_1 + \omega_2(t, \hat{A}) g_2) \left(\frac{d^2}{dt^2} + 1\right)^m \varphi(t)\, dt \right\|$$
$$\leq c \left(\int_a^b \|\omega_1(t, \hat{A}) g_1 + \omega_2(t, \hat{A}) g_2\|^2 \, dt \right)^{\frac{1}{2}} \|\varphi\|_{\overset{\circ}{W}{}_2^{2m}(a,b)},$$

ω is the vector-valued distribution from $\mathcal{D}'(\mathfrak{H}, [a, b])$ corresponding to $y(t)$. By Theorem 5.3 $y(t)$ is a solution inside the interval (a, b), hence a generalized solution, of equation (5.1). □

COROLARY 5.2. *If $\omega \in \mathcal{D}'(\mathfrak{H}, [a, b])$ is a generalized solution of equation* (5.1), *then it is a solution inside the interval* (a, b).

THEOREM 5.5. *A generalized solution of equation* (5.1) *is a solution inside the interval.*

Proof. Let ω be a generalized solution of equation (5.1). Because of the continuity of the imbedding $\mathcal{D}(\mathfrak{H}, [a + \delta, b - \delta]) \subset \mathcal{D}(\mathfrak{H}, (a, b))$ the vector-valued distribution ω belongs to $\mathcal{D}'(\mathfrak{H}, [a + \delta, b - \delta])$ for any sufficiently small $\delta > 0$. According to Corollary 5.2, ω is a solution inside $(a + \delta, b - \delta)$. Since δ is arbitrary, ω is a solution inside (a, b) of equation (5.1). □

5.6. BOUNDARY VALUES OF SOLUTIONS INSIDE AN INTERVAL

It follows from the representation (5.9) and the properties of the functions $\omega_i(t, \lambda)$ $(i = 1, 2)$ listed in subsection 5.3 that any solution inside (a, b) of equation (5.1) is infinitely differentiable inside this interval, and its boundary values at the points a

and b exist in the space $\mathfrak{A}'(A)$. As well as for the equation of the first order, it is possible to describe the spaces of boundary values of solutions inside (a, b) of (5.1) in dependence on their behaviour near the ends of the interval.

Let $\gamma(t)$ be a positive, continuous and integrable function on (a, b). Denote

$$
\begin{aligned}
&G(a,\lambda) = \Big(\int_a^b \gamma(t)e^{-2\lambda(t-a)}\,\mathrm{d}t\Big)^{-\frac{1}{2}}, \quad G(b,\lambda) = \Big(\int_a^b \gamma(t)e^{-2\lambda(b-t)}\,\mathrm{d}t\Big)^{-\frac{1}{2}}, \\
&G_i(\lambda) = \Big(\int_a^b \gamma(t)\omega_i^2(t,\lambda)\,\mathrm{d}t\Big)^{-\frac{1}{2}} \quad (i = 1, 2).
\end{aligned} \tag{5.18}
$$

These functions are positive. In addition they have the following properties:

(i) $G(\lambda) \geq c = \text{const} \quad (\lambda \in [0, \infty))$,

(ii) for any $\alpha > 0$ there exists a constant $c_\alpha > 0$ such that

$$G(\lambda) \leq c_\alpha e^{\alpha\lambda};$$

(iii) $\lim\limits_{\lambda\to\infty} G(\lambda) = \infty$

Set also

$$
\begin{aligned}
&G^\delta(a,\lambda) = \Big(\int_a^{a+\delta} \gamma(t)e^{-2\lambda(t-a)}\,\mathrm{d}t\Big)^{-\frac{1}{2}}, \\
&G^\delta(b,\lambda) = \Big(\int_{b-\delta}^{b} \gamma(t)e^{-2\lambda(b-t)}\,\mathrm{d}t\Big)^{-\frac{1}{2}}, \qquad (0 < \delta < b - a) \\
&G_1^\delta(\lambda) = \Big(\int_a^{a+\delta} \gamma(t)\omega_1^2(t,\lambda)\,\mathrm{d}t\Big)^{-\frac{1}{2}}, \\
&G_2^\delta(\lambda) = \Big(\int_{b-\delta}^{b} \gamma(t)\omega_2^2(t,\lambda)\,\mathrm{d}t\Big)^{-\frac{1}{2}},
\end{aligned} \tag{5.19}
$$

Making use of the properties of the function $\lambda(t)$ we can easily convince ourselves of the validity of the inequalities

$$c_\delta G^\delta(\lambda) \leq G(\lambda) \leq G^\delta(\lambda) \quad (\lambda \in [0,\infty),\ 0 < c_\delta = \text{const}), \tag{5.20}$$

where $G(\lambda)$ denotes any of the functions from (5.18) and $G^\delta(\lambda)$ is the function corresponding by (5.19). Moreover, there exist constants $c_1, c_2 > 0$ such that

$$
\begin{aligned}
&c_1 e^{-\lambda(t-a)} \leq \omega_1(t,\lambda) \leq c_2 e^{-\lambda(t-a)}, \\
&c_1 e^{-\lambda(b-t)} \leq \omega_2(t,\lambda) \leq c_2 e^{-\lambda(b-t)}.
\end{aligned} \tag{5.21}
$$

The first of the two-sided inequalities holds for all $\lambda \in [0,\infty)$, $t \in [a, a+\delta]$, and the second one holds for $\lambda \in [0,\infty)$, $t \in [b-\delta, b]$ $(0 < \delta < \frac{b-a}{4})$.

The inequalities (5.20), (5.21) allow us to conclude that the spaces $\mathfrak{H}_{G^\delta(a,\lambda)}$, $\mathfrak{H}_{G(a,\lambda)}$, $\mathfrak{H}_{G_1}$, $\mathfrak{H}_{G_1^\delta}$ ($\mathfrak{H}_{G^\delta(b,\lambda)}$, $\mathfrak{H}_{G(b,\lambda)}$, $\mathfrak{H}_{G_2}$, $\mathfrak{H}_{G_2^\delta}$, respectively) coincide with each other and that their norms are equivalent. The structure of these spaces depends on the behaviour of the function $\gamma(t)$ in a neighbourhood of the point a (of the point

b): the faster $\gamma(t)$ tends to zero as t approaches a (respectively, b) the narrower they are.

THEOREM 5.6. *Let $\gamma(t)$ be a continuous, positive and integrable function on (a, b) $(-\infty < a < b < \infty)$. The boundary values at a and b of a solution $y(t)$ inside (a, b) of equation (5.1) belong to the spaces $\mathfrak{H}'_{G(a,\lambda)}(A)$ and $\mathfrak{H}'_{G(b,\lambda)}(A)$, respectively, where $G(a, \lambda)$ and $G(b, \lambda)$ are determined as in formula (5.18), if and only if*

$$\int_a^b \gamma(t)\|y(t)\|^2 \, dt < \infty. \tag{5.22}$$

Moreover, formula (5.9) defines a topological isomorphism between the space K_γ of solutions inside (a, b) of the equation considered satisfying (5.22), and the space $\mathfrak{H}'_{G(a,\lambda)} \oplus \mathfrak{H}'_{G(b,\lambda)}$.

Proof. We first show that the set K_γ is a subspace of $L_2(\mathfrak{H}, (a, b), \gamma(t)\,dt)$. Indeed, let $K_\gamma \ni y_n \to y$ in $L_2(\mathfrak{H}, (a, b), \gamma(t)\,dt)$ as $n \to \infty$. Then for a sufficiently small number $\delta_0 > 0$, also $y_n \to y$ $(n \to \infty)$ in the space $L_2(\mathfrak{H}, (a + \delta_0, b - \delta_0), \gamma(t)\,dt)$. By virtue of the positivity and continuity of $\gamma(t)$ on $[a+\delta_0, b-\delta_0]$, the sequence y_n converges to y in the space $L_2(\mathfrak{H}, (a + \delta_0, b - \delta_0))$. By limit transition in the equality

$$\int_{a+\delta_0}^{b-\delta_0} (y_n(t), (-\varphi'' + A^2\varphi)(t)) \, dt = 0,$$

where $\varphi(t)$ is an infinitely differentiable vector-function on $(a + \delta_0, b - \delta_0)$ with compact support and with values in $\mathfrak{D}(A^2)$, we deduce that $y(t)$ is a generalized solution from $L_2(\mathfrak{H}, (a + \delta_0, b - \delta_0))$ of equation (5.1). Therefore it is a solution inside $(a + \delta_0, b - \delta_0)$. Since δ_0 is arbitrary, the vector-function $y(t)$ is a solution inside (a, b) of equation (5.1), i.e. $y \in K_\gamma$.

We now consider the mapping

$$\{g_1, g_2\} \to \omega_1(t, \hat{A})g_1 + \omega_2(t, \hat{A})g_2 \tag{5.23}$$

arising from the expression (5.9). In view of the relations

$$\begin{aligned}
&\int_a^b \gamma(t)\|\omega_1(t, \hat{A})g_1 + \omega_2(t, \hat{A})g_2\|^2 \, dt \\
&\leq 2\Big(\int_a^b \gamma(t)\|\omega_1(t, \hat{A})g_1\|^2 \, dt + \int_a^b \gamma(t)\|\omega_2(t, \hat{A})g_2\|^2 \, dt\Big) \\
&= 2\Big(\int_a^b \gamma(t) \int_0^\infty e^{2\lambda} \, d(\hat{E}_\lambda \omega_1(t, \hat{A})g_1, \omega_1(t, \hat{A})g_1)_{\mathfrak{H}'_e} \, dt + \\
&\quad + \int_a^b \gamma(t) \int_0^\infty e^{2\lambda} \, d(\hat{E}_\lambda \omega_2(t, \hat{A})g_2, \omega_2(t, \hat{A})g_2)_{\mathfrak{H}'_e} \, dt\Big) \\
&= 2\Big(\int_a^b \gamma(t) \int_0^\infty e^{2\lambda}\omega_1^2(t, \lambda) \, d(\hat{E}_\lambda g_1, g_1)_{\mathfrak{H}'_e} \, dt + \\
&\quad + \int_a^b \gamma(t) \int_0^\infty e^{2\lambda}\omega_2^2(t, \lambda) \, d(\hat{E}_\lambda g_2, g_2)_{\mathfrak{H}'_e} \, dt\Big)
\end{aligned}$$

$$= 2\Big(\int_0^\infty e^{2\lambda} G_1^{-2}(\lambda)\, d(\hat{E}_\lambda g_1, g_1)_{\mathfrak{H}'_e} + \int_0^\infty e^{2\lambda} G_2^{-2}(\lambda)\, d(\hat{E}_\lambda g_2, g_2)_{\mathfrak{H}'_e}\Big)$$
$$\le c\big(\|g_1\|^2_{\mathfrak{H}'_{G(a,\lambda)}} + \|g_2\|^2_{\mathfrak{H}'_{G(b,\lambda)}}\big),$$

which hold because of (3.1) and (3.4), this mapping is continuous from $\mathfrak{H}'_{G(a,\lambda)} \oplus \mathfrak{H}'_{G(b,\lambda)}$ into K_γ. In addition, it is a mapping "onto".

In fact, if $y \in K_\gamma$, then $y(t) = \omega_1(t, \hat{A})g_1 + \omega_2(t, \hat{A})g_2$ $(g_1, g_2 \in \mathfrak{A}'(A))$ and

$$\int_a^b \gamma(t)\|\omega_1(t, \hat{A})g_1 + \omega_2(t, \hat{A})g_2\|^2\, dt < \infty.$$

Consequently,

$$\int_a^{\frac{a+b}{2}} \gamma(t)\|\omega_1(t, \hat{A})g_1 + \omega_2(t, \hat{A})g_2\|^2\, dt < \infty,$$
$$\int_{\frac{a+b}{2}}^b \gamma(t)\|\omega_1(t, \hat{A})g_1 + \omega_2(t, \hat{A})g_2\|^2\, dt < \infty.$$

Since the vector-functions $\omega_2(t, \hat{A})g_2$ and $\omega_1(t, \hat{A})g_1$ are continuous on $[a, \frac{a+b}{2}]$ and $[\frac{a+b}{2}, b]$, respectively, we have

$$\omega_2(t, \hat{A})g_2 \in L_2(\mathfrak{H}, (a, \frac{a+b}{2}), \gamma(t)\, dt),\ \ \omega_1(t, \hat{A})g_1 \in L_2(\mathfrak{H}, (\frac{a+b}{2}, b), \gamma(t)\, dt),$$

and

$$\omega_1(t, \hat{A})g_1 \in L_2(\mathfrak{H}, (a, \frac{a+b}{2}), \gamma(t)\, dt),\ \ \omega_2(t, \hat{A})g_2 \in L_2(\mathfrak{H}, (\frac{a+b}{2}, b), \gamma(t)\, dt).$$

Convergence of the integrals $\int_a^b \gamma(t)\|\omega_1(t, \hat{A})g_1\|^2\, dt$ and $\int_a^b \gamma(t)\|\omega_2(t, \hat{A})g_2\|^2\, dt$ follows from this; it is equivalent to $g_1 \in \mathfrak{H}'_{G(a,\lambda)}$, $g_2 \in \mathfrak{H}'_{G(b,\lambda)}$.

By Banach's theorem the mapping (5.23) is a topological isomorphism between K_γ and $\mathfrak{H}'_{G(a,\lambda)} \oplus \mathfrak{H}'_{G(b,\lambda)}$.

The first assertion of the theorem is a direct consequence of Remark 5.1. □

Let now a function $\gamma(t)$ be given on a certain neighbourhood $[a, a+\delta]$ $(0 < \delta < \frac{b-a}{2})$ of the point a. Suppose that $\gamma(t)$ is continuous on this interval and that it also satisfies the conditions 1°, 2° from subsection 3.3. Set

$$G(\lambda) = \Big(\int_a^{a+\delta} \gamma(t) e^{-2\lambda t}\, dt\Big)^{-\frac{1}{2}}.$$

From $G(\lambda)$ we construct the spaces $\mathfrak{H}'_{\{G\}} = \operatorname{limpr}_{\alpha\to 0} \mathfrak{H}'_{G_\alpha}$, $\mathfrak{H}'_{(G)} = \operatorname{limind}_{\alpha\to\infty} \mathfrak{H}'_{G_\alpha}$, where $G_\alpha(\lambda) = G(\alpha\lambda)$.

THEOREM 5.7. *For the boundary value at the point a of a solution $y(t)$ inside (a, b) of equation (5.1) to belong to $\mathfrak{H}'_{\{G\}}(A)$ (respectively, $\mathfrak{H}'_{(G)}(A)$) it is necessary and sufficient that for any $\alpha > 0$ there exists a constant $c = c(\alpha) > 0$ (respectively,* *there exist $\alpha, c > 0$) such that*

$$\|y(t)\|^2 \leq c^{\gamma-1}(\frac{t}{\alpha})$$

if t is close enough to the point a.

Proof. Let $y(t)$ be a solution inside (a,b) of equation (5.1). Then it can be represented in the form (5.9). Since the vector-function $\omega_2(t, \hat{A})g_2$ is continuous on $[a, a+\delta]$ into $\mathfrak{H}$, the existence of the boundary value $y(a)$ of $y(t)$ at the point a in $\mathfrak{H}'_{\{G\}}$ is equivalent to $g_1 \in \mathfrak{H}'_{\{G\}}$. The inequalities (5.21) together with Theorem 3.5 allow us to complete the proof. □

A similar theorem is valid for the boundary value $y(b)$ of a solution $y(t)$ at the point b.

The following statements arise from Theorems 5.3, 3.2, 3.4, 3.6 and estimates (5.21).

COROLLARY 5.3. *A solution $y(t)$ inside (a,b) of equation (5.1) has boundary values $y(a)$ and $y(b)$ in the space $\mathfrak{H}$ if and only if*

$$\|y(t)\| \leq c \quad (c = \text{const}).$$

The boundary values of $y(t)$ belong to the space $\mathfrak{H}^{-\frac{1}{2}}(A)$ if and only if $y \in L_2(\mathfrak{H}, (a,b))$. In order that $y(a) \in \mathfrak{H}^{-\infty}(A)$ (respectively, $y(b) \in \mathfrak{H}^{-\infty}(A)$) it is necessary and sufficient that for some $\alpha > 0$

$$\|y(t)\| \leq c(t-a)^{-\alpha} \quad (\text{respectively, } \|y(t)\| \leq c(b-t)^{-\alpha}) \quad (c = \text{const})$$

near the point a (respectively, near the point b). More precisely, $y(a)$ (respectively, $y(b)$) exists in the space $\mathfrak{H}^{-\alpha}(A)$ $(\alpha > 0)$ if and only if

$$\int_a^b (t-a)^{2\alpha-1}\|y(t)\|^2\,dt < \infty \quad (\text{respectively } \int_a^b (b-t)^{2\alpha-1}\|y(t)\|^2\,dt < \infty).$$

COROLLARY 5.4. *For the boundary value $y(a)$ of a solution $y(t)$ inside (a,b) of equation (5.1) to belong to the space $\mathfrak{G}'_{\{\beta\}}(A)$ (respectively, $\mathfrak{G}'_{(\beta)}(A)$), where $\mathfrak{G}_{\{\beta\}}(A)$ (respectively, $\mathfrak{G}_{(\beta)}(A)$) is the Gevrey class of order $\beta > 1$ of Roumieu (respectively, Beurling) type, it is necessary and sufficient that for any $\alpha > 0$ there exists a constant $c = c(\alpha) > 0$ (respectively, there exist $\alpha, c > 0$) such that*

$$\|y(t)\| \leq c\ \exp(\alpha(t-a)^{-q}), \quad q = \frac{1}{\beta - 1}.$$

Analogously, the estimate

$$\|y(t)\| \leq c\ \exp(\alpha(b-t)^{-q})$$

should be satisfied near the end b.

REMARK 5.3. Everywhere in this section we considered the case that the operator A in equation (5.1) is arbitrary self-adjoint and non-negative. Certainly, the theory is of interest only if A is unbounded. Otherwise all positive and negative spaces constructed from A coincide with $\mathfrak{H}$ and the operator-functions $\omega_i(t, \hat{A})$ $(i = 1, 2)$ are infinitely differentiable on the closed interval $[a, b]$. So, any solution inside the interval (a, b) of equation (5.1) with a bounded A is infinitely differentiable on $[a, b]$ into $\mathfrak{H}$, and among all the theorems proved only Theorem 5.5 is non-trivial for this situation. It now reads: *any generalized solution of equation* (5.1) *with a bounded operator A is infinitely differentiable on the interval $[a, b]$ into $\mathfrak{H}$.*

6. Boundary Values of Periodic Harmonic Functions

In this section the results of the previous one are applied to a concrete situation. Here $\mathfrak{H} = L_2(0, 2\pi)$ and A^2 is the self-adjoint operator generated by the expression $-\frac{d^2}{dx^2}$ and the conditions $y(0) = y(2\pi)$, $y'(0) = y'(2\pi)$, $A = D = \sqrt{A^2}$. Then equation (5.1) becomes the Laplace equation

$$(\Delta u)(x, t) = \frac{\partial^2 u(x, t)}{\partial t^2} + \frac{\partial^2 u(x, t)}{\partial x^2} = 0 \quad (x \in \mathbb{R}^1,\ t > 0).$$

Its solution $u(x, t)$ is a function, 2π-periodic in the variable x and harmonic in the upper half-plane. In this situation the space $\mathfrak{H}_G(D)$ corresponding to a function $G(\lambda)$ is denoted by $(L_2)_G$.

6.1. THE SPACE OF ANALYTIC VECTORS

In this case the space $(L_2)^\alpha = (L_2)_{\lambda^\alpha + 1}$ ($\alpha \geq 0$ is an integer) coincides with the completion of the set of all infinitely differentiable 2π-periodic functions with respect to the norm of the Sobolev space $W_2^\alpha(0, 2\pi)$. It is the set of all $(\alpha - 1)$ times continuously differentiable 2π-periodic functions whose α-th derivative in the sense of the theory of distributions is square integrable over the interval of length 2π.

The space $(L_2)^\infty = \operatorname{limpr}_{\alpha \to \infty} W_2^\alpha(0, 2\pi) = W_2^\infty(0, 2\pi)$ coincides with the set of infinitely differentiable 2π-periodic functions. Convergence of a sequence $\{f_n(x)\}_{n=1}^\infty$ in it means uniform convergence of $\{f_n^{(k)}(x)\}_{n=1}^\infty$ for all $k = 0, 1, \ldots$.

LEMMA 6.1. *The space $(L_2)_{e,\delta} = (L_2)_{e^{\delta\lambda}}$ ($\delta > 0$) coincides with the set of the 2π-periodic functions $f(x)$ that have an analytic continuation as a 2π-periodic in x analytic function $f(x + it)$ to the strip $|t| < \delta$. For every such function*

$$\tfrac{1}{2}\|f\|^2_{(L_2)_{e,\delta}} \leq \sup_{|t| < \delta} \int_0^{2\pi} |f(x) + it)|^2 \, dx \leq \|f\|^2_{(L_2)_{e,\delta}}. \tag{6.1}$$

Proof. Let $f \in (L_2)_{e,\delta}$ ($\delta > 0$). Taking into account that under the isometric mapping $f \to \{c_k(f)\}_{k=-\infty}^\infty$, $c_k(f) = \frac{1}{\sqrt{2\pi}} \int_0^{2\pi} f(x) e^{-ikx} \, dx$, of the space $L_2(0, 2\pi)$ onto l_2 the operator D is transformed to multiplication by $|k|$, we have

$$\|f\|^2_{(L_2)_{e,\delta}} = \sum_{k=-\infty}^{\infty} |c_k(f)|^2 e^{2\delta|k|} < \infty.$$

This implies, by the Cauchy–Buniakowski inequality, that the series

$\frac{1}{\sqrt{2\pi}}\sum_{k=-\infty}^{\infty} c_k(f)e^{ik(x+it)}$ converges uniformly for $|t| \leq \delta'$ (δ' is an arbitrary number less than δ) and defines an analytic continuation $f(x+it)$ of $f(x)$ to the strip $|t| < \delta$.

By applying the Parseval equality to the expansion $f(x+it) = \frac{1}{\sqrt{2\pi}}\sum_{k=-\infty}^{\infty} c_k(f)e^{-kt}e^{ikx}$ ($|t| < \delta$) we find the relations

$$\int_0^{2\pi} |f(x+it)|^2\,dx = \sum_{k=-\infty}^{\infty} |c_k(f)|^2 e^{-2kt} \leq \sum_{k=-\infty}^{\infty} |c_k(f)|^2 e^{2|k|\delta} = \|f\|^2_{(L_2)_{e,\delta}},$$

which imply the right-hand inequality in (6.1).

Conversely, if $f(x)$ has a 2π-periodic (in x) analytic continuation $f(x+it)$ to a certain strip $|t| < \delta$, then on account of the Cauchy theorem and the periodicity of $f(x+it)$,

$$c_k(f) = \frac{1}{\sqrt{2\pi}}\int_0^{2\pi} f(x)e^{-ikx}\,dx = -\frac{1}{\sqrt{2\pi}}\int_0^{2\pi} f(x+it)e^{-ik(x+it)}\,dx \quad (|t| < \delta),$$

whence

$$e^{-kt}c_k(f) = -\frac{1}{\sqrt{2\pi}}\int_0^{2\pi} f(x+it)e^{-ikx}\,dx.$$

Applying the Parseval equality to $f(x \pm it)$ ($|t| < \delta$) we conclude that

$$\sum_{k=-\infty}^{\infty} e^{2|k|t}|c_k(f)|^2 \leq \int_0^{2\pi} |f(x+it)|^2\,dx + \int_0^{2\pi} |f(x-it)|^2\,dx \leq 2\sup_{|t|<\delta}\int_0^{2\pi} |f(x+it)|^2\,dx.$$

After limit transition as $t \to \delta$ we obtain the left-hand inequality in (6.1), and simultaneously the membership $f \in (L_2)_{e,\delta}$. □

This lemma and the definition of $\mathfrak{A}$ as the inductive limit of the spaces $(L_2)_{e,\delta}$ lead to the following result.

THEOREM 6.1. *In the periodic case the space $\mathfrak{A}$ of analytic vectors of the operator D is the set of all 2π-periodic functions which have a 2π-periodic in x analytic continuation $f(x+it)$ to a certain strip containing the real axis.*

Because of the estimates (6.1) we can show that convergence of a sequence $\{f_n(x)\}_{n=1}^{\infty}$ in $\mathfrak{A}$ means that all functions $f_n(x)$ admit 2π-periodic analytic continuations to one and the same strip, in which they converge uniformly.

6.2. FORMAL TRIGONOMETRIC SERIES AND PERIODIC GENERALIZED FUNCTIONS

We denote by $\mathcal{T}$ the set of all trigonometric polynomials

$$P(x) = \sum_{k=-n}^{n} c_k(P)e^{ikx} \quad (x \in \mathbb{R}^1;\ n = 0, 1, \ldots)$$

with coefficients in the field of complex numbers. It is clear that $\mathcal{T}$ is a vector space with respect to the usual addition of polynomials and multiplication of a polynomial by a number.

Let $\mathcal{T}^{(m)}$ $(m = 0, 1, \ldots)$ be the set of all polynomials in $\mathcal{T}$ whose degree (order) does not exceed m. Of course $\mathcal{T} = \bigcup_m \mathcal{T}^{(m)}$. We say that a sequence $P_n \in \mathcal{T}$ converges to P in $\mathcal{T}$ (and write $P_n \xrightarrow{\mathcal{T}} P$) if $P_n \in \mathcal{T}^{(m)}$ for a certain m and $c_k(P_n) \to c_k(P)$, $n \to \infty$, for every k. The space $\mathcal{T}$ is complete relative to this convergence. (Notice that this convergence coincides with convergence in the space $\mathcal{T}$ as the regular inductive limit of the spaces $\mathcal{T}^{(m)}$ equipped with the $C[0, 2\pi]$-topology: $\mathcal{T} = \operatorname{limind}_{m\to\infty} \mathcal{T}^{(m)}$.) The differentiation operation, multiplication and convolution

$$(P * Q)(x) = \frac{1}{2\pi} \int_0^{2\pi} P(s)Q(x - s)\, \mathrm{d}s,$$

$$c_k(P * Q) = c_k(P)c_k(Q)$$

are continuous with respect to it.

Let $\mathcal{T}'$ be the space af all continuous antilinear functionals on $\mathcal{T}$: $\mathcal{T}' = \operatorname{limpr}_{m\to\infty}(\mathcal{T}^{(m)})'$. A sequence $f_n \in \mathcal{T}'$ converges to f in $\mathcal{T}'$ (we write $f_n \xrightarrow{\mathcal{T}'} f$) if $< f_n, P > \to < f, P >$, $n \to \infty$, for an arbitrary $P \in \mathcal{T}$ ($< f, P >$ denotes the action of a functional $f \in \mathcal{T}'$ on an element $P \in \mathcal{T}$). We will call the elements of $\mathcal{T}'$ 2π-periodic generalized functions.

The differentiation operation on the space $\mathcal{T}'$ is defined as follows: the k-th derivative of an element $f \in \mathcal{T}'$ is the unique element $f^{(k)} \in \mathcal{T}'$ for which

$$< f^{(k)}, P > = (-1)^k < f, P^{(k)} > \quad (\forall P \in \mathcal{T}).$$

This operation is continuous on $\mathcal{T}'$. Thus, each element $f \in \mathcal{T}'$ is infinitely differentiable in $\mathcal{T}'$.

The convolution

$$< f * g, P > = < f, \overline{< g_s, \overline{P(s + x)} >} > \quad (f, g \in \mathcal{T}',\ \forall P \in \mathcal{T})$$

(the subscript s in g_s denotes the action of the functional g on $P(s + x)$ in the variable s) is also defined on $\mathcal{T}$. The given definition makes sense because

$$< g_s, P(s + x) > = < g_s, \sum_{k=-n}^{n} c_k(P)e^{ik(s+x)} > = \sum_{k=-n}^{n} c_k(P) < g, e^{iks} > e^{ikx}$$

belongs to $\mathcal{T}$. The operation "$*$" is continuous on $\mathcal{T}'$ in the sense that if $f_n \xrightarrow{\mathcal{T}'} f$ $(n \to \infty)$, then $f_n * g \xrightarrow{\mathcal{T}'} f * g$ $(n \to \infty,\ \forall g \in \mathcal{T}')$. This follows from the definition of the convolution. Moreover,

$$(f * g)^{(k)} = f^{(k)} * g = f * g^{(k)}$$

for arbitrary $f, g \in \mathcal{T}'$.

We would like to note that the space $\mathcal{T}$ is continuously imbedded in $\mathcal{T}'$: if $Q \in \mathcal{T}$, then the element $f_Q \in \mathcal{T}'$ corresponding to Q is defined as

$$< f_Q, P > = \frac{1}{2\pi} \int_0^{2\pi} Q(x)\overline{P(x)}\, dx \quad (\forall P \in \mathcal{T}).$$

This imbedding can clearly be continuously extended to the space $C[0, 2\pi]$ of 2π-periodic continuous functions with the topology of uniform convergence and to the space $L_1(0, 2\pi)$ of all 2π-periodic functions that are integrable over an interval of length 2π. Of course, the differentiation operation is preserved. This means that if $f \in L_1(0, 2\pi)$ is absolutely continuous, then its generalized derivative coincides with the ordinary derivative. Moreover, on the functions in $L_1(0, 2\pi)$ the operation "$*$" on $\mathcal{T}'$ coincides with ordinary convolution.

We now consider the set C^∞ of all sequences $\{c_k\}_{k=-\infty}^{\infty}$ of complex numbers. C^∞ forms a vector space over the field $\mathbb{C}^1$ with respect to the ususal addition of sequences and multiplication of a sequence by a number. Convergence in C^∞ is introduced as follows: a sequence $c_n = \{c_{n,k}\}_{k=-\infty}^{\infty}$ converges to $\{c_k\}_{k=-\infty}^{\infty}$ $(n \to \infty)$ in C^∞ if $c_{n,k} \to c_k \quad (n \to \infty)$ for every k. The space C^∞ with this notion of convergence is complete.

On $\mathcal{T}'$ we define the mapping F by

$$\mathcal{T}' \ni f \to Ff = \{c_k(f) = < f, e^{ikx} >\}_{k=-\infty}^{\infty} \in C^\infty.$$

This is a bijection. In fact, if $f_1 \neq f_2$ $(f_1, f_2 \in \mathcal{T}')$, then there exists a k such that $c_k(f_1) \neq c_k(f_2)$. Otherwise the equalities $< f_1 - f_2, e^{ikx} > = 0 \quad (k = 0, \pm 1, \ldots)$ would imply $< f_1 - f_2, P > = 0 \quad (\forall P \in \mathcal{T})$, which contradicts $f_1 \neq f_2$. Thus, the mapping F is an injection. Let now $\{c_k\}_{k=-\infty}^{\infty} \in C^\infty$. We define a functional f as follows:

$$< f, e^{ikx} > = c_k, \quad < f, \sum_{k=-n}^{n} \alpha_k e^{ikx} > = \sum_{k=-n}^{n} \overline{\alpha}_k c_k.$$

It is understood that this functional is antilinear and continuous, hence $f \in \mathcal{T}'$. Also, convergence of a sequence in $\mathcal{T}'$ is equivalent to coordinatewise convergence of the corresponding sequence in C^∞.

Thus, the mapping F is one-to-one and bi-continuous from $\mathcal{T}'$ onto C^∞. The completeness of $\mathcal{T}'$ relative to convergence in it follows from this.

The bijection F maps the space $\mathcal{T}$ onto the set C_0^∞ of all sequences with a finite number of non-vanishing terms. The differentiation operation is transformed under it to multiplication by ik : $c_k(f') = < f', e^{ikx} > = - < f, e^{ikx} > = ikc_k(f)$, while coordinatewise multiplication corresponds to convolution:

$$c_k(f * g) = < f * g, e^{ikx} > = < f, < g_s, \overline{e^{ik(s+x)}} >>$$
$$= < f, < \overline{g, e^{iks}} > e^{ikx} > = c_k(f)c_k(g).$$

This results in the commutativity and associativity of the convolution on $\mathcal{T}'$.

So, $\mathcal{T}'$ is an algebra with the Dirac δ-function δ, $<\delta, P> = P(0) \quad (\forall P \in \mathcal{T})$, as unit; the image of δ under F in C^∞ is the sequence $\{1\}_{k=-\infty}^{\infty}$.

The series $\sum_{k=-\infty}^{\infty} c_k(f)e^{ikx}$, where $c_k(f) = < f, e^{ikx} >$ are the so-called Fourier coefficients of the function $f \in \mathcal{T}'$, is called the Fourier series of f. If $f \in L_1(0, 2\pi)$, then this series coincides with its ordinary Fourier series.

THEOREM 6.2. *The Fourier series of any 2π-periodic generalized function f converges to f in the space $\mathcal{T}'$. Conversely, the sequence of partial sums of any trigonometric series $\sum_{k=-\infty}^{\infty} c_k e^{ikx}$ converges in $\mathcal{T}'$ to a certain element $f \in \mathcal{T}'$, and the initial series is the Fourier series of f.*

Proof. Let $f \in \mathcal{T}'$ and let $c_k(f)$ be the Fourier coefficients of f. The mapping F transforms the n-th partial sum $S_n(f, x)$ of the Fourier series of f to the sequence $s_n = \{\ldots, 0, c_{-n}(f), \ldots, c_0(f), \ldots, c_n(f), 0 \ldots\}$ which converges in C^∞ to $\{c_k(f)\}_{k=-\infty}^{\infty} = Ff$, whence it follows that $S_n(f, x) \xrightarrow{\mathcal{T}'} f$. Analogously, the sequence of partial sums of an arbitrary trigonometric series $\sum_{k=-\infty}^{\infty} c_k e^{ikx}$ tends in $\mathcal{T}'$ to the generalized element f for which $Ff = \{c_k\}_{k=-\infty}^{\infty} = \{< f, e^{ikx} >\}_{k=-\infty}^{\infty}$. □

COROLLARY 6.1. *The space $\mathcal{T}$ is dense in $\mathcal{T}'$.*

6.3. A CHARACTERIZATION OF TEST FUNCTIONS IN TERMS OF THEIR FOURIER COEFFICIENTS

The spaces $(L_2)^\infty$ and $\mathfrak{A}$ of infinitely differentiable and analytic vectors are, in the situation considered, described as follows.

THEOREM 6.3. *A 2π-periodic function f belongs to the space $(L_2)^\infty$ if and only if for an arbitrary integer $m \geq 0$ there exists a constant $c = c(m) > 0$ such that*

$$|c_k(f)| \leq c|k|^{-m} \quad (k = 0, \pm 1, \ldots). \tag{6.2}$$

Proof. Since $(L_2)^\infty = W_2^\infty(0, 2\pi) = \operatorname{limpr}_{n\in\mathbb{N}} W_2^n(0, 2\pi)$, the membership $f \in (L_2)^\infty$ is equivalent to $f \in W_2^m(0, 2\pi)$ for any $m \geq 0$ or, which is the same, to the inequality

$$\sum_{k=-\infty}^{\infty} (1 + |k|^m)^2 |c_k(f)|^2 < \infty.$$

Because m is arbitrary the last inequality is equivalent to (6.2). □

THEOREM 6.4. *A 2π-periodic function f belongs to the space $\mathfrak{A}$ if and only if there exist constants $c > 0, \delta > 0$ such that*

$$|c_k(f)| \leq ce^{-\delta|k|} \quad (k = 0, \pm 1, \ldots) \tag{6.3}$$

Proof. Since $\mathfrak{A} = \operatorname{limind}_{\delta\to 0}(L_2)_{e,\delta}$ the membership of f to $\mathfrak{A}$ is equivalent to $f \in (L_2)_{e,\delta}$ for some $\delta > 0$, whence

$$\sum_{k=-\infty}^{\infty} |c_k(f)|^2 e^{2\delta|k|} < \infty.$$

This implies the estimate (6.3). Conversely, if the condition (6.3) with certain constants $c > 0, \delta > 0$ is satisfied for $f \in \mathcal{T}'$, then $\sum_{k=-\infty}^{\infty} |c_k(f)|^2 e^{2\delta'|k|} < \infty$ for any $\delta' < \delta$. It follows that $f \in (L_2)_{e,\delta'}$, hence $f \in \mathfrak{A}$. □

Recall that in the case under consideration the Gevrey class $\mathfrak{G}_{\{\beta\}}$ $(\mathfrak{G}_{(\beta)})$ $(\beta > 1)$ of Roumieu (Beurling) type consists of those infinitely differentiable 2π-periodic functions f that satisfy the inequality

$$|f^{(n)}(x)| \leq c\alpha^n n^{n\beta} \quad (n = 0, 1, \ldots) \tag{6.4}$$

with some constants $c > 0, \alpha > 0$ depending only on the function (for any $\alpha > 0$ there exists $c = c(\alpha, f)$). In this concrete situation the elements of $\mathfrak{G}_{\{\beta\}}$ and $\mathfrak{G}_{(\beta)}$ are called 2π-periodic ultradifferentiable functions of order β of the corresponding type. The classes of 2π-periodic ultradifferentiable functions are characterized by the following property.

THEOREM 6.5. *A 2π-periodic function f is ultradifferentiable of order β of Roumieu (Beurling) type if and only if there exist constants $c > 0, \mu > 0$ (for any $\mu > 0$ there exists a constant $c = c(\mu) > 0$) such that*

$$|c_k(f)| \leq c \, \exp(-\mu|k|^{\frac{1}{\beta}}) \quad (k = 0, \pm 1, \ldots). \tag{6.5}$$

Proof. As was shown in subsection 2.6,

$$\mathfrak{G}_{\{\beta\}} = (L_2)_{\{G\}} = \operatorname*{limind}_{\alpha\to 0} (L_2)_{G_\alpha}, \quad \mathfrak{G}_{(\beta)} = (L_2)_{(G)} = \operatorname*{limpr}_{\alpha\to\infty} (L_2)_{G_\alpha},$$

where $G(\lambda) = \exp(\lambda^{\frac{1}{\beta}})$. Let $f \in \mathfrak{G}_{\{\beta\}}$ $(f \in \mathfrak{G}_{(\beta)})$. Since the isometric mapping $f \to \{c_k(f)\}_{k=-\infty}^{\infty}$ from $L_2(0, 2\pi)$ onto l_2 transforms the operator D to multiplication by $|k|$, we have

$$\|f\|^2_{(L_2)_{G_\alpha}} = \sum_{k=-\infty}^{\infty} |c_k(f)|^2 \exp(2\alpha^{\frac{1}{\beta}}|k|^{\frac{1}{\beta}}) < \infty \tag{6.6}$$

for some (any) $\alpha > 0$. Inequality (6.6) implies the estimate (6.5) with $\mu = \alpha^{\frac{1}{\beta}}$.

Conversely, suppose that for the Fourier coefficients of a 2π-periodic function f the estimate (6.5) is valid with some (any) $\mu > 0$. In view of Theorem 6.3, the function f is infinitely differentiable and for an arbitrary ϵ, $0 < \epsilon < \mu$,

$$\|f\|_{(L_2)_{G_{(\mu-\epsilon)^\beta}}} = \sum_{k=-\infty}^{\infty} |c_k(f)|^2 \exp(2(\mu-\epsilon)|k|^{\frac{1}{\beta}}) < \infty,$$

i.e. $f \in (L_2)_{G_\alpha}$, where $\alpha = (\mu-\epsilon)^\beta$. In fact, this means that $f \in \mathfrak{G}_{\{\beta\}}$ $(f \in \mathfrak{G}_{(\beta)}$. □

6.4. A CHARACTERIZATION OF CLASSES OF GENERALIZED FUNCTIONS IN TERMS OF THEIR FOURIER COEFFICIENTS

We now turn to the spaces $(L_2)^{-\infty} = W_2^{-\infty}$, $\mathfrak{A}'$, $\mathfrak{G}'_{\{\beta\}}$, $\mathfrak{G}'_{(\beta)}$, the duals of the test spaces $(L_2)^\infty$, W_2^∞, $\mathfrak{A}$, $\mathfrak{G}_{\{\beta\}}$, $\mathfrak{G}_{(\beta)}$. Their elements are usually called 2π-periodic distributions, hyperfunctions, and ultradistributions of order β of Roumieu and Beurling type, respectively. In terms of formal trigonometric series these spaces can be characterized in the following way.

THEOREM 6.6. *For a trigonometric series $\sum_{k=-\infty}^{\infty} c_k e^{ikx}$ to be the Fourier series of a 2π-periodic distribution f it is necessary and sufficient that there exist a constant $c > 0$ and an integer $m > 0$ such that*

$$|c_k| \le c|k|^m \quad (k = 0, \pm 1, \ldots). \tag{6.7}$$

Proof. Since $W_2^{-\infty}(0,2\pi) = \operatorname{limind}_{n\to\infty} W_2^{-n}(0,2\pi)$, the membership $f \in W_2^{-\infty}(0,2\pi)$ implies the existence of an integer $m > 0$ such that $f \in W_2^{-m}(0,2\pi)$, hence

$$\|f\|^2_{W_2^{-m}} = \sum_{k=-\infty}^{\infty} (1+|k|^m)^{-2} |c_k(f)|^2 < \infty.$$

Inequality (6.7) arises from here.

Conversely, if for the coefficients of a series $\sum_{k=-\infty}^{\infty} c_k e^{ikx}$ the estimate (6.7) is valid, then for the 2π-periodic generalized function $f(x) = \sum_{k=-\infty}^{\infty} c_k e^{ikx}$ we have

$$\sum_{k=-\infty}^{\infty} (1+|k|^{m+1})^2 |c_k|^2 < \infty.$$

Consequently, $f \in W_2^{-(m+1)}(0,2\pi)$, hence $f \in W_2^{-\infty}(0,2\pi)$. □

THEOREM 6.7. *A 2π-periodic generalized function f belongs to the space $\mathfrak{G}'_{\{\beta\}}$ ($\mathfrak{G}'_{(\beta)}$) of ultradistributions if and only if for any $\mu > 0$ there exists a constant $c = c(\mu) > 0$ (there exist constants $\mu > 0, c > 0$) such that*

$$|c_k(f)| \le c \exp(\mu|k|^{\frac{1}{\beta}}) \quad (k = 0, \pm 1, \ldots). \tag{6.8}$$

Proof. Let $f \in \mathfrak{G}'_{\{\beta\}} = \operatorname{limpr}_{\alpha\to 0}(L_2)'_{G_\alpha}$ $(f \in \mathfrak{G}'_{(\beta)} = \operatorname{limind}_{\alpha\to\infty}(L_2)'_{G_\alpha})$, where

$G(\lambda) = \exp(\lambda^{\frac{1}{\beta}})$. Then $f \in (L_2)'_{G_\alpha}$ for any (some) α. Therefore

$$\|f\|^2_{(L_2)'_{G_\alpha}} = \sum_{k=-\infty}^{\infty} \exp(-2\alpha^{\frac{1}{\beta}}|k|^{\frac{1}{\beta}})|c_k(f)|^2 < \infty,$$

whence we obtain (6.8) with $\mu = \alpha^{\frac{1}{\beta}}$.

Conversely, if for $f \in \mathcal{T}'$ (6.8) is satisfied with any (some) μ, then

$$\sum_{k=-\infty}^{\infty} |c_k(f)|^2 \exp(-2(\mu+\epsilon)|k|^{\frac{1}{\beta}}) < \infty,$$

that is, $f \in (L_2)'_{G_\alpha}$ for any (some) $\alpha = (\mu+\epsilon)^\beta$. This implies that $f \in \mathfrak{G}'_{\{\beta\}}$ ($f \in \mathfrak{G}'_{(\beta)}$). □

The next corollary is an immediate consequence of Theorem 6.7 (the Roumieu case, $\beta = 1$).

COROLLARY 6.2. *For a trigonometric series $\sum_{k=-\infty}^{\infty} c_k e^{ikx}$ to be the Fourier series of a 2π-periodic hyperfunction it is necessary and sufficient that for any $\mu > 0$ there exists a constant $c = c(\mu) > 0$ such that*

$$|c_k| \le c \, \exp(\mu|k|) \quad (k = 0, \pm 1, \ldots).$$

REMARK 6.1. Employing the estimates for the Fourier coefficients of 2π-periodic generalized functions from various classes and the Parseval equality, it is possible to obtain a characterization of elements from these classes in terms of the growth of the partial sums $S_n(f,x)$ of their Fourier series. For example, the class of 2π-periodic distributions is characterized by the estimate $|S_n(f,x)| \le cn^m$ ($c = \text{const}; n = 0, 1, \ldots$) with some natural m, while the class $\mathfrak{G}'_{\{\beta\}}$ ($\mathfrak{G}'_{(\beta)}$) is characterized by the inequality $|S_n(f,x)| \le c \, \exp(\mu n^{\frac{1}{\beta}})$ ($n = 0, 1, \ldots$) for any (some) $\mu > 0$.

6.5. BOUNDARY VALUES

Theorem 5.3 and the conclusions made on the basis of it show that any function $u(t,x)$ that is 2π-periodic in x and harmonic in the upper half-plane has a boundary value as t approaches the real axis in the class of periodic hyperfunctions. This boundary value can be described more precisely in dependence on the growth of $u(x,t)$ as $t \to 0$ by the following statements.

THEOREM 6.8. *Let $u(x,t)$ be a function that is 2π-periodic in x and harmonic in the upper half-plane. In order that its boundary value as t approaches the real axis exists in the space $L_2(0,2\pi)$ it is necessary and sufficient that*

$$\int_0^{2\pi} |u(x,t)|^2 \, dx \le c \quad (c = \text{const})$$

for sufficiently small t.

Proof. The proof is a direct consequence of Corollary 5.3. □

THEOREM 6.9. *The boundary value of a function that is 2π-periodic in x and harmonic in the upper half-plane as t approaches the real axis is a 2π-periodic distribution if and only if there exist constants $c > 0, \alpha > 0$ such that*

$$|u(x,t)| \le ct^{-\alpha} \tag{6.9}$$

(t *small enough*). *More exactly, this boundary value belongs to the space $W_2^{-\alpha}(0, 2\pi)$ if and only if*

$$\int_0^b \int_0^{2\pi} t^{2\alpha-1} |u(x,t)|^2 \, dx \, dt < \infty \tag{6.10}$$

for some $b > 0$.

Proof. The second assertion is an immediate consequence of Corollary 5.3. The first one also follows from this corollary. It will be established if we prove that the estimate

$$\int_0^{2\pi} |u(x,t)|^2 \, dx \le c_1 t^{-\alpha'} \tag{6.11}$$

implies an inequality like (6.9) (the fact that (6.9) yields the estimate (6.11) with $\alpha' = 2\alpha$ is obvious).

According to the mean value theorem for harmonic functions,

$$u(x,t) = \frac{1}{2\pi\rho} \int_0^{2\pi} u(x + \rho\cos\varphi, t + \rho\sin\varphi) \, d\varphi \quad (0 < \rho < \frac{t}{2}).$$

Multiplication of this equality by ρ^2 and subsequent integration over the interval $[0, \frac{t}{2}]$ yields

$$\frac{t^3}{24} u(x,t) = \frac{1}{2\pi} \int_{K_{\frac{t}{2}}} u(\xi, \eta) \, d\xi \, d\eta$$

where $K_{\frac{t}{2}}$ is the disc of radius $\frac{t}{2}$ with centre at (x,t). Applying now the Cauchy–Buniakowski inequality and integrating over the rectangle $\Pi_t = [0, 2\pi] \times [\frac{t}{2}, \frac{3t}{2}]$ instead of over $K_{\frac{t}{2}}$, we arrive at the relations

$$\begin{aligned} |u(x,t)|^2 &\le c_2 t^{-4} \int_{K_{\frac{t}{2}}} |u(\xi,\eta)|^2 \, d\xi \, d\eta \le c_2 t^{-4} \int_{\Pi_t} |u(\xi,\eta)|^2 \, d\xi \, d\eta \\ &= c_2 t^{-4} \int_{\frac{t}{2}}^{\frac{3t}{2}} d\eta \int_0^{2\pi} |u(\xi,\eta)|^2 \, d\xi \le c_3 t^{-4} \int_{\frac{t}{2}}^{\frac{3t}{2}} \eta^{-\alpha'} \, d\eta = ct^{-\alpha'-3}, \end{aligned}$$

where c_2, c_3, c are positive constants. Thus, (6.9) with $\alpha = \frac{\alpha'+3}{2}$ is proved. □

THEOREM 6.10. *The boundary value of a function $u(x,t)$ that is 2π-periodic in x and harmonic in the upper half-plane as t approaches the real axis is an ultradistribution of order $\beta = \frac{q+1}{q}$ $(q > 0)$ of Roumieu (Beurling) type if and only if for* *any $\alpha > 0$ there exists a positive constant c (there exist constants $\alpha > 0, c > 0$) such that*

$$\sup_{x\in[0,2\pi]} |u(x,t)| \leq c \exp(\alpha t^{-q}) \tag{6.12}$$

for small $t > 0$.

The proof follows from Corollary 5.4 if one takes into account that the estimate (6.12) for small t is equivalent to the integral estimate

$$\int_0^{2\pi} \exp(-2\alpha' t^{-q})|u(x,t)|^2 \, dx \leq c \quad (c = \text{const})$$

with $\alpha' = \nu\alpha$ $(0 < \nu = \text{const})$. The equivalence is shown in exactly the same way as we have established the equivalence of (6.11) and (6.9) in the previous theorem. □

7. Boundary Values of Harmonic Functions in the Upper Half-Plane That are Square Integrable over Straight Lines Parallel to the Real Axis

In this section the results of section 5 are discussed for the case where $\mathfrak{H} = L_2(\mathbb{R}^1)$, A^2 is the self-adjoint operator that is the closure in $L_2(\mathbb{R}^1)$ of the operator $f(x) \to -f''(x)$ given on the set $C_0^\infty(\mathbb{R}^1)$ of infinitely differentiable functions with compact support, $A = \sqrt{A^2} = D$. Then equation (5.1) turns into the Laplace equation $\Delta u = 0$, while its solution $u(x,t)$ inside $(0,\infty)$ is harmonic in the half-plane $t > 0$ and square integrable over straight lines parallel to the real axis. As was mentioned in subsection 5.6, for such a function a boundary value always exists in the space of coanalytic vectors of the operator D. In this situation we will denote the space $\mathfrak{H}_G$ corresponding to the function $G(\lambda)$ by $(L_2)_G(D)$, or simply by $(L_2)_G$.

7.1. BOUNDARY VALUES OF HARMONIC FUNCTIONS IN THE SPACE $L_2(\mathbb{R}^1)$ AND IN SOBOLEV SPACES

Under the unitary mapping $f(x) \to \tilde{f}(\lambda) = \frac{1}{\sqrt{2\pi}} \int_{-\infty}^{\infty} f(x) e^{-i\lambda x} \, dx$ from the space $L_2(\mathbb{R}^1)$ onto $L_2(\mathbb{R}^1)$ the operator D is transformed to multiplication by $|\lambda|$. Therefore, in this case the space $(L_2)_G$ consists of those functions f for which

$$\|f\|^2_{(L_2)_G} = \int_{-\infty}^{\infty} G^2(|\lambda|)|\tilde{f}(\lambda)|^2 \, d\lambda < \infty.$$

In particular, $(L_2)^\alpha$ $(-\infty < \alpha < \infty)$ coincides with the Sobolev space $W_2^\alpha(\mathbb{R}^1)$. The space $(L_2)^\infty = W_2^\infty(\mathbb{R}^1)$ is the set of all infinitely differentiable functions on $\mathbb{R}^1$ all derivatives of which are square integrable; convergence in $W_2^\infty(\mathbb{R}^1)$ means convergence of functions and all their derivatives in $L_2(\mathbb{R}^1)$; $(L_2)^{-\infty} = W_2^{-\infty}(\mathbb{R}^1)$, the space of all continuous antilinear functionals on $W_2^\infty(\mathbb{R}^1)$, coincides with the space of Schwartz distributions, each of them being a derivative of finite order in the sense of the theory of distributions of a square integrable function on $\mathbb{R}^1$. Corollary 5.3 leads to the following result.

THEOREM 7.1. *In order that the boundary value (as t approaches the real axis) of a function $u(x,t)$ that is harmonic in the upper half-plane and square integrable in x over $\mathbb{R}^1$ exists in the space $L_2(\mathbb{R}^1)$ it is necessary and sufficient that*

$$\int_{-\infty}^{\infty} |u(x,t)|^2 \, dx \le c \quad (c = \text{const})$$

for sufficiently small $t > 0$. This boundary value exists in $W_2^{-\infty}(\mathbb{R}^1)$ if and only if

$$\int_{-\infty}^{\infty} |u(x,t)|^2 \, dx \le ct^{-\alpha} \quad (c = \text{const})$$

for some $\alpha > 0$ and sufficiently small $t > 0$. More precisely, $u(x,0)$ belongs to the space $W_2^{-\alpha}(\mathbb{R}^1)$ $(\alpha > 0)$ if and only if

$$\int_0^b \int_{-\infty}^{\infty} t^{2\alpha-1} |u(x,t)|^2 \, dx \, dt < \infty$$

for some $b > 0$ ($t > 0$ small enough).

7.2. THE FOURIER TRANSFORM OF A FUNCTION ANALYTIC IN A STRIP

Assume that a complex-valued function $f(z) = f(x+it)$ is analytic in the closed strip $|t| \le \delta$ and that

$$\sup_{|t|\le\delta} \int_{-\infty}^{\infty} |f(x+it)|^2 \, dx < \infty. \tag{7.1}$$

According to the Cauchy theorem,

$$f(z) = \frac{1}{2\pi i}\left(\int_{-\rho-i\delta}^{\rho-i\delta} + \int_{\rho-i\delta}^{\rho+i\delta} + \int_{\rho+i\delta}^{-\rho+i\delta} + \int_{-\rho+i\delta}^{-\rho-i\delta} \frac{f(\varsigma)}{\varsigma - z} \, d\varsigma\right)$$

for sufficiently large $\rho > 0$ and $|t| < \delta$. We integrate this equality with respect to ρ over the interval $[\beta, \beta+1]$ with a large β. We have

$$f(z) = \frac{1}{2\pi i}\int_\beta^{\beta+1} d\rho\left(\int_{-\rho-i\delta}^{\rho-i\delta} + \int_{\rho-i\delta}^{\rho+i\delta} + \int_{\rho+i\delta}^{-\rho+i\delta} + \int_{-\rho+i\delta}^{-\rho-i\delta} \frac{f(\varsigma)}{\varsigma - z} \, d\varsigma\right).$$

Further,

$$\begin{aligned}
\left|\frac{1}{2\pi i}\right| \left|\int_\beta^{\beta+1} d\rho \int_{\rho-i\delta}^{\rho+i\delta} \frac{f(\varsigma)}{\varsigma - z} \, d\varsigma\right| &= \frac{1}{2\pi}\left|\int_{-i\delta}^{i\delta} d\varsigma \int_\beta^{\beta+1} \frac{f(\varsigma+\rho)}{\varsigma+\rho-z} \, d\rho\right| \\
&\le \frac{1}{2\pi}\int_{-i\delta}^{i\delta} d\varsigma \left(\int_\beta^{\beta+1} |f(\varsigma+\rho)|^2 \, d\rho \int_\beta^{\beta+1} \frac{1}{|\varsigma+\rho-z|^2} \, d\rho\right)^{\frac{1}{2}} \\
&\le \frac{c}{2\pi}\int_{-\delta}^{\delta} dt \left(\int_\beta^{\beta+1} |f(s+it)|^2 \, ds\right)^{\frac{1}{2}}.
\end{aligned} \tag{7.2}$$

It follows from (7.1) that

$$\int_{\beta}^{\beta+1} |f(s+it)|^2\, ds \le c_1 = \text{const},$$

and

$$\lim_{\beta\to\infty} \int_{\beta}^{\beta+1} |f(s+it)|^2\, ds = 0.$$

By limit transition in (7.2) as $\beta \to \infty$ we obtain

$$\lim_{\beta\to\infty} \frac{1}{2\pi i} \int_{\beta}^{\beta+1} d\rho \int_{\rho-i\delta}^{\rho+i\delta} \frac{f(\varsigma)}{\varsigma - z}\, d\varsigma = 0.$$

Analogously,

$$\lim_{\beta\to\infty} \frac{1}{2\pi i} \int_{\beta}^{\beta+1} d\rho \int_{-\rho-i\delta}^{-\rho+i\delta} \frac{f(\varsigma)}{\varsigma - z}\, d\varsigma = 0.$$

Thus,

$$\begin{aligned}
f(z) &= \lim_{\beta\to\infty} \frac{1}{2\pi i} \int_{\beta}^{\beta+1} d\rho \Big(\int_{-\rho-i\delta}^{\rho-i\delta} + \int_{\rho+i\delta}^{-\rho+i\delta} \frac{f(\varsigma)}{\varsigma - z}\, d\varsigma \Big) \\
&= \lim_{\beta\to\infty} \frac{1}{2\pi i} \int_{\beta}^{\beta+1} d\rho \int_{-\rho}^{\rho} \Big(\frac{f(s-i\delta)}{s-i\delta-z} - \frac{f(s+i\delta)}{s+i\delta-z} \Big)\, ds \\
&= \lim_{\beta\to\infty} \frac{1}{2\pi i} \Big(\int_{-\beta-1}^{-\beta} \frac{(\beta+1+s) f(s-i\delta)}{s-i\delta-z}\, ds - \\
&\quad - \int_{-\beta-1}^{-\beta} \frac{(\beta+1+s) f(s+i\delta)}{s+i\delta-z}\, ds + \int_{-\beta}^{\beta} \frac{f(s-i\delta)}{s-i\delta-z}\, ds - \\
&\quad - \int_{-\beta}^{\beta} \frac{f(s+i\delta)}{s+i\delta-z}\, ds + \int_{\beta}^{\beta+1} \frac{(\beta+1-s) f(s-i\delta)}{s-i\delta-z}\, ds - \\
&\quad - \int_{\beta}^{\beta+1} \frac{(\beta+1-s) f(s+i\delta)}{s+i\delta-z}\, ds \Big).
\end{aligned} \tag{7.3}$$

Next,

$$\begin{aligned}
&\Big| \frac{1}{2\pi i} \int_{\beta}^{\beta+1} \frac{(\beta+1-s) f(s+i\delta)}{s+i\delta-z}\, ds \Big| \\
&\quad \le \frac{1}{2\pi} \Big(\int_{\beta}^{\beta+1} \Big| \frac{\beta+1-s}{s+i\delta-z} \Big|^2 ds \int_{\beta}^{\beta+1} |f(s+i\delta)|^2\, ds \Big)^{\frac{1}{2}} \\
&\quad \le c_2 \Big(\int_{\beta}^{\beta+1} |f(s+i\delta)|^2\, ds \Big)^{\frac{1}{2}} \to 0 \quad (\beta \to \infty).
\end{aligned}$$

A similar argument allows us to eliminate three more terms from (7.3). Consequently,

$$f(z) = \frac{1}{2\pi i} \int_{-\infty}^{\infty} \Big(\frac{f(s-i\delta)}{s-i\delta-z} - \frac{f(s+i\delta)}{s+i\delta-z} \Big)\, ds. \tag{7.4}$$

So, we have proved the following statement.

THEOREM 7.2. *Suppose that a function $f(z) = f(x+it)$ is analytic in the strip $|t| \le \delta$ and that the estimate (7.1) holds in it. Then formula (7.4) is valid for all z lying inside the strip.*

7.3. DESCRIPTION OF ANALYTIC VECTORS

First of all we will describe the spaces $(L_2)_{e,\delta}$ whose inductive limit forms the space of analytic vectors of the operator D.

LEMMA 7.1. *The space $(L_2)_{e,\delta}$ consists of those functions $f(x)$ ($x \in \mathbb{R}^1$) that have an analytic continuation to the strip $|t| < \delta$ such that*

$$\sup_{|t|<\delta} \int_{-\infty}^{\infty} |f(x+it)|^2 \, dx = c < \infty \quad (c = \text{const}), \tag{7.5}$$

and $\|f\|_{(L_2)_{e,\delta}}$ is equivalent to the norm $\left(\sup_{|t|<\delta} \int_{-\infty}^{\infty} |f(x+it)|^2 \, dx\right)^{\frac{1}{2}}$.

Proof. Let $f \in (L_2)_{e,\delta}$. From what we have said in subsection 7.1,

$$\int_{-\infty}^{\infty} e^{2\delta|\lambda|} |\tilde{f}(\lambda)|^2 \, d\lambda < \infty. \tag{7.6}$$

Relation (7.6) and the Cauchy–Buniakowski inequality imply the uniform convergence of the integrals $\int_{-\infty}^{\infty} \lambda^k e^{i\lambda(x+it)} \tilde{f}(\lambda) \, d\lambda$ ($k = 0, 1$) in any strip $|t| \le \delta' < \delta$. Hence, the function

$$f(x+it) = \frac{1}{\sqrt{2\pi}} \int_{-\infty}^{\infty} e^{i\lambda(x+it)} \tilde{f}(\lambda) \, d\lambda$$

is an analytic continuation of $f(x)$ to the strip $|t| < \delta$. Since $f(x+it)$ is the inverse Fourier transform of the function $e^{-t\lambda} \tilde{f}(\lambda)$, we have, taking into account Parseval's equality,

$$\begin{aligned}\int_{-\infty}^{\infty} |f(x+it)|^2 \, dx &= \int_{-\infty}^{\infty} e^{-2\lambda t} |\tilde{f}(\lambda)|^2 \, d\lambda \le \int_{-\infty}^{\infty} e^{2|t||\lambda|} |\tilde{f}(\lambda)|^2 \, d\lambda \\ &< \int_{-\infty}^{\infty} e^{2\delta|\lambda|} |\tilde{f}(\lambda)|^2 \, d\lambda = \|f\|^2_{(L_2)_{e,\delta}},\end{aligned}$$

i.e.

$$\sup_{|t|<\delta} \int_{-\infty}^{\infty} |f(x+it)|^2 \, dx \le \|f\|^2_{(L_2)_{e,\delta}}.$$

Conversely, assume that $f(x)$ has an analytic continuation $f(x+it)$ to the strip $|t| < \delta$ satifying (7.5). According to Theorem 7.2,

$$f(z) = \frac{1}{2\pi i} \int_{-\infty}^{\infty} \left(\frac{f(s-i\delta')}{s - i\delta' - z} - \frac{f(s+i\delta')}{s + i\delta' - z} \right) ds \quad (z = x + it) \tag{7.7}$$

for any z inside the strip $|t| \le \delta'$. Since the boundary value of $f(x+it)$ as $t \to \pm\delta$ belongs to the space $L_2(\mathbb{R}^1)$, limit transition in (7.7) as $\delta' \to \delta$ yields

$$f(x+it) = \frac{1}{2\pi i}\int_{-\infty}^{\infty}\Big(\frac{f(s-i\delta)}{s-i\delta-(x+it)} - \frac{f(s+i\delta)}{s+i\delta-(x+it)}\Big)\,ds \tag{7.8}$$

if $z = x+it$ belongs to the strip $|t| < \delta$.

Further,

$$\Big|\int_{-\infty}^{\infty}\frac{f(s+i\delta)}{s+i\delta-(x+it)}\,ds\Big| \le \int_{-N}^{N} + \int_{C(N)}\frac{|f(s+i\delta)|}{\sqrt{(s-x)^2+(\delta-t)^2}}\,ds,$$

where $C(N) = (-\infty,-N)\cup(N,\infty)$. By applying the Cauchy–Buniakowski inequality we obtain

$$\begin{aligned}\int_{-N}^{N}&\frac{|f(s+i\delta)|}{\sqrt{(s-x)^2+(\delta-t)^2}}\,ds\\ &\le\Big(\int_{-N}^{N}|f(s+i\delta)|^2\,ds\int_{-N}^{N}\frac{ds}{(s-x)^2+(\delta-t)^2}\Big)^{\frac12},\\ \int_{C(N)}&\frac{|f(s+i\delta)|}{\sqrt{(s-x)^2+(\delta-t)^2}}\,ds\\ &\le\Big(\int_{C(N)}|f(s+i\delta)|^2\,ds\int_{C(N)}\frac{1}{(s-x)^2+(\delta-t)^2}\,ds\Big)^{\frac12}.\end{aligned}$$

We choose N so that $\int_{C(N)}|f(s+i\delta)|^2\,ds < \epsilon$ ($\epsilon > 0$ is arbitrary and small). From the relations

$$\int_{C(N)}\frac{1}{(s-x)^2+(\delta-t)^2}\,ds \le \frac{\pi}{\delta-t} \le \frac{\pi}{\delta-\delta'}$$

for $|t| \le \delta' < \delta$ and

$$\int_{-N}^{N}\frac{1}{(s-x)^2+(\delta-t)^2}\,ds = \frac{1}{x^2}\int_{-N}^{N}\frac{1}{(1-\frac{s}{x})^2+(\frac{\delta-t}{x})^2}\,ds \le \frac{4}{|x|}\quad(|x|>2N)$$

we conclude that the second summand in (7.8) tends to zero uniformly in the strip $|t| \le \delta'$ as $|x| \to \infty$. We can say the same about the first term. As a result, $f(x+it) \to 0$ uniformly in t, $|t| \le \delta' < \delta$, as $|x| \to \infty$. Taking into account this fact and the analyticity of $f(x+it)$ in the strip $|t| < \delta$, we deduce on the basis of the Cauchy theorem that

$$\begin{aligned}\tilde f(\lambda) &= \frac{1}{\sqrt{2\pi}}\int_{-\infty}^{\infty}f(x)e^{-i\lambda x}\,dx = \frac{1}{\sqrt{2\pi}}\int_{-\infty}^{\infty}f(x\pm it)e^{-i\lambda(x\pm it)}\,dx\\ &= e^{\pm t\lambda}\cdot\frac{1}{\sqrt{2\pi}}\int_{-\infty}^{\infty}f(x\pm it)e^{-i\lambda x}\,dx,\end{aligned}$$

hence $\tilde f(\lambda)e^{\pm t\lambda}$ is the Fourier transform of $f(x\pm it)$ ($|t| < \delta$). By Parseval's equality we arrive at the inequality

$$\int_{-\infty}^{\infty} e^{2t|\lambda|}|\tilde{f}(\lambda)|^2 \,\mathrm{d}\lambda \leq \int_{-\infty}^{\infty} |f(x+it)|^2 \,\mathrm{d}x + \int_{-\infty}^{\infty} |f(x-it)|^2 \,\mathrm{d}x,$$

which after limit transition and in view of Fatou's theorem implies

$$\|f\|_{(L_2)_{e,\delta}} \leq 2 \sup_{|t|<\delta} \int_{-\infty}^{\infty} |f(x+it)|^2 \,\mathrm{d}x. \qquad \square$$

This lemma makes it possible to describe the space $\mathfrak{A}$ of analytic vectors of the operator D in this case.

THEOREM 7.3. *The space $\mathfrak{A}$ coincides with the set of all functions $f(x)$ ($x \in \mathbb{R}^1$) that have an analytic continuation $f(x+it)$ to a strip $|t| < \delta$ (depending on f) that is square integrable over straight lines parallel to the real axis and is such that $\sup_{|t|<\delta} \int_{-\infty}^{\infty} |f(x+it)|^2 \,\mathrm{d}x < \infty$.*

Finally, note that the space $\mathfrak{A}$ has a ring structure.

THEOREM 7.4. *The space $\mathfrak{A}$ is a topological algebra with respect to ordinary multiplication of functions.*

Proof. We need to prove the two following facts:

a) if $f_1, f_2 \in \mathfrak{A}$, then $f_1 f_2 \in \mathfrak{A}$; b) if $f_{1,n} \to f_1$, $f_{2,n} \to f_2$ in $\mathfrak{A}$ ($n \to \infty$), then $f_{1,n} f_{2,n} \to f_1 f_2$ ($n \to \infty$) in $\mathfrak{A}$.

We verify a). It follows from $f_1, f_2 \in \mathfrak{A}$ that $f_1, f_2 \in (L_2)_{e,\delta}$ for some $\delta > 0$. Then

$$\begin{aligned}
&\int_{-\infty}^{\infty} e^{2(\delta-\epsilon)|\lambda|} \left|(\widetilde{f_1 f_2})(\lambda)\right|^2 \mathrm{d}\lambda \\
&\quad = \int_{-\infty}^{\infty} e^{-2\epsilon|\lambda|} e^{2\delta|\lambda|} \left| \int_{-\infty}^{\infty} \tilde{f}_1(\nu) \tilde{f}_2(\lambda-\nu) \,\mathrm{d}\nu \right|^2 \mathrm{d}\lambda \\
&\quad = \int_{-\infty}^{\infty} e^{-2\epsilon|\lambda|} \left| \int_{-\infty}^{\infty} e^{\delta|\lambda|} \tilde{f}_1(\nu) \tilde{f}_2(\lambda-\nu) \,\mathrm{d}\nu \right|^2 \mathrm{d}\lambda \\
&\quad \leq \int_{-\infty}^{\infty} e^{-2\epsilon|\lambda|} \,\mathrm{d}\lambda \left(\int_{-\infty}^{\infty} e^{\delta|\nu|} |\tilde{f}_1(\nu)| e^{\delta|\nu-\lambda|} |\tilde{f}_2(\lambda-\nu)| \,\mathrm{d}\nu \right)^2 \\
&\quad \leq \int_{-\infty}^{\infty} e^{-2\epsilon|\lambda|} \,\mathrm{d}\lambda \int_{-\infty}^{\infty} e^{2\delta|\nu|} |\tilde{f}_1(\lambda)|^2 \,\mathrm{d}\nu \int_{-\infty}^{\infty} e^{2\delta|\nu-\lambda|} |\tilde{f}_2(\lambda-\nu)|^2 \,\mathrm{d}\nu \\
&\quad = c^2 \|f_1\|^2_{(L_2)_{e,\delta}} \|f_2\|^2_{(L_2)_{e,\delta}} < \infty \quad (c = \text{const})
\end{aligned}$$

for small $\epsilon > 0$. Consequently, $f_1 f_2 \in (L_2)_{e,\delta-\epsilon} \subset \mathfrak{A}$, and also

$$\|f_1 f_2\|_{(L_2)_{e,\delta-\epsilon}} \leq c \|f_1\|_{(L_2)_{e,\delta}} \|f_2\|_{(L_2)_{e,\delta}}. \tag{7.9}$$

We now pass to the proof of b). The assumption $f_{i,n} \xrightarrow{\mathfrak{A}} f_i$ ($n \to \infty$; $i = 1,2$) means that $f_{i,n} \in (L_2)_{e,\delta}$ for a certain δ and $f_{i,n} \xrightarrow{(L_2)_{e,\delta}} f_i$ ($i = 1,2$). Then $f_{1,n} f_{2,n}$ and $f_1 f_2$ belong to the space $(L_2)_{e,\delta-\epsilon}$, where $\epsilon > 0$ is small enough. By virtue of inequality (7.9) we obtain

$$\|f_1f_2 - f_{1,n}f_{2,n}\|_{(L_2)_{e,\delta-\epsilon}} \le \|(f_{1,n} - f_1)f_2\|_{(L_2)_{e,\delta-\epsilon}} + \\ + \|(f_{1,n}(f_{2,n} - f_2)\|_{(L_2)_{e,\delta-\epsilon}} \le c\big(\|f_{1,n} - f_1\|_{(L_2)_{e,\delta}}\|f_2\|_{(L_2)_{e,\delta}} + \\ + \|f_{1,n}\|_{(L_2)_{e,\delta}}\|f_{2,n} - f_2\|_{(L_2)_{e,\delta}}\big) \to 0 \quad (n \to \infty).$$

Hence $f_{1,n}f_{2,n} \to f_1f_2$ in $\mathfrak{A}$ if $n \to \infty$. □

Using Theorem 5.3 we can state with certainty that the boundary value of a function $u(x,t)$ that is harmonic in the upper half-plane and square integrable over straight lines parallel to the real axis belongs to the space $\mathfrak{A}'$ of continuous antilinear functionals on $\mathfrak{A}$.

7.4. BOUNDARY VALUES OF HARMONIC FUNCTIONS IN THE UPPER HALF-PLANE THAT INCREASE EXPONENTIALLY AS t APPROACHES THE REAL AXIS

By definition (see subsection 2.6), the Gevrey class $\mathfrak{G}_{\{\beta\}}$ ($\mathfrak{G}_{(\beta)}$) of order β of Roumieu (Beurling) type consists in the example considered of those infinitely differentiable functions $f(x)$ ($x \in \mathbb{R}^1$) that have the property: there exist constants $\alpha > 0, c > 0$ (for any $\alpha > 0$ there exists a $c = c(\alpha) > 0$) such that

$$\left(\int_{-\infty}^{\infty} |f^{(n)}(x)|^2 \,dx\right)^{\frac{1}{2}} \le c\alpha^n n^{n\beta} \quad (n = 0, 1, \ldots).$$

Since $\mathfrak{G}_{\{\beta\}} = (L_2)_{\{G\}}$, $\mathfrak{G}_{(\beta)} = (L_2)_{(G)}$, where $(L_2)_{\{G\}}$ ($(L_2)_{(G)}$) is the inductive (projective) limit of the Hilbert spaces $(L_2)_{\exp(\alpha\lambda)^{\frac{1}{\beta}}}$ constructed from the operator D, functions in $\mathfrak{G}_{\{\beta\}}$ ($\mathfrak{G}_{(\beta)}$) can be characterized in terms of the behaviour of their Fourier transform as follows:

$$\int_{-\infty}^{\infty} \exp(2(\alpha\lambda)^{\frac{1}{\beta}})|\tilde{f}(\lambda)|^2 \,d\lambda < \infty$$

with some (any) $\alpha > 0$. Sometimes functions in $\mathfrak{G}_{\{\beta\}}$ or $\mathfrak{G}_{(\beta)}$ in the present situation are also called ultradifferentiable functions of order β of corresponding type, while the elements from the dual spaces $\mathfrak{G}'_{\{\beta\}}$ or $\mathfrak{G}'_{(\beta)}$ are called ultradistributions of the same order and type. On the basis of Corollary 5.4 we arrive at the following statement.

THEOREM 7.5. *For the boundary value (as t approaches the real axis) of a function $u(x,t)$ that is harmonic in the upper half-plane and square integrable with respect to $x \in \mathbb{R}^1$ for each $t > 0$ to be an ultradistribution of order $\beta > 1$ of Roumieu (Beurling) type it is necessary and sufficient that for any $\alpha > 0$ there exists a positive constant $c = c(\alpha) > 0$ (there exist constants $\alpha > 0, c > 0$) such that*

$$\int_{-\infty}^{\infty} |u(x,t)|^2 \,dx \le c \exp(\alpha t^{-q}), \quad q = \frac{1}{\beta - 1}$$

for sufficiently small t.

When the growth of $u(x,t)$ as t approaches the real axis is controlled by a function $\gamma(t)$ satisfying the conditions 1°, 2° from subsection 3.3, Theorem 5.7 leads to the following result.

THEOREM 7.6. *The boundary value* (as $t \to 0$) *of a function* $u(x,t)$ *that is harmonic in the upper half-plane and square integrable with respect to* $x \in \mathbb{R}^1$ *belongs to the space* $(L_2)'_{\{G\}} = \operatorname{limpr}_{\alpha\to 0}(L_2)'_{G_\alpha}$ $((L_2)'_{(G)} = \operatorname{limind}_{\alpha\to\infty}(L_2)'_{G_\alpha})$, *where the functions* G *and* γ *are connected by formula* (3.9), *if and only if for an arbitrary* $\alpha > 0$ *there exists a constant* $c = c(\alpha) >$) (*there exist* $\alpha > 0, c > 0$) *such that*

$$\int_{-\infty}^{\infty} |u(x,t)|^2 \, dx \le c\gamma^{-1}\left(\frac{t}{\alpha}\right)$$

for sufficiently small t.

7.5. QUASI-ANALYTIC CLASSES OF FUNCTIONS

Let $\{m_n\}_{n=0}^{\infty}$ be a sequence of positive numbers. Recall that in the case considered an infinitely differentiable function $f(x)$ $(x \in \mathbb{R}^1)$ belongs to the class $C_{\{m_n\}} = C_{\{m_n\}}(D)$ if and only if the estimate

$$\left(\int_{-\infty}^{\infty} |f^{(k)}(x)|^2 \, dx\right)^{\frac{1}{2}} \le c\alpha^k m_k \quad (k = 0, 1, \ldots)$$

holds, where the constants $c > 0, \alpha > 0$ depend only on f.

It is clear that the class $C_{\{n!\}}$ consists of functions analytic on the whole real axis. If the numbers m_n grow faster than $n!$, then it is possible that the class $C_{\{m_n\}}$ contains non-analytic functions.

Under the assumption $\sum_{n=1}^{\infty} \frac{1}{\sqrt[n]{m_n}} = \infty$, the class $C_{\{m_n\}}$ has the following property: if two functions f and g from $C_{\{m_n\}}$ coincide at a certain point x_0 together with all their derivatives, then they coincide identically. For analytic functions this property is well-known.

DEFINITION. If a class of infinitely differentiable functions defined on $\mathbb{R}^1$ is such that coincidence at a certain point of $\mathbb{R}^1$ of two functions in it together with all their derivatives implies coincidence of the functions on $\mathbb{R}^1$, then this class is called quasi-analytic.

The full description of quasi-analytic classes $C_{\{m_n\}}$ is given in the Carleman–Ostrowski theorem (see, for example, Shilov [2, p.352]) as follows.

THEOREM 7.7. *Let*

$$T(\lambda) = \sup_{k\ge 0} \frac{\lambda^k}{m_k}.$$

In order that the class $C_{\{m_n\}}$ *be quasi-analytic it is necessary and sufficient that*

$T(\lambda)$ satisfies the condition

$$\int_1^\infty \frac{\ln T(\lambda)}{\lambda^2}\, d\lambda = \infty. \tag{7.10}$$

For example, if $m_n = n^{n\beta}$ ($\beta > 0$ is fixed), i.e. when $C_{\{m_n\}} = \mathfrak{G}_{\{\beta\}}$ is the Gevrey class of ultradifferentiable functions of order β of Roumieu type, then $T(\lambda) \sim \exp(\lambda^{\frac{1}{\beta}})$ and the integral (7.10) converges for $\beta > 1$ and does not converge if $\beta \leq 1$. Thus, the class $\mathfrak{G}_{\{\beta\}}$ with $\beta \leq 1$ is quasi-analytic (it even consists of analytic functions), while for $\beta > 1$ it is not quasi-analytic.

If the class $C_{\{m_n\}}$ is not quasi-analytic, then for arbitrary disjoint closed intervals $[a_1, b_1]$ and $[a_2, b_2]$ we can find a function from $C_{\{m_n\}}$ which equals one on $[a_1, b_1]$ and vanishes on $[a_2, b_2]$.

7.6. THE NON-QUASI-ANALYTICITY OF THE SPACE $(L_2)_{\{G\}}$

Suppose that a positive, twice continuously differentiable function $\gamma(t)$ on $(0, b)$ satisfies the conditions:

1°) $\gamma(0) = 0$, $\gamma(t_1) > \gamma(t_2)$ if $t_1 > t_2$;
2°) there exist constants $c_0 > 0, \beta_0 : 0 < \beta_0 < 1$ such that

$$t\gamma(t) > c_0\gamma(\beta_0 t);$$

3°) $\frac{t\gamma'(t)}{\gamma(t)} \uparrow \infty$ as $t \to 0$.

Let

$$G(\lambda) = \left(\int_0^b \gamma(t) e^{-2\lambda t}\, dt\right)^{-\frac{1}{2}}$$

(compare with subsection 3.3). From the function $G(\lambda)$ we construct, as has been done in subsection 2.4, the space

$$(L_2)_{\{G\}} = \operatorname*{limind}_{\alpha \to 0}\, (L_2)_{G_\alpha}.$$

It follows from the conditions 1°, 2° that $\gamma(t) \to 0$ faster than powers as $t \to 0$. It is not difficult to infer from this that $G(\lambda) \to \infty$ ($\lambda \to \infty$) faster than powers. Therefore the space $(L_2)_{\{G\}}$ consists of infinitely differentiable functions. If we set, as in subsection 2.4,

$$m_n = \sup_{\lambda \geq 1} \frac{\lambda^n}{G(\lambda)},$$

then $(L_2)_{\{G\}} = C_{\{m_n\}}$.

The following question arises: what additional conditions on the function $\gamma(t)$ have to be imposed in order that the class $(L_2)_{\{G\}}$ be not quasi-analytic? Since $G(\lambda)$ satisfies inequality (2.8), Theorem 7.7 leads to the following statement.

THEOREM 7.8. *The space* $(L_2)_{\{G\}}$ *is not a quasi-analytic class if and only if*

$$\int_1^\infty \frac{\ln G(\lambda)}{\lambda^2}\, d\lambda < \infty.$$

Let

$$N(\lambda) = \sup_t (\ln \gamma(t) - 2\lambda t).$$

LEMMA 7.2. *Under the assumptions* 1°, 3° *on the function* $\gamma(t)$, *the estimate*

$$c_1 \exp\left(-\frac{N(\lambda)}{2}\right) \le G(\lambda) \le c_2 \lambda^{\frac{1}{2}} \exp\left(-\frac{N(\lambda)}{2}\right)$$

$(c_1 > 0, c_2 > 0$ *are constants) is valid for sufficiently large* λ.

Proof. Indeed,

$$G(\lambda) = \left(\int_0^b \gamma(t) e^{-2\lambda t}\, dt\right)^{-\frac{1}{2}} = \left(\int_0^b \exp(\ln \gamma(t) - 2\lambda t)\, dt\right)^{-\frac{1}{2}}$$
$$\ge \frac{1}{\sqrt{b}} \exp\left(-\frac{N(\lambda)}{2}\right).$$

On the other hand, it follows from condition 3° that $\frac{\gamma'(t)}{\gamma(t)} \uparrow \infty$ as $t \to 0$. So the equation $\frac{\gamma'(t)}{\gamma(t)} - 2\lambda = 0$ has a unique solution if $\lambda > \frac{\gamma'(b)}{2\gamma(b)}$. We denote it by $K(\lambda)$. The function $K(\lambda)$ is continuously differentiable on $[\frac{\gamma'(b)}{2\gamma(b)}, \infty)$ and decreases to zero monotonically as $\lambda \to \infty$. Because of the monotonicity of $\frac{\gamma'(t)}{\gamma(t)}$, the function $\frac{\gamma'(t)}{\gamma(t)} - 2\lambda$ changes sign as t passes through the point $K(\lambda)$. Therefore $\ln \gamma(t) - 2\lambda t$ takes its maximum value at $K(\lambda)$, i.e. $N(\lambda) = \ln \gamma(K(\lambda)) - 2\lambda K(\lambda)$. Since $\gamma(t)$ increases on $[0, b]$ monotonically,

$$G(\lambda) \le \left(\gamma(\epsilon) \int_\epsilon^b e^{-2\lambda t}\, dt\right)^{-\frac{1}{2}} = \sqrt{2\lambda} e^{\lambda\epsilon} \left(\gamma(\epsilon)\left(1 - e^{-2\lambda(b-\epsilon)}\right)\right)^{-\frac{1}{2}}$$

for any ϵ, $0 < \epsilon < \frac{b}{2}$. Taking into account that $\left(1 - e^{-2\lambda(b-\epsilon)}\right)^{\frac{1}{2}} > \frac{1}{2}$ if $\lambda > \frac{\ln 2}{b}$, we obtain $G(\lambda) \le 2\sqrt{2\lambda} \exp(-\frac{1}{2}(\ln \gamma(\epsilon) - 2\lambda\epsilon))$. Set $\epsilon = K(\lambda)$. Since $K(\lambda) < \frac{b}{2}$ for sufficiently large λ, we conclude that

$$G(\lambda) \le 2\sqrt{2\lambda} \exp\left(-\frac{N(\lambda)}{2}\right).$$

THEOREM 7.9. *The class* $(L_2)_{\{G\}}$ *is not quasi-analytic if and only if*

$$\int_0^\delta \ln(-\ln \gamma(t))\, dt < \infty \tag{7.11}$$

for sufficiently small $\delta > 0$.

Proof. In view of theorem 7.8 and Lemma 7.2 non-quasi-analyticity of the class $(L_2)_{\{G\}}$ is equivalent to convergence of the integral $\int_1^\infty \frac{N(\lambda)}{\lambda^2}\, d\lambda$, where

$$N(\lambda) = \sup_{t \in [0,\delta]} (\ln \gamma(t) - 2\lambda t) = \ln \gamma(K(\lambda)) - 2\lambda K(\lambda) \quad \left(\lambda > \frac{\gamma'(\delta)}{2\gamma(\delta)} = \lambda_\delta\right).$$

We can choose $\delta > 0$ small such that $\gamma(t) < 1$ when $t \in [0, \delta]$.

Set $M(t) = \ln \gamma(t)$. The change of variable $K(\lambda) = t$ in the integral

$$\int_{\lambda_\delta}^{\infty} -\frac{N(\lambda)}{\lambda^2}\,\mathrm{d}\lambda = \int_{\lambda_\delta}^{\infty} \frac{-\ln \gamma(K(\lambda)) + 2\lambda K(\lambda)}{\lambda^2}\,\mathrm{d}\lambda$$

shows that non-quasi-analyticity of the class $(L_2)_{\{G\}}$ is equivalent to convergence of the integral

$$\int_0^\delta \frac{M(t) - tM'(t)}{(M'(t))^2} M''(t)\,\mathrm{d}t = \int_0^\delta \frac{M(t)M''(t)}{(M'(t))^2}\,\mathrm{d}t + \int_0^\delta -\frac{tM''(t)}{M'(t)}\,\mathrm{d}t.$$

Since $M(t) < 0$, $M'(t) \geq 0$ and $M''(t) \leq 0$ by property 3°, the integrands in the three integrals are non-negative. Therefore, convergence of the integrals is equivalent to boundedness of the corresponding integrals over the interval $[\epsilon, \delta]$ $(\forall \epsilon > 0)$. But

$$\begin{aligned} \int_\epsilon^\delta \frac{M(t)M''(t)}{(M'(t))^2}\,\mathrm{d}t &= \int_\epsilon^\delta \left(1 - \frac{\mathrm{d}}{\mathrm{d}t}\left(\frac{M(t)}{M'(t)}\right)\right)\mathrm{d}t \\ &= \delta - \epsilon - \frac{M(t)}{M'(t)}\Big|_\epsilon^\delta; \\ -\int_\epsilon^\delta \frac{tM''(t)}{M'(t)}\,\mathrm{d}t &= -\int_\epsilon^\delta t(\ln M'(t))'\,\mathrm{d}t \\ &= -t\ln M'(t)\Big|_\epsilon^\delta + \int_\epsilon^\delta \ln M'(t)\,\mathrm{d}t. \end{aligned} \tag{7.12}$$

By the first equality in (7.12) we conclude that

$$\delta - \epsilon - \frac{M(\delta)}{M'(\delta)} + \frac{M(\epsilon)}{M'(\epsilon)} \geq 0,$$

whence

$$-\frac{M(\epsilon)}{M'(\epsilon)} < \delta - \frac{M(\delta)}{M'(\delta)} \tag{7.13}$$

and $\int_0^\delta \frac{M(t)M''(t)}{(M'(t))^2}\,\mathrm{d}t < \infty$. As for the integral $\int_0^\delta \frac{tM''(t)}{M'(t)}\,\mathrm{d}t$, its convergence is equivalent to the convergence of $\int_0^\delta \ln M'(t)\,\mathrm{d}t$. In fact, since $M'(t) \geq 0$ decreases monotonically, the inequality $\int_0^\delta \ln M'(t)\,\mathrm{d}t < \infty$ implies $\epsilon \ln M'(\epsilon) \to 0$ as $\epsilon \to 0$, that is $\int_0^\delta \frac{tM''(t)}{M'(t)}\,\mathrm{d}t < \infty$. The converse follows from the non-negativity of the terms depending on ϵ on the right-hand side of the second equality in (7.12).

Hence, non-quasi-analyticity of the class $(L_2)_{\{G\}}$ is equivalent to convergence of the integral $\int_0^\delta \ln M'(t)\,\mathrm{d}t$. Next we will show that the latter is equivalent to (7.11).

Let $\int_0^\delta \ln(-\ln \gamma(t))\,\mathrm{d}t < \infty$. Then

$$\begin{aligned} \int_0^\delta \ln M'(t)\,\mathrm{d}t &= \int_0^\delta \ln\left(-\frac{M'(t)}{M(t)}(-M(t))\right)\mathrm{d}t \\ &= \int_0^\delta \ln(-\ln \gamma(t))\,\mathrm{d}t - \int_0^\delta \ln\left(-\frac{M(t)}{M'(t)}\right)\mathrm{d}t. \end{aligned} \tag{7.14}$$

From the relations

$$\begin{aligned}\int_\epsilon^\delta \ln(-\ln\gamma(t))\,\mathrm{d}t &= t\ln(-\ln\gamma(t))\big|_\epsilon^\delta + \int_\epsilon^\delta t\Big(-\frac{M'(t)}{M(t)}\Big)\,\mathrm{d}t \\ &\geq t\ln(-\ln\gamma(t))\big|_\epsilon^\delta + \int_\epsilon^\delta \ln\Big(t\Big(-\frac{M'(t)}{M(t)}\Big)\Big)\,\mathrm{d}t \\ &= t\ln(-\ln\gamma(t))\big|_\epsilon^\delta + \int_\epsilon^\delta \ln t\,\mathrm{d}t - \int_\epsilon^\delta \ln\Big(-\frac{M(t)}{M'(t)}\Big)\,\mathrm{d}t,\end{aligned}$$

taking into account the boundedness of the function $t\ln(-\ln\gamma(t))$, which follows from the convergence of the integral (7.11), we obtain the convergence of the last integral in (7.14), hence the convergence of $\int_0^\delta \ln M'(t)\,\mathrm{d}t$.

Assume conversely that

$$\int_0^\delta \ln M'(t)\,\mathrm{d}t < \infty. \tag{7.15}$$

If the integral (7.11) is divergent, then equality (7.14) and the boundedness of the function $-\frac{M(t)}{M'(t)}$, following from the estimate (7.13), would imply divergence of the integral $\int_0^\delta \ln M'(t)\,\mathrm{d}t$, which contradicts (7.15).

8. The Operational Calculus for Certain Classes of Non-Self-Adjoint Operators

If T is a bounded operator on a Hilbert space $\mathfrak{H}$ and $R_z(T)$ is its resolvent, then, as is well-known, the formula

$$f(T) = -\frac{1}{2\pi i}\int_\Gamma f(z)R_z(T)\,\mathrm{d}z,$$

where Γ is the rectifiable Jordan boundary of a domain U containing the spectrum of T, defines the operational calculus on the class of analytic functions $f(z)$ on $U\cup\Gamma$. A similar formula may be also given for a unbounded closed operator (see Chapter 1, subsections 5.1 and 5.2). The theory of boundary values of functions that are harmonic in the upper half-plane and square integrable over straight lines parallel to the real axis set forth in the previous section helps to extend the operational calculus to somewhat broader classes of functions. These classes are determined by the behaviour of the resolvent as z approaches the spectrum.

8.1. THE OPERATIONAL CALCULUS FOR ANALYTIC FUNCTIONS

Let T be a closed linear operator on $\mathfrak{H}$ with spectrum $\sigma(T)$ on the real axis. Suppose that its resolvent $R_z(T)$ $(z\in\rho(T))$ satisfies the condition

$$\int_{-\infty}^{\infty} \big\|R_{x+it}f\big\|^2\,\mathrm{d}x \leq \frac{|f|^2}{\gamma(|t|)} \quad (f\in\mathfrak{H},\ t\neq 0), \tag{8.1}$$

where $\gamma(t)$ is a positive continuous function on a certain interval $(0,b]$. Note that an estimate of such a kind with, for example, $\gamma(t) = \frac{t}{\pi}$ holds in the case of a self-adjoint T.

Denote by $\mathfrak{A}$ the set of functions $f(x)$ ($x \in \mathbb{R}^1$) which have an analytic continuation $f(x+it)$ to a certain strip $|t| < \delta$ (depending on f) that is square integrable over straight lines parallel to the real axis and satisfies

$$\sup_{|t|<\delta} \int_{-\infty}^{\infty} |f(x+it)|^2 \, dx < \infty. \tag{8.2}$$

Clearly, $\mathfrak{A}$ coincides with the class $\mathfrak{A}(D)$ of analytic vectors of the operator D on the space $L_2(\mathbb{R}^1)$ as described in subsection 7.3. It forms a topological algebra with respect to the inductive limit topology of the spaces $(L_2)_{e,\delta}$.

We define a mapping

$$\mathfrak{A} \ni f \to f(T) : f(T)h = -\frac{1}{2\pi i} \int_{\Gamma_\epsilon} f(z) R_z(T) h \, dz \quad (h \in \mathfrak{H}), \tag{8.3}$$

where Γ_ϵ is the boundary of a strip $|t| \le \epsilon$ in which $f(x+it)$ is analytic; the contour Γ_ϵ is assumed to have positive orientation with respect to the strip, as is customary in complex analysis (if T is bounded, it suffices to take for Γ_ϵ the boundary of a rectangle of width 2ϵ and containing $\sigma(T)$). The existence of the integral (8.3) follows from the square integrability of the function $f(x+it)$ over the straight lines $t = \pm\epsilon$ and the estimate (8.1) for the resolvent of the operator T. The integral does not depend on the choice of ϵ, provided that the contour Γ_ϵ is contained in a strip $|t| < \delta$ in which f is analytic. This is established in a standard manner if it is taken into account that $f(x+it) \to 0$ as $|x| \to \infty$ uniformly in t, $|t| \le \epsilon < \delta$ (see the proof of Lemma 7.1); on account of the estimate (8.1) $(R_{x+it}(T)f, g) \to 0$ as $|x| \to \infty$ uniformly in t, $\epsilon \le |t| \le \epsilon' < \delta$.

THEOREM 8.1. *The mapping* (8.3) *is a homomorphism of the algebra $\mathfrak{A}$ into a certain algebra of bounded operators on $\mathfrak{H}$. This mapping is continuous in the sense that if $f_n \xrightarrow{\mathfrak{A}} f$ as $n \to \infty$, then $f_n(T) \to f(T)$ $(n \to \infty)$ in the uniform operator topology. In addition, if $f(x), xf(x) \in \mathfrak{A}$, then $f(T)h \in \mathfrak{D}(T)$ for all $h \in \mathfrak{H}$ and $Tf(T) \supset f(T)T$.*

Proof. It is obvious that $\alpha_1 f_1 + \alpha_2 f_2 \to \alpha_1 f_1(T) + \alpha_2 f_2(T)$ (α_1, α_2 are arbitrary numbers; $f_1, f_2 \in \mathfrak{A}$).

The fact that $f_1 f_2 \to f_1(T) f_2(T)$ can be obtained by contour integration (see, for example, Dunford–Schwartz [1, chapter VII, §3]) using Hilbert's identity for resolvents:

$$R_\lambda(T) - R_\mu(T) = (\lambda - \mu) R_\lambda(T) R_\mu(T),$$

and the Cauchy formula for analytic functions in unbounded domains.

Let $f_n \to f$ in $\mathfrak{A}$. This means that all f_n belong to $(L_2)_{e,\delta}$ for a certain δ and that $f_n \to f$ in $(L_2)_{e,\delta}$. Then

$$\begin{aligned}
\|(f_n(T) - f(T))h\| &= \left\| \int_{\Gamma_\epsilon} (f_n(z) - f(z)) R_z(T) h \, dz \right\| \\
&\leq \int_{-\infty}^{\infty} |f_n(x - i\epsilon) - f(x - i\epsilon)| \|R_{x-i\epsilon}(T)h\| \, dx + \\
&\quad + \int_{-\infty}^{\infty} |f_n(x + i\epsilon) - f(x + i\epsilon)| \|R_{x+i\epsilon}(T)h\| \, dx \\
&\leq \sup_{|t|<\delta} \left(\int_{-\infty}^{\infty} |f_n(x+it) - f(x+it)|^2 \, dx \right)^{\frac{1}{2}} \left(\left(\int_{-\infty}^{\infty} \|R_{x-i\epsilon}(T)h\|^2 \, dx \right)^{\frac{1}{2}} + \right. \\
&\quad \left. + \left(\int_{-\infty}^{\infty} \|R_{x+i\epsilon}(T)h\|^2 \, dx \right)^{\frac{1}{2}} \right) \\
&\leq c_1 \|f_n - f\|_{(L_2)_{e,\delta}} \|h\| \quad (c_1 = \text{const}),
\end{aligned}$$

whence $\|f_n(T) - f(T)\| \to 0$ $(n \to \infty)$.

Finally, the last assertion of the theorem is deduced from the definition of the integral $-\frac{1}{2\pi i} \int_{\Gamma_\epsilon} f(z) R_z(T) h \, dz = f(T)h$ as the limit of integral sums and from the fact that T is a closed operator. □

The space $\mathfrak{A}'$, the dual of $\mathfrak{A}$, coincides with $\operatorname{limpr}_{\delta \to 0} (L_2)'_{e,\delta}$. It consists of the generalized functions that are images under Fourier transformation of all functions that are square integrable with respect to the weight $\exp(-2\delta|\lambda|)$ with any $\delta > 0$. Because of (8.1) the boundary values (as t approaches the real axis) of the real and imaginary parts of the function $(R_z(T)g, h)$, which is analytic in the upper and the lower half-plane, belong to the space $\mathfrak{A}'$ (see subsection 7.3). So, the limits $\lim_{\epsilon \to 0} (R_{x \pm i\epsilon}(T)g, h)$ exist in the space $\mathfrak{A}'$. Set

$$R_{g,h} = \lim_{\epsilon \to 0} \left(\overline{(R_{x+i\epsilon}(T)g, h)} - \overline{(R_{x-i\epsilon}(T)g, h)} \right).$$

We now show that $f(x \pm i\epsilon) \overset{\mathfrak{A}}{\to} f(x)$ as $\epsilon \to 0$ for a function $f \in \mathfrak{A}$ (i.e. $f \in (L_2)_{e,\delta}$ for a certain $\delta > 0$). Indeed, since $\widetilde{f(x \pm i\epsilon)} = \tilde{f}(\lambda) e^{\mp \epsilon \lambda}$ (see the proof of Lemma 7.1), $f(x \pm i\epsilon) \in (L_2)_{e,\delta-\epsilon}$ if ϵ is small enough, so, $f(x \pm i\epsilon) \in (L_2)_{e,\frac{\delta}{2}}$. Furthermore, $\|f(x \pm i\epsilon) - f(x)\|^2_{(L_2)_{e,\frac{\delta}{2}}} = \int_{-\infty}^{\infty} |1 - e^{\mp \epsilon \lambda}|^2 |\tilde{f}(\lambda)|^2 e^{\delta|\lambda|} \, d\lambda$. Taking into account the equality $|1 - e^{\mp \epsilon \lambda}| \leq e^{\delta|\lambda|}$, after limit transition under the integral sign we conclude that $f(x \pm i\epsilon) \to f(x)$ in $(L_2)_{e,\frac{\delta}{2}}$ as $\epsilon \to 0$, hence $f(x \pm i\epsilon) \to f(x)$ $(\epsilon \to 0)$ in $\mathfrak{A}$. All this allows us to pass to the limit as $\epsilon \to 0$ in formula (8.3). As a result we have

$$(f(T)g, h) = (R_{g,h}, f)_{L_2(\mathbf{R}^1)} \tag{8.4}$$

(the expression on the right-hand side is to be understood as the action of the functional $R_{g,h}$ from the negative space $\mathfrak{A}'$ on the element f of the positive space $\mathfrak{A}$ in the chain $\mathfrak{A} \subset L_2 \subset \mathfrak{A}'$).

8.2. THE OPERATIONAL CALCULUS OF OPERATORS WHOSE RESOLVENT HAS POWER GROWTH UPON APPROACHING THE SPECTRUM

Let $\gamma(t)$ in (8.1) be integrable on $(0, b]$. We set as before

$$G(\lambda) = \left(\int_0^b \gamma(t) e^{-2\lambda t} \, dt \right)^{-\frac{1}{2}} \quad (\lambda > 0),$$

and denote by $(L_2)_G$ the space constructed from the function $G(\lambda)$ and the operator D. It was noted in subsection 7.1 that this space consists of all functions $f \in L_2(\mathbb{R}^1)$ that satisfy the condition

$$\|f\|^2_{(L_2)_G} = \int_{-\infty}^{\infty} G^2(|\lambda|) |\tilde{f}(\lambda)|^2 \, d\lambda < \infty.$$

By virtue of Theorem 5.6, applied to the situation of §7, $R_{g,h} \in (L_2)'_G$. Therefore, the expression $(R_{g,h}, f)_{L_2(\mathbb{R}^1)}$ makes sense when $f \in (L_2)_G$. Thus it is possible to define the function $f(T)$ of the operator T by formula (8.4) also for $f \in (L_2)_G$.

The mapping $(L_2)_G \ni f \to f(T)$ is clearly linear. It is also continuous in the sense that

$$\|f(T)\| \le c \|f\|_{(L_2)_G}$$

($0 < c = \text{const}$ does not depend on f). This follows from the relations

$$|(f(T)g, h)| = |(R_{g,h}, f)_{L_2(\mathbb{R}^1)}| \le \|R_{g,h}\|_{(L_2)'_G} \|f\|_{(L_2)_G}$$

and the estimate

$$\begin{aligned} \|R_{g,h}\|^2_{(L_2)'_G} &\le c \int_0^b \gamma(t) \int_{-\infty}^{\infty} |((R_{x+it}(T) - R_{x-it}(T))g, h)|^2 \, dx \, dt \\ &\le c_1 \|g\|^2 \|h\|^2 \quad (c, c_1 = \text{const}) \end{aligned}$$

(see Theorem 5.6). In addition, if $f(x), xf(x) \in (L_2)_G$, then $Tf(T) \subset f(T)T$. This follows from the closedness of T and the fact that f can be approximated in the $(L_2)_G$-metric by functions $f_n \in \mathfrak{A}$ so that $\mathfrak{A} \ni xf_n(x) \to xf(x)$. If $\gamma(t) = t^{2\alpha-1}$ ($\alpha \ge \frac{1}{2}$), i.e. the resolvent of T increases like a power as t approaches the spectrum, then $(L_2)_G = W_2^\alpha(\mathbb{R}^1)$ and the operational calculus can be extended to the functions from the Sobolev space $W_2^\alpha(\mathbb{R}^1)$.

8.3. THE OPERATIONAL CALCULUS OF OPERATORS WHOSE RESOLVENT HAS SUPER-POWER GROWTH UPON APPROACHING THE SPECTRUM

Suppose that $\gamma(t)$ satisfies the conditions 1°-3° from subsection 7.6. From the function $G(\lambda)$ corresponding to $\gamma(t)$ we construct, as in the previous section, the spaces

$$(L_2)_{\{G\}} = \operatorname*{limind}_{\alpha \to 0} (L_2)_{G_\alpha}, \quad (L_2)_{(G)} = \operatorname*{limpr}_{\alpha \to \infty} (L_2)_{G_\alpha}.$$

THEOREM 8.2. *The spaces $(L_2)_{\{G\}}$ and $(L_2)_{(G)}$ are topological algebras with respect to ordinary multiplication.*

Proof. It is obvious that $f_1, f_2 \in (L_2)_{\{G\}}$ ($f_1, f_2 \in (L_2)_{(G)}$) implies $\beta_1 f_1 + \beta_2 f_2 \in (L_2)_{\{G\}}$ ($\beta_1 f_1 + \beta_2 f_2 \in (L_2)_{(G)}$) for arbitrary complex numbers β_1, β_2. We will show that $f_1 f_2 \in (L_2)_{\{G\}}$ ($f_1 f_2 \in (L_2)_{(G)}$). To this end we consider the increasing sequence

$$m_n = \sup_{\lambda \geq 1} \frac{\lambda^n}{G(\lambda)} \quad (n = 0, 1, \ldots).$$

In addition to the properties (i), (ii) mentioned in subsection 2.4, this sequence also satisfies:

(iii) $m_n^2 \leq m_{n-1} m_{n+1}$ for any $n \in \mathbb{N}$ (logarithmic convexity);
(iv) $m_p m_{n-p} \leq m_n$ for any $p < n$.

Assertion (iii) is verified simply:

$$\left(\sup_{\lambda \geq 1} \frac{\lambda^n}{G(\lambda)}\right)^2 = \sup_{\lambda \geq 1} \frac{\lambda^{n-1}\lambda^{n+1}}{G^2(\lambda)} \leq \sup_{\lambda \geq 1} \frac{\lambda^{n-1}}{G(\lambda)} \sup_{\lambda \geq 1} \frac{\lambda^{n+1}}{G(\lambda)} =: m_{n-1} m_{n+1}.$$

In order to obtain (iv), without loss of generality we may set $m_0 = 1$. Then the desired inequality is trivial for $p = 0$. Furthermore, it is sufficient to prove it only for $p \leq n - p - 1$. We use induction on p. Assuming that (iv) is valid for p we prove the inequality $m_{p+1} m_{n-p-1} \leq m_n$. The logarithmic convexity property yields

$$\frac{m_k}{m_{k+1}} \leq \frac{m_{k-1}}{m_k} \leq \cdots \leq \frac{m_l}{m_{l+1}} \quad \text{if } l \leq k.$$

Setting $k = n - p - 1$, $l = p$ we obtain

$$\frac{m_{n-p-1}}{m_{n-p}} \leq \frac{m_p}{m_{p+1}}.$$

Multiplying this inequality and (iv), we arrive at the required assertion.

Let now $C_{\{m_n\}}$ and $C_{(m_n)}$ be the spaces constructed from the sequence $\{m_n\}_{n=0}^{\infty}$ as in subsection 2.4:

$$C_{\{m_n\}} = \operatorname*{limind}_{\alpha \to \infty} C_\alpha < m_n >, \quad C_{(m_n)} = \operatorname*{limpr}_{\alpha \to 0} C_\alpha < m_n >,$$

where $C_\alpha < m_n >$, in the case considered, is the set of all infinitely differentiable functions $\varphi \in L_2(\mathbb{R}^1)$ for which

$$\|\varphi\|_{C_\alpha < m_n >} = \sup_k \frac{\left(\int_{-\infty}^{\infty} |\varphi^{(k)}(x)|^2 \, dx\right)^{\frac{1}{2}}}{\alpha^k m_k} < \infty.$$

According to Theorem 2.2,

$$C_{\{m_n\}} = (L_2)_{\{G\}}, \quad C_{(m_n)} = (L_2)_{(G)}.$$

If $f_1, f_2 \in (L_2)_{\{G\}}$, (respectively, $f_1, f_2 \in (L_2)_{(G)}$) then $f_1 f_2 \in C_\alpha < m_n >$ for some (for any) α. We will show that $f_1 f_2 \in C_\beta < m_n >$ for some (for any) β. In fact,

$$\begin{aligned}
&\int_{-\infty}^{\infty} \left|(f_1 f_2)^{(k)}(x)\right|^2 \mathrm{d}x = \int_{-\infty}^{\infty} \left|(\widetilde{f_1 f_2})(\lambda)\right|^2 \lambda^{2k} \,\mathrm{d}\lambda \\
&= \int_{-\infty}^{\infty} \frac{1}{1+\lambda^2} \left| \int_{-\infty}^{\infty} \tilde{f}_1(\nu) \tilde{f}_2(\lambda-\nu)\,\mathrm{d}\nu \right|^2 (\lambda^{2k} + \lambda^{2k+2})\,\mathrm{d}\lambda \\
&\le \int_{-\infty}^{\infty} \frac{1}{1+\lambda^2} \left| \int_{-\infty}^{\infty} \tilde{f}_1(\nu) \tilde{f}_2(\lambda-\nu)(\lambda^k + \lambda^{k+1})\,\mathrm{d}\nu \right|^2 \mathrm{d}\lambda \\
&\le \int_{-\infty}^{\infty} \frac{2}{1+\lambda^2} \left(\left| \int_{-\infty}^{\infty} \tilde{f}_1(\nu) \tilde{f}_2(\lambda-\nu)\lambda^k\,\mathrm{d}\nu \right|^2 + \right. \\
&\left. + \left| \int_{-\infty}^{\infty} \tilde{f}_1(\nu) \tilde{f}_2(\lambda-\nu)\lambda^{k+1}\,\mathrm{d}\nu \right|^2 \right) \mathrm{d}\lambda.
\end{aligned} \tag{8.5}$$

Further, by virtue of property (iv) of the sequence $\{m_n\}_{n=0}^{\infty}$,

$$\begin{aligned}
&\left| \int_{-\infty}^{\infty} \tilde{f}_1(\nu) \tilde{f}_2(\lambda-\nu)\lambda^k\,\mathrm{d}\nu \right| \le \int_{-\infty}^{\infty} |\tilde{f}_1(\nu)||\tilde{f}_2(\lambda-\nu)|(|\lambda-\nu| + |\nu|)^k\,\mathrm{d}\nu \\
&= \sum_{i=0}^{k} C_k^i \int_{-\infty}^{\infty} |\tilde{f}_1(\nu)||\tilde{f}_2(\lambda-\nu)||\lambda-\nu|^{k-i}|\nu|^i\,\mathrm{d}\nu \\
&\le \sum_{i=0}^{k} C_k^i \left(\int_{-\infty}^{\infty} |\tilde{f}_1(\nu)|^2 |\nu|^{2i}\,\mathrm{d}\nu \right)^{\frac{1}{2}} \left(\int_{-\infty}^{\infty} |\tilde{f}_2(\lambda-\nu)|^2 |\lambda-\nu|^{2(k-i)}\,\mathrm{d}\nu \right)^{\frac{1}{2}} \\
&= \sum_{i=0}^{k} C_k^i \left(\int_{-\infty}^{\infty} |f_1^{(i)}(x)|^2\,\mathrm{d}x \right)^{\frac{1}{2}} \left(\int_{-\infty}^{\infty} |f_2^{(k-i)}(x)|^2\,\mathrm{d}x \right)^{\frac{1}{2}} \\
&\le c_1 \alpha^k \sum_{i=0}^{k} C_k^i m_i m_{k-i} \le c_1 (2\alpha)^k m_k.
\end{aligned} \tag{8.6}$$

Taking into account property (ii) of $\{m_n\}_{n=0}^{\infty}$, we have

$$\begin{aligned}
\int_{-\infty}^{\infty} |(f_1 f_2)^{(k)}(x)|^2\,\mathrm{d}x &\le c^2((2\alpha)^{2k} m_k^2 + (2\alpha)^{2(k+1)} m_{k+1}^2) \\
&\le c^2 m_k^2 (2\alpha)^{2k} (1+h)^{2k} = c' \beta^{2k} m_k^2,
\end{aligned} \tag{8.7}$$

where $\beta = 2\alpha(1+h)$. This implies that $f_1 f_2 \in C_\beta < m_n >$ with $\beta = 2\alpha(1+h)$, hence $f_1 f_2 \in (L_2)_{\{G\}}$ (because β is arbitrary, $f_1 f_2 \in (L_2)_{(G)}$).

Using the relations (8.5)-(8.7) we may also conclude that

$$\|f_1 f_2\|_{C_\beta < m_n >} \le c \|f_1\|_{C_\alpha < m_n >} \|f_2\|_{C_\alpha < m_n >} \quad (c = \mathrm{const}).$$

Thus, the algebras $C_{\{m_n\}} = (L_2)_{\{G\}}$ and $C_{(m_n)} = (L_2)_{(G)}$ are topological. □

Let now the resolvent of the operator T satisfy the condition: There exist constants $\alpha > 0, c > 0$ (for each $\alpha > 0$ there exists a $c = c(\alpha) > 0$) such that

$$\int_{-\infty}^{\infty} \|R_{x+it}(T)h\|^2 \, dx \leq c\gamma^{-1}(\alpha|t|)\|h\|^2 \quad (h \in \mathfrak{H}).$$

Then by Theorem 7.6 the boundary values of the functions $(R_{x+it}g, h)$ (as $t \to 0$) belong to the space $(L_2)'_{(G)}$ (respectively, $(L_2)'_{\{G\}}$), hence $R_{g,h} \in (L_2)'_{(G)}$ (respectively, $R_{g,h} \in (L_2)'_{\{G\}}$) and the following theorem holds.

THEOREM 8.3. *The mapping $f \to f(T) : (f(T)g, h) = (R_{g,h}, f)_{L_2(\mathbb{R}^1)}$ is a homomorphism of the algebra $(L_2)_{\{G\}}$ ($(L_2)_{(G)}$) onto a certain algebra of bounded operators. This mapping is continuous in the sense that if a sequence f_n tends to f in the space $(L_2)_{\{G\}}$ (or $(L_2)_{(G)}$), then $f_n(T) \to f(T)$ in the uniform operator topology.*

Proof. We need only prove that if $f_1, f_2 \in (L_2)_{\{G\}}$, then $(f_1 f_2)(T) = f_1(T) f_2(T)$. This follows from the denseness of $\mathfrak{A}$ in $(L_2)_{\{G\}}$ and the fact that the mapping $f \to f(T)$ is a homomorphism on $\mathfrak{A}$ and is continuous on $(L_2)_G$ in the uniform operator topology. □

8.4. THE SUPPORT OF THE FUNCTIONAL $R_{g,h}$

In this subsection we assume that the operator T is bounded. We also assume that the function $\gamma(t)$ occurring in the estimate (8.1) satisfies the Levinson condition

$$\int_0^{\delta} \ln(-\ln \gamma(t)) \, dt < \infty \quad (\delta > 0 \text{ sufficiently small})$$

Then, according to Theorem 7.9, the class $(L_2)_{\{G\}}$ is not quasi-analytic. So, for any locally finite covering of $\mathbb{R}^1$ there exists a partition of unity by functions from $(L_2)_{\{G\}}$.

DEFINITION. We say that the functional $R_{g,h}$ is concentrated on a closed set Ω if for any function $f \in (L_2)_{\{G\}}$ vanishing in some neighbourhood of Ω we have $(R_{g,h}, f)_{L_2(\mathbb{R}^1)} = 0$.

THEOREM 8.4. *If $\gamma(t)$ satisfies the Levinson condition, then the functional $R_{g,h}$ is concentrated on the spectrum $\sigma(T)$ of the operator T.*

To prove the theorem we need to establish the following fact.

LEMMA 8.1. *Let $\Pi_\delta^\pm$ ($\delta > 0$) be the transformations defined on $L_2(\mathbb{R}^1)$ by*

$$\left(\Pi_\delta^+ f\right)(x) = \frac{1}{2\pi i} \int_{-\infty}^{\infty} \frac{f(\varsigma)}{\varsigma - (x + i\delta)} \, d\varsigma$$

$$\left(\Pi_\delta^- f\right)(x) = \frac{1}{2\pi i} \int_{-\infty}^{\infty} \frac{f(\varsigma)}{\varsigma - (x - i\delta)} \, d\varsigma.$$

Then $\Pi_\delta^+ - \Pi_\delta^- = e^{-D\delta}$.

Proof. Since

$$\Pi_\delta^{\pm} f = \frac{1}{2\pi}(f * p_\delta^{\pm}),$$

where

$$p_\delta^{\pm}(\varsigma) = -i(\varsigma \mp i\delta)^{-1},$$

we have

$$F\left(\Pi_\delta^{+} - \Pi_\delta^{-}\right)F^{-1}f = \frac{1}{2\pi}F(F^{-1}f * p_\delta) \quad (p_\delta = p_\delta^{+} - p_\delta^{-})$$

(here, F and F^{-1} stand for the Fourier transform and its inverse). Thus,

$$F\left(\Pi_\delta^{+} - \Pi_\delta^{-}\right)F^{-1}f = \frac{1}{2\pi}(Fp_\delta f).$$

The identities

$$p_\delta^{\pm}(\varsigma) = 2\pi F^{-1}(\chi(\mp x)e^{\pm\delta x},$$

where

$$\chi(x) = \begin{cases} 1 & \text{if } x > 0, \\ 0 & \text{if } x \leq 0, \end{cases}$$

imply $(Fp_\delta)(x) = 2\pi e^{-\delta|x|}$, whence

$$F\left(\Pi_\delta^{+} - \Pi_\delta^{-}\right)F^{-1}f = e^{-\delta|x|}f. \qquad \square$$

It is now clear that the operator $\Pi_\delta^{+} - \Pi_\delta^{-} = e^{-D\delta}$ is bounded on $L_2(\mathbb{R}^1)$ and maps $L_2(\mathbb{R}^1)$ into $\mathfrak{A}$.

Proof of Theorem 8.4. Suppose $f \in (L_2)_{\{G\}}$ vanishes in a certain neighbourhood U of the spectrum $\sigma(T)$ of the operator T. We observe that the family $f_\delta = e^{-D\delta}f \in \mathfrak{A}$ converges in $(L_2)_{\{G\}}$ to f as $\delta \to 0$. Then

$$(R_{g,h}, f)_{L_2(\mathbb{R}^1)} = (f(T)g, h) = \lim_{\delta\to 0}(f_\delta(T)g, h).$$

But $(\epsilon < \delta)$,

$$\begin{aligned}
(f_\delta(T)g, h) &= -\frac{1}{2\pi i}\int_{\Gamma_\epsilon} f_\delta(z)(R_z(T)g, h)\,\mathrm{d}z \\
&= \frac{1}{4\pi^2}\int_{\Gamma_\epsilon}(R_z(T)g, h)\left(\int_{-\infty}^{\infty} f(\varsigma)\left(\frac{1}{\varsigma - (z + i\delta)} - \frac{1}{\varsigma - (z - i\delta)}\right)\mathrm{d}\varsigma\right)\mathrm{d}z \\
&= \frac{1}{4\pi^2}\int_{\mathbb{R}^1\setminus U} f(\varsigma)\int_{\Gamma_\epsilon}(R_z(T)g, h)\left(\frac{1}{\varsigma - (z + i\delta)} - \frac{1}{\varsigma - (z - i\delta)}\right)\mathrm{d}z\,\mathrm{d}\varsigma \\
&= \frac{1}{2\pi i}\int_{\mathbb{R}^1\setminus U} f(\varsigma)((R_{\varsigma+i\delta}(T) - R_{\varsigma-i\delta}(T))g, h)\,\mathrm{d}\varsigma.
\end{aligned}$$

Taking into account that $|(R_{\varsigma+i\delta}(T)g, h)| \leq \frac{c}{|\varsigma|}$ for large ς and using the analyticity of $(R_z(T)g, h)$ on $\mathbb{R}^1\setminus U$, after limit transition under the last integral we obtain $(R_{g,h}, f)_{L_2(\mathbb{R}^1)} = 0$. $\square$

Under the assumption

$$\|R_{x+it}\| \leq \frac{c}{|t|^k} \quad (c = \text{const}) \tag{8.8}$$

it is possible to determine $f(T)$ for functions $f(x) \in C^l(I)$ (I is an interval containing $\sigma(T)$), where $l = [l] \geq [k] + 1$. In fact, the estimate (8.8) yields inequality (5.22) with $\gamma(t) = t^{2k+2\epsilon-1}$ ($\epsilon > 0$ arbitrarily small). Therefore we can extend the operational calculus of T to functions $f \in C^l(I) \subset W_2^{k+\epsilon}(I)$.

9. The Cauchy Problem for Certain Parabolic Equations

We will consider the Cauchy problem

$$\begin{gathered} \frac{\partial u(x,t)}{\partial t} + (-1)^n \frac{\partial^{2n} u(x,t)}{\partial x^{2n}} = 0 \quad (t \geq 0,\ x \in \mathbb{R}^1) \\ u(x,t)\big|_{t=0} = f(x) \end{gathered} \tag{9.1}$$

If $f \in L_2(\mathbb{R}^1)$, then, as is well-known, the problem has a solution which is infinitely differentiable for $t > 0$, $x \in \mathbb{R}^1$. It turns out that the class of initial data of the problem (9.1) ensuring infinite differentiability in the open upper half-plane of a solution is considerably larger. Here we will describe this class, and distinguish some of its subclasses which possess certain preassigned properties. Moreover, we will study the Cauchy problem for the equation

$$\frac{\partial u(x,t)}{\partial t} = \frac{\partial^2 u(x,t)}{\partial x^2} - x^2 u(x,t). \tag{9.2}$$

It will be shown that the space of initial data guaranteeing the existence of solutions that are smooth for $t > 0$ and square integrable with respect to x, and its subspaces determined by the solutions with given growth (power or exponential) as $t \to 0$, are closely connected with well-known spaces of S type.

9.1. THE SPACE OF INITIAL DATA OF SMOOTH SOLUTIONS

It is not difficult to observe that the problem (9.1) reduces to the Cauchy problem for equation (4.1) if we take for A the positive self-adjoint operator D^{2n} on the space $\mathfrak{H} = L_2(\mathbb{R}^1)$ that is the closure of the operator generated by the expression $(-1)^n \frac{d^{2n}}{dx^{2n}}$ on infinitely differentiable functions with compact support. The space $\mathfrak{A}'(D^{2n})$ serves as the class of initial data f for which the corresponding solution $u(x,t)$ is infinitely differentiable in the upper half-plane ($t > 0$) and square integrable over straight lines parallel to the real axis. In the case considered $\mathfrak{A}(D^{2n})$ can be described as follows.

Denote by Z_β ($\beta > 0$) the set of entire functions $\varphi(x)$ of order not exceeding $\frac{2n}{2n-1}$, of finite type and satisfying the inequality

$$\int_{-\infty}^{\infty} \exp(-\beta|t|^{\frac{2n}{2n-1}}) \int_{-\infty}^{\infty} |\varphi(x+it)|^2 \, dx \, dt < \infty.$$

Clearly, Z_β is a Banach space with respect to the norm

$$\|\varphi\|_{Z_\beta} = \left(\int_{-\infty}^{\infty} \exp(-\beta|t|^{\frac{2n}{2n-1}}) \int_{-\infty}^{\infty} |\varphi(x+it)|^2 \, dx \, dt\right)^{\frac{1}{2}},$$

and also $\|\varphi\|_{Z_{\beta_1}} \geq \|\varphi\|_{Z_{\beta_2}}$ for $\beta_1 < \beta_2$, i.e. the imbedding $Z_{\beta_1} \subseteq Z_{\beta_2}$ is continuous. Let

$$Z = \operatorname*{limind}_{\beta\to\infty} Z_\beta.$$

THEOREM 9.1. *The space $\mathfrak{A}(D^{2n})$ coincides with Z. It consists of those functions $\varphi(x)$ that have an analytic continuation to an entire function $\varphi(x+it)$ of order not exceeding $\frac{2n}{2n-1}$, of finite type and satisfying the inequality*

$$\int_{-\infty}^{\infty} |\varphi(x+it)|^2 \, dx \leq c \exp(\beta|t|^{\frac{2n}{2n-1}}) \tag{9.3}$$

for certain $c > 0, \beta > 0$ depending on φ.

Proof. Let $\varphi \in \mathfrak{A}(D^{2n})$, i.e. there exists a $\delta > 0$ such that $\varphi \in (L_2)_{e,\delta}(D^{2n})$. Since under the unitary transformation $\varphi(x) \to \tilde{\varphi}(\lambda) = \frac{1}{\sqrt{2\pi}} \int_{-\infty}^{\infty} \varphi(x) e^{-i\lambda x} \, dx$ of the space $L_2(\mathbb{R}^1)$ onto itself the operator D^{2n} is transformed to multiplication by λ^{2n},

$$\|\varphi\|^2_{(L_2)_{e,\delta}(D^{2n})} = \int_{-\infty}^{\infty} \exp(2\delta\lambda^{2n}) |\tilde{\varphi}(\lambda)|^2 \, d\lambda < \infty.$$

Then the function

$$\varphi(x+it) = \frac{1}{\sqrt{2\pi}} \int_{-\infty}^{\infty} \tilde{\varphi}(\lambda) e^{i\lambda(x+it)} \, d\lambda = \frac{1}{\sqrt{2\pi}} \int_{-\infty}^{\infty} e^{-\lambda t} \tilde{\varphi}(\lambda) e^{i\lambda x} \, d\lambda$$

is an analytic continuation of $\varphi(x)$ to the complex plane $\mathbb{C}^1$. The application of Parseval's equality to the function $\tilde{\varphi}(\lambda) e^{-\lambda t}$ yields

$$\int_{-\infty}^{\infty} |\varphi(x+it)|^2 \, dx = \int_{-\infty}^{\infty} |\tilde{\varphi}(\lambda)|^2 e^{-2\lambda t} \, d\lambda \leq \int_{-\infty}^{\infty} |\tilde{\varphi}(\lambda)|^2 e^{2|\lambda t|} \, d\lambda.$$

Setting $p = 2n$, $q = \frac{2n}{2n-1}$, $a = (2n\delta)^{\frac{1}{2n-1}} |\lambda|$, $b = |t|$ in the inequality

$$\frac{a^p}{p} + \frac{b^q}{q} \geq ab, \quad \left(\frac{1}{p} + \frac{1}{q} = 1; \ a, b \geq 0\right)$$

yields

$$\begin{aligned} |\lambda t| &\leq \delta\lambda^{2n} + \frac{2n-1}{2n} (2n\delta)^{-\frac{1}{2n-1}} |t|^{\frac{2n}{2n-1}} \\ &= \delta\lambda^{2n} + \frac{\beta}{2} |t|^{\frac{2n}{2n-1}} \qquad \left(\beta = \frac{2n-1}{n} (2n\delta)^{-\frac{1}{2n-1}}\right), \end{aligned}$$

whence

$$\begin{aligned} \int_{-\infty}^{\infty} |\varphi(x+it)|^2 \, dx &\leq \int_{-\infty}^{\infty} |\tilde{\varphi}(\lambda)|^2 \exp(2\delta\lambda^{2n} + \beta|t|^{\frac{2n}{2n-1}}) \, d\lambda \\ &= \exp(\beta|t|^{\frac{2n}{2n-1}}) \|\varphi\|^2_{(L_2)_{e,\delta}(D^{2n})}. \end{aligned} \tag{9.4}$$

Formula (7.4), the Cauchy–Buniakowski inequality and estimate (9.3) just proved imply that the type of the function φ is finite.

Conversely, assume that $\varphi(x)$ $(x \in \mathbb{R}^1)$ admits an analytic continuation $\varphi(x+it)$ to $\mathbb{C}^1$ such that the estimate (9.3) is satisfied. By the same arguments as in subsection 7.3 it is possible to show that $\varphi(x+it) \to 0$ as $|x| \to \infty$, uniformly in any strip $|t| \le \delta$ ($\delta > 0$ is arbitrary). Therefore, because of the Cauchy theorem

$$\int_{-\infty}^{\infty} \varphi(x) e^{i\lambda x}\, dx = \int_{-\infty}^{\infty} \varphi(x+it) e^{-i\lambda(x+it)}\, dx \quad (t \neq 0).$$

Hence, the Fourier transform of the function $\varphi(x+it)$ coincides with $\tilde{\varphi}(\lambda)e^{-\lambda t}$. The Parseval equality and the estimate (9.3) imply

$$\begin{aligned}\int_{-\infty}^{\infty} |\tilde{\varphi}(\lambda)|^2 e^{-2\lambda t}\, d\lambda &= \int_{-\infty}^{\infty} |\varphi(x+it)|^2\, dx \le c\exp\bigl(\beta|t|^{\frac{2n}{2n-1}}\bigr) \\ &= c\exp\bigl(-\beta|t|^{\frac{2n}{2n-1}}\bigr)\exp\bigl(2\beta|t|^{\frac{2n}{2n-1}}\bigr).\end{aligned}$$

By multiplying the last inequality by $\exp\bigl(-2\beta|t|^{\frac{2n}{2n-1}}\bigr)$, integrating with respect to t and then changing the order of integration we obtain

$$\int_{-\infty}^{\infty} |\tilde{\varphi}(\lambda)|^2 \exp(2\delta\lambda^{2n}) \Bigl(\int_{-\infty}^{\infty} \exp\bigl(-2\delta\lambda^{2n} - 2\lambda t - 2\beta|t|^{\frac{2n}{2n-1}}\bigr)\, dt\Bigr)\, d\lambda < \infty,$$

where, for the time being, δ is arbitrary. To prove that $\varphi \in \mathfrak{A}(D^{2n})$ it is now sufficient to show that

$$F_\delta(\lambda) = \int_{-\infty}^{\infty} \exp\bigl(-2\delta\lambda^{2n} - 2\lambda t - 2\beta|t|^{\frac{2n}{2n-1}}\bigr)\, dt \ge c > 0$$

for a certain $\delta > 0$. Since the function $F_\delta(\lambda)$ is positive and continuous on the whole real axis, this inequality needs to be proved only for $|\lambda|$ large.

If $\lambda > 0$, then

$$\begin{aligned}F_\delta(\lambda) &\ge \int_{-\infty}^{0} \exp\bigl(-2\delta\lambda^{2n} - 2\lambda t - 2\beta|t|^{\frac{2n}{2n-1}}\bigr)\, dt \\ &= \int_{0}^{\infty} \exp\bigl(-2\delta\lambda^{2n} + 2\lambda t - 2\beta|t|^{\frac{2n}{2n-1}}\bigr)\, dt.\end{aligned}$$

After the change of variables $t = \xi^{2n-1}$ we have for any $\gamma > 0$,

$$\begin{aligned}F_\delta(\lambda) &\ge (2n-1)\int_{0}^{\infty} \exp(-2\delta\lambda^{2n} + 2\lambda\xi^{2n-1} - 2\beta\xi^{2n})\xi^{2n-2}\, d\xi \\ &\ge (2n-1)\exp(-2\delta\lambda^{2n})\int_{\gamma\lambda-1}^{\gamma\lambda+1} \exp(2\lambda\xi^{2n-1} - 2\beta\xi^{2n})\xi^{2n-2}\, d\xi \\ &\ge 2(2n-1)\exp(-2\delta\lambda^{2n})\exp(2\lambda(\gamma\lambda-1)^{2n-1} - 2\beta(\gamma\lambda+1)^{2n})(\gamma\lambda-1)^{2n-2} \\ &\ge 2(2n-1)\exp(2(-\delta + \gamma^{2n-1} - \beta\gamma^{2n})\lambda^{2n} + P_{2n-1}(\lambda)),\end{aligned}$$

where $P_{2n-1}(\lambda)$ is a polynomial of degree $2n-1$. If we put $\gamma = \frac{2n-1}{2n\beta}$, then the coefficient at λ^{2n} will be equal to $-\delta + \gamma^{2n-1} - \beta\gamma^{2n} = -\delta + \frac{1}{2n}\bigl(\frac{2n-1}{2n\beta}\bigr)^{2n-1}$. It is not difficult to see that δ can be chosen so that this coefficient is positive. Then $F_\delta(\lambda) \geq 2$ when $\lambda > 0$ is sufficiently large. For $\lambda > 0$ the proof is similar.

Let now $\varphi \in (L_2)_{e,\delta}(D^{2n})$. It follows from (9.4) that

$$\|\varphi\|^2_{Z_{\beta_1}} = \int_{-\infty}^{\infty} \exp(-\beta_1 |t|^{\frac{2n}{2n-1}}) \left(\int_{-\infty}^{\infty} |\varphi(x+it)|^2 \, dx\right) dt$$
$$\leq \|\varphi\|^2_{(L_2)_{e,\delta}(D^{2n})} \int_{-\infty}^{\infty} \exp((-\beta_1+\beta)|t|^{\frac{2n}{2n-1}}) \, dt = c^2 \|\varphi\|^2_{(L_2)_{e,\delta}(D^{2n})}$$
$$\text{if } \beta_1 > \beta = \frac{2n-1}{n}(2n\delta)^{-\frac{1}{2n-1}}.$$

Conversely, if $\varphi \in Z_\beta$ for some β, then

$$\|\varphi\|^2_{Z_\beta} = \int_{-\infty}^{\infty} \exp(-\beta|t|^{\frac{2n}{2n-1}}) \left(\int_{-\infty}^{\infty} |\varphi(x+it)|^2 \, dx\right) dt$$
$$= \int_{-\infty}^{\infty} \exp(-\beta|t|^{\frac{2n}{2n-1}}) \left(\int_{-\infty}^{\infty} |\tilde{\varphi}(\lambda)|^2 e^{-2\lambda t} \, d\lambda\right) dt$$
$$= \int_{-\infty}^{\infty} |\tilde{\varphi}(\lambda)|^2 \exp(2\delta\lambda^{2n}) F_\delta(\lambda) \, d\lambda,$$

and, as was established above, for $\delta < \frac{1}{2n}\left(\frac{2n-1}{2n\beta}\right)^{2n-1}$ there exists c^2 such that $F_\delta(\lambda) \geq \frac{1}{c^2}$. Therefore, $\|\varphi\|_{(L_2)_{e,\delta}(D^{2n})} \leq c^2 \|\varphi\|_{Z_\beta}$. So,

$$\operatorname*{limind}_{\beta\to\infty} Z_\beta = \operatorname*{limind}_{\delta\to 0} (L_2)_{e,\delta}(D^{2n}) = \mathfrak{A}(D^{2n}). \qquad \square$$

9.2. BOUNDARY VALUES OF SOLUTIONS IN THE CLASSES OF DISTRIBUTIONS AND ULTRADISTRIBUTIONS

Since in the case considered $\mathfrak{H}^\alpha = W_2^{2n\alpha}(\mathbb{R}^1)$ and, as in subsection 7.1, $\mathfrak{H}^\infty$ coincides with the set of all infinitely differentiable functions all derivatives of which are square integrable on $\mathbb{R}^1$, Corollary 4.2 implies the following statement.

THEOREM 9.2. *In order that the boundary value (as t approaches the real axis) of a solution $u(x,t)$ ($x \in \mathbb{R}^1$, $t > 0$) of equation (9.1) exists in $L_2(\mathbb{R}^1)$ it is necessary and sufficient that*

$$\int_{-\infty}^{\infty} |u(x,t)|^2 \, dx \leq c = \text{const}$$

for sufficiently small $t > 0$. It exists in the Sobolev space $W_2^{-\alpha}(\mathbb{R}^1)$ ($\alpha > 0$) if and only if

$$\int_0^b \int_{-\infty}^{\infty} t^{\frac{\alpha-n}{n}} |u(x,t)|^2 \, dx \, dt < \infty$$

for some $b > 0$. A necessary and sufficient condition for it to belong to $W_2^{-\infty}(\mathbb{R}^1)$ is that

$$\int_{-\infty}^{\infty} |u(x,t)|^2 \, dx \leq ct^{-\alpha}$$

for certain $c > 0, \alpha > 0$ and sufficiently small t.

In view of the well-known inequality

$$\|\varphi^{(m)}\|_{L_2(\mathbb{R}^1)} \le \|\varphi\|_{L_2(\mathbb{R}^1)}^{\frac{n-m}{n}} \|\varphi^{(n)}\|_{L_2(\mathbb{R}^1)}^{\frac{m}{n}} \quad (m \le n),$$

which is valid for any function $\varphi \in W_2^n(\mathbb{R}^1)$ (see, for example, Glazman [1, p. 132]), the space $\mathfrak{G}_{\{\beta\}}(D^{2n})$ (respectively, $\mathfrak{G}_{(\beta)}(D^{2n})$) coincides with the class of ultradifferentiable functions of order $\frac{\beta}{2n}$ of Roumieu (respectively, Beurling) type, i.e. with $\mathfrak{G}_{\{\frac{\beta}{2n}\}}(D)$ (respectively, $\mathfrak{G}_{(\frac{\beta}{2n})}(D)$. For this reason Corollary 4.3 leads in the situation discussed to the following statement.

THEOREM 9.3. *For the boundary value (as $t \to 0$) of a solution $u(x,t)$ ($x \in \mathbb{R}^1, t > 0$) of equation (9.1) to be an ultradistribution of order $\beta > \frac{1}{2n}$ of Roumieu (Beurling) type it is necessary and sufficient that for any $\alpha > 0$ there exists a constant $c = c(\alpha) > 0$ (there exist $c > 0, \alpha > 0$) such that*

$$\int_{-\infty}^{\infty} |u(x,t)|^2 \, dx \le c \exp\left(\alpha t^{-\frac{1}{2n\beta-1}}\right)$$

if $t > 0$ is sufficiently small.

9.3. SPACES OF S TYPE

Spaces of S type consist of infinitely differentiable functions on $\mathbb{R}^1$ with a certain decrease at the infinity and a certain growth of derivatives as their order increases. The respective conditions are expressed by the inequality

$$|x^m f^{(n)}(x)| \le c_{mn} \quad (m, n \in \mathbb{N}),$$

where $\{c_{mn}\}_{m,n\in\mathbb{N}}$ is a double sequence of positive numbers. If these numbers and f are changed arbitrarily, then we obtain the Schwartz space S of rapidly decreasing functions. Imposing various restrictions on a sequence $\{c_{mn}\}_{m,n=1}^{\infty}$ we arrive at specific spaces, called spaces of S type; these are contained in S.

We define three series of spaces of S type, which we will need in the sequel.

For any $\alpha \ge 0$, any $\beta \ge 0$ we set

$$S_\alpha = \left\{f \in S : \exists h > 0 \, \forall n \in \mathbb{N} \sup_{m\in\mathbb{N}} \left(\|x^m f^{(n)}(x)|_{L_2(\mathbb{R}^1)} h^{-m} m^{-m\alpha}\right) < \infty\right\}, \tag{9.5}$$

$$S^\beta = \left\{f \in S : \exists h > 0 \, \forall m \in \mathbb{N} \sup_{n\in\mathbb{N}} \left(\|x^m f^{(n)}(x)|_{L_2(\mathbb{R}^1)} h^{-n} n^{-n\beta}\right) < \infty\right\}, \tag{9.6}$$

$$S_\alpha^\beta = \left\{f \in S : \exists h > 0 \sup_{m,n\in\mathbb{N}} \left(\|x^m f^{(n)}(x)|_{L_2(\mathbb{R}^1)} h^{-(m+n)} m^{-m\alpha} n^{-n\beta}\right) < \infty\right\}. \tag{9.7}$$

We would like to note that the inequalities (9.5) - (9.7) are equivalent to analogous ones with $\|\cdot\|_{L_2(\mathbb{R}^1)}$ replaced by $\sup_{x\in\mathbb{R}^1} |\cdot|$ because m, n run through all natural numbers while h is an arbitrary positive number. These spaces of S type can be characterized as follows (see Gelfand, Shilov [1, chapt. IV]).

S_α ($\alpha > 0$) consists of all infinitely differentiable functions that satisfy the estimate

$$|f^{(n)}(x)| \le c_n \exp(-a|x|^{\frac{1}{\alpha}}) \quad (c_n > 0, \ a > 0; \ n = 0, 1, \ldots).$$

S_0 coincides with the Schwartz space $\mathcal{D}(\mathbb{R}^1)$.

Each of the functions from S^1 has an analytic continuation to a certain strip $|\operatorname{Im} z| < \delta$.

For $0 < \beta < 1$, $f \in S^\beta$ if and only if f has an analytic continuation as an antire function satisfying the estimate

$$|x^m f(x+it)| \leq c_m \exp\left(b|t|^{\frac{1}{1-\beta}}\right) \quad (m \in \mathbb{N},\ 0 < b = \text{const}).$$

S^0 coincides with the space of entire functions of order 1 that decrease on the real axis faster than $\frac{1}{|x|^k}$ (for all $k > 0$) as $|x| \to \infty$.

The space S_α^β is non-trivial if: 1) $\alpha + \beta \geq 1$, $\alpha > 0$, $\beta > 0$; 2) $\alpha = 0$, $\beta > 1$; 3) $\beta = 0$, $\alpha > 1$; in each case indicated it is dense in $L_2(\mathbb{R}^1)$.

The space S_α^1 ($\alpha > 0$ arbitrary) consists of the functions $f(x)$ having an analytic continuation $f(x+it)$ to a certain strip $|t| < \delta$, depending on $f(x)$, for which

$$|f(x+it)| \leq c \exp\left(-a|x|^{\frac{1}{\alpha}}\right)$$

($c > 0$, $a > 0$ are constants).

If $0 < \beta < 1$ and $\alpha > 1 - \beta$, then S_α^β consists of the functions $f(x)$ that have an analytic continuation as an entire function $f(x+it)$ satisfying the inequality

$$|f(x+it)| \leq c \exp\left(-a|x|^{\frac{1}{\alpha}} + b|t|^{\frac{1}{1-\beta}}\right) \quad (a > 0, b > 0).$$

The Schwartz space corresponds formally to the symbol S_∞^∞.

One of the principal advantages of spaces of S type is the possibility of using freely Fourier transformation: the Fourier transformation maps a space of a certain S type onto a space of another S type; namely: $\widetilde{S}_\alpha = S^\alpha, \widetilde{S}_\beta = S_\beta, \widetilde{S_\alpha^\beta} = S_\beta^\alpha$.

It can be seen from the definition of the space S_α^β that $S_\alpha^\beta \subseteq S_\alpha \cap S^\beta$. The converse inclusion is also true.

THEOREM 9.4. *For any $\alpha \geq 0$ and any $\beta \geq 0$ we have*

$$S_\alpha \cap S^\beta = S_\alpha^\beta.$$

Proof. Suppose first that $\alpha + \beta < 1$. Since S_α^β is trivial in this case, in order to prove

$$S_\alpha \cap S^\beta \subseteq S_\alpha^\beta$$

it suffices to establish that an arbitrary function $f \in S_\alpha \cap S^\beta$ is identical zero.

If $\alpha = 0$, $0 < \beta < 1$ and $f \in S_0 \cap S^\beta$, then $f(x)$ has compact support and it can be analytically continued as an entire function. By virtue of the uniqueness theorem for analytic functions, $f(x) \equiv 0$. The case $0 \leq \alpha < 1$, $\beta = 0$ and $f \in S_\alpha \cap S^0$ reduces to the previous one using Fourier transformation.

If $\alpha, \beta > 0$ and $\alpha + \beta < 1$, then $f \in S_\alpha \cap S^\beta$ has an analytic continuation as an entire function $f(x+it)$ satisfying the inequalities

$$|f(x)| \leq c_0 \exp\left(-a|x|^{\frac{1}{\alpha}}\right),$$
$$|f(x+it)| \leq c_0 \exp\left(b|t|^{\frac{1}{1-\beta}}\right) \quad (a > 0, b > 0).$$

Since $\alpha+\beta<1$, $\frac{1}{\alpha}>\frac{1}{1-\beta}$ and either $0<\alpha<\frac{1}{2}$ or $0<\beta<\frac{1}{2}$. If $0<\beta<\frac{1}{2}$, then the function $F(z)=f(z)f(iz)$ has exponential order of growth $\frac{1}{1-\beta}<2$ and tends to zero on the coordinate axes as $z\to\infty$. According to the Phragmén–Lindelöf theorem (see, for example, Evgrafov [1, §6, Chapter 8]) for each sector $\frac{(k-1)\pi}{2}<\arg z<\frac{k\pi}{2}$ $(k=1,2,3,4)$, we deduce that $F(z)$ is bounded on the whole complex plane. Since $F(z)\to 0$ on the coordinate axes as $z\to\infty$, we have, by the Liouville theorem, $F(z)\equiv 0$, whence $f(z)\equiv 0$. The case $0<\alpha<\frac{1}{2}$ is reduced to the previous one using Fourier transformation.

Assume now that $\alpha+\beta\geq 1$. According to the definition it is possible to find, for any function $f\in S_\alpha\cap S^\beta$, $c>0$ and $h>0$ such that

$$\|x^m f(x)\|_{L_2(\mathbf{R}^1)}\leq ch^m m^{m\alpha},$$
$$\|f^{(n)}(x)\|_{L_2(\mathbf{R}^1)}\leq ch^n n^{n\beta}.$$

Then, from Leibniz' formula on differentiation of a product, the Cauchy-Buniakowski inequality and estimates for the binomial coefficients we obtain

$$\begin{aligned}
\|x^m f^{(n)}(x)\|^2_{L_2(\mathbf{R}^1)} &= \left(x^m f^{(n)}(x), x^m f^{(n)}(x)\right)_{L_2(\mathbf{R}^1)}\\
&= \left|\left(\frac{d^n}{dx^n}(x^{2m}f^{(n)}(x)), f(x)\right)_{L_2(\mathbf{R}^1)}\right|\\
&= \left|\sum_{k=0}^{r} C_n^k \frac{(2m)!}{(2m-k)!}\left(x^{2m-k}f^{(2n-k)}(x), f(x)\right)_{L_2(\mathbf{R}^1)}\right|\\
&\leq \sum_{k=0}^{r}\frac{(2m)!n!}{k!(2m-k)!(n-k)!}\left\|x^{2m-k}f(x)\right\|_{L_2(\mathbf{R}^1)}\left\|f^{(2n-k)}(x)\right\|_{L_2(\mathbf{R}^1)}\\
&\leq 2^{m+n}\sum_{k=0}^{r} k!c^2h^{2(m+n-k)}(2m-k)^{\alpha(2m-k)}(2n-k)^{\beta(2n-k)}\\
&\leq c^2\left(2^{2(\alpha+\beta+\frac{1}{2})}h^2\right)^{m+n}\sum_{k=0}^{r} r^{k(\alpha+\beta)}m^{2\alpha m-\alpha k}n^{2\beta n-\beta k}\\
&\leq c^2h_1^{2(m+n)}\sum_{k=0}^{r}(2m)^{\alpha k}n^{\beta k}m^{2\alpha m-\alpha k}n^{2\beta n-\beta k}\\
&\leq c^2 2^{\alpha r}(r+1)h_1^{2(m+n)}m^{2\alpha m}n^{2\beta n}\\
&\leq c^2h_2^{2(m+n)}m^{2\alpha m}n^{2\beta n},
\end{aligned}$$

where $r=\min\{2m,n\}$, $h_1=2^{\alpha+\beta+\frac{1}{2}}h$, $h_2=2^{\frac{1}{2}(\alpha+1)}h_1$ $(m,n\in\mathbb{N})$. This enables us to conclude that

$$\left\|x^m f^{(n)}(x)\right\|_{L_2(\mathbf{R}^1)}\leq ch_2^{m+n}m^{m\alpha}n^{n\beta},$$

i.e. $f\in S_\alpha^\beta$. □

9.4. SPACES OF S TYPE AS SUBCLASSES OF INFINITELY DIFFERENTIABLE VECTORS OF THE HARMONIC OSCILLATOR

We consider the positive self-adjoint operator A on the space $L_2(\mathbb{R}^1)$ that is the closure of the operator generated by $-\frac{d^2}{dx^2} + x^2$ on the set $C_0^\infty(\mathbb{R}^1)$ of infinitely differentiable functions with compact support. The spectrum of this operator is discrete, its eigenvalues are $\lambda_n(A) = 2n+1$ $(n = 0, 1, \ldots)$ and the corresponding eigenfunctions

$$h_n(x) = (2^n n!)^{-\frac{1}{2}}(-1)^n \pi^{-\frac{1}{4}} e^{\frac{x^2}{2}} (e^{-x^2})^{(n)}$$
$$= \pi^{-\frac{3}{4}} 2^{\frac{n}{2}} (n!)^{-\frac{1}{2}} \int_{-\infty}^{\infty} (x + i\xi)^n \exp(-\frac{x^2}{2} - \xi)\, d\xi$$

(the Hermite functions) form an orthonormal basis in $L_2(\mathbb{R}^1)$.

Denote by B the operator of multiplication by x^2 on $L_2(\mathbb{R}^1) : (Bf)(x) = x^2 f(x)$. As before, D^2 will stand for the closure of the operation $-\frac{d^2}{dx^2}$ on $C_0^\infty(\mathbb{R}^1)$. These are positive self-adjoint operators, and their spectra fill the whole semi-axis $[0, \infty)$.

It is not difficult to observe that $S_{\frac{\beta}{2}} \subseteq \mathfrak{S}_{\{\beta\}}(B)$. Moreover, as was noted in subsection 9.2, $\mathfrak{S}_{\{\beta\}}(D^2) = \mathfrak{S}_{\frac{\beta}{2}}(D)$, whence the inclusion $\mathfrak{S}_{\{\beta\}}(D^2) \supseteq S^{\frac{\beta}{2}}$ follows. Therefore

$$S_{\frac{\beta}{2}} \cap S^{\frac{\beta}{2}} \subseteq \mathfrak{S}_{\{\beta\}}(B) \cap \mathfrak{S}_{\{\beta\}}(D^2).$$

On the other hand, it can be seen from the proof of Theorem 9.4 that, in particular, $\mathfrak{S}_{\{\beta\}}(B) \cap \mathfrak{S}_{\{\beta\}}(D^2) \subseteq S_{\frac{\beta}{2}}^{\frac{\beta}{2}} = S_{\frac{\beta}{2}} \cap S^{\frac{\beta}{2}}$. Hence,

$$S_{\frac{\beta}{2}}^{\frac{\beta}{2}} = \mathfrak{S}_{\{\beta\}}(B) \cap \mathfrak{S}_{\{\beta\}}(D^2).$$

THEOREM 9.5. *For $\beta \geq 1$, $\mathfrak{S}_{\{\beta\}}(A) = S_{\frac{\beta}{2}}^{\frac{\beta}{2}}$.*

To prove this theorem we need the following statement.

LEMMA 9.1. *If $f(x) \in S$, then*

$$(A^n f)(x) = \sum_{0 \leq p+q \leq 2n} C_{p,q}^{(n)} x^p f^{(q)}(x), \tag{9.8}$$

where the coefficients $C_{p,q}^{(n)}$ satisfy the estimate

$$|C_{p,q}^{(n)}| \leq 10^n n^{n-\frac{1}{2}(p+q)}. \tag{9.9}$$

Proof. We proceed by induction on n. For $n = 1$ the representation (9.8) is obvious. Suppose it is true for $n = m$. Then

$$(A^{m+1}f)(x) = \left(-\frac{\mathrm{d}^2}{\mathrm{d}x^2} + x^2\right) \sum_{0\le p+q\le 2m} C_{p,q}^{(m)} x^p f^{(q)}(x)$$
$$= \sum_{0\le p+q\le 2(m+1)} C_{p,q}^{(m+1)} x^p f^{(q)}(x),$$
$$C_{p,q}^{(m+1)} = -(p+2)(p+1)C_{p+2,q}^{(m)} - 2(p+1)C_{p+q,q-1}^{(m)} - C_{p,q-2}^{(m)} + C_{p-2,q}^{(m)}$$

(the coefficients $C_{p,q}^{(m)}$ with $p+q > 2m$, $p < 0$ or $q < 0$ are assumed to be equal to zero). By virtue of the induction assumption, we have by using (9.9) and the inequalities

$$p+1 \le 2m+1 < 2(m+1),$$
$$(p+2)(p+1) \le (2m+2)(2m+1) < 4(m+1)^2,$$

the following estimate for $C_{p,q}^{(m+1)}$:

$$\left|C_{p,q}^{(m+1)}\right| \le (p+2)(p+1)10^m m^{m-\frac{1}{2}(p+q+2)} + 2(p+1)10^m m^{m-\frac{1}{2}(p+q)} +$$
$$+ 10^m m^{m-\frac{1}{2}(p+q-2)} + 10^m m^{m-\frac{1}{2}(p+q-2)}$$
$$< 10^{m+1}(m+1)^{m+1-\frac{1}{2}(p+q)}.$$

Hence, formula (9.8) has been established, □

Proof of theorem 9.5. We wish to show first the inclusion

$$\mathfrak{G}_{\{\beta\}}(A) \subseteq S_{\frac{\beta}{2}}^{\frac{\beta}{2}} = \mathfrak{G}_{\{\beta\}}(B) \cap \mathfrak{G}_{\{\beta\}}(D^2). \tag{9.10}$$

Because of Corollaries 2.2, 2.3 and Remark 2.2, it suffices to prove

$$e^A \ge \mu e^{\tau B}, \quad e^A \ge \mu_1 e^{\tau_1 D^2} \tag{9.11}$$

for some μ, μ_1, τ, $\tau_1 > 0$. Since under Fourier transformation the operator B becomes D^2 while A is invariant, the first of the inequalities (9.11) implies the second with $\mu_1 = \mu_2$, $\tau_1 = \tau_2$. So, let us prove the inequality

$$\mu^{-1}e^{-\tau B} \ge e^{-A},$$

which is equivalent to (9.11).

The operator e^{-A} is an integral operator with kernel

$$K(x,y) = \sum_{n=0}^{\infty} e^{-(2n+1)} h_n(x)h_n(y).$$

Take $0 < \epsilon < \frac{1}{2}$. Since

$$|(x+i\xi)^n \exp(-\epsilon(x^2+\xi^2))| = (x^2+\xi^2)^{\frac{n}{2}} \exp(-\epsilon(x^2+\xi^2))$$
$$\le \epsilon^{-\frac{n}{2}} \sup_{u>0}(u^{\frac{n}{2}}e^{-u}) \le \left(\frac{n}{2\epsilon}\right)^{\frac{n}{2}}$$

and $\int_{-\infty}^{\infty} \exp(-(1-\epsilon)\xi^2)\,d\xi = \sqrt{\frac{\pi}{1-\epsilon}}$, by means of the Stirling formula we get

$$|h_n(x)| \le \frac{2^{\frac{1}{4}}}{e^{\frac{1}{4}}\sqrt{\pi(1-\epsilon)}}\left(\frac{e}{\epsilon}\right)^{\frac{n}{2}} e^{-(\frac{1}{2}-\epsilon)x^2}.$$

If ϵ has been taken from the interval $(\frac{1}{e}, \frac{1}{2})$ then

$$\begin{aligned} K(x,y) &\le \frac{\sqrt{2}}{\sqrt{e}(1-\epsilon)\pi}\sum_{n=0}^{\infty} e^{-2n-1}\left(\frac{e}{\epsilon}\right)^n e^{-(\frac{1}{2}-\epsilon)(x^2+y^2)} \\ &= \frac{\sqrt{2}}{e^{\frac{3}{2}}\pi(1-\epsilon)} e^{-(\frac{1}{2}-\epsilon)(x^2+y^2)} \sum_{n=0}^{\infty}(\epsilon e)^{-n} \\ &= \frac{\sqrt{2}\epsilon e^{-(\frac{1}{2}-\epsilon)(x^2+y^2)}}{e^{\frac{1}{2}}\pi(1-\epsilon)(\epsilon e-1)}. \end{aligned}$$

This estimate and the Cauchy–Buniakowski inequality enable us to obtain the relations

$$\begin{aligned} \left(e^{-A}f,f\right)_{L_2(\mathbf{R}^1)} &\le \int_{-\infty}^{\infty}\int_{-\infty}^{\infty} K(x,y)|f(x)||f(y)|\,dx\,dy \\ &\le \frac{\sqrt{2}\epsilon}{e^{\frac{1}{2}}\pi(1-\epsilon)(\epsilon e-1)}\int_{-\infty}^{\infty}\int_{-\infty}^{\infty} e^{-(\frac{1}{2}-\epsilon)(x^2+y^2)}|f(x)||f(y)|\,dx\,dy \\ &\le \frac{\sqrt{2}\epsilon}{e^{\frac{1}{2}}\pi(1-\epsilon)(\epsilon e-1)}\left(\int_{-\infty}^{\infty} e^{-(\frac{1}{2}-\epsilon)x^2}|f(x)|\,dx\right)^2 \\ &\le \mu^{-1}\int_{-\infty}^{\infty} e^{-(\frac{1}{2}-\epsilon)x^2}|f(x)|^2\,dx \\ &= \mu^{-1}\left(e^{-(\frac{1}{2}-\epsilon)B}f,f\right)_{L_2(\mathbf{R}^1)}, \end{aligned}$$

where $\mu^{-1} = \frac{\sqrt{2}\epsilon}{e^{\frac{1}{2}}\pi(1-\epsilon)(\epsilon e-1)}\int_{-\infty}^{\infty} e^{-(\frac{1}{2}-\epsilon)x^2}\,dx = \frac{2\epsilon}{\sqrt{\pi e}(1-\epsilon)(\epsilon e-1)(1-2\epsilon)}$, which proves (9.10). We establish the converse inclusion.

Let $f \in S_{\frac{\beta}{2}}^{\frac{\beta}{2}}$ $(\beta \ge 1)$, i.e. there exist $c > 0, h \ge 1$ such that

$$\|x^m f^{(n)}(x)\|_{L_2(\mathbf{R}^1)} \le c h^{m+n} m^{\frac{\beta}{2}m} n^{\frac{\beta}{2}n}. \tag{9.12}$$

By the representation (9.8) and the inequality (9.12)

$$\begin{aligned} \|A^n f\|_{L_2(\mathbf{R}^1)} &\le \sum_{0\le p+q\le 2n} |C_{p,q}^{(n)}|\,\|x^p f^{(q)}(x)\|_{L_2(\mathbf{R}^1)} \\ &\le c\sum_{0\le p+q\le 2n} 10^n n^{n-\frac{1}{2}(p+q)} h^{p+q} p^{\frac{\beta}{2}p} q^{\frac{\beta}{2}q} \\ &\le c\cdot 10^n h^{2n}(2n)^{\beta n}(2n+1)^2 \\ &= c(30\cdot 2^\beta h^2)^n n^{n\beta}. \end{aligned}$$

So, $f \in \mathfrak{G}_{\{\beta\}}(A)$. □

It can be seen from the above-mentioned arguments that the Schwartz space $S = S_\infty^\infty$ coincides with $(L_2)^\infty(A)$.

We would also like to note that it is possible, according to Theorem 9.5, to introduce on S_α^α $(\alpha \geq \frac{1}{2})$ the topology of the space $\mathfrak{G}_{\{2\alpha\}}(A)$. It can be proved that the topology introduced in this way coincides with the generally accepted topology of the space S_α^α (see, for example, Gelfand and Shilov [1]). Therefore, $\mathfrak{G}'_{\{2\alpha\}}(A) = (S_\alpha^\alpha)'$. In particular, $(L_2)^{-\infty}(A) = S'$, where S' is the space of generalized functions of slow growth.

Thus, on the basis of Theorem 4.3 and Corollaries 4.2 and 4.3 we arrive at the conclusion: 1) a solution $u(x,t)$ of equation (9.2) that is square integrable with respect to x has a boundary value in the space $\left(S_{\frac{1}{2}}^{\frac{1}{2}}\right)'$; the solution can be uniquely recovered from its boundary value. In other words, $\left(S_{\frac{1}{2}}^{\frac{1}{2}}\right)'$ is the space of initial data of smooth solutions of the Cauchy problem for equation (9.2) that are square integrable with respect to x; 2) the space $\left(S_\alpha^\alpha\right)'$ with $\alpha > \frac{1}{2}$ is the set of boundary values of all solutions $u(x,t)$ of the Cauchy problem for (9.2) that satisfy the estimate

$$\int_{-\infty}^{\infty} |u(x,t)|^2 \, dx \leq c \exp\left(at^{-\frac{1}{2\alpha-1}}\right)$$

for sufficiently small $t > 0$ (any $a > 0, c = c(a)$); 3) S' coincides with the space of initial data of the problem considered which guarantee power growth of a solution $u(x,t)$ $(t > 0)$ as t approaches the real axis, i.e. $\int_{-\infty}^{\infty} |u(x,t)|^2 \, dx \leq ct^{-\alpha}$ $(t \to 0)$ (certain $\alpha > 0$ and $c > 0$, depending on u).

CHAPTER 3

Extensions of Symmetric Operators

1. Dissipative Extensions and Boundary Value Problems

This section deals with the theory of dissipative (in particular, self-adjoint) extensions of a symmetric operator with equal deficiency numbers. Unlike the traditional approach used in the theory of self-adjoint extensions of symmetric operators, in this section we describe the extensions in terms of abstract boundary conditions which are convenient to use in boundary value problems for differential equations. In this connection, the notions of a dissipative linear relation and a boundary value space are of principal importance.

1.1. SYMMETRIC OPERATORS

DEFINITION. A linear operator A on a Hilbert space $\mathfrak{H}$ with dense domain $\mathfrak{D}(A)$ is called symmetric if $A \subset A^*$, i.e. if for every $f, g \in \mathfrak{D}(A)$,

$$(Af, g) = (f, Ag). \tag{1.1}$$

Since an adjoint operator is always closed, the relation $A \subset A^*$ implies that a symmetric operator is closeable. The symmetric operators considered below are assumed to be closed.

EXAMPLE. Let $\mathfrak{H} = L_2(0,1)$, $Af = -\frac{d^2 f}{dt^2}$, $\mathfrak{D}(A) = \overset{\circ}{W}{}_2^2(0,1)$. It follows from the definition of $\overset{\circ}{W}{}_2^2(0,1)$ and the continuity of the imbedding $\overset{\circ}{W}{}_2^2(0,1) \subset C^1[0,1]$ that the operator A is closed. A is called the minimal operator generated on $L_2(0,1)$ by the differential expression

$$(l[f])(t) = -\frac{d^2 f(t)}{dt^2}.$$

For every $f, g \in W_2^2(0,1)$,

$$\int_0^1 (l[f])(t)\overline{g(t)}\,dt - \int_0^1 f(t)\overline{(l[g])(t)}\,dt$$
$$= (f(1)\overline{g'(1)} - f'(1)\overline{g(1)}) - (f(0)\overline{g'(0)} - f'(0)\overline{g(0)}), \tag{1.2}$$

so (1.1) holds for $f, g \in \mathfrak{D}(A)$. Hence the operator A is symmetric.

Let us find A^*. Set $M : Mf = -f''$, $\mathfrak{D}(M) = W_2^2(0,1)$. According to (1.2), for $f \in \mathfrak{D}(A)$, $g \in \mathfrak{D}(M)$,

$$(Af, g) - (f, Mg) = 0,$$

i.e. $M \subseteq A^*$. On the other hand, if $g \in \mathfrak{D}(A^*)$, $h = A^*g$ and $g_0 \in W_2^2(0,1)$ is an arbitrary solution of the equation $l[x] = h$, then for all $f \in \mathfrak{D}(A)$ we have, by (1.2),

$$(f, h) = (f, l[g_0]) = (l[f], g_0) = (Af, g_0).$$

Moreover, by the definition of adjoint operator, $(f, h) = (f, A^*g) = (Af, g)$. Therefore,

$$(Af, g - g_0) = 0$$

for all $f \in \mathcal{D}(0,1)$, whence it follows that $g - g_0$ is a generalized solution of the equation

$$\frac{d^2}{dt^2}(g - g_0) = 0.$$

Thus (see Remark 5.3, Chapter 2), $g(t) = g_0(t) + \alpha + \beta t$ for almost all $t \in [0,1]$. Consequently, $g \in W_2^2(0,1) = \mathfrak{D}(M)$, $A^*g = h = l[g_0] = l[g]$.

So, $A^* = M$. The set $\mathfrak{D}(A)$ is distinguished from $\mathfrak{D}(A^*)$ by the boundary conditions $f(0) = f'(0) = f(1) = f'(1) = 0$. By posing boundary conditions in a certain way, it is possible to determine proper extensions of the operator A, i.e. operators $\widetilde{A}$ for which $A \subset \widetilde{A} \subset A^*$. Later we shall see that the condition $f(0) = f(1) = 0$, for example, gives a self-adjoint extension of A. On the other hand, we will see that broad classes of extensions of A can be described in terms of boundary conditions. Moreover, the notion of a boundary condition, generalized in a suitable way, will turn out to be a universal means for describing extensions of general symmetric operators.

Now we return to the general case. Let A be an arbitrary symmetric operator on $\mathfrak{H}$ and λ any non-real complex number. Put $\mathfrak{M}_\lambda = \mathfrak{R}(A - \lambda E)$. Then the set $\mathfrak{N}_\lambda = \mathfrak{H} \ominus \mathfrak{M}_\lambda$ coincides with the eigenspace of the operator A^* corresponding to the eigenvalue $\overline{\lambda}$.

Indeed, if $g \in \mathfrak{N}_\lambda$, then for any vector $f \in \mathfrak{D}(A)$,

$$(Af - \lambda f, g) = 0, \tag{1.3}$$

that is,

$$(Af, g) = (f, \overline{\lambda} g). \tag{1.4}$$

This means, by the definition of the operator A^*, that $g \in \mathfrak{D}(A^*)$ and $A^*g = \overline{\lambda} g$. Conversely, if $A^*g = \overline{\lambda} g$, then for every vector $f \in \mathfrak{D}(A)$ the equality (1.4), which is equivalent to (1.3), holds, so $g \in \mathfrak{N}_\lambda$.

DEFINITION. Linear sets $\mathfrak{M}_1, \mathfrak{M}_2, \ldots, \mathfrak{M}_n$ are called linearly independent if the equality

$$f_1 + f_2 + \cdots + f_n = 0, \quad f_k \in \mathfrak{M}_k \quad (k = 1, \ldots, n)$$

is possible only for $f_1 = f_2 = \cdots = f_n = 0$.

THEOREM 1.1. *The linear sets* $\mathfrak{D}(A)$, $\mathfrak{N}_{\overline{\lambda}}$, $\mathfrak{N}_\lambda$ *are linearly independent and their direct sum coincides with* $\mathfrak{D}(A^*)$:

$$\mathfrak{D}(A^*) = \mathfrak{D}(A) \dotplus \mathfrak{N}_{\overline{\lambda}} \dotplus \mathfrak{N}_\lambda. \tag{1.5}$$

Proof. We first prove the linear independence part. Let

$$f + g + h = 0, \quad f \in \mathfrak{D}(A), \quad g \in \mathfrak{N}_{\overline{\lambda}}, \quad h \in \mathfrak{N}_\lambda. \tag{1.6}$$

Letting the operator $A^* - \overline{\lambda}E$ act on both sides of the equation in (1.6) we obtain

$$(A - \overline{\lambda}E)f + (\lambda - \overline{\lambda})g = 0. \tag{1.7}$$

But $(A - \overline{\lambda}E)f \in \mathfrak{M}_{\overline{\lambda}}$, $(\lambda - \overline{\lambda})g \in \mathfrak{N}_{\overline{\lambda}}$. Since these subspaces are orthogonal, (1.7) is possible only if

$$(A - \overline{\lambda}E)f = 0, \quad (\lambda - \overline{\lambda})g = 0. \tag{1.8}$$

The first of these equalities shows that $\overline{\lambda}(f, f) = (Af, f)$. Because of the symmetry of A the number (Af, f) is real, so $f = 0$. The second equality from (1.8) gives $g = 0$. In view of (1.6) we conclude that $h = 0$ too.

We now prove formula (1.5). Clearly, $\mathfrak{D}(A^*) \supset \mathfrak{D}(A) \dotplus \mathfrak{N}_{\overline{\lambda}} \dotplus \mathfrak{N}_\lambda$. It remains to show that every vector $u \in \mathfrak{D}(A^*)$ can be represented in the form

$$u = f + g + h, \quad f \in \mathfrak{D}(A), \quad g \in \mathfrak{N}_{\overline{\lambda}}, \quad h \in \mathfrak{N}_\lambda. \tag{1.9}$$

First of all we verify that the set $\mathfrak{M}_{\overline{\lambda}}$ is closed. For any $\varphi \in \mathfrak{D}(A)$,

$$\begin{aligned} \|(A - \overline{\lambda}E)\varphi\|^2 &= ((A - \operatorname{Re}\lambda \cdot E)\varphi + i \operatorname{Im}\lambda \cdot \varphi,\ (A - \operatorname{Re}\lambda \cdot E)\varphi + i \operatorname{Im}\lambda \cdot \varphi) \\ &= \|(A - \operatorname{Re}\lambda \cdot E)\varphi\|^2 + i \operatorname{Im}\lambda(\varphi, (A - \operatorname{Re}\lambda \cdot E)\varphi) - \\ &\quad - i \operatorname{Im}\lambda((A - \operatorname{Re}\lambda \cdot E)\varphi, \varphi) + (\operatorname{Im}\lambda)^2\|\varphi\|^2 \\ &= \|(A - \operatorname{Re}\lambda \cdot E)\varphi\|^2 + (\operatorname{Im}\lambda)^2\|\varphi\|^2, \end{aligned}$$

whence

$$\|(A - \overline{\lambda}E)\varphi\|^2 \geq (\operatorname{Im}\lambda)^2\|\varphi\|^2 \quad (\varphi \in \mathfrak{D}(A)). \tag{1.10}$$

If a sequence $\varphi_n \in \mathfrak{H}$ is such that $(A - \overline{\lambda}E)\varphi_n \to \psi \in \mathfrak{H}$ as $n \to \infty$, then, on account of (1.10) and the completeness of the space $\mathfrak{H}$, we have $\varphi_n \to \varphi_0$, where φ_0 is a certain vector from $\mathfrak{H}$. Since the operator A is closed, we have $\varphi_0 \in \mathfrak{D}(A)$, $\psi = (A - \overline{\lambda}E)\varphi_0$. That means that $\mathfrak{M}_{\overline{\lambda}}$ is closed.

Due to the definition of $\mathfrak{N}_\lambda$ and the closedness of $\mathfrak{M}_{\overline{\lambda}}$, the orthogonal decomposition

$$\mathfrak{H} = \mathfrak{M}_{\overline{\lambda}} \oplus \mathfrak{N}_{\overline{\lambda}}.$$

holds. In particular,

$$(A^* - \overline{\lambda}E)u = v' + v'', \quad v' \in \mathfrak{M}_{\overline{\lambda}}, \quad v'' \in \mathfrak{N}_{\overline{\lambda}}.$$

Since $v' = (A - \overline{\lambda}E)f$ $(f \in \mathfrak{D}(A))$ and v'' can be written as $v'' = (\lambda - \overline{\lambda})g$ $(g \in \mathfrak{N}_{\overline{\lambda}})$, we get

$$(A^* - \overline{\lambda}E)u = (A - \overline{\lambda}E)f + (\lambda - \overline{\lambda})g = (A^* - \overline{\lambda}E)(f + g),$$

because $A^*f = Af$, $A^*g = \lambda g$. This implies that $(A^* - \overline{\lambda}E)(u - f - g) = 0$ hence, $h = u - f - g \in \mathfrak{N}_\lambda$. Thus, $u = f + g + h$, where $f \in \mathfrak{D}(A)$, $g \in \mathfrak{N}_{\overline{\lambda}}$, $h \in \mathfrak{N}_\lambda$, and the representation (1.9) has been proved. □

Theorem 1.1 gives a complete description of the operator A^*, since

$$A^*u = Af + \lambda g + \overline{\lambda}h.$$

It is possible to prove (see, for example, Naimark [2, p. 175]) that the numbers $n_+ = \dim \mathfrak{N}_\lambda$, $n_- = \dim \mathfrak{N}_{\overline{\lambda}}$ do not change when λ runs through the half-plane $\operatorname{Im}\lambda > 0$. The pair (n_+, n_-) is called the deficiency index of the operator A, while the numbers n_+ and n_- are called its deficiency numbers.

The following corollary directly follows from Theorem 1.1.

COROLLARY 1.1. *A closed symmetric operator is self-adjoint if and only if its deficiency index is equal to* $(0,\ 0)$.

1.2. DISSIPATIVE OPERATORS

DEFINITION. A linear operator A on $\mathfrak{H}$ with dense domain $\mathfrak{D}(A)$ is called dissipative if

$$\operatorname{Im}(Af, f) \geq 0 \quad \text{for all} \quad f \in \mathfrak{D}(A). \tag{1.11}$$

A linear operator A is called accumulative if

$$\operatorname{Im}(Af, f) \leq 0 \quad \text{for all} \quad f \in \mathfrak{D}(A).$$

Since a linear operator A is accumulative if and only if $-A$ is dissipative, all results concerning dissipative operators can be immediately transferred to accumulative operators.

A dissipative (accumulative) operator is called maximal dissipative (maximal accumulative) if it has no non-trivial, that is, different from A itself, dissipative (accumulative) extensions.

A dissipative operator is always closeable. In fact, let $f_n \to 0$ ($f_n \in \mathfrak{D}(A)$), $Af_n \to g$ ($n \to \infty$). For any $f \in \mathfrak{D}(A)$ and any complex number α,

$$\operatorname{Im}(A(f + \alpha f_n), f + \alpha f_n) \geq 0,$$

from which follows, by limit transition,

$$\operatorname{Im}(Af, f) + \operatorname{Im}\alpha(g, f) \geq 0.$$

This inequality can hold for arbitrary complex α if and only if $(g, f) = 0$. By virtue of the denseness of $\mathfrak{D}(A)$ in $\mathfrak{H}$, we have $g = 0$.

It is obvious that the closure of a dissipative operator is also dissipative. A maximal dissipative operator is always closed.

THEOREM 1.2. *Every dissipative operator has a maximal dissipative extension. A dissipative operator A is maximal dissipative if and only if $\mathfrak{R}(A-\lambda E)$ coincides with the whole space for any λ with* $\operatorname{Im}\lambda < 0$.

Proof. Let A be a closed dissipative operator and $\operatorname{Im}\lambda < 0$. Then $\mathfrak{R}(A-\lambda E)$ is closed. In fact, inequality (1.11) implies $\operatorname{Im}((A-\lambda E)f, f) \geq -\operatorname{Im}\lambda(f, f)$. But $\operatorname{Im}((A-\lambda E)f, f) \leq \|(A-\lambda E)f\|\|f\|$, i.e.

$$-\operatorname{Im}\lambda \cdot \|f\| \leq \|(A-\lambda E)f\|,$$

whence follows that $\mathfrak{R}(A-\lambda E)$ is closed.

Now there are two possibilities. If $\mathfrak{R}(A-\lambda E) = \mathfrak{H}$ for some λ ($\operatorname{Im}\lambda < 0$), then the operator A is maximal dissipative (otherwise for a non-trivial extension $\widetilde{A}$ of it we could find an element $f_0 \neq 0$ such that $(\widetilde{A}-\lambda E)f_0 = 0$, hence $\operatorname{Im}(\widetilde{A}f_0, f_0) = \operatorname{Im}\lambda(f_0, f_0) < 0$ which contradicts the dissipativeness of $\widetilde{A}$). If $\mathfrak{R}(A-\lambda E) \neq \mathfrak{H}$, then the operator A has a non-trivial dissipative extension $\widetilde{A}$; it can be constructed as follows: having the λ we set $\mathfrak{N} = \mathfrak{H} \ominus \mathfrak{R}(A-\lambda E)$, $\mathfrak{D}(\widetilde{A}) = \mathfrak{D}(A) \dotplus \mathfrak{N}$, $\widetilde{A}(f+u) = Af + \bar{\lambda}u$, $f \in \mathfrak{D}(A)$, $u \in \mathfrak{N}$. The operator $\widetilde{A}$ is defined correctly: if $f + u = 0$, then $(Af, f) = (Au, u) = (u, A^*u) = (u, \bar{\lambda}u) = \lambda(u, u)$ and $\operatorname{Im}(Af, f) \leq 0$, whence $u = 0$ and therefore $f = 0$.

The operator $\widetilde{A}$ is dissipative:

$$\begin{aligned}(\widetilde{A}(f+u), f+u) &= (Af, f) + \bar{\lambda}(u, u) + (Af, u) + \bar{\lambda}(u, f)\\ &= (Af, f) + \bar{\lambda}(u, u) + (f, A^*u) + \bar{\lambda}(u, f)\\ &= (Af, f) + \bar{\lambda}(u, u) + \lambda(f, u) + \bar{\lambda}(u, f),\end{aligned}$$

so

$$\operatorname{Im}(\widetilde{A}(f+u), f+u) = \operatorname{Im}(Af, f) + \operatorname{Im}\bar{\lambda}(u, u) \geq 0.$$

Finally, it is not difficult to check that $\mathfrak{R}(\widetilde{A}-\lambda E) = \mathfrak{H}$. As shown above, the operator $\widetilde{A}$ is maximal dissipative. □

We would like to note that a symmetric operator is dissipative and accumulative at the same time. It can be seen from Theorem 1.2 and Corollary 1.1 that an operator is simultaneously maximal dissipative and maximal accumulative if and only if it is self-adjoint. A symmetric operator which is maximal dissipative or maximal accumulative is called maximal symmetric. According to Theorem 1.2 a necessary and sufficient condition for a symmetric operator A to be maximal symmetric is that one of its deficiency numbers be equal to zero. It is possible to prove (see, for example, Naimark [2, p.168]) that an operator A is maximal symmetric if and only if it has no non-trivial symmetric extensions.

THEOREM 1.3. *Let $\widetilde{A}$ be a dissipative (accumulative) extension of a symmetric operator A. Then $\widetilde{A} \subset A^*$.*

Proof. Suppose that $\widetilde{A} \supset A$ is a dissipative extension. By Theorem 1.2 we may assume the operator $\widetilde{A}$ to be maximal dissipative.

Let us consider the operators

$$B = (A - iE)(A + iE)^{-1}, \quad \widetilde{B} = (\widetilde{A} - iE)(\widetilde{A} + iE)^{-1}. \tag{1.12}$$

It is not difficult to see that the number $-i$ cannot be an eigenvalue of a dissipative (hence, symmetric) operator. Therefore, formulas (1.12) make sense. The operator B isometrically maps $\mathfrak{M}_{-i}$ onto $\mathfrak{M}_i$: if $f \in \mathfrak{M}_{-i}$, i.e. $f = (A + iE)g$ $(g \in \mathfrak{D}(A))$, then

$$\begin{aligned} \|Bf\|^2 &= \|(A - iE)g\|^2 = ((A - iE)g, (A - iE)g) \\ &= (Ag, Ag) - i(g, Ag) + i(Ag, g) + (g, g); \\ \|f\|^2 &= \|(A + iE)g\|^2 = ((A + iE)g, (A + iE)g) \\ &= (Ag, Ag) + i(g, Ag) - i(Ag, g) + (g, g); \end{aligned}$$

the equality $(Ag, g) = (g, Ag)$ implies $\|Bf\| = \|f\|$.

It is proved analogously that the operator $\widetilde{B}$ (defined on the whole of $\mathfrak{H}$ according to Theorem 1.2) is a contraction, i.e. $\|\widetilde{B}f\| \leq \|f\|$. It follows from (1.12) that

$$A = -i(B + E)(B - E)^{-1}, \quad \widetilde{A} = -i(\widetilde{B} + E)(\widetilde{B} - E)^{-1}.$$

Clearly, $\widetilde{B}$ is an extension of the operator B. Let $u \in \mathfrak{N}_{-i}$. Then $u \perp \mathfrak{D}(B)$. For any vector $v \in \mathfrak{D}(B)$ and any complex number ξ,

$$\|\xi v + u\|^2 - \|\widetilde{B}(\xi v + u)\|^2 \geq 0. \tag{1.13}$$

Taking into account that $\|\widetilde{B}v\| = \|Bv\| = \|v\|$, we obtain from (1.13)

$$\|u\|^2 - \|\widetilde{B}u\|^2 - 2\,\mathrm{Re}(\xi(\widetilde{B}v, \widetilde{B}u)) \geq 0.$$

Since ξ is arbitrary, we have $(\widetilde{B}v, \widetilde{B}u) = 0$. Thus, $\widetilde{B}u \perp \widetilde{B}v = Bv$ for all $v \in \mathfrak{D}(B)$, i.e. $\widetilde{B}u \perp \mathfrak{M}_i$, or, which is the same, $\widetilde{B}u \in \mathfrak{N}_i$. This means that $\widetilde{B} = B \oplus C$, where C is a contraction from $\mathfrak{N}_{-i}$ into $\mathfrak{N}_i$.

Let g be an arbitrary vector from $\mathfrak{D}(\widetilde{A})$. Set $f = (\widetilde{A} + iE)g$. Then $\widetilde{B}f = (\widetilde{A} - iE)g$. It follows from the two last equalities that

$$\begin{aligned} g &= \frac{1}{2i}(f - \widetilde{B}f) = \frac{1}{2i}(f_1 + f_2 - Bf_1 - Cf_2) = \\ &= \frac{1}{2i}(f_1 - Bf_1) + \frac{1}{2i}(f_2 - Cf_2), \end{aligned}$$

where $f_1 \in \mathfrak{M}_{-i}$, $f_2 \in \mathfrak{N}_{-i}$. By the definition of the operator B we have $f_1 - Bf_1 \in \mathfrak{D}(A)$. Further, from what we have proved above, $f_2 - Cf_2 \in \mathfrak{N}_{-i} \dotplus \mathfrak{N}_i$. Hence, $g \in \mathfrak{D}(A^*)$ and

$$A^*g = \frac{1}{2i}A(f_1 - Bf_1) + \tfrac{1}{2}(f_2 + Cf_2) = \tfrac{1}{2}(f_1 + Bf_1) + \tfrac{1}{2}(f_2 + Cf_2).$$

On the other hand,

$$\widetilde{A}g = \tfrac{1}{2}(f + \widetilde{B}f) = \tfrac{1}{2}(f_1 + Bf_1) + \tfrac{1}{2}(f_2 + Cf_2),$$

so $\widetilde{A}g = A^*g$. □

1.3. LINEAR RELATIONS

By a linear relation in a Hilbert space $\mathfrak{H}$ we mean an arbitrary linear subset $\theta \in \mathfrak{H} \oplus \mathfrak{H}$. If θ_1, θ_2 are linear relations and $\theta_1 \subset \theta_2$, then θ_2 is called an extension of θ_1.

DEFINITION. A linear relation θ is called dissipative (accumulative, symmetric) if for any $\{x, x'\} \in \theta$ we have $\operatorname{Im}(x', x) \geq 0$ (respectively, if $\operatorname{Im}(x', x) \leq 0$, or $\operatorname{Im}(x', x) = 0$).

A dissipative (accumulative, symmetric) relation is called maximal dissipative (maximal accumulative, maximal symmetric) if it has no non-trivial dissipative (accumulative, symmetric) extensions. A symmetric relation is called Hermitian (or self-adjoint) if it is maximal dissipative and maximal accumulative at the same time.

It is useful to associate with a dissipative relation θ the operator U_θ determined as follows:

$$\mathfrak{D}(U_\theta) = \{x' + ix : \{x, x'\} \in \theta\}, \quad U_\theta(x' + ix) = x' - ix.$$

The operator U_θ (called the Cayley transform of the relation θ) is correctly defined: if $\{x, x'\} \in \theta$, $\{y, y'\} \in \theta$ and $x' + ix = y' + iy$, then $\{x - y, x' - y'\} \in \theta$, $x' - y' = -i(x - y)$. On the other hand,

$$0 \leq \operatorname{Im}(x' - y', x - y) = \operatorname{Im}(-i(x - y), x - y) = -\|x - y\|^2,$$

whence $x = y$, $x' = y'$.

Observing that

$$\|U_\theta(x' + ix)\|^2 = \|x'\|^2 + \|x\|^2 - 2\operatorname{Im}(x', x),$$
$$\|x' + ix\|^2 = \|x'\|^2 + \|x\|^2 + 2\operatorname{Im}(x', x),$$

and because θ is dissipative, we have

$$\|U_\theta(x' + ix)\| \leq \|x' + ix\|, \quad \{x, x'\} \in \theta. \tag{1.14}$$

If the relation θ is symmetric, then equality holds in (1.14).

The following theorem describes the structure of linear relations belonging to the classes introduced above.

THEOREM 1.4. *The linear relations determined by the equations*

$$(K - E)x' + i(K + E)x = 0, \tag{1.15}$$
$$(K - E)x' - i(K + E)x = 0, \tag{1.16}$$

where K is a contraction on $\mathfrak{H}$, are maximal dissipative and maximal accumulative, respectively. Conversely, any maximal dissipative (maximal accumulative) relation can be represented in the form of (1.15) *(or* (1.16)*), where a contraction K is uniquely determined by a relation. A maximal dissipative (maximal accumulative) relation is maximal symmetric if and only if the operator K in* (1.15) *(in* (1.16)*) is* *isometric. The general form of Hermitian relations is given by the formula* (1.15) *or* (1.16) *where* K *is a unitary operator on* $\mathfrak{H}$.

Proof. Consider the linear relation θ determined by (1.15). Let $\{x, x'\} \in \theta$. Then

$$K(x' + ix) = x' - ix,$$
$$\|K(x' + ix)\|^2 = \|x'\|^2 + \|x\|^2 - 2\,\mathrm{Im}(x', x);$$
$$\|x' + ix\|^2 = \|x'\|^2 + \|x\|^2 + 2\,\mathrm{Im}(x', x).$$

By subtracting we obtain

$$4\,\mathrm{Im}(x', x) = \|x' + ix\|^2 - \|K(x' + ix)\|^2 \geq 0. \tag{1.17}$$

Thus, the relation θ is dissipative.

Next, for a vector $u \in \mathfrak{H}$ we set

$$x' = \tfrac{1}{2}(u + Ku), \quad x = \frac{1}{2i}(u - Ku).$$

Then $\{x, x'\} \in \theta$, $x' + ix = u$, $x' - ix = Ku$. This means that $\mathfrak{D}(U_\theta) = \mathfrak{H}$, $U_\theta = K$.

If $\widetilde{\theta}$ is a dissipative extension of the relation θ, then $U_{\widetilde{\theta}} \supset U_\theta$ which is possible only if $U_{\widetilde{\theta}} = U_\theta$. The equality $\widetilde{\theta} = \theta$ follows from this. Hence we have proved that θ is maximal dissipative.

Now, let θ be an arbitrary maximal dissipative relation and let U_θ be its Cayley transform. Then $\mathfrak{D}(U_\theta) = \mathfrak{H}$. Indeed, if we assume the opposite we will be able to continuously extend U_θ to $\overline{\mathfrak{D}(U_\theta)}$. If $\overline{\mathfrak{D}(U_\theta)} \neq \mathfrak{H}$, then we extend U_θ to $\mathfrak{H}$, putting it equal to zero on $\mathfrak{H} \ominus \overline{\mathfrak{D}(U_\theta)}$. As a result we obtain a contraction $K \supset U_\theta$ defined on the whole space. Consider the relation $\widetilde{\theta}$ corresponding to the equation (1.15) with the K just constructed. As was proved, $\widetilde{\theta}$ is dissipative; it is clear that $\widetilde{\theta} \supset \theta$. Since the θ is maximal dissipative, $\widetilde{\theta} = \theta$, whence $U_{\widetilde{\theta}} = U_\theta$ and we obtain a contradiction.

Thus, U_θ is a contraction on $\mathfrak{H}$. By the definition of the Cayley transform,

$$U_\theta(x' + ix) = x' - ix \tag{1.18}$$

for all $\{x, x'\} \in \theta$. As has already been mentioned, equality (1.18) defines a maximal dissipative relation $\widetilde{\theta} \supset \theta$. Because θ is maximal dissipative, we have $\theta = \widetilde{\theta}$, i.e. the relation θ is determined by a formula of the kind (1.15).

It can be seen from (1.17) that a relation θ is maximal symmetric if and only if K is an isometric operator.

Let θ be a maximal accumulative relation. Then the relation

$$\theta_1 = \{\{-x, x'\} : \{x, x'\} \in \theta\} \tag{1.19}$$

is maximal dissipative. As has been shown, $\{x, x'\} \in \theta_1$ (or, which is the same, $\{-x, x'\} \in \theta$) if and only if

$$(K - E)x' + i(K + E)(-x) = 0,$$

which is equivalent to (1.16). The converse assertion is proved in a similar way.

Let, finally, θ be a Hermitian relation. Then the equations

$$(K_1 - E)x' + i(K_1 + E)x = 0,$$
$$(K_2 - E)x' - i(K_2 + E)x = 0$$

(K_1, K_2 are isometries on $\mathfrak{H}$) are equivalent. We deduce from them that

$$K_1K_2(x' - ix) = x' - ix, \quad K_2K_1(x' + ix) = x' + ix.$$

Since $\{x'+ix \mid \{x,x'\} \in \theta\} = \{x'-ix \mid \{x,x'\} \in \theta\} = \mathfrak{H}$, we have $K_1K_2 = K_2K_1 = E$, i.e. K_1 and K_2 are unitary. The converse is trivial. □

REMARK 1.1. It is not difficult to verify that a maximal symmetric relation θ is either maximal dissipative or maximal accumulative.

In fact, let us consider the Cayley transform U_θ of the relation θ and that of a relation θ_1 of the kind (1.19), U_{θ_1}. They are isometric operators. Since $\mathfrak{D}(U_\theta) = \mathfrak{R}(U_{\theta_1})$ and $\mathfrak{R}(U_\theta) = \mathfrak{D}(U_{\theta_1})$, at least one of the operators U_θ, U_{θ_1} is defined on the whole of $\mathfrak{H}$ (otherwise we could construct an isometric extension of one of them to the whole space $\mathfrak{H}$, hence a symmetric extension of the corresponding linear relation, contradicting its being maximal symmetric). Thus, θ is determined by one of the equations (1.15) or (1.16).

REMARK 1.2. Equation (1.15) with a unitary operator K may be also written as

$$(\cos C)x - (\sin C)x' = 0,$$

where C is a self-adjoint operator on $\mathfrak{H}$ connected with the operator K by the equality $K = e^{-2iC}$.

REMARK 1.3. Arguments similar to those in the proof of Theorem 1.4 enable us to represent any (generally speaking, not maximal) dissipative (accumulative) relation in $\mathfrak{H}$ in the form

$$K(x' + ix) = x' - ix, \quad x' + ix \in \mathfrak{D}(K)$$

(respectively,

$$K(x' - ix) = x' + ix, \quad x' - ix \in \mathfrak{D}(K)),$$

where K is a linear operator with the property $\|Kf\| \leq \|f\|$ for all $f \in \mathfrak{D}(K)$. A dissipative (accumulative) relation in $\mathfrak{H}$ is symmetric if and only if the operator K associated with it is isometric.

1.4. BOUNDARY VALUE SPACES AND A DESCRIPTION OF DISSIPATIVE EXTENSIONS

Let A be a closed symmetric operator on $\mathfrak{H}$ with equal, finite or infinite, deficiency numbers.

DEFINITION. A triple $(\mathcal{H}, \Gamma_1, \Gamma_2)$, where $\mathcal{H}$ is a Hilbert space and Γ_1, Γ_2 are linear mappings of $\mathfrak{D}(A^*)$ into $\mathcal{H}$, is called a boundary value space of the operator A if:

1) for any $f, g \in \mathfrak{D}(A^*)$,

$$(A^* f, g) - (f, A^* g) = (\Gamma_1 f, \Gamma_2 g)_{\mathcal{H}} - (\Gamma_2 f, \Gamma_1 g)_{\mathcal{H}};$$

2) for any $F_1, F_2 \in \mathcal{H}$ there exists a vector $f \in \mathfrak{D}(A^*)$ such that $\Gamma_1 f = F_1$, $\Gamma_2 f = F_2$.

It follows from the definition of a boundary value space that $f \in \mathfrak{D}(A)$ if and only if $\Gamma_1 f = \Gamma_2 f = 0$. In fact, according to 2) it is possible to select for a given $f \in \mathfrak{D}(A)$ a vector $g \in \mathfrak{D}(A^*)$ so that $\Gamma_1 g = -\Gamma_2 f$, $\Gamma_2 g = \Gamma_1 f$. Then

$$0 = (A^* f, g) - (f, A^* g) = (\Gamma_1 f, \Gamma_2 g)_{\mathcal{H}} - (\Gamma_2 f, \Gamma_1 g)_{\mathcal{H}} = \|\Gamma_1 f\|^2_{\mathcal{H}} + \|\Gamma_2 f\|^2_{\mathcal{H}},$$

whence $\Gamma_1 f = \Gamma_2 f = 0$. Conversely, if $\Gamma_1 f = \Gamma_2 f = 0$, then $(A^* f, g) = (f, A^* g)$ for any $g \in \mathfrak{D}(A^*)$. This means that $f \in \mathfrak{D}(A^{**}) = \mathfrak{D}(A)$.

EXAMPLE. Let A be the operator on $L_2(0,1)$ considered in the previous example (see subsection 1.1). A boundary value space of this operator can be chosen as follows:

$$\mathcal{H} = \mathbb{C}^2, \quad \Gamma_1 f = \{-f(0), f(1)\}, \quad \Gamma_2 f = \{f'(0), f'(1)\}.$$

The form of the mappings Γ_1 and Γ_2 in this example justifies the term "boundary value space". In what follows we construct boundary value spaces of more general differential operators.

Notice that the dimension of the space $\mathcal{H}$ in the example happens to be equal to each of the deficiency numbers of A, as a matter of fact, the deficiency index of A is $(2,2)$.

THEOREM 1.5. *For any symmetric operator with deficiency index $(n,n)(n \leq \infty)$ there exists a boundary value space $(\mathcal{H}, \Gamma_1, \Gamma_2)$ with* dim $\mathcal{H} = n$.

Proof. Formula (1.5) with $\lambda = i$ yields

$$\mathfrak{D}(A^*) = \mathfrak{D}(A) \dotplus \mathfrak{N}_{-i} \dotplus \mathfrak{N}_i. \tag{1.20}$$

Let P_{-i}, P_i be the projectors of $\mathfrak{D}(A^*)$ onto $\mathfrak{N}_{-i}$ and $\mathfrak{N}_i$ which correspond to the decomposition (1.20). Since dim $\mathfrak{N}_{-i} =$ dim $\mathfrak{N}_i$, there exists an isometric mapping from $\mathfrak{N}_i$ onto $\mathfrak{N}_{-i}$. We denote it by U.

Set $\mathcal{H} = \mathfrak{N}_{-i}$ (with inner product induced by that in $\mathfrak{H}$), $\Gamma_1 = P_{-i} + UP_i$, $\Gamma_2 = -iP_{-i} + iUP_i$. We verify that $(\mathcal{H}, \Gamma_1, \Gamma_2)$ is a boundary value space of A. According to (1.20), if $f, g \in \mathfrak{D}(A^*)$, then

$$f = f_0 + P_{-i} f + P_i f, \quad g = g_0 + P_{-i} g + P_i g,$$

where $f_0, g_0 \in \mathfrak{D}(A)$. Taking into account the equalities $A^* P_i = -iP_i$, $A^* P_{-i} =$

iP_{-i} and the symmetry of A, we obtain

$$(A^*f,g)-(f,A^*g)=2i((P_{-i}f,P_{-i}g)-(P_if,P_ig)).$$

On the other hand, a consequence of the isometry of U is that

$$(\Gamma_1 f,\Gamma_2 g)_{\mathcal{H}}-(\Gamma_2 f,\Gamma_1 g)_{\mathcal{H}}=2i((P_{-i}f,P_{-i}g)-(P_if,P_ig)).$$

So, condition 1) in the definition of a boundary value space is satisfied.

Further, if $F_1,F_2\in\mathcal{H}$, then we take $f\in\mathfrak{D}(A^*)$, setting

$$f=f_0+f_{-i}+f_i,$$

where f_0 is any vector from $\mathfrak{D}(A)$, $f_{-i}=\frac{1}{2i}(iF_1-F_2)\in\mathfrak{N}_{-i}$, $f_i=\frac{1}{2i}U^{-1}(iF_1+F_2)\in\mathfrak{N}_i$. It is simple to check that $\Gamma_1 f=F_1$, $\Gamma_2 f=F_2$. □

Let now $(\mathcal{H},\Gamma_1,\Gamma_2)$ be an arbitrary boundary value space of A.

THEOREM 1.6. *If K is a contraction on $\mathfrak{H}$, then the restriction of the operator A^* to the set of vectors $f\in\mathfrak{D}(A^*)$ satisfying the condition*

$$(K-E)\Gamma_1 f+i(K+E)\Gamma_2 f=0 \tag{1.21}$$

or

$$(K-E)\Gamma_1 f-i(K+E)\Gamma_2 f=0 \tag{1.22}$$

is a maximal dissipative, respectively, a maximal accumulative extension of A. Conversely, any maximal dissipative (maximal accumulative) extension of A is the restriction of A^* to the set of vectors $f\in\mathfrak{D}(A^*)$ satisfying (1.21) ((1.22)), where a contraction K is uniquely determined by an extension. The maximal symmetric extensions of an operator A on $\mathfrak{H}$ are described by the conditions (1.21) ((1.22)), where K is an isometric operator. These conditions define a self-adjoint extension if K is unitary. In the last case (1.21), (1.22) are equivalent to

$$(\cos C)\Gamma_2 f-(\sin C)\Gamma_1 f=0,$$

where C is a self-adjoint operator on $\mathfrak{H}$. The general form of dissipative (accumulative) extensions of A is given by the condition

$$K(\Gamma_1 f+i\Gamma_2 f)=\Gamma_1 f-i\Gamma_2 f,\quad \Gamma_1 f+i\Gamma_2 f\in\mathfrak{D}(K), \tag{1.23}$$

(respectively,

$$K(\Gamma_1 f-i\Gamma_2 f)=\Gamma_1 f+i\Gamma_2 f,\quad \Gamma_1 f-i\Gamma_2 f\in\mathfrak{D}(K)), \tag{1.24}$$

where K is a linear operator satisfying $\|Kf\|\le\|f\|$ $(f\in\mathfrak{D}(K))$, while symmetric extensions are described by the formulas (1.23) and (1.24), where K is an isometric operator.

Proof. We present the proof only for maximal dissipative extensions (the other cases are proved analogously).

Let $\widetilde{A}$ be a maximal dissipative extension of A. By Theorem 1.3, $\widetilde{A} \subset A^*$. Set $\theta = \{\{\Gamma_2 f, \Gamma_1 f\} : f \in \mathfrak{D}(\widetilde{A})\}$, where $(\mathcal{H}, \Gamma_1, \Gamma_2)$ is a boundary value space of A. Then, by virtue of the property 1) for $(\mathcal{H}, \Gamma_1, \Gamma_2)$, θ is a dissipative relation in $\mathcal{H}$. If $\widetilde{\theta} \supset \theta$ and $\widetilde{\theta}$ is a dissipative relation, then the operator $\approx{A}$, defined as the restriction of A^* to $\mathfrak{D}(\approx{A}) = \{f \in \mathfrak{D}(A^*) : \{\Gamma_2 f, \Gamma_1 f\} \in \widetilde{\theta}\}$, is a dissipative extension of the operator $\widetilde{A}$. Hence, $\approx{A} = \widetilde{A}$. If $\{x, x'\} \in \widetilde{\theta}$, then $x = \Gamma_2 f$, $x' = \Gamma_1 f$ with a certain $f \in \mathfrak{D}(A^*)$ (according to the definition of a boundary value space). Here $f \in \mathfrak{D}(\approx{A})$, hence $f \in \mathfrak{D}(\widetilde{A})$, i.e. $\{x, x'\} \in \theta$. Thus, $\widetilde{\theta} = \theta$ and θ is a maximal dissipative relation. The required result follows from Theorem 1.4.

Suppose that $\widetilde{A}$ is the restriction of A^* to the set $\mathfrak{D}(\widetilde{A})$ of vectors satisfying (1.21). We consider the relation

$$\theta = \{\{\Gamma_2 f, \Gamma_1 f\} : f \in \mathfrak{D}(\widetilde{A})\}.$$

According to Theorem 1.4, the relation θ is maximal dissipative. This fact and property 1) for $(\mathcal{H}, \Gamma_1, \Gamma_2)$ imply that $\widetilde{A}$ is dissipative. We prove that it is maximal dissipative.

Assume that $\approx{A}$ is a dissipative extension of $\widetilde{A}$. Denote

$$\widetilde{\theta} = \{\{\Gamma_2 f, \Gamma_1 f\} : f \in \mathfrak{D}(\approx{A})\}.$$

Since $\widetilde{\theta}$ is a dissipative extension of the relation θ, $\widetilde{\theta} = \theta$. If now $f \in \mathfrak{D}(\approx{A})$, then for some $g \in \mathfrak{D}(\widetilde{A})$, we have $\Gamma_1 f = \Gamma_1 g$, $\Gamma_2 f = \Gamma_2 g$. This implies that $f - g \in \mathfrak{D}(A)$, hence $f \in \mathfrak{D}(\widetilde{A})$. So, $\widetilde{A} = \approx{A}$, and $\widetilde{A}$ is maximal dissipative. □

2. Positive Definite Symmetric Operators and Solvable Extensions of Them

In this section we describe, in terms of abstract boundary conditions, all solvable extensions of a positive definite symmetric operator A that are restrictions of A^*.

2.1. THE FRIEDRICHS EXTENSION

Recall that a symmetric operator A on a Hilbert space $\mathfrak{H}$ is called positive definite with lower bound $\mu > 0$ if $(Af, f) \geq \mu \|f\|^2$ for all $f \in \mathfrak{D}(A)$. For instance, the operator A discussed in the example of subsection 1.1 is positive definite.

In fact, integration by parts yields for it:

$$(Af, f) = \int_0^1 |f'(t)|^2 \, dt \tag{2.1}$$

for any function $f \in \mathfrak{D}(A)$. Since $f(0) = 0$ we have $f(t) = \int_0^t f'(\tau)\,d\tau$, whence

$$|f(t)|^2 \leq \left(\int_0^t |f'(\tau)|\,d\tau\right)^2 \leq \int_0^t d\tau \int_0^t |f'(\tau)|^2\,d\tau$$
$$= t\int_0^t |f'(\tau)|^2\,d\tau \leq t\int_0^1 |f'(\tau)|^2\,d\tau \quad (0 \leq t \leq 1).$$

By integrating with respect to t from zero to one, we find that

$$\int_0^1 |f(t)|^2\,dt \leq \tfrac{1}{2}\int_0^1 |f'(t)|^2\,dt. \tag{2.2}$$

The inequality $(Af, f) \geq 2\|f\|^2$ follows from (2.1) and (2.2). Thus, the operator A is positive definite with lower bound 2.

In subsection 1.4 we have constructed a boundary value space of A and have, in particular, described (see Theorem 1.6) its self-adjoint extensions. If, for example, $K = -E$, then we obtain the boundary condition $f(0) = f(1) = 0$, which defines a certain self-adjoint extension $A' \supset A$. The foregoing arguments show that

$$(A'f, f) \geq 2\|f\|^2.$$

So, A' is a positive definite extension with a lower bound that is the same as that of A. The following theorem establishes a possibility of constructing positive definite self-adjoint extensions while preserving a lower bound for an arbitrary positive definite symmetric operator.

THEOREM 2.1. *A positive definite symmetric operator A has a positive definite self-adjoint extension A' with a lower bound that is the same as that of A.*

Proof. Complete $\mathfrak{D}(A)$ with respect to the norm

$$\|u\|_A = (Au, u)^{1/2}$$

which corresponds to the inner product $(u, v)_A = (Au, v)$. We denote this completion by $\mathfrak{H}_A$. The positive definiteness of the operator A implies the inclusion $\mathfrak{H}_A \subset \mathfrak{H}$.

We define the operator A' as the restriction A^* to the set

$$\mathfrak{D}(A') = \mathfrak{D}(A^*) \cap \mathfrak{H}_A.$$

It is obvious that A' is an extension of A. We show that A' is symmetric and that it has a lower bound equal to μ.

If $u, v \in \mathfrak{D}(A')$, then there exist sequences $\{u_n\}_{n=1}^{\infty}$ and $\{v_n\}_{n=1}^{\infty}$ such that $u_n, v_n \in \mathfrak{D}(A)$, $\|u - u_n\|_A \to 0$, $\|v - v_n\|_A \to 0$, $n \to \infty$. Hence, the expression $(u_n, v_m)_A = (Au_n, v_m) = (u_n, Av_m)$ has a limit as $n, m \to \infty$. We calculate this limit:

$$\lim_{n\to\infty}\lim_{m\to\infty}(Au_n, v_m) = \lim_{n\to\infty}(Au_n, v) = \lim_{n\to\infty}(u_n, A^*v)$$
$$= \lim_{n\to\infty}(u_n, A'v) = (u, A'v).$$

On the other hand,

$$\lim_{n\to\infty}\lim_{m\to\infty}(Au_n, v_m) = \lim_{n\to\infty}\lim_{m\to\infty}(u_n, Av_m) = \lim_{m\to\infty}(u, Av_m)$$
$$= \lim_{m\to\infty}(A'u, v_m) = (A'u, v).$$

Consequently, $(A'u, v) = (u, A'v)$. Next,

$$(A'u, u) = \lim_{n\to\infty}(A'u, u_n) = \lim_{n\to\infty}(A^*u, u_n) = \lim_{n\to\infty}(u, Au_n)$$
$$= \lim_{n\to\infty}(u, u_n)_A = (u, u)_A \geq \mu\|u\|^2.$$

It remains to prove that the operator A' is self-adjoint. It suffices to verify that its range coincides with $\mathfrak{H}$. If this has been proved, then for $g \in \mathfrak{D}(A'^*)$ there exists a vector $h \in \mathfrak{D}(A')$ such that $A'h = A'^*g$. Observing that

$$(A'f, g) = (f, A'^*g) = (f, A'h) = (A'f, h)$$

for any $f \in \mathfrak{D}(A')$, i.e. $g - h \perp A'f$ for all $f \in \mathfrak{D}(A')$, we will obtain $g = h \in \mathfrak{D}(A')$, hence $A'^* = A'$.

So, let v be an arbitrary element from $\mathfrak{H}$. As a result of the inequalities

$$|(u, v)| \leq \|u\|\|v\| \leq \frac{1}{\sqrt{\mu}}\|v\|\|u\|_A, \quad \forall u \in \mathfrak{D}(A)$$

the expression (u, v) is a continuous linear functional on the dense subset $\mathfrak{D}(A)$ of the space $\mathfrak{H}_A$. According to the Riesz theorem there exists a unique element $v^* \in \mathfrak{H}_A$ such that $(u, v) = (u, v^*)_A = (Au, v^*)$ for all $u \in \mathfrak{D}(A)$. This implies that $v^* \in \mathfrak{D}(A^*) \cap \mathfrak{H}_A$ and $A^*v^* = A'v^* = v$. The proof is complete. □

The extension A' constructed in the process of proving Theorem 2.1 is called the Friedrichs extension of the operator A.

It is not difficult to check that the extension A' of the concrete operator A from subsection 1.1 is the Friedrichs extension of this operator.

2.2. SOLVABLE EXTENSIONS

DEFINITION. Let A be a positive definite symmetric operator. An operator $\widetilde{A}$ satisfying the condition $A \subset \widetilde{A} \subset A^*$ and for which $\widetilde{A}^{-1}$ exists as a bounded operator with domain $\mathfrak{H}$ is called a proper solvable extension of the operator A.

Theorem 2.1 shows that a positive definite symmetric operator has always proper solvable extensions. We will give a complete description of them in terms of abstract boundary conditions.

We fix a positive definite extension $\widetilde{A}$ of A (usually the Friedrichs extension). The decomposition

$$\mathfrak{D}(A^*) = \mathfrak{D}(\widetilde{A}) \dotplus \operatorname{Ker} A^* \tag{2.3}$$

holds. Indeed, assume $f \in \mathfrak{D}(A^*)$ and $g = \widetilde{A}^{-1}A^*f$, $h = f - g$. Then $g \in \mathfrak{D}(\widetilde{A})$, $A^*h = A^*f - \widetilde{A}g = A^*f - A^*f = 0$, i.e. $h \in \operatorname{Ker} A^*$ and $f = g + h$.

Denote by $\widetilde{P}$ and P_0 the projectors of $\mathfrak{D}(A^*)$ onto $\mathfrak{D}(\widetilde{A})$ and $\operatorname{Ker} A^*$, respectively, corresponding to the decomposition (2.3).

DEFINITION. A triple $(\mathcal{H}, \Gamma_1\Gamma_2)$, where $\mathcal{H}$ is a Hilbert space, Γ_1 and Γ_2 are linear mappings $\mathfrak{D}(A^*) \to \mathcal{H}$, is called a positive boundary value space of the operator A corresponding to the decomposition (2.3) if

1) $(A^*f, g) = (\widetilde{A}\widetilde{P}f, \widetilde{P}g) + (\Gamma_1 f, \Gamma_2 g)_{\mathcal{H}}$ for any $f, g \in \mathfrak{D}(A^*)$;
2) for any $F_1, F_2 \in \mathcal{H}$ there exists a vector $f \in \mathfrak{D}(A^*)$ such that $\Gamma_1 f = F_1$, $\Gamma_2 f = F_2$.

It can be easily seen that a positive boundary value space is also a boundary value space in the meaning of the definition from subsection 1.4.

Let P be the orthoprojector from $\mathfrak{H}$ onto $\operatorname{Ker} A^*$.

LEMMA 2.1. *A positive boundary value space of the operator A may be constructed as follows**

$$\mathcal{H}^0 = \operatorname{Ker} A^*, \quad \Gamma_1^0 = P\widetilde{A}\widetilde{P}, \quad \Gamma_2^0 = P_0. \tag{2.4}$$

Proof. If $f, g \in \mathfrak{D}(A^*)$, then

$$\begin{aligned}(A^*f, g) &= (A^*(\widetilde{P}f + P_0 f), \widetilde{P}g + P_0 f), \widetilde{P}g + P_0 g) = (\widetilde{A}\widetilde{P}f, \widetilde{P}g) + (\widetilde{A}\widetilde{P}f, P_0 g)\\ &= (\widetilde{A}\widetilde{P}f, \widetilde{P}g) + (\Gamma_1^0 f, \Gamma_2^0 g)_{\mathcal{H}^0}.\end{aligned}$$

Setting $f = \widetilde{A}^{-1}F_1 + F_2$ for $F_1, F_2 \in \mathcal{H}^0$ we obtain

$$\Gamma_1^0 f = P\widetilde{A}\widetilde{A}^{-1}F_1 = PF_1 = F_1, \quad \Gamma_2^0 f = P_0 F_2 = F_2. \qquad \square$$

The following lemma, disclosing, in particular, a connection between an arbitray positive boundary value space and the boundary value space (2.4), plays an essential part in what follows.

LEMMA 2.2. *Assume that* $(\mathcal{H}^1, \Gamma_1^1, \Gamma_2^1)$ *and* $(\mathcal{H}^2, \Gamma_1^2, \Gamma_2^2)$ *are positive boundary value spaces of the operator A corresponding to the decomposition* (2.3). *Then*

$$\Gamma_1^1 = UX_1\Gamma_1^2, \quad \Gamma_2^1 = UX_2\Gamma_2^2,$$

where U is an isometric mapping of $\mathcal{H}^2$ onto $\mathcal{H}^1$, $X_i (i = 1, 2)$ is a bounded operator on $\mathcal{H}^2$ having a bounded inverse and $X_2^* X_1 = E$.

Proof. First of all we wish to prove that $\dim \mathcal{H}^1 = \dim \mathcal{H}^2$. In fact, the mapping $\Gamma_1^1 \oplus \Gamma_2^1 : \mathfrak{D}(A^*) \to \mathcal{H}^1 \oplus \mathcal{H}^1$ induces a bijection

$$\mathfrak{D}(A^*)/\mathfrak{D}(A) \to \mathcal{H}^1 \oplus \mathcal{H}^1,$$

---

* Here and in the sequel, $\operatorname{Ker} A^*$ is regarded as a Hilbert space, with inner product induced by that of $\mathfrak{H}$

whence $\dim(\mathcal{H}^1 \oplus \mathcal{H}^1) = \dim(\mathfrak{D}(A^*)/\mathfrak{D}(A))$. Analogously, $\dim(\mathcal{H}^2 \oplus \mathcal{H}^2) = \dim(\mathfrak{D}(A^*)/\mathfrak{D}(A))$. Therefore $\dim(\mathcal{H}^1 \oplus \mathcal{H}^1) = \dim(\mathcal{H}^2 \oplus \mathcal{H}^2)$, hence $\dim \mathcal{H}^1 = \dim \mathcal{H}^2$.

Let U be an isometric mapping of $\mathcal{H}^2$ onto $\mathcal{H}^1$. Set

$$\widetilde{\Gamma}_1^2 = U^{-1}\Gamma_1^1, \quad \widetilde{\Gamma}_2^2 = U^{-1}\Gamma_2^1.$$

Clearly, $(\mathcal{H}^2, \widetilde{\Gamma}_1^2, \widetilde{\Gamma}_2^2)$ is a positive boundary value space. We construct on $\mathcal{H}^2 \oplus \mathcal{H}^2$ a linear operator X as follows. For any $Y_1, Y_2 \in \mathcal{H}^2$ there exists a vector $y \in \mathfrak{D}(A^*)$ such that $\Gamma_1^2 y = Y_1$, $\Gamma_2^2 y = Y_2$. Set $Z_1 = \widetilde{\Gamma}_1^2 y$, $Z_2 = \widetilde{\Gamma}_2^2 y$. We define the operator X by the formula

$$X\{Y_1, Y_2\} = \{Z_1, Z_2\}.$$

Note that the vector y is determined by Y_1, Y_2 uniquely up to a summand from $\mathfrak{D}(A)$; the vectors Z_1, Z_2 do not depend on this term, so the operator X is correctly defined.

It follows from the definition of a positive boundary value space that

$$(Z_1, Z_2)_{\mathcal{H}^2} = (Y_1, Y_2)_{\mathcal{H}^2}, \tag{2.5}$$

i.e. the operator X leaves invariant the form $(Y_1, Y_2)_{\mathcal{H}^2}$ ($\{Y_1, Y_2\} \in \mathcal{H}^2 \oplus \mathcal{H}^2$). In particular, the form

$$2\,\mathrm{Re}(Y_1, Y_2)_{\mathcal{H}^2} = (J\{Y_1, Y_2\}, \{Y_1, Y_2\})_{\mathcal{H}^2 \oplus \mathcal{H}^2},$$

where $J = \begin{pmatrix} 0 & E \\ E & 0 \end{pmatrix}$, is invariant under X. Since the operator X maps $\mathcal{H}^2 \oplus \mathcal{H}^2$ onto $\mathcal{H}^2 \oplus \mathcal{H}^2$, it is J-unitary. In view of Theorem 4.1 from Chapter 1 it is bounded.

We represent X by the matrix

$$X = \begin{pmatrix} X_{11} & X_{12} \\ X_{21} & X_{22} \end{pmatrix},$$

where the X_{ij} are bounded operators on $\mathcal{H}^2$. Then formula (2.5) turns into

$$\begin{aligned}(Y_1, Y_2)_{\mathcal{H}^2} = (X_{11}Y_1, X_{21}Y_1)_{\mathcal{H}^2} + (X_{11}Y_1, X_{22}Y_2)_{\mathcal{H}^2} + \\ + (X_{12}Y_2, X_{21}Y_1)_{\mathcal{H}^2} + (X_{12}Y_2, X_{22}Y_2)_{\mathcal{H}^2}.\end{aligned} \tag{2.6}$$

Setting here $Y_1 = 0$ and taking into account that $Y_2 \in \mathcal{H}^2$ is arbitrary we find that

$$X_{12}^* X_{22} = 0. \tag{2.7}$$

Treating Y_2 in the same way we arrive at the equality

$$X_{21}^* X_{11} = 0. \tag{2.8}$$

Now (2.6) takes the form

$$(Y_1, Y_2)_{\mathcal{H}^2} = (X_{11}Y_1, X_{22}Y_2)_{\mathcal{H}^2} + (X_{12}Y_2, X_{21}Y_1)_{\mathcal{H}^2}. \tag{2.9}$$

Substituting λY_1 ($\mathrm{Im}\,\lambda \neq 0$) for Y_1 in (2.9) we get

$$(\lambda Y_1, Y_2)_{\mathcal{H}^2} = \lambda(Y_1, Y_2)_{\mathcal{H}^2} = \lambda(X_{11}Y_1, X_{22}Y_2)_{\mathcal{H}^2} + \overline{\lambda}(X_{12}Y_2, X_{21}Y_1)_{\mathcal{H}^2}.$$

On the other hand,

$$(\lambda Y_1, Y_2)_{\mathcal{H}^2} = \lambda(X_{11}Y_1, X_{22}Y_2)_{\mathcal{H}^2} + \lambda(X_{12}Y_2, X_{21}Y_1)_{\mathcal{H}^2}.$$

Comparison of the last equalities shows that

$$X_{21}^* X_{12} = 0. \tag{2.10}$$

Therefore formula (2.9) becomes

$$(Y_1, Y_2)_{\mathcal{H}^2} = (X_{11}Y_1, X_{22}Y_2)_{\mathcal{H}^2},$$

whence

$$X_{22}^* X_{11} = E. \tag{2.11}$$

It can be seen from (2.11) that $\operatorname{Ker} X_{11} = \operatorname{Ker} X_{22} = \{0\}$. According to (2.7), (2.8),

$$X_{21}^* \restriction \mathfrak{R}(X_{11}) = 0, \quad X_{12}^* \restriction \mathfrak{R}(X_{22}) = 0. \tag{2.12}$$

Because of the surjectivity of the operator X, for any $Z_1, Z_2 \in \mathcal{H}^2$ there exist Y_1, Y_2 such that

$$X_{11}Y_1 + X_{12}Y_2 = Z_1, \quad X_{21}Y_1 + X_{22}Y_2 = Z_2.$$

This implies that, in particular,

$$\mathfrak{R}(X_{11}) + \mathfrak{R}(X_{12}) = \mathcal{H}^2.$$

But $\mathfrak{R}(X_{12}) \subset \operatorname{Ker} X_{21}^*$ (see (2.10)), i.e. $\mathfrak{R}(X_{11}) + \operatorname{Ker} X_{21}^* = \mathcal{H}^2$. By (2.12), $X_{21}^* = 0$. Similarly, $X_{12}^* = 0$, i.e. $X_{12} = X_{21} = 0$. Then $\mathfrak{R}(X_{11}) = \mathfrak{R}(X_{22}) = \mathcal{H}^2$. Putting $X_1 = X_{11}$, $X_2 = X_{22}$ and taking into account (2.11) we arrive at the desired result. □

Let now $(\mathcal{H}, \Gamma_1, \Gamma_2)$ be an arbitrary positive boundary value space.

THEOREM 2.2. *For any bounded linear operator B on $\mathfrak{H}$ the restriction of the operator A^* to the set of vectors $f \in \mathfrak{D}(A^*)$ satisfying the condition*

$$\Gamma_2 f = B\Gamma_1 f \tag{2.13}$$

is a proper solvable extension of A. Conversely, any proper solvable extension of an operator A is the restriction of A^* to the set of vectors $f \in \mathfrak{D}(A^*)$ satisfying (2.13), *where a bounded linear operator B on $\mathcal{H}$ is uniquely determined by an extension.*

Proof. Let A_B be the restriction of A^* to the set of all vectors $f \in \mathfrak{D}(A^*)$ satisfying (2.13). Clearly, $A \subset A_B \subset A^*$. We prove that the operator A_B has a bounded inverse.

We first establish the equality

$$A_B^* = A_{B^*}, \tag{2.14}$$

where A_{B^*} is the extension corresponding to the boundary condition $\Gamma_2 f = B^*\Gamma_1 f$. If $f \in \mathfrak{D}(A_B)$, $g \in \mathfrak{D}(A^*)$, then

$$\begin{aligned}(A_B f, g) - (f, A^* g) &= (\Gamma_1 f, \Gamma_2 g)_{\mathcal{H}} - (\Gamma_2 f, \Gamma_1 g)_{\mathcal{H}} \\ &= (\Gamma_1 f, \Gamma_2 g)_{\mathcal{H}} - (B\Gamma_1 f, \Gamma_1 g)_{\mathcal{H}} = (\Gamma_1 f, (\Gamma_2 - B^*\Gamma_1) g)_{\mathcal{H}},\end{aligned}$$

whence it is seen that $g \in \mathfrak{D}(A_B^*)$ if and only if $g \in \mathfrak{D}(A_{B^*})$. Since $A_B^* \subset A^*$, $A_{B^*} \subset A^*$, equality (2.14) is true.

It implies that $A_B = (A_{B^*})^*$, i.e. A_B is closed. If $A_B f = 0$, then, assuming $f = g + h$, $g \in \mathfrak{D}(\widetilde{A})$, $h \in \operatorname{Ker} A^*$, we find that $A_B f = A^* f = \widetilde{A} g = 0$, whence $g = 0$ and $f \in \operatorname{Ker} A^*$. By Lemma 2.2,

$$\Gamma_1 = U X_1 \Gamma_1^0, \quad \Gamma_2 = U X_2 \Gamma_2^0, \tag{2.15}$$

where U is an isometry of $\operatorname{Ker} A^*$ onto $\mathfrak{H}$ and X_1, X_2 are bounded operators in $\operatorname{Ker} A^*$. In particular, $\Gamma_1 f = U X_1 P \widetilde{A} \widetilde{P} f = 0$. Then $\Gamma_2 f = B\Gamma_1 f = 0$ too, i.e. $f \in \mathfrak{D}(A)$, $Af = A_B f = 0$. Since A is positive definite, $f = 0$.

So we have shown that $\operatorname{Ker} A_B = \{0\}$.

We now prove that $\mathfrak{R}(A_B) = \mathfrak{H}$. For an arbitrary $h \in \mathfrak{H}$ we set $f = \widetilde{A}^{-1} h + X_2^{-1} U^{-1} B U X_1 P h$. It is obvious that $f \in \mathfrak{D}(A^*)$ and $A^* f = h$. On the other hand, according to (2.15),

$$\begin{aligned}\Gamma_1 f &= U X_1 P \widetilde{A} \widetilde{P} f = U X_1 P h, \\ \Gamma_2 f &= U X_2 P_0 f = B U X_1 P h.\end{aligned}$$

Consequently, $\Gamma_2 f = B\Gamma_1 f$, i.e. $f \in \mathfrak{D}(A_B)$ and $A_B f = A^* f = h$.

Thus, the operator A_B^{-1} exists, it is defined on the whole space $\mathfrak{H}$ and closed. By Banach's theorem A_B^{-1} is bounded, and the extension A_B is thus solvable.

Conversely, let $\widetilde{A}_1$ be a proper solvable extension of the operator A. We use (2.15). For $f \in \mathfrak{D}(\widetilde{A}_1)$,

$$\begin{aligned}\Gamma_1 f &= U X_1 P \widetilde{A} \widetilde{P} f, \\ \Gamma_2 f &= U X_2 P_0 f = U X_2 P_0 \widetilde{A}_1^{-1} A^* (\widetilde{P} f + P_0 f) = U X_2 P_0 \widetilde{A}_1^{-1} \widetilde{A} \widetilde{P} f.\end{aligned}$$

Since $E - P$ is the orthoprojector onto $\mathfrak{R}(A)$, $(\widetilde{A}_1^{-1} - \widetilde{A}^{-1})(E - P) = 0$, whence

$$\begin{aligned}\Gamma_2 f &= U X_2 P_0 ((\widetilde{A}_1^{-1} - \widetilde{A}^{-1}) + \widetilde{A}^{-1}) \widetilde{A} \widetilde{P} f = U X_2 (\widetilde{A}_1^{-1} - \widetilde{A}^{-1}) \widetilde{A} \widetilde{P} f \\ &= U X_2 (\widetilde{A}_1^{-1} - \widetilde{A}^{-1}) P \widetilde{A} \widetilde{P} f = U X_2 (\widetilde{A}_1^{-1} - \widetilde{A}^{-1}) X_1^{-1} U^{-1} \Gamma_1 f,\end{aligned}$$

so $\Gamma_2 f = B\Gamma_1 f$, where

$$B = U X_2 (\widetilde{A}_1^{-1} - \widetilde{A}^{-1}) X_1^{-1} U^{-1}.$$

Thus, we have proved that $\widetilde{A}_1 \subset A_B$. Since the operators $\widetilde{A}_1$ and A_B are boundedly invertible and their inverses are defined on the whole space $\mathfrak{H}$ we have $\widetilde{A}_1 = A_B$. □

COROLLARY 2.1. *The formula*

$$B = UX_2(A_B^{-1} - \widetilde{A}^{-1})X_1^{-1}U^{-1}$$

holds, where U is an isometry of Ker A^* *onto $\mathcal{H}$ and X_1 and X_2 are bounded and boundedly invertible operators on* Ker A^* *such that $X_2^*X_1 = E$.*

3. Spectral Properties of Extensions

In this section we establish a connection between spectral properties of the maximal dissipative (maximal accumulative) extension A_K of a symmetric operator A associated with the boundary condition (1.21) ((1.22)) and properties of the operator K. The theorems collected here are of comparative nature: the operator A_{K_2} is compared with an operator A_{K_1} that possesses definite properties, such as: the spectrum is discrete, the resolvent belongs to the von Neumann–Schatten class $\mathfrak{S}_p$, the eigenvalues have certain asymptotic behaviour. It is established that if the operators K_1 and K_2 are "close" in a certain sense, then A_{K_2} has the same properties as A_{K_1}.

3.1. THE BOUNDARY MAPPINGS

In the definition of a boundary value space of a symmetric operator A (see subsection 1.4) it was not required that the boundary mappings $\Gamma_1, \Gamma_2 : \mathfrak{D}(A^*) \to \mathcal{H}$ have any continuity properties. Nevertheless they appear automatically.

Let $(\mathcal{H}, \Gamma_1, \Gamma_2)$ be an arbitrary boundary value space of the operator A. Consider the mapping $\gamma : \mathfrak{D}(A^*) \to \mathcal{H} \oplus \mathcal{H}$ given by $\gamma f = \{\Gamma_2 f, \Gamma_1 f\}$. We define on $\mathfrak{D}(A^*)$ a new scalar product by

$$\langle f, g\rangle = (f, g) + (A^* f, A^* g). \tag{3.1}$$

As was mentioned in Chapter 1, subsection 4.1, $\mathfrak{D}(A^*)$ is a Hilbert space relative to $\langle\cdot,\cdot\rangle$. One can immediately check that the direct decomposition (1.20) is orthogonal with respect to the scalar product (3.1):

$$\mathfrak{D}(A^*) = \mathfrak{D}(A) \oplus \mathfrak{N}_{-i} \oplus \mathfrak{N}_i.$$

Moreover,

$$\langle f, f\rangle = 2(f, f) = \pm i((\Gamma_1 f, \Gamma_2 f)_{\mathcal{H}} - (\Gamma_2 f, \Gamma_1 f)_{\mathcal{H}}) \tag{3.2}$$

for any $f \in \mathfrak{N}_{\pm i}$.

LEMMA 3.1. *γ is a continuous one-to-one mapping of $\mathfrak{N}_{-i} \oplus \mathfrak{N}_i$ onto $\mathcal{H} \oplus \mathcal{H}$.*

Proof. If $\gamma f = 0$, then $f \in \mathfrak{D}(A)$. So the mapping γ is one-to-one on $\mathfrak{N}_{-i} \oplus \mathfrak{N}_i$. The surjectivity of γ follows from the definition of a boundary value space.

We will now show that the set $M_i = \gamma(\mathfrak{N}_i)$ is closed in $\mathcal{H} \oplus \mathcal{H}$. Let $f_n \in \mathfrak{N}_i$, $\gamma f_n = \{F_n^2, F_n^1\}$. Suppose that the sequence $\{F_n^2, F_n^1\}$ converges in $\mathcal{H} \oplus \mathcal{H}$ to a certain $\{F^2, F^1\}$. It follows from (3.2) that the sequence f_n is fundamental, hence, it converges to a certain element $f \in \mathfrak{N}_i$. Let us show that $\{\Gamma_1 f, \Gamma_2 f\} = \{F^1, F^2\}$.

By the definition of a boundary value space there exists an element $g \in \mathfrak{N}_{-i} \oplus \mathfrak{N}_i$ such that $\gamma g = \{F^2, F^1\}$. For arbitrary $y \in \mathfrak{N}_{-i}$ we have

$$\langle g - f, y\rangle = i \lim_{n\to\infty} ((F_n^1 - F^1, \Gamma_2 y)_{\mathcal{H}} - (F_n^2 - F^2, \Gamma_1 y)_{\mathcal{H}}) = 0,$$

whence $g - f \in \mathfrak{N}_i$, hence $g \in \mathfrak{N}_i$. Because of (3.2),

$$\langle g - f, g - f\rangle = \lim_{n\to\infty} \langle g - f_n, g - f_n\rangle = -2 \lim_{n\to\infty} \operatorname{Im}(F_n^1 - F^1, F_n^2 - F^2)_{\mathcal{H}} = 0.$$

The last equality implies $g = f$. Consequently, $\{\Gamma_1 f, \Gamma_2 f\} = \{F^1, F^2\}$. It is established analogously that the set $M_{-i} = \gamma(\mathfrak{N}_{-i})$ is closed in $\mathcal{H} \oplus \mathcal{H}$. It follows from (3.2) that the mapping $\gamma^{-1} : M_{\pm i} \to \mathfrak{N}_{\pm i}$ is continuous. The closedness of M_i and M_{-i} implies the continuity of the mapping γ. □

Denote by P_1, P_2 the operators acting from $\mathcal{H} \oplus \mathcal{H}$ into $\mathcal{H}$ according to the formulas $P_1\{F^1, F^2\} = F^1$, $P_2\{F^1, F^2\} = F^2$.

LEMMA 3.2. *The equality $P_1(M_{\pm i}) = P_2(M_{\pm i}) = \mathcal{H}$ is valid.*

Proof. We will prove this only for P_2 and M_i (in the other cases the proof is the same). Let an element $F_0 \in \mathcal{H}$ be such that $(F_0, F)_{\mathcal{H}} = 0$ for all $F \in P_2(M_i)$. Denote by f_0 the element from $\mathfrak{N}_{-i} \oplus \mathfrak{N}_i$ with the property $\Gamma_1 f_0 = 0$, $\Gamma_2 f_0 = F_0$. Then for arbitrary $f \in \mathfrak{N}_i$ we have

$$(A^* f_0, f) - (f_0, A^* f) = (\Gamma_1 f_0, \Gamma_2 f)_{\mathcal{H}} - (\Gamma_2 f_0, \Gamma_1 f)_{\mathcal{H}} = -(F_0, P_2 \gamma f)_{\mathcal{H}} = 0.$$

On the other hand,

$$(A^* f_0, f) - (f_0, A^* f) = -i(A^* f_0, A^* f) - i(f_0, f) = -i\langle f_0, f\rangle.$$

So, $f_0 \in \mathfrak{N}_{-i}$. By virtue of (3.2) $f_0 = 0$. Thus, $P_2(M_i)$ is dense in $\mathcal{H}$.

Assume that a sequence $X_n \in P_2(M_i)$ converges to $X \in \mathcal{H}$. Let $x_n \in \mathfrak{N}_i$, $\Gamma_1 x_n = X_n$, $\Gamma_2 x_n = X_n'$. The sequence X_n' is bounded. Otherwise, without loss of generality we may assume that $\lim_{n\to\infty} \|X_n'\|_{\mathcal{H}} = \infty$. Then, as follows from (3.2),

$$\lim_{n\to\infty} \|X_n'\|_{\mathcal{H}}^{-1} \langle x_n, x_n\rangle^{1/2} = 0.$$

According to Lemma 3.1, the sequence $\|X_n'\|_{\mathcal{H}}^{-1}\{X_n', X_n\}$ converges to zero in $\mathcal{H} \oplus \mathcal{H}$, which is impossible.

The boundedness of $\|X_n'\|_{\mathcal{H}}$ and (3.2) show that the sequence $x_n \in \mathfrak{N}_i$ is fundamental, and therefore it converges to a certain element $x \in \mathfrak{N}_i$. In view of Lemma 3.1, $\lim_{n\to\infty}\{X_n', X_n\}$ exists in $\mathcal{H} \oplus \mathcal{H}$ and equals γx. Hence, $X \in P_2(M_i)$ and $P_2(M_i) = \mathcal{H}$. □

The lemma and formula (3.2) allow us to conclude that the operators Γ_1, Γ_2 map the spaces $\mathfrak{N}_{\pm i}$ one-to-one onto $\mathcal{H}$. So, if $\{X', X\} = \gamma f = \{\Gamma_2 f, \Gamma_1 f\} (f \in \mathfrak{N}_i)$, then $X' = CX$, where $C = \Gamma_2 \Gamma_1^{-1}$ (Γ_1^{-1} is the inverse of the restriction of Γ_1 to $\mathfrak{N}_i$)

is a continuous linear operator defined on the whole of $\mathcal{H}$ and having a bounded inverse defined also on the whole of $\mathcal{H}$.

Conversely, if a pair $\{X', X\} \in \mathcal{H} \oplus \mathcal{H}$ satisfies the relation $X' = CX$, where $C = \Gamma_2\Gamma_1^{-1}$, then $\{X', X\} = \gamma f$ with $f = \Gamma_1^{-1}X \in \mathfrak{N}_i$. If now $f \in \mathfrak{N}_i$, $g \in \mathfrak{N}_{-i}$ are arbitrary, then the equality

$$\begin{aligned} 0 &= (A^*f, g) - (f, A^*g) = (\Gamma_1 f, \Gamma_2 g)_{\mathcal{H}} - (\Gamma_2 f, \Gamma_1 g)_{\mathcal{H}} \\ &= (\Gamma_1 f, \Gamma_2 g - C^*\Gamma_1 g)_{\mathcal{H}} \end{aligned}$$

implies $C^*\Gamma_1 g = \Gamma_2 g$. We conclude from this that $\{X', X\} = \gamma g = \{\Gamma_2 g, \Gamma_1 g\}$ with $g \in \mathfrak{N}_{-i}$ if and only if $X' = C^*X$.

LEMMA 3.3. *The operator $C - C^*$ has a bounded inverse.*

Proof. Since the operator $i(C - C^*)$ is self-adjoint, it is sufficient to show that zero belongs neither to the point spectrum nor to the continuous spectrum of $C - C^*$.

Let us assume the contrary, i.e. there exists a sequence $X_n \in \mathcal{H}$, $\|X_n\|_{\mathcal{H}} = 1$ such that $(C - C^*)X_n \to 0$ as $n \to \infty$. Set

$$X_n' = CX_n, \quad Z_n' = C^*X_n.$$

Then $Z_n' = \varepsilon_n + X_n'$, where $\lim_{n\to\infty} \|\varepsilon_n\|_{\mathcal{H}} = 0$. Applying property (1) of a boundary value space to elements $f_n \in \mathfrak{N}_i$ and $g_n \in \mathfrak{N}_{-i}$ for which $\gamma f_n = \{X_n', X_n\}$, $\gamma g_n = \{Z_n', X_n\}$, we obtain

$$(X_n', X_n)_{\mathcal{H}} - (X_n, X_n')_{\mathcal{H}} - (X_n, \varepsilon_n)_{\mathcal{H}} = 0,$$

whence

$$\lim_{n\to\infty} ((X_n', X_n)_{\mathcal{H}} - (X_n, X_n')_{\mathcal{H}}) = 0.$$

Taking into account $\{X_n', X_n\} \in M_i$ we deduce on the base of (3.2) and the continuity of the mapping γ that $X_n \to 0$ as $n \to \infty$. This contradicts the equality $\|X_n\|_{\mathcal{H}} = 1$. □

3.2. COMPARABILITY IN THE SENSE OF RESOLVENTS

Let A_K denote the maximal dissipative (maximal accumulative) extension of the operator A determined by the abstract boundary condition

$$(K-E)\Gamma_1 f + i(K+E)\Gamma_2 f = 0 \text{ (respectively, } (K-E)\Gamma_1 f - i(K+E)\Gamma_2 f = 0),$$

where K is a contraction on $\mathcal{H}$.

THEOREM 3.1. *Let K_1, K_2 be contractions on $\mathcal{H}$ and let $\mu \in \rho(A_{K_1}) \cap \rho(A_{K_2})$. In order that $R_\mu(A_{K_1}) - R_\mu(A_{K_2}) \in \mathfrak{S}_p$ it is necessary and sufficient that $K_1 - K_2 \in \mathfrak{S}_p$ $(1 \le p \le \infty)$.*

Proof. We will prove this for maximal accumulative extensions (the case of maximal dissipative extensions is similar). Let

$$F_{K_j} = (A_{K_j} + iE)(A_{K_j} - iE)^{-1} \restriction \mathfrak{N}_i \quad (j = 1, 2).$$

By arguments analogous to those in the proof of Theorem 1.3 we verify that $\mathfrak{R}(F_{K_j}) \subset \mathfrak{N}_{-i}$, the operator F_{K_j} is a contraction operator and the domain of the operator A_{K_j} consists of the vectors of the form $y = y_0 + F_{K_j}x - x$ ($y_0 \in \mathfrak{D}(A)$, $x \in \mathfrak{N}_i$). Denote by P_i the orthoprojector on $\mathfrak{H}$ onto $\mathfrak{N}_i$. It is obvious that for all $f \in \mathfrak{H}$,

$$\begin{aligned}(F_{K_1} - F_{K_2})P_i f &= ((A_{K_1} + iE)(A_{K_1} - iE)^{-1} - (A_{K_2} + iE)(A_{K_2} - iE)^{-1})P_i f \\ &= (A^* + iE)(R_i(A_{K_1}) - R_i(A_{K_2}))P_i f.\end{aligned}$$

Since $(A^* - iE)(R_i(A_{K_1}) - R_i(A_{K_2})) = 0$ and for any $g = (A - iE)h$ ($h \in \mathfrak{D}(A)$) from $\mathfrak{M}_i$, $(R_i(A_{K_1}) - R_i(A_{K_2}))g = (R_i(A_{K_1})(A_{K_1} - iE) - R_i(A_{K_2})(A_{K_2} - iE))h = 0$, we get

$$(F_{K_1} - F_{K_2})P_i f = 2i(R_i(A_{K_1}) - R_i(A_{K_2}))f,$$

i.e.

$$R_i(A_{K_1}) - R_i(A_{K_2}) = \frac{1}{2i}(F_{K_1} - F_{K_2})P_i. \tag{3.3}$$

Set $\widetilde{F}_{K_j} = \gamma F_{K_j}\gamma^{-1} : M_i = \gamma(\mathfrak{N}_i) \to M_{-i} = \gamma(\mathfrak{N}_{-i})$ ($j = 1, 2$). We denote by G_{K_j} the linear operator on $\mathfrak{H}$ defined by the formula

$$\widetilde{F}_{K_j}\{X, CX\} = \{G_{K_j}X, C^*G_{K_j}X\}, X \in \mathcal{H} \quad (j = 1, 2) \tag{3.4}$$

and also introduce the following bounded operators on $\mathcal{H}$:

$$S_{K_j} = C^*G_{K_j} - C - i(G_{K_j} - E) \tag{3.5}$$

$$T_{K_j} = C^*G_{K_j} - C + i(G_{K_j} - E), \quad j = 1, 2. \tag{3.6}$$

Let us consider the sets

$$N_j = \{\{G_{K_j}X - X, C^*G_{K_j}X - CX\}\},\ X \in \mathfrak{H} \quad (j = 1, 2).$$

According to (3.4)

$$\begin{aligned}N_j &= \{\widetilde{F}_{K_j}\{X, CX\} - \{X, CX\} : X \in \mathcal{H}\} = \{\widetilde{F}_{K_j}\varphi - \varphi : \varphi \in M_i\} \\ &= \{\gamma F_{K_j}x - \gamma x : x \in \mathfrak{N}_i\} = \{\gamma y : y \in \mathfrak{D}(A_{K_j})\}.\end{aligned}$$

As a consequence of Theorems 1.6 and 1.4 it follows from A_{K_j} being maximal accumulative that N_j is a maximal accumulative relation.

According to the theory of linear relations (see subsection 1.3) the operator S_{K_j} maps $\mathcal{H}$ onto $\mathcal{H}$ and

$$\|S_{K_j}X\| \geq \|T_{K_j}X\| \quad (j = 1, 2)$$

for any $X \in \mathcal{H}$. It has a bounded inverse: if $S_{K_j}X = 0$, then $T_{K_j}X = 0$, whence $G_{K_j}X = X$, $C^*G_{K_j}X = CX$; by virtue of the linear independence of M_i and M_{-i}, $X = 0$ and it remains to use Banach's theorem.

We see from what has been proved that

$$K_j = T_{K_j}S_{K_j}^{-1} \quad (j = 1, 2). \tag{3.7}$$

Multiplying the equalities (3.5), (3.6) on the right by $S_{K_j}^{-1}$ and taking into account (3.7), we obtain after some elementary manipulations

$$(K_j + E) + iC(K_j - E) = 2(C^* - C)G_{K_j}S_{K_j}^{-1}, \tag{3.8}$$

$$(K_j + E) + iC^*(K_j - E) = 2(C^* - C)S_{K_j}^{-1} \quad (j = 1, 2). \tag{3.9}$$

The equalities (3.8), (3.9) and the invertibility of the operator $C^* - C$ enable us to conclude that

$$K_j = 2i(G_{K_j} - E)S_{K_j}^{-1} + E, \quad j = 1, 2,$$

whence

$$\begin{aligned} K_1 - K_2 &= 2i((G_{K_1}S_{K_1}^{-1} - G_{K_2}S_{K_2}^{-1}) - (S_{K_1}^{-1} - S_{K_2}^{-1})) \\ &= 2i((G_{K_1} - G_{K_2})S_{K_1}^{-1} + (G_{K_2} - E)(S_{K_1}^{-1} - S_{K_2}^{-1})). \end{aligned} \tag{3.10}$$

But, because of (3.5)

$$S_{K_1}^{-1} - S_{K_2}^{-1} = S_{K_1}^{-1}(S_{K_2} - S_{K_1})S_{K_2}^{-1} = S_{K_1}^{-1}(C^* - iE)(G_{K_2} - G_{K_1})S_{K_2}^{-1}.$$

Therefore,

$$\begin{aligned} K_1 - K_2 = 2i((G_{K_1} - G_{K_2})S_{K_1}^{-1} - \\ - (G_{K_2} - E)S_{K_1}^{-1}(C^* - iE)(G_{K_1} - G_{K_2})S_{K_2}^{-1}. \end{aligned} \tag{3.11}$$

In addition, (3.9) gives

$$S_{K_1}^{-1} - S_{K_2}^{-1} = \tfrac{1}{2}(C^* - C)^{-1}(E + iC^*)(K_2 - K_1).$$

This and (3.10) imply

$$\begin{aligned} G_{K_1} - G_{K_2} = -\tfrac{1}{2}(iE + (G_{K_2} - E)(C^* - C)^{-1}(E + iC^*)) * \\ * (K_1 - K_2)S_{K_1}. \end{aligned} \tag{3.12}$$

On account of (3.11), (3.12) we can state that $K_1 - K_2 \in \mathfrak{S}_p$ if and only if $G_{K_1} - G_{K_2} \in \mathfrak{S}_p$. We show next that $(F_{K_1} - F_{K_2})P_i \in \mathfrak{S}_p$ if and only if $G_{K_1} - G_{K_2} \in \mathfrak{S}_p$.

The operator $F_{K_1} - F_{K_2}$ acts from $\mathfrak{N}_i$ into $\mathfrak{N}_{-i}$. Its adjoint $F_{K_1}^* - F_{K_2}^*$ maps $\mathfrak{N}_{-i}$ into $\mathfrak{N}_i$. Since

$$((F_{K_1} - F_{K_2})P_i)^*(F_{K_1} - F_{K_2})P_i = (F_{K_1} - F_{K_2})^*(F_{K_1} - F_{K_2})P_i,$$

the s-numbers of the operators $((F_{K_1} - F_{K_2})^*(F_{K_1} - F_{K_2}))^{1/2}$ and $(F_{K_1} - F_{K_2})P_i$ coincide (perhaps with the exception of zero). Since γ is a continuous and one-to-one mapping of $\mathfrak{N}_{-i} \oplus \mathfrak{N}_i$ onto $\mathcal{H} \oplus \mathcal{H}$,

$$\begin{aligned} a_1 s_j(((F_{K_1} - F_{K_2})^*(F_{K_1} - F_{K_2}))^{1/2}) &\le s_j(((\widetilde{F}_{K_1} - \widetilde{F}_{K_2})^*(\widetilde{F}_{K_1} - \widetilde{F}_{K_2}))^{1/2}) \\ &\le a_2 s_j(((F_{K_1} - F_{K_2})^*(F_{K_1} - F_{K_2}))^{1/2}), \quad 0 < a_1, a_2 < \infty. \end{aligned}$$

It follows from (3.4) that the operator $(\widetilde{F}_{K_1} - \widetilde{F}_{K_2})(\widetilde{F}_{K_1} - \widetilde{F}_{K_2})^*$ acts on M_i by

$$\begin{aligned} &(\widetilde{F}_{K_1} - \widetilde{F}_{K_2})(\widetilde{F}_{K_1} - \widetilde{F}_{K_2})^*\{X, CX\} \\ &= \{L^{-1}(N^{1/2}(G_{K_1} - G_{K_2}))^*(N^{1/2}(G_{K_1} - G_{K_2}))X, \\ &\qquad CL^{-1}(N^{1/2}(G_{K_1} - G_{K_2}))^*(N^{1/2}(G_{K_1} - G_{K_2}))X\}, \end{aligned}$$

where $X \in \mathcal{H}$, $L = E + C^*C$, $N = E + CC^*$. This and the properties (6.7), (6.8) (Chapter 1) of s-numbers imply the inequalities

$$a_1' s_j(((\widetilde{F}_{K_1} - \widetilde{F}_{K_2})^*(\widetilde{F}_{K_1} - \widetilde{F}_{K_2}))^{\frac{1}{2}}) \leq s_j(((G_{K_1} - G_{K_2})^*(G_{K_1} - G_{K_2}))^{\frac{1}{2}}) \leq a_2' s_j(((\widetilde{F}_{K_1} - \widetilde{F}_{K_2})^*(\widetilde{F}_{K_1} - \widetilde{F}_{K_2}))^{\frac{1}{2}}).$$

Thus, $(F_{K_1} - F_{K_2})P_i \in \mathfrak{S}_p$ if and only if $G_{K_1} - G_{K_2} \in \mathfrak{S}_p$ $(1 \leq p \leq \infty)$. For $p < \infty$ this is derived from the last estimate for s-numbers, for $p = \infty$ it is obtained immediately from the relations for $F_{K_1} - F_{K_2}$, $\widetilde{F}_{K_1} - \widetilde{F}_{K_2}$, $G_{K_1} - G_{K_2}$.

Taking into account (3.3) we find that $R_i(A_{K_1}) - R_i(A_{K_2}) \in \mathfrak{S}_p$ if and only if $K_1 - K_2 \in \mathfrak{S}_p$. It remains to prove that $R_\mu(A_{K_1}) - R_\mu(A_{K_2}) \in \mathfrak{S}_p$ if and only if $R_i(A_{K_1}) - R_i(A_{K_2}) \in \mathfrak{S}_p$. We now establish this fact. Indeed,

$$\begin{aligned} R_\mu(A_{K_1}) - R_\mu(A_{K_2}) &= (R_\mu(A_{K_1}) - R_i(A_{K_1})) + (R_i(A_{K_1}) - R_i(A_{K_2})) + \\ &\quad + (R_i(A_{K_2}) - R_\mu(A_{K_2})) \\ &= (\mu - i)R_\mu(A_{K_1})R_i(A_{K_1}) + (R_i(A_{K_1}) - R_i(A_{K_2})) + \\ &\quad + (i - \mu)R_i(A_{K_2})R_\mu(A_{K_2}) \\ &= (R_i(A_{K_1}) - R_i(A_{K_2})) + (\mu - i)(R_\mu(A_{K_1})(R_i(A_{K_1}) - \\ &\quad - R_i(A_{K_2})) + (R_\mu(A_{K_1}) - R_\mu(A_{K_2}))R_i(A_{K_2})) \\ &= (E + (\mu - i)R_\mu(A_{K_1}))(R_i(A_{K_1}) - R_i(A_{K_2})) + \\ &\quad + (\mu - i)(R_\mu(A_{K_1}) - R_\mu(A_{K_2}))R_i(A_{K_2}). \end{aligned}$$

After transferring the second summand at the right-hand side of the equality to the left-hand side, we have

$$\begin{aligned} &(R_\mu(A_{K_1}) - R_\mu(A_{K_2}))(E - (\mu - i)R_i(A_{K_2})) \\ &\quad = (E + (\mu - i)R_\mu(A_{K_1}))(R_i(A_{K_1}) - R_i(A_{K_2})). \end{aligned} \tag{3.13}$$

It is readily verified that

$$\begin{aligned} (E - (\mu - i)R_i(A_{K_2}))^{-1} &= E + (\mu - i)R_\mu(A_{K_2}), \\ (E + (\mu - i)R_\mu(A_{K_1}))^{-1} &= E - (\mu - i)R_i(A_{K_1}). \end{aligned}$$

Then (3.13) yields

$$\begin{aligned} R_\mu(A_{K_1}) - R_\mu(A_{K_2}) &= (E + (\mu - i)R_\mu(A_{K_1}))(R_i(A_{K_1}) - \\ &\quad - R_i(A_{K_2}))(E + (\mu - i)R_\mu(A_{K_2})), \end{aligned} \tag{3.14}$$

$$\begin{aligned} R_i(A_{K_1}) - R_i(A_{K_2}) &= (E - (\mu - i)R_i A_{K_1}))(R_\mu(A_{K_1}) - \\ &\quad - R_\mu(A_{K_2}))(E - (\mu - i)R_i(A_{K_2})). \end{aligned} \tag{3.15}$$

The relations (3.14), (3.15) and properties (6.7), (6.8) (Chapter 1) of s-numbers result in the required assertion. □

COROLLARY 3.1. *If K_1 and K_2 are unitary and $K_1 - K_2 \in \mathfrak{S}_\infty$, then the essential spectra of the operators A_{K_1} and A_{K_2} coincide.*

Theorem 3.1 and Corollary 3.1 show that a "small" change in the operator K occurring in a boundary condition results in comparatively small changes of spectral properties of the extension A_K.

3.3. THE ASYMPTOTIC BEHAVIOUR OF S-NUMBERS OF EXTENSIONS

The connection found above between properties of the operator K occurring in the abstract boundary condition and properties of the resolvent of the extension A_K permits us to obtain information about the s-number asymptotics (eigenvalue asymptotics in a self-adjoint case) of the operator A_K, if we know the asymptotic behaviour of the s-numbers of the operator K.

THEOREM 3.2. *Let the resolvent $R_{\mu_1}(A_{K_1})$ of a certain maximal dissipative (maximal accumulative) extension A_{K_1} of the operator A ($\mu_1 \in \rho(A_{K_1})$) be a completely continuous operator and let*

$$\lim_{n\to\infty} n^\alpha s_n(R_{\mu_1}(A_{K_1})) = a \quad (\alpha > 0). \tag{3.16}$$

If A_{K_2} is another maximal dissipative (maximal accumulative) extension of A and $\mu_2 \in \rho(A_{K_2})$, then for the equality

$$\lim_{n\to\infty} n^\alpha s_n(R_{\mu_2}(A_{K_2})) = a \tag{3.17}$$

to be valid it is sufficient that $K_1 - K_2$ be a completely continuous operator and that

$$\lim_{n\to\infty} n^\alpha s_n(K_1 - K_2) = 0. \tag{3.18}$$

Proof. By Theorem 3.1 the operator $R_{\mu_2}(A_{K_2})$ is completely continuous. Let us prove (3.17). We first assume that $\mu_1 = \mu_2 = i$, and we will consider only maximal accumulative extensions (the case of maximal dissipative ones is proved in a similar way). On the basis of the properties (6.7), (6.8) (Chapter 1) of s-numbers and formulas (3.12), (3.18) one can conclude that

$$\lim_{n\to\infty} n^\alpha s_n(G_{K_1} - G_{K_2}) = 0.$$

From here by repeating the arguments contained in the proof of Theorem 3.1 we arrive at the equality

$$\lim_{n\to\infty} n^\alpha s_n((F_{K_1} - F_{K_2})P_i) = 0.$$

The required statement follows from (3.3) and property (6.23) (Chapter 1) of s-numbers.

Let now μ_1 and μ_2 be arbitrary. We wish to show that the equalities (3.16), (3.17) hold if and only if they are valid for $\mu_1 = \mu_2 = i$. Assume, for example, that equation (3.17) holds when $\mu_2 = i$. Using the resolvent identity

$$R_{\mu_2}(A_{K_2}) - R_i(A_{K_2}) = (\mu_2 - i)R_{\mu_2}(A_{K_2})R_i(A_{K_2}) \tag{3.19}$$

and property (6.7) (Chapter 1) of s-numbers, we deduce that

$$n^\alpha s_n(R_{\mu_2}(A_{K_2})R_i(A_{K_2})) < a_1.$$

Because of (6.14) from Chapter 1, $n^\alpha s_n(R_{\mu_2}(A_{K_2})) < a_2$. Taking into account (6.15) (Chapter 1), we obtain

$$n^{2\alpha} s_n(R_{\mu_2}(A_{K_2})R_i(A_{K_2})) < a_3.$$

$(a_1, a_2, a_3 < \infty)$. The desired equality

$$\lim_{n\to\infty} n^\alpha s_n(R_{\mu_2}(A_{K_2})) = a$$

follows from (3.19) and property (6.23) of s-numbers (Chapter 1). □

THEOREM 3.3. *Let the resolvent $R_{\mu_1}(A_{K_1})$ of a certain maximal dissipative (maximal accumulative) extension A_{K_1} of the operator A $(\mu_1 \in \rho(A_{K_1}))$ be completely continuous and let*

$$\lim_{n\to\infty} n^\beta s_n(R_{\mu_1}(A_{K_1})) = 0 \quad (\beta > 0). \tag{3.20}$$

If A_{K_2} is another maximal dissipative (maximal accumulative) extension of A, $\mu_2 \in \rho(A_{K_2})$ and the operator $K_1 - K_2$ is completely continuous, then the inequality

$$b_1 \le n^\delta s_n(R_{\mu_2}(A_{K_2})) \le b_2 \quad (0 < b_1, b_2 < \infty) \tag{3.21}$$

holds for $0 < \delta \le \beta$ and all $n \in \mathbb{N}$ if and only if

$$c_1 \le n^\delta s_n(K_1 - K_2) \le c_2 \quad (0 < c_1, c_2 < \infty). \tag{3.22}$$

Proof. As before we will consider the case of maximal accumulative extensions. According to Theorem 3.1, the operator $R_{\mu_2}(A_{K_2})$ is completely continuous. Formulas (3.3), (3.20) and property (6.23) of s-numbers (Chapter 1) imply that inequality (3.21) is valid if and only if

$$d_1 \le n^\delta s_n((F_{K_1} - F_{K_2})P_i) \le d_2 \quad (0 < d_1, d_2 < \infty). \tag{3.23}$$

Arguing as in the proof of Theorem 3.1 we arrive at the conclusion that (3.23) holds if and only if

$$d_1' \le n^\delta s_n(G_{K_1} - G_{K_2}) \le d_2' \quad (0 < d_1', d_2' < \infty).$$

It follows from (3.11), (3.12) and properties (6.7), (6.8) of s-numbers (Chapter 1) that the last inequality is equivalent to (3.22).

If $\mu_1 \ne i$ or $\mu_2 \ne i$, then the stated fact is obtained by word-for-word repetition of the final part of the proof of Theorem 3.2. □

3.4. COMPLETELY SOLVABLE EXTENSIONS

The theorems of subsections 3.2, 3.3 rely on the formulas which connect the operator occurring in the abstract boundary condition with the resolvent of the corresponding extension. Similar kinds of formulas exist for the proper solvable extension A_B of a positive definite operator A associated with the boundary condition

$\Gamma_2 f = B\Gamma_1 f$ (see Theorem 2.2; the boundary value space $(\mathcal{H}, \Gamma_1, \Gamma_2)$ is assumed to be positive, corresponding to a proper solvable extension $\widetilde{A}$). The expression for the operator B in terms of A_B^{-1} was given in Corollary 2.1.

LEMMA 3.4. *The formula*

$$A_B^{-1} = \widetilde{A}^{-1} + U'X_2'B(X_1')^{-1}(U')^{-1}P \tag{3.24}$$

holds, where U' is an isometry of $\mathcal{H}$ onto Ker A^*, *X_1', X_2' are bounded operators on $\mathcal{H}$ having bounded inverses and are such that $(X_2')^* X_1' = E$, and P is the orthoprojector on $\mathfrak{H}$ onto* Ker A^*.

Proof. According to Lemmas 2.1 and 2.2 there exist bounded operators X_1', X_2' on $\mathcal{H}$ such that $(X_2')^* X_1' = E$,

$$P\widetilde{A}\widetilde{P} = U'X_1'\Gamma_1, \quad P_0 = U'X_2'\Gamma_2 \tag{3.25}$$

(the operators $\widetilde{P}$ and P_0 are defined in subsection 2.2).

Let $A_B y = h$. Then

$$h = A^* y = A^*(\widetilde{P}y + P_0 y) = A^*\widetilde{P}y = \widetilde{A}\widetilde{P}y,$$

whence $\widetilde{P}y = \widetilde{A}^{-1}h$. By virtue of (3.25),

$$\begin{aligned} P_0 y &= U'X_2'\Gamma_2 y = U'X_2'B\Gamma_1 y = U'X_2'B(X_1')^{-1}(U')^{-1}P\widetilde{A}\widetilde{P}y \\ &= U'X_2'B(X_1')^{-1}(U')^{-1}Ph. \end{aligned} \tag{3.26}$$

Since $y = \widetilde{P}y + P_0 y = \widetilde{A}^{-1}h + P_0 y$, we have $P_0 y = A_B^{-1}h - \widetilde{A}^{-1}h$. Comparing this equality with (3.26) one obtains (3.24). □

DEFINITION. The extension A_B is called completely solvable if $A_B^{-1} \in \mathfrak{S}_\infty$.

The next result immediately follows from Corollary 2.1 and Lemma 3.4.

THEOREM 3.4. *Let $\widetilde{A}^{-1} \in \mathfrak{S}_p$ $(1 \le p \le \infty)$. In order that $A_B^{-1} \in \mathfrak{S}_p$ it is necessary and sufficient that $B \in \mathfrak{S}_p$.*

3.5. POSITIVE DEFINITE EXTENSIONS

Let A be a positive definite symmetric operator on $\mathfrak{H}$. By Theorem 2.1 the operator A has a positive definite self-adjoint extension. We will describe all such extensions in terms of abstract boundary conditions.

Assume that $(\mathcal{H}, \Gamma_1, \Gamma_2)$ is a positive boundary value space of the operator A corresponding to a positive definite self-adjoint extension $\widetilde{A} \supset A$ (for example, the Friedrichs extension).

THEOREM 3.5. *The restriction A_B of the operator A^* to the set of vectors $f \in \mathfrak{D}(A^*)$ satisfying the condition*

$$\Gamma_2 f = B\Gamma_1 f, \tag{3.27}$$

where B is a non-negative bounded operator on $\mathcal{H}$, is a positive definite self-adjoint extension of A. If $\widetilde{A}$ is the Friedrichs extension, then any positive definite self-adjoint extension of A is given by condition (3.27) with a non-negative bounded operator B.

Proof. Let B be a non-negative bounded operator on $\mathcal{H}$. According to (2.14) the operator A_B is self-adjoint. By the definition of a positive boundary value space,

$$(A_B f, f) = (\widetilde{A}\widetilde{P}f, \widetilde{P}f) + (\Gamma_1 f, \Gamma_2 f)_{\mathcal{H}} - (\widetilde{A}\widetilde{P}f, \widetilde{P}f) + (\Gamma_1 f, B\Gamma_1 f)_{\mathcal{H}} \geq 0$$

for all $f \in \mathfrak{D}(A_B)$.
On the other hand, in view of Theorem 2.2 the operator A_B has a bounded inverse. This implies that A_B is positive definite.

Assume that $\widetilde{A}$ is the Friedrichs extension and that A' is an arbitrary positive definite self-adjoint extension of A. Theorem 2.2 and equality (2.14) show that A' is given by condition (3.27) with a self-adjoint operator B. We prove that B is non-negative.

Suppose there exists a vector $F \in \mathcal{H}$ such that $(BF, F)_{\mathcal{H}} = -\epsilon < 0$. Take $f \in \mathfrak{D}(A^*)$ so that $\Gamma_1 f = F$, $\Gamma_2 f = BF$. Then $f \in \mathfrak{D}(A_B)$. It can be seen from the construction of the Friedrichs extension (see the proof of Theorem 2.1) that there exists a vector $g \in \mathfrak{D}(A)$ such that

$$(\widetilde{A}(g - \widetilde{P}f), g - \widetilde{P}f) < \frac{\epsilon}{2}.$$

Set $h = g - f$. Then $h \in \mathfrak{D}(A_B)$ and

$$(A_B h, h) = (\widetilde{A}\widetilde{P}h, \widetilde{P}h) + (\Gamma_1 h, \Gamma_2 h)_{\mathcal{H}} = (\widetilde{A}(g - \widetilde{P}f), g - \widetilde{P}f) + (F, BF)_{\mathcal{H}} < -\frac{\epsilon}{2}$$

which contradicts the positive definiteness of the operator A_B. □

4. Boundary Value Problems for the Sturm–Liouville Equation with Bounded Operator Potential

We consider a differential expression of the form

$$(l[y])(t) = -y''(t) + q(t)y(t) \quad (t \in [a, b],\ -\infty < a < b < \infty), \tag{4.1}$$

where, for each $t \in [a, b]$, $q(t)$ is a bounded self-adjoint operator on $\mathfrak{H}$. It is also assumed that the operator-function $q(t)$ is continuous (in the strong sense) on $[a, b]$. This section is devoted to a description of all maximal dissipative (maximal accumulative), among them self-adjoint, boundary-value problems for expression (4.1) and also to an investigation of the structure of their spectrum.

4.1. THE MAXIMAL AND MINIMAL OPERATOR. EXTENSIONS OF THE MINIMAL OPERATOR

Let $\mathfrak{D}_0'$ be the set of all vector-functions of the form $\sum_{k=1}^n \varphi_k(t) f_k$ where $f_k \in \mathfrak{H}$, $\varphi_k(t) \in \overset{\circ}{W}{}_2^2(a,b)$. Expression (4.1) generates on $L_2(\mathfrak{H},(a,b))$ the operator L_0' : $L_0' y = l[y]$, $\mathfrak{D}(L_0') = \mathfrak{D}_0'$. Evidently, the domain of L_0' is dense in $L_2(\mathfrak{H},(a,b))$, and L_0' is symmetric. The symmetry follows from the formula for integration by parts:

$$\int_a^b ((l[y])(t), z(t))\,dt - \int_a^b (y(t), (l[z])(t))\,dt$$

$$= -((y'(t), z(t)) - (y(t), z'(t)))\big|_a^b. \tag{4.2}$$

So, L_0' is closeable. Its closure $L_0 = \overline{L_0'}$ is called the minimal operator generated by expression (4.1). The operator L_0^*, the adjoint of L_0, is called the maximal operator.

Since the operator $q : (qy)(t) = q(t)y(t)$ is bounded on $L_2(\mathfrak{H},(a,b))$, the presence of the coefficient $q(t)$ in expression (4.1) is not essential in the study of the domains of L_0, L_0^* and of extensions of L_0. So, one can set $q(t) \equiv 0$ and consider the expression $l[y] = -y''$.

THEOREM 4.1. *The domain $\mathfrak{D}(L_0^*)$ of the maximal operator L_0^* coincides with the space $W_2^2(\mathfrak{H},(a,b))$, i.e. $\mathfrak{D}(L_0^*)$ consists of those vector-functions $y(t)$ for which $y'(t)$ is absolutely continuous on $[a,b]$ and $y'' \in L_2(\mathfrak{H},(a,b))$. Furthermore, $L_0^* y = l[y]$.*

Proof. It follows from formula (4.2) that any vector-function $y(t)$ possessing the properties mentioned in the theorem belongs to $\mathfrak{D}(L_0^*)$.

Conversely, let $y \in \mathfrak{D}(L_0^*)$, i.e. for any $z \in \mathfrak{D}(L_0')$,

$$\int_a^b ((L_0' z)(t), y(t))\,dt = \int_a^b (z(t), h(t))\,dt$$

with a certain $h \in L_2(\mathfrak{H},(a,b))$. The vector-function $y_1(t) = \int_a^t (t-s)h(s)\,ds$ has an absolutely continuous derivative on $[a,b]$ and $y_1''(t) = h(t) \in L_2(\mathfrak{H},(a,b))$. Therefore $y_1 \in \mathfrak{D}(L_0^*)$. Moreover, if $y_0(t) = y_1(t) - y(t)$, then

$$\int_a^b ((L_0' z)(t), y_0(t))\,dt = 0,$$

i.e. $y_0(t)$ is a generalized solution of the equation $-y'' = 0$. In view of Remark 5.3 from Chapter 2, the vector-function $y_0(t)$ is infinitely differentiable on $[a,b]$. Consequently, $y(t) = y_1(t) - y_0(t)$ has an absolutely continuous derivative and $y''(t) \in L_2(\mathfrak{H},(a,b))$. Taking into account (4.2) and the denseness of $\mathfrak{D}_0'$ in $L_2(\mathfrak{H},(a,b))$ we conclude that $L_0^* y = l[y]$. □

For arbitrary vector-functions $y(t), z(t) \in \mathfrak{D}(L_0^*)$ formula (4.2) holds. We write it in the form

$$(L_0^* y, z)_{L_2(\mathfrak{H},(a,b))} - (y, L_0^* z)_{L_2(\mathfrak{H},(a,b))} = (\hat{Y}, \hat{Z}')_{\mathfrak{H}\oplus\mathfrak{H}} - (\hat{Y}', \hat{Z})_{\mathfrak{H}\oplus\mathfrak{H}}, \tag{4.3}$$

where

$$\hat{Y} = \{-y(a), y(b)\}, \quad \hat{Y}' = \{y'(a), y'(b)\}.$$

The following statement is a consequence of this formula.

COROLLARY 4.1. *The domain $\mathfrak{D}(L_0)$ of the minimal operator L_0 generated by expression (4.1) consists of those vector-functions $y(t) \in \mathfrak{D}(L_0^*)$ for which $y(a) = y(b) = y'(a) = y'(b) = 0$, i.e. $\mathfrak{D}(L_0) = \overset{\circ}{W}{}_2^2(\mathfrak{H}, (a, b))$.*

We now consider the triple $(\mathfrak{H} \oplus \mathfrak{H}, \Gamma_1, \Gamma_2)$, where $\Gamma_1 y = \hat{Y}$, $\Gamma_2 y = \hat{Y}'$ are operators from $\mathfrak{D}(L_0^*)$ into $\mathfrak{H}\oplus\mathfrak{H}$. If $F = \{f_1, f_2\}$, $G = \{g_1, g_2\}$ are arbitrary vectors from $\mathfrak{H} \oplus \mathfrak{H}$, then the vector-function $y(t) = \alpha_1(t) f_1 + \alpha_2(t) g_1 + \beta_1(t) f_2 + \beta_2(t) g_2$, where the $\alpha_i(t), \beta_i(t) \in C^\infty[a, b]$ $(i = 1, 2)$ satisfy the conditions

$$\begin{aligned}
\alpha_1(a) &= -1, \quad \alpha_1'(a) = \alpha_1(b) = \alpha_1'(b) = 0,\\
\alpha_2'(a) &= \ \ 1, \quad \alpha_2(a) = \alpha_2(b) = \alpha_2'(b) = 0,\\
\beta_1(b) &= \ \ 1, \quad \beta_1(a) = \beta_1'(a) = \beta_1'(b) = 0,\\
\beta_2'(b) &= \ \ 1, \quad \beta_2(a) = \beta_2(b) = \beta_2'(a) = 0,
\end{aligned}$$

belongs to $\mathfrak{D}(L_0^*)$ and $\Gamma_1 y = F$, $\Gamma_2 y = G$. Taking into account (4.3) we conclude that the triple $(\mathfrak{H} \oplus \mathfrak{H}, \Gamma_1, \Gamma_2)$ is a boundary value space of the minimal operator in the sense of the definition from subsection 1.4. Now, by Theorem 1.6 we have the possibility to describe some classes of extensions of the operator L_0 in terms of boundary conditions.

THEOREM 4.2. *Let K be a contraction on the space $\mathfrak{H}\oplus\mathfrak{H}$. The extensions L_K of the minimal operator L_0 generated by expression (4.1) and the boundary conditions*

$$(K - E)\hat{Y} + i(K + E)\hat{Y}' = 0, \tag{4.4}$$

$$(K - E)\hat{Y} - i(K + E)\hat{Y}' = 0 \tag{4.5}$$

are maximal dissipative and maximal accumulative, respectively. Conversely, any maximal dissipative (maximal accumulative) extension of L_0 is generated by expression (4.1) and a boundary condition of the kind (4.4), (respectively, (4.5)), where a contraction K on $\mathfrak{H} \oplus \mathfrak{H}$ is uniquely determined by an extension. Maximal symmetric extensions of the minimal operator are described by the conditions (4.4) and (4.5) with an isometric operator K. These conditions give a self-adjoint extension if and only if K is unitary. In the last case (4.4) and (4.5) are equivalent to

$$(\cos C)\hat{Y}' - (\sin C)\hat{Y} = 0, \tag{4.6}$$

where C is a self-adjoint operator on $\mathfrak{H} \oplus \mathfrak{H}$.

4.2. ON THE SPECTRA OF EXTENSIONS OF THE MINIMAL OPERATOR

We formulate the results below only for maximal dissipative extensions, because they are analogous in the accumulative case.

An important role in what follows is taken by the extension L_D generated by the expression $l[y]$ and the boundary condition $y(a) = y(b) = 0$ which corresponds to $K = -E$ or $C = \frac{\pi}{2}E$. Since $\mathfrak{D}(L_D)$ coincides with $W_2^2(\mathfrak{H}, (a,b)) \cap \overset{0}{W}{}_2^1(\mathfrak{H}, (a,b))$ the extension L_D is the Friedrichs extension of the minimal operator L_0. It is not difficult to convince ourselves that the spectrum of this operator when $q(t) \equiv 0$ consists of the eigenvalues $\lambda_k(L_D) = \pi^2 k^2/(b-a)^2$ $(k \in \mathbb{N})$, and the corresponding eigenfunctions are $\sin \frac{\pi k(t-a)}{b-a} f$, where the element $f \in \mathfrak{H}$ is arbitrary. Thus, the multiplicity of each eigenvalue coincides with $\dim \mathfrak{H}$. Hence the spectrum of L_D is discrete if and only if $\dim \mathfrak{H} < \infty$. In this case $L_D^{-1} \in \mathfrak{S}_p$ $(p > \frac{1}{2})$, the operator $K + E$ is finite dimensional and, according to Theorem 3.1, the spectrum of any maximal dissipative extension L_K of the minimal operator is discrete. In addition, $R_\lambda(L_K^{-1}) \in \sigma_p$ $(p > \frac{1}{2})$.

Since

$$\lim_{n\to\infty} n^2 s_n(L_D^{-1}) = \lim_{n\to\infty} n^2 \lambda_n(L_D^{-1}) = m^2 \frac{(b-a)^2}{\pi^2} \qquad (m = \dim \mathfrak{H})$$

and

$$\lim_{n\to\infty} n^2 s_n(K+E) = 0,$$

the s-numbers of any maximal dissipative extension L_K have by Theorem 3.2 aymptotics $\frac{\pi^2 n^2}{m^2(b-a)^2}$ $(n \to \infty)$ (this fact is written as $s_n(L_K) \sim \frac{\pi^2 n^2}{m^2(b-a)^2}$). In particular, if L_K is a self-adjoint extension, then it is lower semi-bounded and its eigenvalues have asymptotics that are the same as those of the operator L_D.

In the case when $q(t) \not\equiv 0$, we use the resolvent formula

$$R_\lambda(L_K) = R_\lambda(L_K^0) - R_\lambda(L_K^0) q R_\lambda(L_K), \tag{4.7}$$

where L_K^0 is the maximal dissipative extension generated by the expression $-y''$ and the boundary condition (4.4). This formula is derived from the equality

$$L_K = L_K^0 + q$$

by subtracting λE from both sides of it and subsequent multiplication on the left by $R_\lambda(L_K^0)$ and on the right by $R_\lambda(L_K)$. It follows from (4.7) that the spectrum of any maximal dissipative extension L_K of the operator L_0 is discrete. Taking into account property (6.23) of s-numbers (Chapter 1), we conclude that $\lim_{n\to\infty} n^{-2} s_n(L_K) = \frac{\pi^2}{m^2(b-a)^2}$. The above-said enables us to formulate the following result.

THEOREM 4.3. *If* $\dim \mathfrak{H} = m$ $(m < \infty)$, *then the spectrum of any boundary value problem* (4.4) *for the extension* (4.1) *is discrete and its s-numbers* s_n *have asymptotic* $(n \to \infty)$ *behaviour* $\frac{\pi^2 n^2}{m^2(b-a)^2}$. *In particular, the asymptotic behaviour of the eigenvalues of problem* (4.1), (4.6), *associated with a self-adjoint extension* L_C *of the minimal operator, is determined by*

$$\lambda_n(L_C) \sim \frac{\pi^2 n^2}{m^2(b-a)^2} \quad (n \to \infty).$$

THEOREM 4.4. *In case* $\dim \mathfrak{H} = \infty$ *the minimal operator has no extensions with discrete spectrum.*

Proof. Let f_n $(n \in \mathbb{N})$ be an orthonormal basis in the space $\mathfrak{H}$ and let $\varphi(t)$ $(\|\varphi\|_{L_2(a,b)} = 1)$ be a scalar infinitely-differentiable function on (a,b) with compact support. Then the vector-functions $y_n(t) = \varphi(t) f_n$ belong to $\mathfrak{D}(L_0')$ and $\|y_n\|_{L_2(\mathfrak{H},(a,b))} = 1$. For any $\lambda \in \mathbb{C}^1$ we have

$$\begin{aligned}\|(L_0 - \lambda E) y_n\|_{L_2(\mathfrak{H},(a,b))} &= \| -\varphi''(t) f_n - \lambda \varphi(t) f_n + \varphi(t) q(t) f_n \|_{L_2(\mathfrak{H},(a,b))} \\ &\le c(\lambda),\end{aligned}$$

that is, at fixed λ the sequence $z_n = (L_0 - \lambda E) y_n$ is bounded.

Suppose now that the operator L_0 has an extension $\widetilde{L}$ with discrete spectrum. This means that there exists $\lambda \in \mathbb{C}^1$ such that $(\widetilde{L} - \lambda E)^{-1} \in \mathfrak{S}_\infty$. So, one can select from the sequence $y_n = (\widetilde{L} - \lambda E)^{-1} z_n$ a convergent subsequence. But this contradicts the fact that $\|y_n - y_m\|_{L_2(\mathfrak{H},(a,b))} = \sqrt{2} \quad (n \neq m)$. □

5. Boundary Value Problems for the Sturm–Liouville Equation of Hyperbolic Type with Unbounded Operator Potential

We will consider the differential expression

$$(l[y])(t) = y''(t) + A^2 y(t) + q(t) y(t) \quad (t \in [a,b], -\infty < a < b < \infty), \tag{5.1}$$

where A is a lower semi-bounded self-adjoint operator on $\mathfrak{H}$, $q(t) = q^*(t)$ is a strongly-continuous operator-function taking values in the set of bounded operators on $\mathfrak{H}$. Since in a concrete situation (for instance, $\mathfrak{H} = L_2(\mathbb{R}^n)$, $A^2 = -\Delta$) expression (5.1) is a partial differential expression of hyperbolic type, we will call the considered operator differential expression to be of hyperbolic type.

In this section we describe all maximal dissipative (maximal accumulative), among them self-adjoint, boundary value problems for the expression (5.1) of hyperbolic type. We will also study the spectra of such problems. In contrast to the case of a bounded A, the minimal operator generated by (5.1) with an unbounded A can have self-adjoint extensions with discrete spectrum. In what follows we will give a description of these extensions. For simplicity we set $a = 0$.

5.1. THE MINIMAL AND MAXIMAL OPERATORS. THE STRUCTURE OF THEIR DOMAINS

On the dense set $\mathfrak{D}_0'$ of elements in $L_2(\mathfrak{H},(0,b))$ of the form $\sum_{k=1}^n \varphi_k(t) f_k$, $f_k \in \mathfrak{D}(A^2)$, $\varphi_k(t) \in \overset{\circ}{W}{}_2^2(0,b)$, we define the operator L_0' by $L_0' y = l[y]$. The formula for integration by parts,

$$\int_0^b ((l[y])(t), z(t))\,dt - \int_0^b (y(t), (l[z])(t))\,dt = ((y'(t), z(t)) - (y(t), z'(t)))\big|_0^b$$

applied to expression (5.1) when $y(t), z(t) \in \mathfrak{D}_0'$, shows that the operator L_0' is symmetric. The closure L_0 of the operator L_0' is called the minimal operator generated by expression (5.1), and L_0^* is called the maximal operator.

Because the operator $q : (qy)(t) = q(t)y(t)$ is bounded, the presence of the coefficient $q(t)$ in (5.1) is not essential in the study of the domains of the minimal and the maximal operator, and of extensions of L_0. In this connection we shall set $q(t) \equiv 0$, when such a kind of question is discussed. Without loss of generality we can also suppose that $A \geq E$.

DEFINITION. A vector-function $y(t)$ taking values in $\mathfrak{H}$ is called a weak solution of the equation

$$y''(t) + A^2 y(t) = 0 \quad (t \in [0, b]), \tag{5.2}$$

if it is twice weakly differentiable into $\mathfrak{H}^{-2}(A)$ (i.e. the scalar function $(y(t), f)$ is twice differentiable for arbitrary $f \in \mathfrak{H}^2 = \mathfrak{D}(A^2)$) and

$$\frac{d^2}{dt^2}(y(t), f) + (y(t), A^2 f) = 0.$$

If

$$\hat{l}[y] = y'' + \hat{A}^2 y,$$

where $\hat{A}$ is the extension of A constructed in Chapter 2, subsection 3.1, then a $\mathfrak{H}$-valued vector-function $y(t)$ is a weak solution of equation (5.2) if and only if it is twice weakly differentiable into $\mathfrak{H}^{-2}$ and satisfies the equation $\hat{l}[y] = 0$.

We would also like to recall that a vector-function $y(t)$ on $[0, b]$ that is twice strongly differentiable into $\mathfrak{H}$ is an ordinary solution of equation (5.2) if $y(t) \in \mathfrak{D}(A^2)$ at every $t \in [0, b]$ and $l[y] = 0$.

It is not hard to convince ourselves (see Chapter 1, subsection 5.4) that the vector-functions $\cos A(t - t_0)f$, $\frac{\sin A(t-t_0)}{A} g$ ($\forall t_0 \in [0, b]$, $f \in \mathfrak{D}(A^2)$, $g \in \mathfrak{D}(A)$) are ordinary solutions of equation (5.2). But if $f \in \mathfrak{H}$, $g \in \mathfrak{H}^{-1}$, then these vector-functions are weak solutions of the same equation. In addition,

$$\cos A\,(t - t_0) f\big|_{t=t_0} = f, \quad (\cos A\,(t - t_0) f)'\big|_{t=t_0} = 0,$$
$$\frac{\sin A\,(t - t_0)}{A} g\Big|_{t=t_0} = 0, \quad \Big(\frac{\sin \hat{A}\,(t - t_0)}{\hat{A}} g\Big)'\Big|_{t=t_0} = g.$$

Therefore, the Cauchy problem of finding an ordinary (weak) solution $y(t)$ of equation (5.2) satisfying the conditions $y(t_0) = f$, $y'(t_0) = g$ is solvable if $f \in \mathfrak{D}(A^2) = \mathfrak{H}^2$, $g \in \mathfrak{D}(A) = \mathfrak{H}^1$ (if $f \in \mathfrak{H}$, $g \in \mathfrak{H}^{-1}$). This solution is given by the formula

$$y(t) = \cos A\,(t - t_0) f + \frac{\sin \hat{A}\,(t - t_0)}{\hat{A}} g. \tag{5.3}$$

LEMMA 5.1. *A weak solution of the Cauchy problem*

$$y(t_0) = f, \quad y'(t_0) = g \quad (f \in \mathfrak{H},\ g \in \mathfrak{H}^{-1})$$

for equation (5.2) *is unique.*

Proof. It is sufficient to show that if $v(t)$ is a weak solution of equation (5.2) satisfying the conditions $v(t_0) = 0$, $v'(t_0) = 0$, then $v(t) \equiv 0$. Multiplying, in the sense of the inner product, the identity

$$\frac{d^2}{dt^2}\Big(\frac{\sin \hat{A}\,(t-s)}{\hat{A}}h\Big) + A^2\frac{\sin \hat{A}\,(t-s)}{\hat{A}}h = 0 \quad (h \in \mathfrak{D}(A),\ \forall s \in [0,b])$$

by $v(t)$ and subsequently integrating with respect to t along the interval $[t_0, s]$ we obtain

$$\int_{t_0}^{s}\Big(v(t), \Big(\frac{\sin A\,(t-s)}{A}h\Big)''\Big)\,dt + \int_{t_0}^{s}\Big(v(t), A^2\frac{\sin A\,(t-s)}{A}h\Big)\,dt = 0.$$

Integration by parts yields

$$\int_{t_0}^{s}\Big((\hat{l}[v])(t), \frac{\sin A\,(t-s)}{A}h\Big)\,dt$$
$$= \Big((v(t), \cos A\,(t-s)h) - \Big(v'(t), \frac{\sin A\,(t-s)}{A}h\Big)\Big)\Big|_{t_0}^{s}.$$

Since $\hat{l}[v] = 0$, $v(t_0) = v'(t_0) = 0$, $\cos A\,(t-s)h|_{t=s} = h$, $\frac{\sin A\,(t-s)}{A}h\big|_{t=s} = 0$, we have $(v(s), h) = 0$ for an arbitrary $h \in \mathfrak{D}(A)$. Since $\mathfrak{D}(A)$ is dense in $\mathfrak{H}$ and s is arbitrary, $v(t) \equiv 0$ on $[0, b]$. □

LEMMA 5.2. *The domain $\mathfrak{D}(L_0^*)$ of the maximal operator L_0^* consists of those vector-functions $y(t) \in L_2(\mathfrak{H}, (0,b))$ that can be represented as*

$$y(t) = \cos A\,(t-t_0)f + \frac{\sin \hat{A}\,(t-t_0)}{\hat{A}}g + \int_{t_0}^{t}\frac{\sin A\,(t-s)}{A}h(s)\,ds \tag{5.4}$$
$$(f = y(t_0) \in \mathfrak{H},\ g = y'(t_0) \in \mathfrak{H}^{-1},\ h = L_0^*y \in L_2(\mathfrak{H}, (0,b)),\ \forall t_0 \in [0,b]).$$

Proof. Let $y(t) \in \mathfrak{D}(L_0^*)$. Then

$$(L_0 z, y)_{L_2(\mathfrak{H},(0,b))} = (z, h)_{L_2(\mathfrak{H},(0,b))} \quad (h = L_0^* y \in L_2(\mathfrak{H}, (0,b)),\ \forall z \in \mathfrak{D}(L_0)).$$

Suppose first that $h = 0$. Since the vector-function $z(t) = \varphi(t)f$, where $\varphi(t) \in \overset{\circ}{W}{}_2^2(0,b)$, $f \in \mathfrak{D}(A^2)$ belongs to $\mathfrak{D}_0'$,

$$\int_0^b \varphi''(t)(f, y(t))\,dt = -\int_0^b \varphi(t)(A^2 f, y(t))\,dt. \tag{5.5}$$

Consider the function

$$w(t) = -\int_0^t (t-s)\hat{A}^2 y(s)\,ds.$$

It has the properties: $w(t) \in \mathfrak{H}^{-2}$ for every t, its derivative is absolutely continuous into $\mathfrak{H}^{-2}$ and $w''(t) \in L_2(\mathfrak{H}^{-2}, (0, b))$. Integration by parts of the right-hand side of (5.5) yields

$$\int_0^b \varphi''(t)(f, y_1(t))\, dt = 0, \tag{5.6}$$

where $y_1(t) = y(t) - w(t) - g_1 - g_2 t$ ($g_1, g_2 \in \mathfrak{H}^{-2}$ are arbitrary). Select g_1, g_2 so that $\psi(b) = \psi'(b) = 0$ for the function

$$\psi(t) = \int_0^t (t-s) y_1(s)\, ds$$

(these conditions determine g_1, g_2 uniquely). Setting in (5.6) $\varphi(t) = (\psi(t), f)$ we have $\int_0^b |(y_1(t), f)|^2\, dt = 0$ ($\forall f \in \mathfrak{H}^2$). This implies that $y_1(t) = 0$ almost everywhere on $[0, b]$. Therefore the vector-function $y(t) = w(t) + g_1 + g_2 t$, taking by assumption values in $\mathfrak{H}$, has on $[0, b]$ a derivative that is absolutely continuous into $\mathfrak{H}^{-2}$. By integration by parts we conclude from (5.5) that $\hat{l}[y] = 0$, i.e. $y(t)$ is a weak solution of equation (5.2). Since $y(t) \in L_2(\mathfrak{H}, (0, b))$, $y''(t) \in L_2(\mathfrak{H}^{-2}, (0, b))$, according to the intermediate derivative theorem (Theorem 7.2 from Chapter 1), $y'(t) \in L_2(\mathfrak{H}^{-1}, (0, b))$. Consequently, there exists a point $\bar{t}_0 \in [0, b]$ such that $y(\bar{t}_0) \in \mathfrak{H}$, $y'(\bar{t}_0) \in \mathfrak{H}^{-1}$.

Let

$$z(t) = \cos A\, (t - \bar{t}_0) y(\bar{t}_0) + \frac{\sin \hat{A}\, (t - \bar{t}_0)}{\hat{A}} y'(\bar{t}_0).$$

The vector-function $z(t)$ is the weak solution of equation (5.2) for which $z(\bar{t}_0) = y(\bar{t}_0)$, $z'(\bar{t}_0) = y'(\bar{t}_0)$. By Lemma 5.1, $z(t) \equiv y(t)$. Taking into account the continuity of $z(t)$ into $\mathfrak{H}$ and its continuous differentiability into $\mathfrak{H}^{-1}$ on $[0, b]$, we arrive at the representation

$$y(t) = \cos A\, (t - t_0) f + \frac{\sin \hat{A}\, (t - t_0)}{\hat{A}} g \tag{5.7}$$

$$(f = y(t_0) \in \mathfrak{H}, \quad g = y'(t_0) \in \mathfrak{H}^{-1}).$$

Now we assume that $h \neq 0$. We set

$$v(t) = \int_{t_0}^t \frac{\sin A\, (t-s)}{A} h(s)\, ds.$$

It is not difficult to verify that the vector-function $v(t)$ has a derivative that is continuous into $\mathfrak{H}$ and also that $v'(t)$ is absolutely continuous into $\mathfrak{H}^{-1}$. Moreover, $\hat{l}[v] = h$, $v(t_0) = 0$, $v'(t_0) = 0$. By integration by parts we obtain

$$(L_0 z, v)_{L_2(\mathfrak{H}, (0,b))} = (z, h)_{L_2(\mathfrak{H}, (0,b))} \quad (z \in \mathfrak{D}_0'),$$

hence, $(L_0 z, y - v)_{L_2(\mathfrak{H}, (0,b))} = 0$. But, as was proved, $y - v$ can be represented in the form of (5.7). So, $y(t)$ has a representation (5.4).

The converse assertion of the lemma can be immediately checked. $\square$

THEOREM 5.1. *The domain $\mathfrak{D}(L_0^*)$ of the maximal operator L_0^* consists of the vector-functions $y(t)$ that satisfy the conditions:* a) *$y(t)$ is continuous on $[0,b]$ into $\mathfrak{H}$;* b) *on $[0,b]$, $y'(t)$ is continuous into $\mathfrak{H}^{-1}$ and absolutely continuous into $\mathfrak{H}^{-2}$;* c) *$\hat{l}[y] \in L_2(\mathfrak{H},(0,b))$. Furthermore, $L_0^* y = \hat{l}[y]$.*

Proof. If $y(t) \in \mathfrak{D}(L_0^*)$, then $y(t)$ can be represented by (5.4). This implies that the conditions a), b), c) of the theorem are satisfied. The converse can be simply verified. The equality $L_0^* y = \hat{l}[y]$ follows from the formula for integration by parts, applied to the vector-function $z(t) \in \mathfrak{D}_0'$, while $y(t)$ satisfies the conditions a), b), c). □

We denote by $\mathfrak{D}'$ the set of all vector-functions $y(t)$ on $[0,b]$ that are continuous into $\mathfrak{H}^2$ and twice continuously differentiable into $\mathfrak{H}$. Let L' be the restriction of L_0^* to $\mathfrak{D}'$.

COROLLARY 5.1. *The operator L_0^* coincides with the closure of L' in the space $L_2(\mathfrak{H},(0,b))$.*

Proof. Let $y(t) \in \mathfrak{D}(L_0^*)$ and $L_0^* y = h$. Then $y(t)$ can be represented in the form (5.4). Choose sequences of vectors f_n, $g_n \in \mathfrak{H}^2$ convergent to f, g in the space $\mathfrak{H}$ and $\mathfrak{H}^{-1}$, respectively. Take also a sequence of vector-functions $h_n(t)$ that are strongly continuously differentiable into $\mathfrak{H}$ which converges to $h(t)$ in $L_2(\mathfrak{H},(0,b))$. Then the sequence of vector-functions

$$y_n(t) = \cos A\,(t-t_0) f_n + \frac{\sin \hat{A}\,(t-t_0)}{\hat{A}} g_n + \int_{t_0}^{t} \frac{\sin A\,(t-s)}{A} h_n(s)\,ds$$

is such that $\mathfrak{D}' \ni y_n \to y$, $L' y_n = h_n \to h$ as $n \to \infty$ in $L_2(\mathfrak{H},(0,b))$. □

THEOREM 5.2. *The domain $\mathfrak{D}(L_0)$ of the minimal operator L_0 consists of the vector-functions $y(t) \in \mathfrak{D}(L_0^*)$ that satisfy the conditions*

$$y(0) = y'(0) = y(b) = y'(b) = 0.$$

Proof. Suppose, that $y(t) \in \mathfrak{D}(L_0)$. Since $L_0 = L_0^{**} = (L')^*$,

$$\int_0^b ((l[z])(t), y(t))\,dt = \int_0^b (z(t), y^*(t))\,dt \quad (y^* = L_0^* y) \tag{5.8}$$

for any $z(t) \in \mathfrak{D}(L')$. Since $\mathfrak{D}(L_0) \subset \mathfrak{D}(L_0^*)$, the vector-function $y(t)$ satisfies the conditions a), b), c) from Theorem 5.1. Therefore, for $y(t)$ and $z(t) \in \mathfrak{D}'$ one can apply the formula for integration by parts. Taking into account equality (5.8) we get

$$(z(b), y'(b)) - (z'(b), y(b)) - (z(0), y'(0)) + (z'(0), y(0)) = 0.$$

Setting $z(t) = \psi(t) f$ ($f \in \mathfrak{H}^2$, $\psi(t) \in C^2[0,b]$, $\psi(0) - \psi'(0) = \psi'(b) = 0$, $\psi(b) = 1$) we obtain from the last relation that $(f, y'(b)) = 0\ (\forall f \in \mathfrak{H}^2)$, i.e. $y'(b) = 0$. The rest is proved in a similar way.

The converse statement of the theorem follows from the formula for integration by parts. □

COROLLARY 5.2. *Any vector-function $y(t) \in \mathfrak{D}(L_0)$ is strongly continuous on $[0,b]$ into the space $\mathfrak{H}^1$. On this interval its derivative $y'(t)$ is strongly continuous into $\mathfrak{H}$ and absolutely continuous into $\mathfrak{H}^{-1}$.*

Proof. The proof follows from the fact that if $y(t) \in \mathfrak{D}(L_0)$, then in its representation (5.4) with t_0 replaced by 0, we have $f = g = 0$. □

COROLLARY 5.3. *If $y(t) \in \mathfrak{D}(L_0)$ and $\varphi(t) \in W_2^2(0,b)$, then the vector-function $z(t) = \varphi(t)y(t)$ belongs to $\mathfrak{D}(L_0)$.*

Proof. Clearly $z(t) \in L_2(\mathfrak{H},(0,b))$, $(\hat{l}[z])(t) = \varphi''(t)y(t) + 2\varphi'(t)y'(t) + \varphi(t)(\hat{l}[y])(t) \in L_2(\mathfrak{H},(0,b))$ and $z(0) = z'(0) = z(b) = z'(b) = 0$. It remains to apply Theorem 5.2 and Corollary 5.2. □

We conclude the subsection with the following result, which is a consequence of the representation (5.4).

LEMMA 5.3. *Let $y_n(t) \in \mathfrak{D}(L_0^*)$ and $y_n \to y$ $(n \to \infty)$ in the sense of the graph of the operator L_0^*. Then $y_n(t) \to y(t)$, $A^{-1}y_n'(t) \to A^{-1}y'(t)$ uniformly.*

Proof. Assume that a sequence of vector-functions $y_n(t)$ tends to a vector-function $y(t)$ in the sense of the graph of L_0^*, i.e. $y_n \to y$ and $L_0^*y_n \to L_0^*y$ in $L_2(\mathfrak{H},(0,b))$. The representation (5.4) holds for $y_n(t)$; namely

$$y_n(t) = \cos At\, f_n + \frac{\sin \hat{A}t}{\hat{A}} g_n + \int_0^t \frac{\sin A(t-s)}{A} h_n(s)\, ds \tag{5.9}$$

$$(f_n = y_n(0) \in \mathfrak{H},\quad \hat{A}^{-1}g_n = \hat{A}^{-1}y_n'(0) \in \mathfrak{H},\quad h_n = L_0^*y_n \in L_2(\mathfrak{H},(0,b))).$$

Certainly, the third term in this equality tends uniformly to $\int_0^t \frac{\sin A(t-s)}{A}(L_0^*y)(s)\, ds$. Therefore the sequence $\cos At\, f_n + \frac{\sin \hat{A}t}{\hat{A}} g_n$ converges in $L_2(\mathfrak{H},(0,b))$ to a certain vector-function $z(t)$. This implies that

$$\int_0^{t_0} \cos^2 At\, f_n\, dt + \int_0^{t_0} \cos At \sin At \cdot A^{-1}g_n\, dt \to \int_0^{t_0} \cos At\, z(t)\, dt,$$

$$\int_0^{t_0} \cos At \sin At\, f_n\, dt + \int_0^{t_0} \sin^2 At \cdot A^{-1}g_n\, dt \to \int_0^{t_0} \sin At\, z(t)\, dt$$

$$(\forall t_0 \in [0,b])$$

in $\mathfrak{H}$. Since the determinant of the matrix

$$\begin{pmatrix} \int_0^{t_0} \cos^2 At\, dt & \int_0^{t_0} \cos At \sin At\, dt \\ \int_0^{t_0} \cos At \sin At\, dt & \int_0^{t_0} \sin^2 At\, dt \end{pmatrix}$$

is a continuous and continuously invertible operator on $\mathfrak{H}$, the sequences f_n and $\hat{A}^{-1}g_n$ converge to certain elements f and $A^{-1}g$ of $\mathfrak{H}$. Now the representation (5.9) implies the uniform convergence of $y_n(t)$ to $y(t)$ and $\hat{A}^{-1}y_n'(t)$ to $\hat{A}^{-1}y'(t)$ in $\mathfrak{H}$. □

5.2. MAXIMAL DISSIPATIVE EXTENSIONS OF THE MINIMAL OPERATOR, A DESCRIPTION OF THEM IN TERMS OF BOUNDARY CONDITIONS

The representation (5.4) shows that any vector-function $y(t) \in \mathfrak{D}(L_0^*)$ takes its values at the ends of the interval $[0, b]$ in $\mathfrak{H}$ while the values of its derivative at the points 0 and b lie in $\mathfrak{H}^{-1}$. Let us consider what properties $f_1, g_1 \in \mathfrak{H}$, $f_2, g_2 \in \mathfrak{H}^{-1}$ must have in order that a vector-function $y(t) \in \mathfrak{D}(L_0^*)$ satisfying

$$y(0) = f_1, \quad y'(0) = f_2; \quad y(b) = g_1, \quad y'(b) = g_2 \tag{5.10}$$

exists.

LEMMA 5.4. *Let $f_1, g_1 \in \mathfrak{H}$, $f_2, g_2 \in \mathfrak{H}^{-1}$. In order that a vector-function $y(t) \in \mathfrak{D}(L_0^*)$ satisfying* (5.10) *exists it is necessary and sufficient that the vectors $\cos \hat{A}bg_2 + \hat{A}\sin Abg_1 - f_2$ and $-\sin \hat{A}bg_2 + \hat{A}\cos Abg_1 - \hat{A}f_1$ belong to $\mathfrak{H}$.*

Proof. Suppose that a vector-function $y(t) \in \mathfrak{D}(L_0^*)$ satisfies (5.10). Setting in representation (5.4) $t_0 = b$ and substituting it for $y(t)$ in $y(0) = f_1$, $y'(0) = f_2$ we obtain

$$-\sin \hat{A}bg_2 + \hat{A}\cos Abg_1 - \hat{A}f_1 = -\int_0^b \sin Ash(s)\, ds \in \mathfrak{H},$$

$$\cos \hat{A}bg_2 + \hat{A}\sin Abg_1 - f_2 = -\int_0^b \cos Ash(s)\, ds \in \mathfrak{H}.$$

Conversely, let the vectors mentioned in the formulation belong to $\mathfrak{H}$. We look for the vector-function $y(t) \in \mathfrak{D}(L_0^*)$ required in the form

$$y(t) = \cos A(t-b)g_1 + \frac{\sin \hat{A}(t-b)}{\hat{A}}g_2 - \int_t^b \frac{\sin A(t-s)}{A}h(s)\, ds.$$

Of course, $y(b) = g_1$, $y'(b) = g_2$ and it is necessary to choose $h(s) \in L_2(\mathfrak{H}, (0, b))$ so that

$$\begin{aligned} \int_0^b \frac{\sin As}{A}h(s)\, ds &= \quad f_1 - \cos Abg_1 + \frac{\sin \hat{A}b}{\hat{A}}g_2, \\ \int_0^b \cos Ash(s)\, ds &= -f_2 + \hat{A}\sin Abg_1 + \cos \hat{A}bg_2. \end{aligned} \tag{5.11}$$

Denote the right-hand sides of these equalities by f and g. Obviously, $f \in \mathfrak{H}^1$, $g \in \mathfrak{H}$. Our task is to find a vector-function $z(t) \in \mathfrak{D}(L_0^*)$ such that $z(b) = z'(b) = 0$, $z(0) = f$, $z'(0) = g$.

Take $z_1(t) = \cos Atf + \frac{\sin At}{A}g$. It is clear that on $[0, b]$, $z_1(t)$ is strongly continuous into $\mathfrak{H}^1$ and $z_1'(t)$ is continuous into $\mathfrak{H}$. In addition, $z_1(0) = f$, $z_1'(0) = g$. So, the vector-function $z(t) = \varphi(t)z_1(t)$ where $\varphi(t) \in C^2[0, b]$, $\varphi(t) = 1$ in a neighbourhood of zero and $\varphi(t)$ vanishes in a neighbourhood of the point b, is the required function. □

For a vector-function $y(t) \in \mathfrak{D}(L_0^*)$ we put

$$
\begin{aligned}
y_0 &= \frac{1}{\sqrt{2}}(-\sin Ab\hat{A}^{-1}y'(b) + \cos Aby(b) + y(0)),\\
y_b &= \frac{1}{\sqrt{2}}(-\cos Ab\hat{A}^{-1}y'(b) - \sin Aby(b) - \hat{A}^{-1}y'(0)),\\
y_0' &= \frac{1}{\sqrt{2}}(\ \cos \hat{A}by'(b) + \hat{A}\sin Aby(b) - y'(0)),\\
y_b' &= \frac{1}{\sqrt{2}}(-\sin \hat{A}by'(b) + \hat{A}\cos Aby(b) - \hat{A}y(0)),
\end{aligned}
\tag{5.12}
$$

$$
Y = \{y_0, y_b\}, \quad Y' = \{y_0', y_b'\}. \tag{5.13}
$$

Lemma 5.4 shows that the vectors Y, Y' belong to the space $\mathfrak{H} \oplus \mathfrak{H}$. The next lemma follows from the unique solvability of the system (5.12) relative to $y(0)$, $y(b)$, $y'(0)$, $y'(b)$ (its determinant equals E) and Lemma 5.4.

LEMMA 5.5 *For arbitrary vectors Y, Y' from $\mathfrak{H} \oplus \mathfrak{H}$ there exists a vector-function $y(t) \in \mathfrak{D}(L_0^*)$ such that $Y = \{y_0, y_b\}$, $Y' = \{y_0', y_b'\}$.*

By means of a direct calculation we find that formally

$$
\begin{aligned}
(Y'Z)_{\mathfrak{H}\oplus\mathfrak{H}} - (Y, Z')_{\mathfrak{H}\oplus\mathfrak{H}} &= (y_0', z_0) + (y_b', z_b) - (y_0, z_0') - (y_b, z_b')\\
&= (y'(b), z(b)) - (y'(0), z(0)) - (y(b), z'(b)) + (y(0), z'(0))\\
&= ((y'(t), z(t)) - (y(t), z'(t)))\Big|_0^b.
\end{aligned}
\tag{5.14}
$$

Because of the formula for integration by parts, which is valid for arbitrary vector-functions $y(t), z(t) \in \mathfrak{D}(L')$, and taking into account (5.14) we write

$$
(L_0^*y, z)_{L_2(\mathfrak{H},(0,b))} - (y, L_0^*z)_{L_2(\mathfrak{H},(0,b))} = (Y', Z)_{\mathfrak{H}\oplus\mathfrak{H}} - (Y, Z')_{\mathfrak{H}\oplus\mathfrak{H}}. \tag{5.15}
$$

Let now $y(t)$ be a vector-function from $\mathfrak{D}(L_0^*)$. Representing it in the form (5.4) with $t_0 = 0$ and using (5.11) we find

$$
Y = \frac{1}{\sqrt{2}}\begin{pmatrix} 2y(0) - \int_0^b \frac{\sin As}{A}h(s)\,ds \\ -A^{-1}(2y'(0)) + \int_0^b \cos Ash(s)\,ds \end{pmatrix}
$$

$$
Y' = \frac{1}{\sqrt{2}}\begin{pmatrix} \int_0^b \cos Ash(s)\,ds \\ -\int_0^b \sin Ash(s)\,ds \end{pmatrix},
$$

whence it is seen, in view of Lemma 5.3, that the mapping $y \to \{Y, Y'\}$ acts continuously from $\mathfrak{D}(L_0^*)$ (with the norm of the graph of the operator L_0^*) onto $(\mathfrak{H} \oplus \mathfrak{H}) \oplus (\mathfrak{H} \oplus \mathfrak{H})$. Consequently, we can pass, on the basis of Corollary 5.1, to the limit in formula (5.15) and extend this formula to arbitrary vector-functions $y(t), z(t) \in \mathfrak{D}(L_0^*)$.

Consider the triple $(\mathfrak{H} \oplus \mathfrak{H}, \Gamma_1\Gamma_2)$ where $\Gamma_1 y = Y'$, $\Gamma_2 y = Y$ are operators from $\mathfrak{D}(L_0^*)$ onto $\mathfrak{H} \oplus \mathfrak{H}$ (see Lemma 5.5). Formula (5.15) enables us to state that the triple $(\mathfrak{H} \oplus \mathfrak{H}, \Gamma_1, \Gamma_2)$ is a boundary value space of L_0. According to Theorem 1.6 we obtain the following description of maximal dissipative and other extensions of the minimal operator.

THEOREM 5.3. *Let K be a contraction on the space $\mathfrak{H} \oplus \mathfrak{H}$. The boundary conditions*

$$(K - E)Y' + i(K + E)Y = 0, \tag{5.16}$$

$$(K - E)Y' - i(K + E)Y = 0 \tag{5.17}$$

define on $L_2(\mathfrak{H}, (0, b))$ a maximal dissipative and a maximal accumulative extension of the operator L_0, respectively. Conversely, a maximal dissipative (a maximal accumulative) extension on $L_2(\mathfrak{H}, (0, b))$ of the operator L_0 is generated by the operation $\hat{l}[y]$ and a boundary condition of the kind (5.16), (respectively (5.17)), where K is a contraction on $\mathfrak{H} \oplus \mathfrak{H}$. Maximal symmetric extensions of L_0 on $L_2(\mathfrak{H}, (0, b))$ are described by the conditions (5.16) and (5.17) where K is an isometric operator on $\mathfrak{H} \oplus \mathfrak{H}$. These conditions give a self-adjoint extension of L_0 on $L_2(\mathfrak{H}, (0, b))$ if and only if K is a unitary operator on $\mathfrak{H} \oplus \mathfrak{H}$.

It should be noted that the set of vectors $\{Y, Y'\}$ satisfying the condition (5.16) (respectively (5.17)) coincides with the set of vectors of the form

$$Y = (K - E)G, \quad Y' = -i(K + E)G$$

(respectively,

$$Y = (K - E)G, \quad Y' = i(K + E)G), \tag{5.18}$$

where G runs through the whole space $\mathfrak{H} \oplus \mathfrak{H}$. Indeed, if Y, Y' satisfy the boundary condition (5.16) (respectively, (5.17)) then they can be expressed by the formulas (5.18) with $G = -\frac{1}{2}(Y' - iY)$. The converse can be checked immediately.

It is also possible to describe self-adjoint extensions of the operator L_0 using the boundary condition

$$(\cos C)Y - (\sin C)Y' = 0,$$

where C is a self-adjoint operator on $\mathfrak{H} \oplus \mathfrak{H}$.

We stress once more that the extensions discussed above are actually given by boundary conditions, because the vectors Y, Y' occurring in the formulas (5.16), (5.17) are determined, as can be seen from (5.12), (5.13), by the values of vector-functions from the domain of an extension and their derivatives at the ends of the interval. We will call the vectors Y, Y' the regularized boundary values of the vector-function $y(t) \in \mathfrak{D}(L_0^*)$.

5.3. THE SPECTRUM OF A MAXIMAL DISSIPATIVE EXTENSION

Our purpose now is to study the spectrum of a maximal dissipative extension L_K of the minimal operator. In this connection we consider the eigenvalue problem

$$\begin{aligned} &y''(t) + A^2 y(t) - \lambda y(t) = h(t), \\ &(K - E)Y' + i(K + E)Y = 0. \end{aligned} \tag{5.19}$$

A vector-function $y(t) \in \mathfrak{D}(L_0^*)$ is a solution of the first of these equations if and only if it has a representation

$$y(t) = \cos\sqrt{A^2 - \lambda E}(t-b) f_1 + \frac{\sin\sqrt{\hat{A}^2 - \lambda E}(t-b)}{\sqrt{\hat{A}^2 - \lambda E}} f_2 - \int_t^b \frac{\sin\sqrt{A^2 - \lambda E}(t-s)}{\sqrt{A^2 - \lambda E}} h(s)\, ds \tag{5.20}$$
$$(f_1 \in \mathfrak{H},\ f_2 \in \mathfrak{H}^{-1},\ h = (L_0^* - \overline{\lambda} E) y \in L_2(\mathfrak{H}, (0, b))),$$

which is obtained in the same way as the representation (5.4) if we replace A^2 by $A^2 - \lambda E$ and set $t_0 = b$. The substitution of this representation for $y(t)$ in the boundary condition (5.19) yields

$$((K-E)\Omega_1(\lambda) + i(K+E)\Omega_2(\lambda))F = Q\widetilde{h}, \tag{5.21}$$

where

$$\Omega_1(\lambda) = \frac{1}{\sqrt{2}} \begin{pmatrix} A\sin Ab - \sqrt{A^2 - \lambda E}\sin\sqrt{A^2 - \lambda E}\, b & A\cos Ab - A\cos\sqrt{A^2 - \lambda E}\, b \\ A\cos Ab - A\cos\sqrt{A^2 - \lambda E}\, b & -A\sin Ab + A^2 \frac{\sin\sqrt{A^2 - \lambda E}\, b}{\sqrt{A^2 - \lambda E}} \end{pmatrix},$$

$$\Omega_2(\lambda) = \frac{1}{\sqrt{2}} \begin{pmatrix} \cos Ab + \cos\sqrt{A^2 - \lambda E}\, b & -\sin Ab - A\frac{\sin\sqrt{A^2 - \lambda E}\, b}{\sqrt{A^2 - \lambda E}} \\ -\sin Ab - A^{-1}\sqrt{A^2 - \lambda E}\sin\sqrt{A^2 - \lambda E}\, b & -\cos Ab - \cos\sqrt{A^2 - \lambda E}\, b \end{pmatrix},$$

$$Q = (K-E)\begin{pmatrix} -E & 0 \\ 0 & E \end{pmatrix} + i(K+E)\begin{pmatrix} 0 & -A^{-1} \\ -A^{-1} & 0 \end{pmatrix}, \quad F = \{f_1, \hat{A}^{-1} f_2\},$$

$$\widetilde{h}(\lambda) = \frac{1}{\sqrt{2}} \begin{pmatrix} \int_0^b \cos\sqrt{A^2 - \lambda E}\, s h(s)\, ds \\ A\int_0^b \frac{\sin\sqrt{A^2 - \lambda E}\, s}{\sqrt{A^2 - \lambda E}} h(s)\, ds \end{pmatrix}. \tag{5.22}$$

It is not difficult to see that $\Omega_1(\lambda)$, $\Omega_2(\lambda)$ are entire operator-functions, Q is a bounded operator on $\mathfrak{H} \oplus \mathfrak{H}$ and $\widetilde{h}(\lambda)$ is an entire vector-function whose values lie in $\mathfrak{H} \oplus \mathfrak{H}$.

Let $\mathcal{D}_K(\lambda) = (K-E)\Omega_1(\lambda) + i(K+E)\Omega_2(\lambda)$. The operator-function $\mathcal{D}_K(\lambda)$, taking values in $[\mathfrak{H} \oplus \mathfrak{H}]$, is entire with respect to λ.

THEOREM 5.4. *A point λ belongs to the spectrum of the operator L_K if and only if zero belongs to the spectrum of the operator $\mathcal{D}_K(\lambda)$. Moreover, if λ is an eigenvalue of multiplicity m (a point in the continuous spectrum, a point in the residual spectrum) of the operator L_K, then 0 is an eigenvalue of multiplicity m (a point in the continuous spectrum, a point in the residual spectrum) of the operator $\mathcal{D}_K(\lambda)$.*

Proof. Let λ be an eigenvalue of L_K. Then there exists a vector-function $y(t) \not\equiv 0$

from $\mathfrak{D}(L_K)$ such that $L_K y = \lambda y$. Furthermore,

$$y(t) = \cos\sqrt{A^2 - \lambda E}\,(t-b) f_1 + \frac{\sin\sqrt{\hat{A}^2 - \lambda E}\,(t-b)}{\sqrt{\hat{A}^2 - \lambda E}} f_2 \tag{5.23}$$
$$(0 \neq F = \{f_1, \hat{A}^{-1} f_2\} \in \mathfrak{H} \oplus \mathfrak{H}).$$

The substitution of this expression for $y(t)$ in the boundary condition (5.16) gives $\mathcal{D}_K(\lambda)F = 0$, i.e. F is an eigenvector of $\mathcal{D}_K(\lambda)$ with 0 as an eigenvalue.

Conversely, assume that there exists a vector $F = \{f_1, \hat{A}^{-1} f_2\} \in \mathfrak{H} \oplus \mathfrak{H}$, $F \neq 0$, such that $\mathcal{D}_K(\lambda)F = 0$. Then the vector-function $y(t)$ defined by (5.23) is an eigenvector of L_K corresponding to the eigenvalue λ

It can be seen from this proof that the multiplicity of the eigenvalue λ of the operator L_K is equal to that of the eigenvalue zero of the operator $\mathcal{D}_K(\lambda)$.

Now we prove that if zero is a regular point of $\mathcal{D}_K(\lambda)$, then λ is a regular point of L_K. It is sufficient to show that the equation $(L_K - \lambda E)y = h$ is solvable for any $h \in L_2(\mathfrak{H}, (0, b))$. But it is not difficult to observe that the vector-function (5.20), where $F = \{f_1, \hat{A}^{-1} f_2\}$ is found from equation (5.21) and $\widetilde{h}$ is related to h by formula (5.22), is a solution of this equation.

Suppose that zero is a point in the residual spectrum of $\mathcal{D}_K(\lambda)$. Then there exists an element $G \neq 0$ in $\mathfrak{H} \oplus \mathfrak{H}$ such that

$$(\mathcal{D}_K(\lambda)F, G)_{\mathfrak{H}\oplus\mathfrak{H}} = 0 \quad (\forall F \in \mathfrak{H} \oplus \mathfrak{H}).$$

Choose $z(t) \in \mathfrak{D}(L_0^*)$ so that the vectors Z, Z' constructed according to formula (5.13) are

$$Z = (K^* - E)G, \quad Z' = i(K^* + E)G \tag{5.24}$$

(this is possible by virtue of Lemma 5.5). If we take in (5.15) $y(t)$ as in (5.23) and the $z(t)$ just selected, then

$$-(y, (L_0^* - \overline{\lambda}E)z)_{L_2(\mathfrak{H},(0,b))} = (\mathcal{D}_K(\lambda)F, G)_{\mathfrak{H}\oplus\mathfrak{H}} = 0. \tag{5.25}$$

It can be seen from the construction of $z(t)$ that

$$(K^* - E)Z' - i(K^* + E)Z = 0. \tag{5.26}$$

Since the boundary condition (5.26) generates the maximal accumulative extension $L_{K^*} = L_K^*$ of L_0, we have $z(t) \in \mathfrak{D}(L_K^*)$. Equality (5.25) shows that the vector-function $(L_K - \overline{\lambda}E)z$ is orthogonal to the deficiency subspace $\mathfrak{N}_{\overline{\lambda}} = L_2(\mathfrak{H}, (0, b)) \ominus \mathfrak{M}_{\overline{\lambda}}$, $\mathfrak{M}_\lambda = (L_0 - \lambda E)\mathfrak{D}(L_0)$, of L_0, whence

$$(L_K^* - \overline{\lambda}E)z = (L_0 - \overline{\lambda}E)z_0 \quad (z_0 \in \mathfrak{D}(L_0)).$$

If $\overline{\lambda}$ is not an eigenvalue of L_K^*, then $z = z_0 \in \mathfrak{D}(L_0)$, hence $Z' = Z = 0$. So $G = 0$, which is impossible. Thus $\overline{\lambda}$ is an eigenvalue of L_K^*. Therefore λ is either an eigenvalue or a point in the residual spectrum of L_K. But λ cannot be an eigenvalue of L_K, so λ belongs to the residual spectrum of L_K.

Conversely, let λ be a point in the residual spectrum of L_K. Then $\overline{\lambda}$ is an eigenvalue of L_K^*, i.e. there exists a vector-function $z \notin \mathfrak{D}(L_0)$ from $\mathfrak{D}(L_K^*)$ which is a solution of the equation $(L_K^* - \overline{\lambda}E)z = 0$. Consequently, the vectors Z and Z' corresponding to this vector-function according to the formulas (5.12), (5.13) have a representation (5.24) with $G \neq 0$. Applying formula (5.15) to $z(t)$ and a vector-function $y(t)$ of the form (5.23), we obtain $(\mathcal{D}_K(\lambda)F, G)_{\mathfrak{H}\oplus\mathfrak{H}} = 0$, hence zero is a point in the residual spectrum of $\mathcal{D}_K(\lambda)$.

We now assume that zero is a point in the continuous spectrum of $\mathcal{D}_K(\lambda)$ and show that λ belongs to the continuous spectrum of L_K. Indeed, if it is not true, then λ is a regular point of L_K. By assumption there exists a sequence $F_n = \{f_1^n, \hat{A}^{-1}f_2^n\} \in \mathfrak{H}\oplus\mathfrak{H}$ with $\|F_n\|_{\mathfrak{H}\oplus\mathfrak{H}} = 1$ such that $\mathcal{D}_K(\lambda)F_n \to 0$ as $n \to \infty$. Let us denote by y_n the solution of the equation

$$L_K^* y - \overline{\lambda}y = \cos\sqrt{A^2 - \lambda E}\,(t-b)f_1^n + \frac{\sin\sqrt{\hat{A}^2 - \lambda E}\,(t-b)}{\sqrt{\hat{A}^2 - \lambda E}} f_2^n. \tag{5.27}$$

Since $y_n \in \mathfrak{D}(L_K^*)$ satisfies the condition

$$(K^* - E)Y_n' - i(K^* + E)Y_n = 0,$$

where Y_n, Y_n' can be represented in the form (5.18), we have

$$Y_n = (K^* - E)G_n, \quad Y_n' = i(K^* + E)G_n \quad (G_n \in \mathfrak{H}\oplus\mathfrak{H}).$$

Using formula (5.15) for $y_n(t)$ and the vector-function $z_n(t)$ equal to the right-hand side of (5.27), we find

$$(\mathcal{I}(\lambda)F_n, F_n)_{\mathfrak{H}\oplus\mathfrak{H}} = -(G_n, \mathcal{D}_K(\lambda)F_n),$$

where

$$\mathcal{I}(\lambda) = \begin{pmatrix} \int_0^b \cos\sqrt{A^2 - \lambda E}\,(t-b)\cos\sqrt{A^2 - \overline{\lambda}E}\,(t-b)\,dt & A\int_0^b \cos\sqrt{A^2 - \overline{\lambda}E}\,(t-b)\frac{\sin\sqrt{A^2-\lambda E}\,(t-b)}{\sqrt{A^2-\lambda E}}\,dt \\ A\int_0^b \cos\sqrt{A^2 - \lambda E}\,(t-b)\frac{\sin\sqrt{A^2-\overline{\lambda}E}\,(t-b)}{\sqrt{A^2-\overline{\lambda}E}}\,dt & A^2\int_0^b \frac{\sin\sqrt{A^2-\lambda E}\,(t-b)}{\sqrt{A^2-\lambda E}}\frac{\sin\sqrt{A^2-\overline{\lambda}E}\,(t-b)}{\sqrt{A^2-\overline{\lambda}E}}\,dt \end{pmatrix}$$

is a continuous and continuously invertible operator on $\mathfrak{H}\oplus\mathfrak{H}$. If we could establish the boundedness of the sequence G_n, then on account of $\mathcal{D}_K(\lambda)F_n \to 0$ $(n \to \infty)$ we would arrive at $F_n \to 0$, contradicting $\|F_n\|_{\mathfrak{H}\oplus\mathfrak{H}} = 1$.

So, it remains to show that the sequence G_n is bounded. The vector-function $y_n(t)$, being a solution of equation (5.27), can be represented as

$$y_n(t) = \cos\sqrt{A^2 - \lambda E}\,(t-b)e_n^1 + \frac{\sin\sqrt{\hat{A}^2 - \lambda E}\,(t-b)}{\sqrt{\hat{A}^2 - \lambda E}}e_n^2 -$$
$$- \int_t^b \frac{\sin\sqrt{A^2 - \lambda E}\,(t-s)}{\sqrt{A^2 - \lambda E}} z_n(s)\, ds \quad (e_n^1 \in \mathfrak{H},\ e_n^2 \in \mathfrak{H}^{-1}).$$

Since the sequences $z_n(t)$ ($\|F_n\|_{\mathfrak{H}\oplus\mathfrak{H}} = 1$) and $y_n = (L_K^* - \bar{\lambda}E)^{-1}z_n$ are bounded in $L_2(\mathfrak{H},(0,b))$, the sequence $\cos\sqrt{A^2 - \lambda E}\,(t-b)e_n^1 + \frac{\sin\sqrt{\hat{A}^2-\lambda E}\,(t-b)}{\sqrt{\hat{A}^2-\lambda E}}e_n^2$ is also bounded in this space; because $\mathcal{I}(\lambda)$ has a continuous inverse, the sequence $\{e_n^1, \hat{A}^{-1}e_n^2\}$ is bounded in $\mathfrak{H}\oplus\mathfrak{H}$. But the mapping $y = \{Y, Y'\}$ acts continuously from $\mathfrak{D}(L_0^*)$ (with the norm of the graph of the operator L_0^*) onto $(\mathfrak{H}\oplus\mathfrak{H})\oplus(\mathfrak{H}\oplus\mathfrak{H})$. One may conclude from this that the sequences Y_n, Y_n' are bounded in $\mathfrak{H}\oplus\mathfrak{H}$. Therefore the sequence G_n is also bounded. □

COROLLARY 5.4. *The maximal dissipative (maximal accumulative) extension L_K of the operator L_0 is solvable if and only if the operator $(K+E)^{-1}$ exists and is defined on the whole space $\mathfrak{H}\oplus\mathfrak{H}$.*

The proof follows from Theorem 5.4 and the fact that

$$\mathcal{D}_K(0) = \sqrt{2}\,i(K+E)\begin{pmatrix} \cos Ab & -\sin Ab \\ -\sin Ab & -\cos Ab \end{pmatrix}.$$

Since the operator $\begin{pmatrix} \cos Ab & -\sin Ab \\ -\sin Ab & -\cos Ab \end{pmatrix}$ is continuously invertible on $\mathfrak{H}\oplus\mathfrak{H}$, the invertibility of $K+E$ is necessary and sufficient for the continuous invertibility of $\mathcal{D}_K(0)$. □

Let now a vector-function $y(t)$ belong to $\mathfrak{D}(L_K)$. Since λ ($\operatorname{Im}\lambda < 0$) is a regular point for the maximal dissipative extension L_K, the operator $\mathcal{D}_K^{-1}(\lambda)$ exists and is defined on the whole space $\mathfrak{H}\oplus\mathfrak{H}$. Then the vector $F = \{f_1, \hat{A}^{-1}f_2\}$ occurring in the representation (5.20) can be found from equality (5.21) as follows:

$$F = \mathcal{D}_K^{-1}(\lambda)Q\tilde{h}(\lambda) = \mathcal{D}_K^{-1}(\lambda)\left((K-E)\begin{pmatrix} -E & 0 \\ 0 & E \end{pmatrix} + i(K+E)\begin{pmatrix} 0 & -A^{-1} \\ -A^{-1} & 0 \end{pmatrix}\right)\tilde{h}(\lambda).$$

Substituting this expression for F in (5.20) we derive the formula

$$\begin{aligned}(R_\lambda(L_K)h)(t) &= y(t) \\ &= \left[\cos\sqrt{A^2-\lambda E}\,(t-b),\ A\frac{\sin\sqrt{A^2-\lambda E}\,(t-b)}{\sqrt{A^2-\lambda E}}\right]\mathcal{D}_K^{-1}(\lambda)\times \\ &\quad \times\left((K-E)\begin{pmatrix} -E & 0 \\ 0 & E \end{pmatrix} + i(K+E)\begin{pmatrix} 0 & -A^{-1} \\ -A^{-1} & 0 \end{pmatrix}\right)\tilde{h}(\lambda) - \\ &\quad - \int_t^b \frac{\sin\sqrt{A^2-\lambda E}\,(t-s)}{\sqrt{A^2-\lambda E}}h(s)\, ds \end{aligned} \tag{5.28}$$

which connects the resolvent of the maximal dissipative extension L_K with the operator K involved in the boundary condition determining L_K.

5.4. EXTENSIONS WITH PURELY DISCRETE SPECTRUM

Unlike the case considered in section 4, when the operator A in the differential expression was bounded, the minimal operator L_0 generated by the expresion (5.1) with unbounded A can have maximal dissipative extensions (among them self-adjoint) whose spectrum is purely discrete. This is possible if and only if the spectrum of the operator A is discrete. Indeed, if the spectrum of A is not discrete, then there exist a bounded sequence of numbers λ_n ($\operatorname{Im}\lambda_n = 0$) and an orthonormal system of vectors f_n from $\mathfrak{D}(A^2)$ such that $(A^2 - \lambda_n E)f_n \to 0$ as $n \to \infty$. If L_0 had an extension with purely discrete spectrum, then just as in the proof of Theorem 4.4 we could select a convergent subsequence from the sequence $y_n(t) = \varphi(t)f_n$, where $\varphi(t) \in \mathcal{D}(0,b)$, $\|\varphi\|_{L_2(0,b)} = 1$. This contradicts $\|y_n - y_m\|_{L_2(\mathfrak{H},(0,b))} = \sqrt{2}$, $\forall n \neq m \in \mathbb{N}$.

Under the assumption that the spectrum of A is discrete, the minimal operator L_0 has maximal dissipative extensions with purely discrete spectrum. A description of them is given in the following theorem.

THEOREM 5.5. *Let $A^{-1} \in \mathfrak{S}_\infty$. In order that the spectrum of the extension L_K be discrete it is necessary and sufficient for the operator $K - E$ to be completely continuous.*

Proof. Suppose that $K - E \in \mathfrak{S}_\infty$ and L_K is the maximal dissipative extension of L_0 corresponding to K. Formula (5.28) holds for $R_\lambda(L_K)$ ($\operatorname{Im}\lambda < 0$). Since the mapping $B_1 : h \to \tilde{h}$ is continuous from $L_2(\mathfrak{H},(0,b))$ into $\mathfrak{H}\oplus\mathfrak{H}$ (λ is fixed), the operator $B : h \to \int_t^b \frac{\sin\sqrt{A^2-\lambda E}\,(t-s)}{\sqrt{A^2-\lambda E}} h(s)\,ds$ is completely continuous from $L_2(\mathfrak{H},(0,b))$ into $\mathfrak{H}$ as an integral operator with continuous operator kernel $\frac{\sin\sqrt{A^2-\lambda E}\,(t-s)}{\sqrt{A^2-\lambda E}}$ on $[0,b]\times[0,b]$ taking values in $\mathfrak{S}_\infty$ (see Chapter 1, subsection 3.6), and the operator $B_2 : f = \{f_1, f_2\} \to \cos\sqrt{A^2-\lambda E}\,(t-b)f_1 + A\frac{\sin\sqrt{\hat{A}^2-\lambda E}\,(t-b)}{\sqrt{\hat{A}^2-\lambda E}} f_2$ is continuous from $\mathfrak{H}\oplus\mathfrak{H}$ into $L_2(\mathfrak{H},(0,b))$, we conclude that $R_\lambda(L_K) \in \mathfrak{S}_\infty$.

Conversely, assuming $R_\lambda(L_K) \in \mathfrak{S}_\infty$ ($\operatorname{Im}\lambda < 0$) we obtain on the basis of (5.28) $B_2\mathcal{D}_K^{-1}(\lambda)(K-E)\begin{pmatrix}-E & 0\\ 0 & E\end{pmatrix}B_1 \in \mathfrak{S}_\infty$ in the space $[L_2(\mathfrak{H},(0,b))]$. Then $B_1B_2\mathcal{D}_K^{-1}(\lambda)(K-E)\begin{pmatrix}-E & 0\\ 0 & E\end{pmatrix}B_1B_2 = \mathcal{I}(\lambda)\mathcal{D}_K^{-1}(\lambda)(K-E)\begin{pmatrix}-E & 0\\ 0 & E\end{pmatrix}\mathcal{I}(\lambda) \in \mathfrak{S}_\infty$ in $[\mathfrak{H}\oplus\mathfrak{H}]$. Taking into account that the operators $\mathcal{I}(\lambda)$ and $\mathcal{D}_K^{-1}(\lambda)$ have continuous inverses, we find $K - E \in \mathfrak{S}_\infty$ in $[\mathfrak{H}\oplus\mathfrak{H}]$. □

REMARK 5.1. The assertiom of Theorem 5.5 is true not only for maximal dissipative, but also for maximal accumulative extensions of the minimal operator. The proof is similar.

REMARK 5.2. Let $A^{-1} \in \mathfrak{S}_p$ in $[\mathfrak{H}]$ and $B \in \mathfrak{S}_p$ $(p \geq 1)$ in $[L_2(\mathfrak{H}, (0, b))$. By arguments as in the proof of Theorem 5.5 it is possible to show that the membership $K - E \in \mathfrak{S}_\infty$ in $[\mathfrak{H} \oplus \mathfrak{H}]$ is a necessary and sufficient condition for $R_\lambda(L_K) \in \mathfrak{S}_p$ in $[L_2(\mathfrak{H}, (0, b))]$.

We do not know conditions guaranteeing the membership $B \in \mathfrak{S}_p$ when p is distinct from 2. However, if $p = 2$, then the following statement holds.

LEMMA 5.6. *The operator*

$$(Bh)(t) = \int_t^b \frac{\sin\sqrt{A^2 - \lambda E}\,(t-s)}{\sqrt{A^2 - \lambda E}} h(s)\,ds$$

is a Hilbert–Schmidt operator on the space $L_2(\mathfrak{H}, (0, b))$ if and only if A^{-1} is a Hilbert–Schmidt operator on $\mathfrak{H}$.

Proof. Set $B_J = \frac{1}{2i}(B - B^*)$. Obviously,

$$(B_J h)(t) = \int_0^b G_\lambda(t, s) h(s)\,ds,$$

where

$$G_\lambda(t,s) = \begin{cases} \frac{1}{2i} \frac{\sin\sqrt{A^2 - \lambda E}\,(t-s)}{\sqrt{A^2 - \lambda E}} & \text{if } s \geq t \geq 0 \\ \frac{1}{2i} \frac{\sin\sqrt{A^2 - \bar{\lambda} E}\,(t-s)}{\sqrt{A^2 - \bar{\lambda} E}} & \text{if } s \leq t \leq b. \end{cases}$$

Since B is a Volterra operator, by Theorem 6.6, Chapter 1, $B \in \mathfrak{S}_2$ is equivalent to $B_J \in \mathfrak{S}_2$. So it suffices to show that $B_J \in \mathfrak{S}_2$ if and only if A^{-1} is a Hilbert–Schmidt operator.

Let $\varphi_i(s)$ be an orthonormal basis in $L_2(0, b)$. Denote by $\{e_k\}_{k=1}^\infty$ an orthonormal basis in $\mathfrak{H}$ consisting of eigenvectors of A corresponding to its eigenvalues μ_k. Then $\varphi_i(s) e_k$ $(i, k \in \mathbb{N})$ is an orthonormal basis in $L_2(\mathfrak{H}, (0, b))$. Setting

$$\widetilde{G}_{\mu_k}(t,s) = \begin{cases} \frac{1}{2i} \frac{\sin\sqrt{\mu_k^2 - \lambda}\,(t-s)}{\sqrt{\mu_k^2 - \lambda}} & \text{if } s \geq t \geq 0, \\ \frac{1}{2i} \frac{\sin\sqrt{\mu_k^2 - \bar{\lambda}}\,(t-s)}{\sqrt{\mu_k^2 - \bar{\lambda}}} & \text{if } s \leq t \leq b, \end{cases}$$

we have

$$\sum_{i,k=1}^\infty \left\| B_J \varphi_i(\cdot) e_k \right\|^2_{L_2(\mathfrak{H},(0,b))} = \sum_{i,k=1}^\infty \int_0^b \left\| \int_0^b G_\lambda(t,s)\varphi_i(s) e_k\,ds \right\|^2 dt$$

$$= \sum_{i,k=1}^\infty \int_0^b \left\| \int_0^b \widetilde{G}_{\mu_k}(t,s)\varphi_i(s)\,ds\, e_k \right\|^2 dt = \sum_{i,k=1}^\infty \int_0^b \left| \int_0^b \widetilde{G}_{\mu_k}(t,s)\varphi_i(s)\,ds \right|^2 dt$$

$$= \sum_{k=1}^\infty \int_0^b \int_0^b \left| \widetilde{G}_{\mu_k}(t,s) \right|^2 ds\,dt = \frac{1}{4} \sum_{k=1}^\infty \int_0^b \int_0^b \left| \frac{\sin\sqrt{\mu_k - \lambda}\,(t-s)}{\sqrt{\mu_k - \lambda}} \right|^2 ds\,dt$$

$$= \frac{1}{4}\sum_{k=1}^{\infty}\frac{1}{|\mu_k^2-\lambda|}\int_0^b\int_0^b \left|\sin\sqrt{\mu_k^2-\lambda}\,(t-s)\right|^2 ds\,dt.$$

A direct computation yields

$$\int_0^b\int_0^b |\sin\sqrt{\mu_k^2-\lambda}\,(t-s)|^2\,ds\,dt = \frac{b^2}{2}\left(\frac{\operatorname{sh}^2\tau_k b}{b^2\tau_k^2}-\frac{\sin^2\sigma_k b}{b^2\sigma_k^2}\right),$$

where $\sqrt{\mu_k^2-\lambda}=\sigma_k+i\tau_k$. Since $\mu_k\to\infty$ as $k\to\infty$, $\sigma_k\to\infty$ and $\tau_k\to 0$. Therefore $\lim_{k\to\infty}\int_0^b\int_0^b|\sin\sqrt{\mu_k-\lambda}\,(t-s)|^2\,ds\,dt=\frac{b^2}{2}$. Consequently, convergence of the series $\sum_{i,k=1}^{\infty}\left\|B_J\varphi_i(\cdot)e_k\right\|^2_{L_2(\mathfrak{H},(0,b))}$ is equivalent to convergence of the series $\sum_{k=1}^{\infty}\left|\mu_k^2-\lambda\right|^{-1}$, which is equivalent to $A^{-1}\in\mathfrak{S}_2$. □

Using this lemma we find, by repeating the arguments of Theorem 5.5, the following result.

THEOREM 5.6. *Let $A^{-1}\in\mathfrak{S}_2$. The resolvent $R_\lambda(L_K)$ of the maximal dissipative (maximal accumulative) extension L_K of L_0 belongs to the ideal $\mathfrak{S}_2$ if and only if $K-E\in\mathfrak{S}_2$.*

REMARK 5.3. Theorems 5.5, 5.6 are also true if $q(t)\not\equiv 0$ in expression (5.1). In this case, as in subsection 4.2, one must use the resolvent formula (4.7).

5.5. EXTENSIONS, SMOOTH UP TO THE BOUNDARY

Denote by $C(\mathfrak{H}^\alpha,[0,b])$ the space of vector-functions continuous on $[0,b]$ and taking values in $\mathfrak{H}^\alpha$. As has been shown in subsection 5.1, $\mathfrak{D}(L_0^*)\subset C(\mathfrak{H},[0,b])$, $\mathfrak{D}(L_0)\subset C(\mathfrak{H}^1,[0,b])$.

DEFINITION. A maximal dissipative (maximal accumulative) extension L of the operator L_0 is called α-smooth ($0<\alpha\le 1$), if $\mathfrak{D}(L)\subset C(\mathfrak{H}^\alpha,[0,b])$.

The vector-function $\cos A(t-t_0)f\left(\frac{\sin A(t-t_0)}{A}g\right)$ is continuous in $\mathfrak{H}^\alpha$ ($\mathfrak{H}^{\alpha-1}$) if and only if $f\in\mathfrak{H}^\alpha$ ($g\in\mathfrak{H}^{\alpha-1}$). Then the representation (5.4) implies that a vector-function $y(t)\in\mathfrak{D}(L_0^*)$ belongs to $C(\mathfrak{H}^\alpha,[0,b])$ if and only if $y(t_0)\in\mathfrak{H}^\alpha$, $y'(t_0)\in\mathfrak{H}^{\alpha-1}$ for at least one point $t_0\in[0,b]$.

Let us clarify under what conditions on a contraction K the maximal dissipative or maximal accumulative extension corresponding to K is α-smooth. As follows immediately from (5.12),

$$\begin{pmatrix} y(0)\\ -\hat{A}^{-1}y'(0)\end{pmatrix}=\frac{1}{\sqrt{2}}\left(Y+\begin{pmatrix}0 & -A^{-1}\\ A^{-1} & 0\end{pmatrix}Y'\right),$$

for any vector-function $y(t)\in\mathfrak{D}(L_0^*)$. Therefore, for the extension L_K to be smooth it is necessary and sufficient that the vector Y, constructed according to the formulas (5.12), (5.13), belongs to $\mathfrak{H}^\alpha\oplus\mathfrak{H}^\alpha$ for an arbitrary vector-function $y(t)\in\mathfrak{D}(L_K)$.

Taking into account the representation (5.18) for the vector Y we arrive at the following conclusion.

THEOREM 5.7. *In order that the maximal dissipative or maximal accumulative extension L_K of the operator L_0 be α-smooth it is necessary and sufficient that the operator* $\begin{pmatrix} A^\alpha & 0 \\ 0 & A^\alpha \end{pmatrix}$ $(K - E)$, *acting on $\mathfrak{H} \oplus \mathfrak{H}$, be continuous.*

COROLLARY 5.5. *If the spectrum of the operator A is discrete, then the spectrum of any α-smooth maximal dissipative or maximal accumulative extension of the minimal operator is also discrete.*

5.6. THE CASE OF ONE FIXED END POINT

The results of the previous subsections are considerably simplified if the minimal operator L_0 is generated by the expression (5.1) with $q(t) \equiv 0$ and a boundary condition at one of the ends of the interval, for example,

$$y(b) = 0. \tag{5.29}$$

Then the minimal operator is defined as the closure of the symmetric operator $L_0' : y \to l[y]$ given on the set $\mathfrak{D}_0'$ of vector-functions $\sum_{k=1}^n \varphi_k(t) f_k$, where $f_k \in \mathfrak{D}(A^2)$, while $\varphi_k \in W_2^2(0,b)$ satisfies the conditions $\varphi_k(0) = \varphi_k'(0) = \varphi(b) = 0$. The maximal operator is defined as L_0^*. In this case the domain of the maximal operator consists of the vector-functions that have a representation

$$y(t) = \frac{\sin \hat{A}\,(t-b)}{\hat{A}} f - \int_t^b \frac{\sin A\,(t-s)}{A} h(s)\,ds \tag{5.30}$$

$$(f = y'(b) \in \mathfrak{H}^{-1}, \quad h = L_0^* y \in L_2(\mathfrak{H}, (0,b)).$$

Using this representation one can show without difficulty that $\mathfrak{D}(L_0^*)$ coincides with the set of all vector-functions $y(t)$ having the following properties: a) $y(t)$ is continuous on $[0,b]$ into the space $\mathfrak{H}$; b) on $[0,b]$, $y'(t)$ is continuous into $\mathfrak{H}^{-1}$ and absolutely continuous into $\mathfrak{H}^{-2}$; c) $y(b) = 0$; d) $\hat{l}[y] \in L_2(\mathfrak{H}, (0,b))$.

If we denote by L' the operator $y \to l[y]$; $\mathfrak{D}(L') = \{y(t) \in C^2(\mathfrak{H}, [0,b]) \cap C^1(\mathfrak{H}^2, [0,b]),\ y(b) = 0\}$, then $L_0^* = \overline{L'}$. This implies that $\mathfrak{D}(L_0)$ consists precisely of the vector-functions $y(t) \in \mathfrak{D}(L_0^*)$ for which $y(0) = y'(0) = y(b) = 0$. This implies the inclusion

$$\mathfrak{D}(L_0) \subset C(\mathfrak{H}^1, [0,b]);$$

furthermore, the scalar functions $\varphi(t) \in W_2^2(0,b)$ such that $\varphi(b) = 0$ act as multipliers in $\mathfrak{D}(L_0)$.

With the representation (5.30) at hand and by word-for-word repetition of the proof of Lemma 5.4 we arrive at the following result.

LEMMA 5.7. *Let $f_1 \in \mathfrak{H}$, $f_2 \in \mathfrak{H}^{-1}$. For the existence of a vector-function $y(t) \in \mathfrak{D}(L_0^*)$ with the properties $y(0) = f_1$, $y'(0) = f_2$ it is necessary and sufficient* *that the vector* $\sin \hat{A}bf_2 + \hat{A}\cos Abf_1$ *belongs to* $\mathfrak{H}$.

For any vector-function $y(t) \in \mathfrak{D}(L_0^*)$ we set

$$\begin{aligned} y_0 &= \sin \hat{A}by'(0) + \hat{A}\cos Aby(0) \in \mathfrak{H}, \\ y_0' &= A^{-1}\cos \hat{A}by'(0) - \sin Aby(0) \in \mathfrak{H}. \end{aligned} \tag{5.31}$$

Then for arbitrary $y(t)$, $z(t)$ from $\mathfrak{D}(L')$,

$$(y_0, z_0') - (y_0', z_0) = (y(0), z'(0)) - (y'(0), z(0)). \tag{5.32}$$

The formula for integration by parts,

$$(L_0^* y, z)_{L_2(\mathfrak{H},(0,b))} - (y, L_0^* z)_{L_2(\mathfrak{H},(0,b))} = (y(0), z'(0)) - (y'(0), z(0)),$$

valid for $y, z \in \mathfrak{D}(L')$, leads together with (5.32) to

$$(L_0^* y, z)_{L_2(\mathfrak{H},(0,b))} - (y, L_0^* z)_{L_2(\mathfrak{H},(0,b))} = (y_0, z_0') - (y_0', z_0). \tag{5.33}$$

Assume now $y \in \mathfrak{D}(L_0^*)$. Representing it in the form (5.30) we can write the regularized boundary values y_0 and y_0' of the vector-function $y(t)$ as follows

$$\begin{aligned} y_0 &= -\int_0^b \sin A\,(b-s)h(s)\,ds, \\ y_0' &= \hat{A}^{-1}f - A^{-1}\int_0^b \cos A\,(b-s)h(s)\,ds. \end{aligned} \tag{5.34}$$

Next, we take a sequence of vector-functions $y_n(t)$ of the type (5.30) with $f_n \in \mathfrak{D}(A^2) : f_n \to f$ in $\mathfrak{H}^{-1}$ and $h_n \in C(\mathfrak{H}^2, [0,b]) : h_n \to h$ in $L_2(\mathfrak{H}, (0,b))$. It can be seen from (5.34) that $(y_n)_0 \to y_0$, $(y_n)_0' \to y_0'$ in $\mathfrak{H}$. The representation (5.30) shows that $y_n \to y$, $L_0^* y_n \to L_0^* y$ in $L_2(\mathfrak{H}, (0,b))$. Therefore formula (5.33) can be extended to arbitrary vector-functions $y(t), z(t) \in \mathfrak{D}(L_0^*)$.

It follows from Lemma 5.7 and formula (5.33) that the triple $(\mathfrak{H}, \Gamma_1, \Gamma_2)$, where $\Gamma_1 y = y_0$, $\Gamma_2 y = y_0'$, is a boundary value space of the operator L_0. So, all maximal dissipative and maximal accumulative extensions of the minimal operator generated by expression (5.1) and the boundary condition (5.29) are described by the boundary conditions

$$\begin{aligned} (K-E)y_0 + i(K+E)y_0' &= 0, \\ (K-E)y_0 - i(K+E)y_0' &= 0, \end{aligned} \tag{5.35}$$

respectively. Here, as before, K is a contraction on $\mathfrak{H}$. In the self-adjoint case the equations (5.35) are equivalent to

$$(\cos C)y_0' - (\sin C)y_0 = 0, \tag{5.36}$$

where C is a self-adjoint operator on $\mathfrak{H}$.

As in Theorem 5.4 the spectrum of the extension L_K can be investigated using the entire function

$$\mathcal{D}_K(\lambda) = (K-E)\Omega_2(\lambda) \pm i(K+E)\Omega_1(\lambda)$$

(the sign "+" corresponds to the dissipative case, and "−" to the accumulative case), where

$$\Omega_1(\lambda) = A\sin Ab\frac{\sin\sqrt{A^2-\lambda E}\,b}{\sqrt{A^2-\lambda E}} + \cos Ab\cos\sqrt{A^2-\lambda E}\,b,$$

$$\Omega_2(\lambda) = A\sin Ab\cos\sqrt{A^2-\lambda E}\,b - A^2\cos Ab\frac{\sin\sqrt{A^2-\lambda E}\,b}{\sqrt{A^2-\lambda E}}$$

are entire operator-functions with values in $[\mathfrak{H}]$. Namely, a point λ belongs to the spectrum of L_K if and only if zero is a point in the spectrum of $\mathcal{D}_K(\lambda)$. Moreover, if λ is an eigenvalue of multiplicity m, a point in the continuous or residual spectrum of L_K, then zero is an eigenvalue of multiplicity m, a point in the continuous or residual spectrum of $\mathcal{D}_K(\lambda)$, respectively. It is not difficult to conclude from this that the resolvent of the maximal dissipative (maximal accumulative) extension L_K can be represented by

$$\begin{aligned}(R_\lambda(L_K)h)(t) = {} & \frac{\sin\sqrt{A^2-\lambda E}\,(t-b)}{\sqrt{A^2-\lambda E}}A\mathcal{D}_K^{-1}(\lambda)(K-E)\int_0^b \mathcal{A}_2(s,\lambda)h(s)\,ds\pm \\ & \pm i\frac{\sin\sqrt{A^2-\lambda E}\,(t-b)}{\sqrt{A^2-\lambda E}}\mathcal{D}_K^{-1}(\lambda)(E+K)\int_0^b \mathcal{A}_1(s,\lambda)h(s)\,ds- \\ & -\int_t^b \frac{\sin\sqrt{A^2-\lambda E}\,(t-s)}{\sqrt{A^2-\lambda E}}h(s)\,ds, \end{aligned} \tag{5.37}$$

where

$$\mathcal{A}_1(s,\lambda) = \cos Ab\cos\sqrt{A^2-\lambda E}\,s + A\sin Ab\frac{\sin\sqrt{A^2-\lambda E}\,s}{\sqrt{A^2-\lambda E}},$$

$$\mathcal{A}_2(s,\lambda) = -A\cos Ab\frac{\sin\sqrt{A^2-\lambda E}\,s}{\sqrt{A^2-\lambda E}} + \sin Ab\cos\sqrt{A^2-\lambda E}\,s.$$

If the spectrum of the operator A appearing in expression (5.1) is discrete, then the second and third summands at the right-hand side of (5.37) are completely continuous operators on the space $L_2(\mathfrak{H},(0,b))$. Arguing as in Theorem 5.5 with regard to the invertibility on $\mathfrak{H}$ of the operator $A^2\int_0^b \frac{\sin\sqrt{A^2-\lambda E}\,s}{\sqrt{A^2-\lambda E}}\cdot\frac{\sin\sqrt{A^2-\bar\lambda E}\,s}{\sqrt{A^2-\bar\lambda E}}\,ds$, we deduce that the spectrum of the extension L_K is discrete if and only if $K-E\in\mathfrak{S}_\infty$ (in the self-adjoint case, $\sin C\in\mathfrak{S}_\infty$). Under the assumption $A^{-1}\in\mathfrak{S}_2$ the resolvent $R_\lambda(L_K)$ belongs to the ideal $\mathfrak{S}_2$ if and only if $K-E$ is a Hilbert–Schmidt operator.

Because $\mathcal{D}_K(0) = -i(K+E)$, the extension L_K is solvable if and only if the operator $K+E$ (in the self-adjoint case, $\cos C$) has a bounded inverse defined on the whole space $\mathfrak{H}$. For L_K to be α-smooth in the situation considered it is necessary and sufficient that the operator $A^\alpha(K-E)$ be continuous on $\mathfrak{H}$.

5.7. THE DIRICHLET PROBLEM

We denote by L_D the closure on $L_2(\mathfrak{H}, (0,b))$ of the operator L'_D generated by the differential expression (5.1) ($q(t) \equiv 0$) on the set of vector-functions $y(t)$ on $[0,b]$ that are twice continuously differentiable into $\mathfrak{H}^2$ and satisfy the conditions $y(0) = y(b) = 0$. Since $L_2(\mathfrak{H}, (0,b)) = L_2(0,b) \otimes \mathfrak{H}$ and the operator L_D admits separation of variables: $L_D = l_D \otimes E + E \otimes A^2$, where l_D is the self-adjoint operator on $L_2(0,b)$ generated by the expression φ'' and the boundary conditions $\varphi(0) = \varphi(b) = 0$, the operator L_D is self-adjoint (see Chapter 1, subsection 4.6). Being the extension of the minimal operator generated by expression (5.1) and condition (5.29), it is given by the boundary condition (5.36) with a certain self-adjoint operator C.

It can be easily seen that for $y(t) \in \mathfrak{D}(L_0^*)$ the equality

$$A^{-1} \cos Ab y_0 - \sin Ab y'_0 = y(0)$$

is valid. Therefore, $y(t) \in \mathfrak{D}(L_D)$ if and only if

$$\cos Ab y_0 - A \sin Ab y'_0 = 0,$$

whence

$$\cos C = \sin Ab(A^{-2} \cos^2 Ab + \sin^2 Ab)^{-1/2}$$
$$\sin C = A^{-1} \cos Ab(A^{-1} \cos^2 Ab + \sin^2 Ab)^{-1/2}.$$

These formulas show that the operator L_D is invertible, i.e. the equation $L_D y = 0$ has a unique solution, only when the points $\pi m/b$ ($m \in \mathbb{N}$) are not eigenvalues of A. The extension L_D is solvable if and only if the set $\{\mu^{-1} \operatorname{ctg} \mu b : \mu \in \sigma(A)\}$ is bounded. Under the assumption that the spectrum $\{\mu_k\}_{k=1}^{\infty}$ of A is discrete, the spectrum of L_D is discrete if and only if infinity is the unique limit point of the set $\{\mu_k \operatorname{tg} \mu_k\}_{k=1}^{\infty}$.

5.8. BOUNDARY VALUE PROBLEMS FOR PARTIAL DIFFERENTIAL EQUATIONS

1. Let Ω be a bounded domain in the space $\mathbb{R}^n$ ($n \geq 1$) with sufficiently smooth boundary $\partial\Omega$. Denote by Ξ a cylinder $\Omega \times [0,b]$ ($b < \infty$) and by S its side surface, i.e. the set $\partial\Omega \times [0,b]$. On the space $L_2(\Xi) = L_2(\mathfrak{H}, (0,b))$ ($\mathfrak{H} = L_2(\Omega)$) we consider the operator generated by the differential expression

$$(l[u])(x,t) = \frac{\partial^2 u(x,t)}{\partial t^2} + \sum_{|\alpha| \leq 2m} a_\alpha(x) D_x^\alpha u(x,t) \tag{5.38}$$
$$(x \in \Omega,\ 0 \leq t \leq b)$$

and Dirichlet boundary conditions on the side surface of the cylinder:

$$\left.\frac{\partial^i u}{\partial \nu^i}\right|_S = 0 \quad (i = 0, 1, \ldots, m-1;\ m \geq 1) \tag{5.39}$$

(ν is the outward normal to S). Let us assume that the coefficients of (5.38) are sufficiently smooth and such that the operator A^2 generated by the expression

$M = \sum_{|\alpha|\le 2m} a_\alpha(x) D_x^\alpha$ and boundary conditions $\frac{\partial^i u}{\partial \nu^i}\big|_{\partial\Omega} = 0$ $(i = 0, 1, \dots, m-1)$ is positive and self-adjoint (for this it is sufficient to require the expression M to be formally self-adjoint and strongly elliptic (see, for example, Berezansky [4, Chapters 3,6])). In the case considered, $\mathfrak{D}(A^2) = W_2^{2m}(\Omega) \cap \overset{\circ}{W}{}_2^m(\Omega)$.

Using (5.38) and the boundary conditions (5.39) we construct on the space $L_2(\Xi)$ the minimal operator L_0 as the closure of the operator L_0':

$$L_0'u = l[u], \quad \mathfrak{D}(L_0') = \Big\{u(x,t) \in C^\infty(\Xi) : \frac{\partial^i u}{\partial \nu^i}\Big|_S = 0 \ (i = 0, 1, \dots, m-1);$$
$$u(x,0) = \frac{\partial u}{\partial t}(x,0) = u(x,b) = \frac{\partial u}{\partial t}(x,b) = 0\Big\},$$

as well as the maximal operator L_0^*. The operators L_0 and L_0^* coincide, respectively, with the minimal and maximal ones generated by the differential operator expression $l[y] = y'' + A^2 y$ on the space $L_2(\mathfrak{H}, (0,b))$.

The deficiency numbers of the operator L_0 are infinite. Consequently, it has self-adjoint extensions on $L_2(\Xi)$. Theorem 5.3 shows that each of them is defined by two homogeneous boundary conditions which connect the values of a function $u(x,t)$ and its derivative with respect to t on the bases of the cylinder. More exactly, every self-adjoint extension of the operator L_0 on $L_2(\Xi)$ is given by conditions of the form

$$\begin{aligned} a_1 u(x,0) + a_2 u(x,b) + a_3 \frac{\partial u(x,0)}{\partial t} + a_4 \frac{\partial u(x,b)}{\partial t} &= 0, \\ b_1 u(x,0) + b_2 u(x,b) + b_3 \frac{\partial u(x,0)}{\partial t} + b_4 \frac{\partial u(x,b)}{\partial t} &= 0, \end{aligned} \tag{5.40}$$

where the a_i, b_i $(i = 1,2,3,4)$ are operators acting on $L_2(\Omega)$. In particular, the operators L_D and L_N generated by expression (5.38), the Dirichlet boundary conditions on the side surface of the cylinder, and Dirichlet $(u(x,0) = u(x,b) = 0)$ or Neuman $\left(\frac{\partial u}{\partial t}(x,0) = \frac{\partial u}{\partial t}(x,b) = 0\right)$ conditions, respectively, on the bases are self-adjoint extensions of the minimal operator. It should be noted that the spectrum of these extensions is not always discrete, even if the spectrum of A^2 is discrete. However, if we know that an arbitrary function $u(x,t)$ from the domain of L_D, L_N, or any other self-adjoint extension of the operator L_0, belongs at fixed $t_0 \in [0,b]$ to the space $\mathfrak{H}^\alpha$ $(0 < \alpha \le 1)$, then according to Corollary 5.5 the spectrum of such an extension is discrete.

The problem of the existence of a solvable extension of the minimal operator generated by a general partial differential expression has not been solved in the theory of partial differential equations up to now. When the coefficients of an expression are constant the existence of such an extension has been proved (see, for example, Hörmander [1]). But the question whether it is possible to give it using boundary conditions remains open. Corollary 5.4 and Theorem 5.5 state that the minimal operator generated by the differential expression (5.38) in the cylinder Ξ always has self-adjoint completely solvable extensions, and all these extensions can be given by boundary conditions of the form (5.39) on the side surface of the cylinder, and conditions of the form (5.40) on its bases. For instance, the extension corresponding to the operator $K = E$ is completely solvable because $K - E = 0 \in \mathfrak{S}_\infty$ and the operator $K + E = 2E$ is invertible. Moreover, according to Theorem 5.6, if $m > \frac{n}{2}$, then among the self-adjoint solvable extensions there are extensions whose inverses are Hilbert–Schmidt operators. This follows from the asymptotic relation $\mu_k \sim c_k k^{m/n}$ for the eigenvalues μ_k of the operator A (see, for example, Birman, Solomyak [1]).

2. Let us consider a special case of the previous example, when Ω is a parallelepiped $0 \le x_j \le a_j$ ($j = 1, \dots, n$) and

$$M = -\sum_{j=1}^{n} \frac{\partial^2}{\partial x_j^2}.$$

In this case the spectrum of the operator A^2 is discrete and $\pi^2 \sum_{j=1}^{n} \frac{k_j^2}{a_j^2}$ ($k_j \in \mathbb{N}$) are its eigenvalues. We discuss only the Dirichlet problem corresponding to the extension L_D for the expression

$$l = \frac{\partial^2}{\partial t^2} - \sum_{j=1}^{n} \frac{\partial^2}{\partial x_j^2}.$$

As has been shown in subsection 5.7, a necessary and sufficient condition for the unique solvability of this problem is the non-solvability of the equation $b^2 \sum_{j=1}^{n} \frac{k_j^2}{a_j^2} = m^2$ in the class of natural numbers (a solution of this equation is understood to be a set $\{k_1, \dots, k_n;\ m\}$).

In particular, in the case $n = 1$ a solution of the Dirichlet problem for the string equation in the rectangle $0 \le t \le b$, $0 \le x \le a$ is unique if and only if the ratio $\frac{b}{a}$ is irrational. We will discuss what additional properties the numbers a and b need in this case for the operator L_D to be solvable. According to the results of subsection 5.7, a necessary and sufficient condition for this is the boundedness of the set

$$\left\{ \frac{\cos \frac{\pi k b}{a}}{\frac{\pi k b}{a} \sin \frac{\pi k b}{a}} \right\}_{k=1}^{\infty},$$

which is equivalent to the fact that the numbers $(\pi k b/a)\sin(\pi k b/a)$ should not have a limit point at zero as $k \to \infty$. The question on approaching zero by numbers of the form of $(\pi k b/a)\sin(\pi k b/a)$ (with irrational $\frac{b}{a}$) reduces to the question of the rate of approximation of zero by the numbers $\frac{\pi k b}{a} - m\pi$ ($m \in \mathbb{N}$) as $k \to \infty$. In turn, the latter is equivalent to the question of the rate of approximation of the irrational number b/a by numbers m/k ($k \to \infty$). The answer depends on the properties of the continued fraction expansion of b/a.

So, let

$$\frac{b}{a} = a_0 + \cfrac{1}{a_1 + \cfrac{1}{a_2 + \cdots}}. \tag{5.41}$$

If the sequence $\{a_i\}_{i=0}^{\infty}$ is bounded, there exists $\epsilon_0 > 0$ such that the inequality $\left|\frac{b}{a} - \frac{m}{k}\right| \geq \frac{\epsilon_0}{k^2}$ holds for all natural m, k (see, for example, Khinchin [1]). Hence, $\left|\pi k \frac{b}{a} - m\pi\right| \geq \frac{\pi\epsilon_0}{k}$. Therefore, if zero is a limit point of the sequence $\left\{\sin \pi k \frac{b}{a}\right\}_{k=1}^{\infty}$, this sequence tends to 0 with rate not exceeding const$\cdot \frac{1}{k}$. Now it is not difficult to conclude that the sequence of numbers $\left\{\pi k \frac{b}{a} \sin \frac{\pi k b}{a}\right\}_{k=1}^{\infty}$ which we are interested in has no limit point at zero, i.e. the operator L_D is solvable. If the sequence $\{a_i\}_{i=0}^{\infty}$ is not bounded, then for $\epsilon > 0$ one can find an infinite set of natural numbers m and k such that $\left|\frac{b}{a} - \frac{m}{k}\right| < \frac{\epsilon}{k^2}$. Hence, $\left|\pi k \frac{b}{a} - m\pi\right| < \frac{\pi\epsilon}{k}$. This implies that the sequence of numbers $\left\{\sin \pi k \frac{b}{a}\right\}_{k=1}^{\infty}$ has a limit point at zero and tends to it with rate not less than $\frac{\epsilon}{k}$. Hence, the sequence $\left\{\pi k \frac{b}{a} \sin \pi k \frac{b}{a}\right\}_{k=1}^{\infty}$ which is the object of our investigation, has distance from zero at most $\frac{\pi b \epsilon}{a}$ for large k. Since ϵ is arbitrary, zero is a limit point of this sequence. In this case the operator L_D is not solvable.

So, if the number b/a is irrational and the sequence $\{a_k\}_{k=0}^{\infty}$ appearing in its continued fraction expansion (5.41) is bounded, then the operator L_D is solvable. In all other cases zero belongs to the spectrum L_D.

3. Consider for expression (5.1) with $q(t) \equiv 0$ the problem

$$y'(0) = y(b) = 0. \tag{5.42}$$

It is associated with a certain extension L_N of the minimal operator L_0 constructed in subsection 5.6. Since the operator L_N admits separation of variables, it is self-adjoint. Because of

$$A^{-1} \sin Aby_0 + \cos Aby_0' = A^{-1} y'(0),$$

the equalities (5.42) are equivalent to the boundary conditions

$$A^{-1} \sin Aby_0 + \cos Aby_0' = 0, \quad y(b) = 0.$$

If we denote by C a self-adjoint operator with spectrum located in $\left[-\frac{\pi}{2}, \frac{\pi}{2}\right]$ for which

$$\cos C = \cos Ab\left(A^{-2}\sin^2 Ab + \cos^2 Ab\right)^{-1/2},$$
$$\sin C = -A^{-1}\sin Ab\left(A^{-2}\sin^2 Ab + \cos^2 Ab\right)^{-1/2},$$

then L_N coincides with the extension L_C of the operator L_0. Therefore, the operator L_N is invertible, i.e. the equation $L_N y = 0$ has a unique solution, if and only if the points $\frac{\pi}{b}(m + \frac{1}{2})$ are not eigenvalues of the operator A. For the extension L_N to be solvable it is necessary and sufficient that the set $\{\mu^{-1}\text{tg}\,\mu b : \mu \in \sigma(A)\}$ be bounded. Under the assumption that the spectrum $\{\mu_k\}_{k=1}^{\infty}$ of A is discrete, the spectrum of L_N is discrete if and only if the set $\{\mu_k \text{ctg}\,\mu_k b\}_{k=1}^{\infty}$ has infinity as its unique limit point.

If we take for A^2 the positive definite self-adjoint operator generated on $L_2(0, a)$ $(a < \infty)$ by the expression $-\frac{d^2}{dx^2}$ and the boundary condition $\varphi(0) = \varphi(a) = 0$, then the operator L_N is generated on the space $L_2([0,a]\times[0,b])$ by the expression

$$(l[u])(x,t)=\frac{\partial^2 u(x,t)}{\partial t^2}-\frac{\partial^2 u(x,t)}{\partial x^2} \tag{5.43}$$

and boundary conditions

$$u(0,t)=u(a,t)=u(x,b)=u_t'(x,0)=0. \tag{5.44}$$

Since the spectrum of A coincides with the set $\left\{\mu_k=\frac{k\pi}{a}\right\}_{k=1}^{\infty}$, problem (5.44) for the equation $l[u]=0$ has a unique solution if and only if the equation $\frac{m+\frac{1}{2}}{b}=\frac{k}{a}$ is not solvable in the class of integers. The latter holds if and only if b/a is irrational or if $\frac{b}{a}=\frac{p}{q}$, where p,q are coprime integers and in addition q is odd. In the latter case the set $\left\{\frac{a}{\pi k}\operatorname{tg}\frac{\pi k b}{a}\right\}_{k=1}^{\infty}$ is bounded; consequently, the operator L_N is solvable. Moreover, it follows from the complete continuity of the operator $\sin C$ that the operator L_N is completely solvable. Since $\alpha_k=\left(1+\frac{k^2\pi^2}{a^2}\operatorname{ctg}^2\frac{\pi k b}{a}\right)^{-1/2}$ $(k=1,2,\ldots)$ are the eigenvalues of $\sin C$, the operator $\sin C$ belongs to $\mathfrak{S}_2$, hence $L_N^{-1}\in\mathfrak{S}_2$.

Thus, if $\frac{b}{a}=\frac{p}{q}$ (p,q are coprime integers and q is odd), then L_N^{-1} is a Hilbert–Schmidt operator.

CHAPTER 4

Boundary Value Problems for a Second-Order Elliptic-Type Operator Differential Equation

1. Dissipative Boundary Value Problems

We will consider the differential expression

$$(l[y])(t) = -y''(t) + A^2 y(t) + q(t)y(t) \quad (t \in [a,b],\ -\infty < a < b < \infty), \tag{1.1}$$

where A is a lower semi-bounded self-adjoint operator, and for any fixed $t \in [a,b]$, $q(t)$ is a bounded self-adjoint operator on a Hilbert space $\mathfrak{H}$. We assume that the operator-function $q(t)$ is strongly measurable and that $\|q(t)\|$, as a function of t, is bounded on $[a,b]$.

Let us define the operator L_0' by $L_0'y = l[y]$ on the set $\mathfrak{D}_0'$ of elements of the form $y(t) = \sum_{k=1}^{m} \varphi_k(t) f_k$, $\varphi_k(t) \in \overset{\circ}{W}{}_2^2(a,b)$, $f_k \in \mathfrak{D}(A^2)$. It follows from the formula for integration by parts,

$$\int_\alpha^\beta ((l[y])(t), z(t))\, dt - \int_\alpha^\beta (y(t), (l[z])(t))\, dt = ((y(t), z'(t)) - (y'(t), z(t)))\Big|_\alpha^\beta$$
$$(y, z \in \mathfrak{D}_0',\ a \le \alpha < \beta \le b), \tag{1.2}$$

that the operator L_0' is symmetric on the space $L_2(\mathfrak{H}, (a,b))$. Its closure L_0 is called the minimal operator generated by expression (1.1). The operator L_0^*, the adjoint of L_0, is called the maximal operator.

In this section we study the domains of the minimal and the maximal operator generated by expression (1.1). In terms of boundary conditions we describe all maximal dissipative and maximal accumulative, in particular self-adjoint, extensions of L_0.

1.1. THE DOMAIN OF THE MAXIMAL OPERATOR

Since the operators L_0 and L_0^* generated by expression (1.1) differ from those for expression (1.1) with $q(t) \equiv 0$ and $A \ge E$ by a bounded operator, we only need to consider

$$(l[y])(t) = -y''(t) + A^2 y(t) \quad (A^2 \ge E)$$

if we wish to investigate the domain of the minimal or the maximal operator, as well as the domain of a maximal dissipative (maximal accumulative) extension of L_0.

We set

$$(\hat{l}[y])(t) = -y''(t) + \hat{A}^2 y(t),$$

where $\hat{A}$ is the extension of A as defined in Chapter 2, subsection 3.1.

LEMMA 1.1. *The linear operators*

$$f \to e^{-\hat{A}(t-a)} f, \quad f \to e^{-\hat{A}(b-t)} f$$

are continuous from $\mathfrak{H}^{-\frac{1}{2}}(A)$ into $L_2(\mathfrak{H}, (a,b))$.

Proof. Since the vector-functions $e^{-\hat{A}(t-a)} f$ and $e^{-\hat{A}(b-t)} f$ $(f \in \mathfrak{H}^{-\frac{1}{2}})$ are infinitely differentiable into $\mathfrak{H}$ for $t \in (a,b)$ (see Theorem 3.1, Chapter 2) and, by virtue of Corollary 3.1, Chapter 2, $f = \hat{A}^{\frac{1}{2}} g$ $(g \in \mathfrak{H})$, we have

$$\left\| e^{-\hat{A}(t-a)} f \right\|^2_{L_2(\mathfrak{H},(a,b))} = \tfrac{1}{2} \int_1^{\infty} (1 - e^{-2\lambda(b-a)}) \, \mathrm{d}(E_\lambda g, g) \le \|g\|^2 = \|f\|^2_{\mathfrak{H}^{-\frac{1}{2}}}.$$

A similar estimate can be given for $\left\| e^{-\hat{A}(b-t)} f \right\|_{L_2(\mathfrak{H},(a,b))}$. □

LEMMA 1.2. *The linear operators*

$$f(t) \to \int_a^b e^{-\hat{A}(t-a)} f(t) \, \mathrm{d}t, \quad f(t) \to \int_a^b e^{-\hat{A}(b-t)} f(t) \, \mathrm{d}t$$

are continuous from $L_2(\mathfrak{H}, (a,b))$ into $\mathfrak{H}^{\frac{1}{2}}(A)$.

Proof. The operators considered are adjoint to the operators of Lemma 1.1. This implies the desired result. □

Lemma 1.2 shows that the vector-function $\int_a^b e^{-\hat{A}|t-s|} h(s) \, \mathrm{d}s$ $(h(t) \in L_2(\mathfrak{H}, (a,b)))$ is continuous into the space $\mathfrak{H}^{\frac{1}{2}}$ on the closed interval $[a,b]$.

The Cauchy–Buniakowski inequality

$$|(f,g)| \le \|f\|_{\mathfrak{H}^{-\alpha}} \|g\|_{\mathfrak{H}^{\alpha}} \tag{1.3}$$

(see Chapter 2, subsection 1.1) implies that a scalar function $(y(t), z(t))$ is continuous on $[a,b]$ provided that a vector-function $y(t)$ is strongly (weakly) continuous into the space $\mathfrak{H}^{-\alpha}$ while $z(t)$ is weakly (strongly) continuous into $\mathfrak{H}^{\alpha}$. Moreover, the following lemma is valid.

LEMMA 1.3. *The formula*

$$(y''(t), z(t)) - (y(t), z''(t)) = \frac{\mathrm{d}}{\mathrm{d}t}\big((y'(t), z(t)) - (y(t), z'(t))\big) \tag{1.4}$$

is true under the following assumptions on $y(t), z(t)$:

a) *the $\mathfrak{H}$-valued function $y(t)$ is twice weakly differentiable into $\mathfrak{H}^{-2}$ while the $\mathfrak{H}^2$-valued function $z(t)$ is strongly continuously differentiable into $\mathfrak{H}^2$ and twice strongly differentiable into $\mathfrak{H}$;*

b) *the $\mathfrak{H}$-valued function $y(t)$ is weakly differentiable into $\mathfrak{H}^{-1}$ and twice weakly differentiable into $\mathfrak{H}^{-2}$, while the vector-function $z(t)$ with values in $\mathfrak{H}^2$ has a strongly continuous derivative into $\mathfrak{H}^1$ and the second strong derivative into $\mathfrak{H}$.*

Proof. Taking into account (1.3) we verify by means of a direct calculation that in both cases

$$\begin{aligned}
\frac{\mathrm{d}}{\mathrm{d}t}(y'(t), z(t)) &= \lim_{\Delta t\to 0}\frac{(y'(t+\Delta t), z(t+\Delta t)) - (y'(t), z(t))}{\Delta t}\\
&= \lim_{\Delta t\to 0}\Big(\frac{y'(t+\Delta t) - y'(t)}{\Delta t}, z(t+\Delta t)\Big) +\\
&\quad + \lim_{\Delta t\to 0}\Big(y'(t), \frac{z(t+\Delta t) - z(t)}{\Delta t}\Big)\\
&= (y''(t), z(t)) + (y'(t), z'(t));\\
\frac{\mathrm{d}}{\mathrm{d}t}(y(t), z'(t)) &= \lim_{\Delta t\to 0}\frac{(y(t+\Delta t), z'(t+\Delta t)) - (y(t), z'(t))}{\Delta t}\\
&= \lim_{\Delta t\to 0}\Big(\frac{y(t+\Delta t) - y(t)}{\Delta t}, z'(t+\Delta t)\Big) +\\
&\quad + \lim_{\Delta t\to 0}\Big(y(t), \frac{z'(t+\Delta t) - z'(t)}{\Delta t}\Big)\\
&= (y'(t), z'(t)) + (y(t), z''(t)),
\end{aligned}$$

whence (1.4) follows. □

If now the two vector-functions $y(t), z(t)$ satisfy condition a) or b) of the lemma, then

$$\int_a^b ((\hat{l}[y])(t), z(t))\,\mathrm{d}t - \int_a^b (y(t), (\hat{l}[z])(t))\,\mathrm{d}t = \big((y(t), z'(t)) - (y'(t), z(t))\big)\big|_a^b. \tag{1.5}$$

LEMMA 1.4. *The domain $\mathfrak{D}(L_0^*)$ of the maximal operator L_0^* generated by expression (1.1) consists precisely of the vector-functions $y(t)$ that have a representation*

$$y(t) = e^{-\hat{A}(t-a)}f_1 + e^{-\hat{A}(b-t)}f_2 + \tfrac{1}{2}\int_a^b e^{-A|t-s|}A^{-1}h(s)\,\mathrm{d}s \tag{1.6}$$

$$(f_1, f_2 \in \mathfrak{H}^{-\frac{1}{2}}(A), \quad h = L_0^* y \in L_2(\mathfrak{H}, (a,b)).$$

Proof. Suppose $y(t)$ has a representation (1.6). According to Lemma 1.1 $y \in L_2(\mathfrak{H}, (a,b))$. The first and the second summands in (1.6) are infinitely differentiable inside (a,b) into $\mathfrak{H}$. They satisfy the equation $l[y] = 0$. The third summand is a vector-function that is continuously differentiable into $\mathfrak{H}$. By Lemma 1.2 its second derivative exists almost everywhere into $\mathfrak{H}^{-\frac{1}{2}}$. This vector-function satisfies the equation $\hat{l}[y] = h$. Since (1.5) holds for a vector-function $y(t)$ of the form (1.6) and $z(t) \in \mathfrak{D}_0'$, we obtain $(L_0'z, y)_{L_2(\mathfrak{H},(a,b))} = (z, h)_{L_2(\mathfrak{H},(a,b))}$, i.e. $y \in \mathfrak{D}(L_0^*)$ and $L_0^* y = h$.

Conversely, let $y \in \mathfrak{D}(L_0^*)$. First $y(t)$ is assumed to be such that

$$(L_0^* z, y)_{L_2(\mathfrak{H},(a,b))} = (z, 0)_{L_2(\mathfrak{H},(a,b))} = 0 \quad (\forall z \in \mathfrak{D}_0'). \tag{1.7}$$

Substituting for $z(t)$ in (1.7) vector-functions $z(t) = \varphi(t)f$ ($f \in \mathfrak{H}^\infty(A)$, $\varphi(t) \in \mathcal{D}(a,b)$), we conclude that the vector-function $y(t)$ is a generalized solution in the sense of Chapter 2, subsection 5.2 of the equation $l[y] = 0$. By Theorem 5.5, Chapter 2, $y(t)$ is a solution inside (a,b) of this equation. But, $y(t) \in L_2(\mathfrak{H},(a,b))$. Taking into account that A is positive definite by Theorem 5.2 and Corollary 5.3, Chapter 2, we obtain

$$y(t) = e^{-\hat{A}(t-a)} f_1 + e^{-\hat{A}(b-t)} f_2 \quad (f_1, f_2 \in \mathfrak{H}^{-\frac{1}{2}}). \tag{1.8}$$

Let now $y \in \mathfrak{D}(L_0^*)$ be arbitrary. For any $z \in \mathfrak{D}_0'$,

$$(L_0' z, y)_{L_2(\mathfrak{H},(a,b))} = (z, h)_{L_2(\mathfrak{H},(a,b))},$$

where $h = L_0^* y \in L_2(\mathfrak{H},(a,b))$. Using formula (1.5) we can directly verify that if

$$y^*(t) = \frac{1}{2} \int_a^b e^{-A|t-s|} A^{-1} h(s)\, ds \in L_2(\mathfrak{H},(a,b))$$

and $z(t) \in \mathfrak{D}_0'$, then

$$(L_0 z, y^*)_{L_2(\mathfrak{H},(a,b))} = (z, h)_{K_2(\mathfrak{H},(a,b))},$$

that is, $\tilde{y}(t) = y(t) - y^*(t) \in \mathfrak{D}(L_0^*)$ satisfies (1.7). Therefore, $\tilde{y}(t)$ has a representation (1.8), which implies the representation (1.6) for $y(t)$. □

THEOREM 1.1. *A vector-function $y(t)$ belongs to the domain $\mathfrak{D}(L_0^*)$ of the maximal operator L_0^* generated on $L_2(\mathfrak{H},(a,b))$ by expression (1.1) if and only if: a) $y'(t)$ exists and is an absolutely continuous function on (a,b) into $\mathfrak{H}^{-2}$; b) $\hat{l}[y] \in L_2(\mathfrak{H},(a,b))$. The operator L_0^* acts on $\mathfrak{D}(L_0^*)$ as $L_0^* y = \hat{l}[y]$.*

Proof. Representation (1.6) shows that a vector-function $y(t) \in \mathfrak{D}(L_0^*)$ satisfies the conditions a) and b) of the theorem. The converse assertion follows from (1.5). According to Lemma 1.3, this formula is valid for vector-functions $z(t) \in \mathfrak{D}_0'$ and $y(t)$ satisfying the conditions of the theorem. □

Representation (1.6) and Theorem 1.1 lead to the following assertions.

COROLLARY 1.1. *If a scalar function $\varphi(t) \in W_2^2(a,b)$ is constant in neighbourhoods of two points a, b, then*

$$(y \in \mathfrak{D}(L_0^*)) \Rightarrow (z(t) = \varphi(t) y(t) \in \mathfrak{D}(L_0^*)).$$

Proof. Obviously, the vector-function $z'(t)$ is absolutely continuous into $\mathfrak{H}^{-2}$ on (a,b). Furthermore, $(\hat{l}[z])(t) = \varphi(t)(-y''(t) + A^2 y(t)) - \varphi''(t) y(t) - 2\varphi'(t) y'(t) \in L_2(\mathfrak{H},(a,b))$. This follows from the continuity into $\mathfrak{H}$ of the vector-functions $y(t)$ and $y'(t)$ on any interval $[a_1, b_1] \subset (a,b)$, which is a consequence of representation (1.6).

We denote by $\mathfrak{D}(L')$ the set of vector-functions $y(t)$ satisfying the following conditions; a) on the interval $[a,b]$, $y(t)$ is twice continuously differentiable (in the strong sense) into $\mathfrak{H}$; b) $y(t) \in \mathfrak{D}(A^2)$ at any $t \in [a,b]$; c) $l[y] \in L_2(\mathfrak{H},(a,b))$. Formula (1.2) shows that $\mathfrak{D}(L') \subset \mathfrak{D}(L_0^*)$ and $L_0^* y = l[y]$ on $\mathfrak{D}(L')$. It is also obvious that the set of vector-functions from $\mathfrak{D}(L')$ with compact support inside (a,b) is contained in $\mathfrak{D}(L_0)$. Denote by L' the restriction of the operator L_0^* to $\mathfrak{D}(L')$.

COROLLARY 1.2. *The operator L_0^* coincides with the closure in $L_2(\mathfrak{H},(a,b))$ of the operator L'.*

Proof. Let $y \in \mathfrak{D}(L_0^*)$. The vector-function $y(t)$ can be represented in the form (1.6). Since $\overline{\mathfrak{D}(A^2)} = \mathfrak{H}$, there exist sequences $f_{1,n}, f_{2,n} \in \mathfrak{H}^2$ convergent to f_1, f_2, respectively, in $\mathfrak{H}^{-\frac{1}{2}}$, and a sequence $h_n(t)$ of continuous $\mathfrak{H}^2$-valued vector-functions convergent to $h(t)$ in $L_2(\mathfrak{H},(a,b))$. Then the vector-functions

$$y_n(t) = e^{-\hat{A}(t-a)} f_{1,n} + e^{-\hat{A}(b-t)} f_{2,n} + \frac{1}{2}\int_a^b e^{-A|t-s|} A^{-1} h_n(s)\, ds$$

belong to $\mathfrak{D}(L')$, and also $y_n \to y$, $L' y_n = h_n \to h$ in $L_2(\mathfrak{H},(a,b))$. □

1.2. REGULARIZED BOUNDARY VALUES AND GREEN'S FORMULA

Three terms are involved in the representation (1.6) of a vector-function $y(t)$ in the domain of the maximal operator. It is not difficult to verify that the third term is continuously differentiable on $[a,b]$ into $\mathfrak{H}$. As for the other terms, they are infinitely differentiable inside (a,b) into $\mathfrak{H}$, which follows from Theorem 3.1, Chapter 2. Since $f_1, f_2 \in \mathfrak{H}^{-\frac{1}{2}}$, their boundary values at the ends of the interval belong to the space $\mathfrak{H}^{-\frac{1}{2}}$; the boundary values of the derivatives of these summands belong to $\mathfrak{H}^{-\frac{3}{2}}$.

Let now f_a, f_b be arbitrary vectors in $\mathfrak{H}^{-\frac{1}{2}}$ and $f_a', f_b' \in \mathfrak{H}^{-\frac{3}{2}}$. The following question arises: are there any vector-functions $y(t), z(t) \in \mathfrak{D}(L_0^*)$ such that

$$\begin{aligned} y(a) &= f_a, & y'(a) &= f_a', \\ z(b) &= f_b, & z'(b) &= f_b'. \end{aligned} \tag{1.9}$$

The answer is "yes", but not always.

LEMMA 1.5. *Assume that $f_a, f_b \in \mathfrak{H}^{-\frac{1}{2}}$; $f_a', f_b' \in \mathfrak{H}^{-\frac{3}{2}}$. A necessary and sufficient condition for the existence of vector-functions $y(t), z(t) \in \mathfrak{D}(L_0^*)$ satisfying (1.9) is*

$$\hat{A} f_a + f_a' \in \mathfrak{H}^{\frac{1}{2}}, \quad \hat{A} f_b - f_b' \in \mathfrak{H}^{\frac{1}{2}}.$$

Proof. We present the proof only for the point a, since it is the same for the point b.

Let $f_a \in \mathfrak{H}^{-\frac{1}{2}}$, $f_a' \in \mathfrak{H}^{-\frac{3}{2}}$, and let $y(t) \in \mathfrak{D}(L_0^*)$ be a function such that $y(a) = f_a$,

$y'(a) = f'_a$. Substituting for y in these equations its representation (1.6) we obtain

$$f_1 + e^{-\hat{A}(b-a)} f_2 + \tfrac{1}{2} \int_a^b e^{-A(s-a)} A^{-1} h(s)\, ds = f_a,$$

$$-\hat{A} f_1 + \hat{A} e^{-\hat{A}(b-a)} f_2 + \tfrac{1}{2} \int_a^b e^{-A(s-a)} h(s)\, ds = f'_a,$$

whence

$$f_1 = \tfrac{1}{2}(f_a - \hat{A}^{-1} f'_a),$$

$$2e^{-\hat{A}(b-a)} f_2 + A^{-1} \int_a^b e^{-A(s-a)} h(s)\, ds = f_a + \hat{A}^{-1} f'_a.$$

Since $e^{-\hat{A}(b-a)} f_2 \in \mathfrak{H}^{\frac{3}{2}}$ and, because of Lemma 1.2, $A^{-1} \int_a^b e^{-A(s-a)} h(s)\, ds \in \mathfrak{H}^{\frac{3}{2}}$, we have $f_a + \hat{A}^{-1} f'_a \in \mathfrak{H}^{\frac{3}{2}}$, i.e. $\hat{A} f_a + f'_a \in \mathfrak{H}^{\frac{1}{2}}$.

Conversely, let $\hat{A} f_a + f'_a \in \mathfrak{H}^{\frac{1}{2}}$ for $f_a \in \mathfrak{H}^{-\frac{1}{2}}$, $f'_a \in \mathfrak{H}^{-\frac{3}{2}}$. According to Lemma 1.1 the vector-function $h(s) = e^{-\hat{A}(s-a)} f_0$, $f_0 = 2\left(E - e^{-2\hat{A}(b-a)}\right)^{-1} \hat{A}(\hat{A} f_a + f'_a) \in \mathfrak{H}^{-\frac{1}{2}}$, belongs to $L_2(\mathfrak{H}, (a,b))$. Setting in representation (1.6) $f_1 = \frac{1}{2}(f_a - \hat{A}^{-1} f'_a) \in \mathfrak{H}^{-\frac{1}{2}}$, $f_2 = 0$, $h(s) = e^{-\hat{A}(s-a)} f_0$, we obtain the vector-function $y(t) = e^{-\hat{A}(t-a)} f_1 + \frac{1}{2} \int_a^b e^{-A|t-s|} A^{-1} h(s)\, ds$, which belongs, by Lemma 1.4, to $\mathfrak{D}(L_0^*)$. One can immediately check that for this function $y(a) = f_a$, $y'(a) = f'_a$. □

LEMMA 1.6. *If $y(t) \in \mathfrak{D}(L_0^*)$, then on the half-open interval $[a,b)$ (respectively, $(a,b]$) the vector-function $\hat{A} y(t) + y'(t)$ (respectively, $\hat{A} y(t) - y'(t)$) is continuous into the space $\mathfrak{H}^{\frac{1}{2}}$.*

Proof. Let us compute the expressions $\hat{A} y(t) \pm y'(t)$ using representation (1.6):

$$\hat{A} y(t) + y'(t) = 2e^{-\hat{A}(b-t)} \hat{A} f_2 + \int_t^b e^{-A(s-t)} h(s)\, ds, \tag{1.10}$$

$$\hat{A} y(t) - y'(t) = 2e^{-\hat{A}(t-a)} \hat{A} f_1 + \int_a^t e^{-A(t-s)} h(s)\, ds. \tag{1.11}$$

It is evident that on $[a,b)$ the first term at the right-hand side of (1.10) is continuous into $\mathfrak{H}^{\frac{1}{2}}$. In view of Lemma 1.2, the second summand is continuous on $[a,b]$ into $\mathfrak{H}^{\frac{1}{2}}$. So, $\hat{A} y(t) + y'(t)$ is a continuous $\mathfrak{H}^{\frac{1}{2}}$-valued function on $(a,b]$. The continuity into the space $\mathfrak{H}^{\frac{1}{2}}$ of $\hat{A} y(t) - y'(t)$ on $(a,b]$ is proved in exactly the same way. □

THEOREM 1.2. *The Green formula*

$$(L_0^* y, z)_{L_2(\mathfrak{H},(a,b))} - (y, L_0^* z)_{L_2(\mathfrak{H},(a,b))} = \left((\overline{y}_t, \overline{z}'_t) - (\overline{y}'_t, \overline{z}_t)\right)\Big|_a^b, \tag{1.12}$$

where

$$\begin{aligned} \overline{y}_a &= y(a), \quad \overline{y}'_a = y'(a) + \hat{A} y(a), \\ \overline{y}_b &= y(b), \quad \overline{y}'_b = y'(b) - \hat{A} y(b), \end{aligned} \tag{1.13}$$

holds for any $y(t), z(t) \in \mathfrak{D}(L_0^*)$.

Proof. It follows from representation (1.6) that the vector-functions $y(t), z(t)$ are continuously differentiable inside the interval (a, b), into $\mathfrak{H}^{\frac{1}{2}}$ while their derivatives are absolutely continuous there into $\mathfrak{H}^{-\frac{1}{2}}$. According to Lemma 1.3 formula (1.5) holds for $y(t), z(t)$ on the interval $[\alpha, \beta]$ $(a < \alpha < \beta < b)$:

$$\begin{aligned}\int_\alpha^\beta ((\hat{l}[y])(t), z(t))\,\mathrm{d}t - \int_\alpha^\beta (y(t), (\hat{l}[z])(t))\,\mathrm{d}t \\ = (y(\beta), z'(\beta)) - (y'(\beta), z(\beta)) - (y(\alpha), z'(\alpha)) + (y'(\alpha), z(\alpha)) \\ = (y(\beta), z'(\beta) - \hat{A}z(\beta)) - (y'(\beta) - \hat{A}y(\beta), z(\beta)) - \\ -(y(\alpha), z'(\alpha) + \hat{A}z(\alpha)) + (y'(\alpha) + \hat{A}y(\alpha), z(\alpha))\end{aligned} \tag{1.14}$$

Taking into account Lemma 1.6 and the continuity of the vector-functions $y(t), z(t)$ into $\mathfrak{H}^{-\frac{1}{2}}$ on the closed interval $[a, b]$, by limit transition in (1.14) as $\alpha \to a$, $\beta \to b$ we arrive at (1.12). □

1.3. THE DOMAIN OF THE MINIMAL OPERATOR

THEOREM 1.3. *The domain* $\mathfrak{D}(L_0)$ *of the minimal operator* L_0 *generated by expression* (1.1) *consists of those vector-functions* $y(t) \in \mathfrak{D}(L_0^*)$ *for which* $y(a) = y(b) = y'(a) = y'(b) = 0$.

Proof. Let $y(t) \in \mathfrak{D}(L_0)$. Then

$$(L_0 y, z)_{L_2(\mathfrak{H},(a,b))} = (y, L_0^* z)_{L_2(\mathfrak{H},(a,b))}$$

for an arbitrary $z(t) \in \mathfrak{D}(L_0^*)$. Because of (1.12),

$$(\overline{y}'_b, \overline{z}_b) - (\overline{y}_b, \overline{z}'_b) - (\overline{y}'_a, \overline{z}_a) + (\overline{y}_a, \overline{z}'_a) = 0.$$

Taking $z(t)$ of the form $z(t) = \varphi(t)f$, where $f \in \mathfrak{D}(A^2)$, $\varphi(t) \in W_2^2(a, b)$ and $\varphi(a) = \varphi(b) = \varphi'(b) = 0$, $\varphi'(a) = 1$, we obtain $\overline{y}_a = 0$. Similarly, by choosing suitable vector-functions $z(t)$ we may conclude that $y(a) = y(b) = y'(a) = y'(b) = 0$.

Conversely, if $y(t) \in \mathfrak{D}(L_0^*)$ has the property $y(a) = y(b) = y'(a) = y'(b) = 0$, then on account of (1.12) and (1.13), $y(t) \in \mathfrak{D}(L_0^{**}) = \mathfrak{D}(L_0)$. □

COROLLARY 1.3. *If* $y(t) \in \mathfrak{D}(L_0^*)$, *while the scalar function* $\varphi(t)$ *belongs to* $\overset{\circ}{W}{}_2^2(a, b) \cap C^2[a, b]$, *then* $z(t) = \varphi(t)y(t) \in \mathfrak{D}(L_0)$.

Proof. Since $z(a) = z(b) = z'(a) = z'(b) = 0$, it remains to show that $z \in \mathfrak{D}(L_0^*)$, i.e.

$$\hat{l}[z] = \varphi\hat{l}[y] - \varphi'' y - 2\varphi' y' \in L_2(\mathfrak{H}, (a, b)). \tag{1.15}$$

Obviously, the first and second summands in (1.15) belong to the space $L_2(\mathfrak{H}, (a, b))$. Let us prove that $\varphi' y' \in L_2(\mathfrak{H}, (a, b))$.

Representation (1.6) yields

$$y'(t) = -\hat{A}e^{-\hat{A}(t-a)}f_1 + \hat{A}e^{-\hat{A}(b-t)}f_2 - \tfrac{1}{2}\int_a^t e^{-A(t-s)}h(s)\,ds +$$
$$+\tfrac{1}{2}\int_t^b e^{-A(s-t)}h(s)\,ds \quad (f_1, f_2 \in \mathfrak{H}^{-\frac{1}{2}},\ h = L_0^* y).$$

This implies that $y'(t)$ is continuous inside (a,b), while at the ends of the interval it has singularities that are not square integrable because of the presence of the first and the second summands. Therefore the proof will be completed if we show that $\varphi'(t)\hat{A}e^{-\hat{A}(t-a)}f_1$ and $\varphi'(t)\hat{A}e^{-\hat{A}(b-t)}f_2$ belong to $L_2(\mathfrak{H},(a,b))$. We verify this, for example, for the first of these vector-functions. Namely,

$$\int_a^b \|\varphi'(t)\hat{A}e^{-\hat{A}(t-a)}f_1\|^2\,dt = \int_a^b |\varphi''(\tau)(t-a)|^2\|\hat{A}e^{-\hat{A}(t-a)}f_1\|^2\,dt$$
$$\le c\int_a^b (t-a)^2\int_1^\infty e^{-2\mu(t-a)}\mu^3\,d(E_\mu g_1, g_1)\,dt,$$

where $a < \tau < t$, $c = \sup_{t\in[a,b]}|\varphi''(t)|$, $g_1 = \hat{A}^{-\frac{1}{2}}f_1 \in \mathfrak{H}$. Interchanging the order of integration in the last integral with subsequent integration by parts yields the desired result. □

Lemma 1.4 and Theorem 1.3 imply the following statement.

COROLLARY 1.4. *The representation*

$$y(t) = \tfrac{1}{2}\int_a^b e^{-A|t-s|}A^{-1}h(s)\,ds,$$

where $h = L_0^* y = \hat{l}[y]$ *satisfies the conditions*

$$\int_a^b e^{-A(s-a)}h(s)\,ds = 0, \quad \int_a^b e^{-A(b-s)}h(s)\,ds = 0,$$

is valid for $y \in \mathfrak{D}(L_0)$.

As a direct consequence of Lemmas 1.2, 1.4 and Corollary 1.4 we obtain the following statement.

COROLLARY 1.5. *If a vector-function* $y(t)$ *belongs to* $\mathfrak{D}(L_0)$, *then on* $[a,b]$ *it is continuous into* $\mathfrak{H}^{\frac{3}{2}}$, *continuously differentiable into* $\mathfrak{H}^{\frac{1}{2}}$ *and* $y'(t)$ *is absolutely continuous into* $\mathfrak{H}^{-\frac{1}{2}}$.

1.4. DESCRIPTION OF DISSIPATIVE AND ACCUMULATIVE EXTENSIONS OF THE MINIMAL OPERATOR

As was shown in subsection 1.2, $\overline{y}_a = y(a) \in \mathfrak{H}^{-\frac{1}{2}}$, $\overline{y}_b = y(b) \in \mathfrak{H}^{-\frac{1}{2}}$ and

$$\overline{y}'_a = y'(a) + \hat{A}y(a) \in \mathfrak{H}^{\frac{1}{2}}, \quad \overline{y}'_b = y'(b) - \hat{A}y(b) \in \mathfrak{H}^{\frac{1}{2}}$$

for every vector-function $y(t) \in \mathfrak{D}(L_0^*)$. Let

$$Y = \{-y_a, y_b\}, \quad Y' = \{y'_a, y'_b\}, \tag{1.16}$$

where

$$y_a = \hat{A}^{-\frac{1}{2}}\overline{y}_a, \quad y_b = \hat{A}^{-\frac{1}{2}}\overline{y}_b,$$
$$y'_a = \hat{A}^{\frac{1}{2}}\overline{y}'_a, \quad y'_b = \hat{A}^{\frac{1}{2}}\overline{y}'_b$$

are regularized boundary values of $y(t)$ and its derivative. The vectors Y and Y' belong to the space $\mathfrak{H} \oplus \mathfrak{H}$. With the notations (1.16) the Green formula turns into

$$\left(L_0^* y, z\right)_{L_2(\mathfrak{H},(a,b))} - \left(y, L_0^* z\right)_{L_2(\mathfrak{H},(a,b))} = \left(Y, Z'\right)_{\mathfrak{H}\oplus\mathfrak{H}} - \left(Y', Z\right)_{\mathfrak{H}\oplus\mathfrak{H}}. \tag{1.17}$$

LEMMA 1.7. *For an arbitrary pair of vectors $Y, Y' \in \mathfrak{H} \oplus \mathfrak{H}$ there exists a vector-function $y(t) \in \mathfrak{D}(L_0^*)$ such that Y, Y' can be represented as in* (1.16).

Proof. Let $Y = \{y_1, y_2\}$, $Y' = \{y'_1, y'_2\}$, where $y_1, y_2, y'_1, y'_2 \in \mathfrak{H}$. Set

$$f_a = -\hat{A}^{\frac{1}{2}} y_1, \quad f_b = \hat{A}^{\frac{1}{2}} y_2;$$
$$f'_a = \hat{A}^{\frac{1}{2}}(\hat{A}^{-1}y'_1 + \hat{A}y_1), \quad f'_b = \hat{A}^{\frac{1}{2}}(\hat{A}^{-1}y'_2 + \hat{A}y_2).$$

Since $f_a, f_b \in \mathfrak{H}^{-\frac{1}{2}}$, $f'_a, f'_b \in \mathfrak{H}^{-\frac{3}{2}}$ and $\hat{A}f_a + f'_a = \hat{A}^{-\frac{1}{2}}y'_1 \in \mathfrak{H}^{\frac{1}{2}}$, $\hat{A}f_b - f'_b = \hat{A}^{-\frac{1}{2}}y'_2 \in \mathfrak{H}^{\frac{1}{2}}$, there exist, by Lemma 1.5, vector-functions $z_1(t), z_2(t) \in \mathfrak{D}(L_0^*)$ such that $z_1(a) = f_a$, $z'_1(a) = f'_a$, $z_2(b) = f_b$, $z'_2(b) = f'_b$. Because of Corollary 1.1 the vector-function $y(t) = \varphi_1(t)z_1(t) + \varphi_2(t)z_2(t)$, where $\varphi_1(t), \varphi_2(t) \in W_2^2(a,b)$, $\varphi_1(t) = 1$, $\varphi_2(t) = 0$ in a neighbourhood of the point a, $\varphi_1(t) = 0$, $\varphi_2(t) = 1$ in a neighbourhood of b, belongs to $\mathfrak{D}(L_0^*)$. Direct verification gives that $y(t)$ is the vector-function required. □

Take now the triple $(\mathfrak{H} \oplus \mathfrak{H}, \Gamma_1, \Gamma_2)$, where $\Gamma_1 y = Y$, $\Gamma_2 y = Y'$ are operators acting from $\mathfrak{D}(L_0^*)$ into $\mathfrak{H} \oplus \mathfrak{H}$; Y and Y' are determined by (1.16). Formula (1.17) and Lemma 1.7 enable us to state that this triple $(\mathfrak{H} \oplus \mathfrak{H}, \Gamma_1, \Gamma_2)$ is a boundary value space for the symmetric operator L_0 in the sense of Chapter 3, subsection 1.4. On the basis of Theorem 1.6, Chapter 3, we arrive at the following description of maximal dissipative and other extensions of the minimal operator.

THEOREM 1.4. *Let K be a contraction on the space $\mathfrak{H} \oplus \mathfrak{H}$. The extension L_K of the minimal operator L_0 generated on $L_2(\mathfrak{H}, (a,b))$ by the operation $\hat{l}[y]$ and the* *boundary condition*

$$(K-E)Y+i(K+E)Y'=0, \tag{1.18}$$

or

$$(K-E)Y-i(K+E)Y'=0, \tag{1.19}$$

is maximal dissipative, respectively, maximal accumulative. Conversely, any maximal dissipative (maximal accumulative) extension on $L_2(\mathfrak{H},(a,b))$ of the operator L_0 is generated by the expression $\hat{l}[y]$ and a boundary condition of the form (1.18) ((1.19)), where a contraction K is uniquely determined by an extension. Maximal symmetric extensions of the minimal operator are completely described by the conditions (1.18) and (1.19) with an isometric K. These conditions give a self-adjoint extension if and only if K is unitary.

REMARK 1.1. For self-adjoint extensions the boundary condition (1.18) is equivalent to

$$(\cos C)Y'-(\sin C)Y=0, \tag{1.20}$$

which is derived from (1.18) if one takes into account that a unitary operator K can be written as $K=e^{-2iC}$, where C is a self-adjoint operator on $\mathfrak{H}\oplus\mathfrak{H}$. The operator e^{-2iC} is uniquely determined by the corresponding self-adjoint extension.

If we set in (1.20) $C=\frac{\pi}{2}E$, then we get the self-adjoint extension of the minimal operator which is given by the boundary condition $y(a)=y(b)=0$ (the Dirichlet problem). If $C=\begin{pmatrix}-\operatorname{arctg}A^2 & 0\\ 0 & -\operatorname{arctg}A^2\end{pmatrix}$, then we obtain the self-adjoint extension corresponding to the boundary condition $y'(a)=y'(b)=0$ (the Neumann problem). In what follows we shall often encounter these extensions and we denote them by L_D and L_N.

2. Some Classes of Extensions of the Minimal Operator

If the spectrum of the operator A is discrete, then, as we will show below, next to extensions of the minimal operator with discrete spectrum there are extensions whose spectrum fills the entire real axis. So, it is natural to select in terms of boundary conditions the extensions that have a purely discrete spectrum. This is done below. Moreover, we describe the self-adjoint extensions having certain other special properties.

2.1. DESCRIPTION OF THE MAXIMAL DISSIPATIVE EXTENSIONS WITH DISCRETE SPECTRUM

THEOREM 2.1. *If $A^{-1}\in\mathfrak{S}_\infty$, then for any closed set Ω on the real axis there exists a self-adjoint extension L of the minimal operator L_0 whose spectrum contains Ω.*

Proof. For simplicity we regard $a = 0$, $q(t) \equiv 0$. We denote by $\{\mu_k\}_{k=1}^{\infty}$ the spectrum of the operator A^2. The extension required will be looked for among the self-adjoint extensions L_φ which are given by condition (1.20) with an operator C of the form $C = \begin{pmatrix} \frac{\pi}{2}E & 0 \\ 0 & \varphi(A^2) \end{pmatrix}$, where $\varphi(\mu)$ is a Borel–measurable function on $(1, \infty)$. This is a boundary condition of the form

$$y(0) = 0, \quad \hat{A} \cos \varphi(A^2)\overline{y}'_b - \sin \varphi(A^2)\overline{y}_b = 0. \tag{2.1}$$

Using Lemma 1.4 with A^2 replaced by $A^2 - \lambda E$, substituting for $y(t)$ in (2.1) its new representation and taking into account (1.13), we arrive at the following conclusion: a point λ is an eigenvalue of L_φ if and only if zero is an eigenvalue of the operator

$$K^{\varphi}_{\lambda}(A^2) = A \cos \varphi(A^2)\Big(\operatorname{ch}\sqrt{A^2 - \lambda E}\, b - A\frac{\operatorname{sh}\sqrt{A^2 - \lambda E}\, b}{\sqrt{A^2 - \lambda E}}\Big)e^{-\sqrt{A^2-\lambda E}\, b} -$$
$$- \sin \varphi(A^2)\frac{\operatorname{sh}\sqrt{A^2 - \lambda E}\, b}{\sqrt{A^2 - \lambda E}}e^{-\sqrt{A^2-\lambda E}\, b}.$$

Let now Ω be a closed set on the real axis and let $\{\lambda_k\}_{k=1}^{\infty}$ be a set that is everywhere dense in Ω. We define the function $\varphi(\mu) = \omega_k$ if $\mu_k \le \mu < \mu_{k+1}$, where ω_k is found from the equation

$$\operatorname{tg}\omega_k = \sqrt{\mu_k(\mu_k - \lambda_k)}\operatorname{cth}\sqrt{\mu_k - \lambda_k} - \mu_k.$$

Since $K^{\varphi}_{\lambda_k}(\mu_k) = 0$, zero is an eigenvalue of the operator $K^{\varphi}_{\lambda_k}(A^2)$, hence λ_k is an eigenvalue of the operator L_φ. Thus, the set $\{\lambda_k\}_{k=1}^{\infty}$, hence Ω, belongs to the spectrum of L_φ. □

Theorem 2.1 shows that if the operator A^{-1} is completely continuous, then the minimal operator has self-adjoint extensions whose spectrum fills the whole real axis. On the other hand, there exist self-adjoint extensions of the operator L_0 with discrete spectrum. For instance, the extensions L_D and L_N have a purely discrete spectrum.

Indeed, suppose that $q(t) \equiv 0$. Since $L_2(\mathfrak{H}, (a, b)) = L_2(a, b) \otimes \mathfrak{H}$, $L_D = l_D \otimes E + E \otimes A^2$, $L_N = l_N \otimes E + E \otimes A^2$, where the self-adjoint extension $l_D(L_N)$ of the minimal operator generated on the space $L_2(a, b)$ by the expression $-y''$ and the boundary condition $y(a) = y(b) = 0$ (respectively, $y'(a) = y'(b) = 0$) has purely discrete spectrum $\nu_k(l_D)$ (respectively, $\nu_k(L_N)$), the spectrum of the operator L_D (L_N) is discrete (see Chapter 1, subsection 4.7) and its eigenvalues are expressed by the formula

$$\lambda_{k,m} = \{\nu_k + \mu_m\}_{k,m=1}^{\infty} \tag{2.2}$$

(here $\lambda_{k,m}, \nu_k$ stand either for $\lambda_{k,m}(L_D)$, $\nu_k(l_D)$ or for $\lambda_{k,m}(L_N)$, $\nu_k(l_N)$; μ_m are the eigenvalues of A^2).

If $q(t) \not\equiv 0$ and L_K is the maximal dissipative extension of the minimal operator corresponding to the boundary condition (1.18), then

$$L_K = L_K^0 + q,$$

where L_K^0 is the extension generated by expression (1.1) with $q(t) \equiv 0$ and the same boundary condition. It remains to apply the resolvent formula (4.7) from Chapter 3,

$$R_\lambda(L_K) = R_\lambda(L_K^0) - R_\lambda(L_K^0) q R_\lambda(L_K).$$

Since the set of completely continuous operators forms a two-sided ideal in the ring of all bounded operators, the resolvent formula allows us to establish that the spectra of the operators L_D and L_N in the case $q(t) \equiv 0$ are discrete.

It turns out that the condition $A^{-1} \in \mathfrak{S}_\infty$ is necessary and sufficient for the existence of extensions of L_0 with purely discrete spectrum. In fact, if the spectrum of the operator A were not discrete, then one could find an infinite orthonormal system of vectors $f_n \in \mathfrak{D}(A^2)$ and a bounded sequence of numbers λ_n ($\operatorname{Im} \lambda_n = 0$) such that $(A^2 - \lambda_n E) f_n \to 0$ as $n \to \infty$. Taking $y_n = \varphi(t) f_n$ as in the proof of Theorem 4.4 (Chapter 3) we would be able to select a convergent subsequence from $\{y_n\}_{n=1}^\infty$ with $\|y_n - y_m\| = \sqrt{2}$ ($\forall m, n \in \mathbb{N}$).

Let now L_K be the maximal dissipative (maximal accumulative) extension of the minimal operator associated with the boundary condition (1.18) ((1.19)). Setting $K_1 = -E$ ($L_{K_1} = L_D$) in Theorem 3.1 from Chapter 3, we arrive at the following description of maximal dissipative (maximal accumulative) extensions of the operator L_0 with discrete spectrum.

THEOREM 2.2. *Suppose that $A^{-1} \in \mathfrak{S}_\infty$. The spectrum of the maximal dissipative (maximal accumulative) extension L_K of the minimal operator is discrete if and only if $K + E$ is completely continuous.*

COROLLARY 2.1. *Let $A^{-1} \in \mathfrak{S}_\infty$. In order that the spectrum of the self-adjoint extension L_C corresponding to the boundary condition (1.20) be discrete it is necessary and sufficient that the operator $\cos C$ be completely continuous.*

2.2. EXTENSIONS WITH RESOLVENT BELONGING TO THE IDEAL $\mathfrak{S}_p$

Recall that $\mathfrak{S}_p$ ($0 < p < \infty$) denotes the ideal in the algebra of all bounded linear operators on a Hilbert space which consists precisely of the completely continuous operators B satisfying $\sum_{i=1}^\infty \lambda_i^p(\sqrt{B^* B}) < \infty$, where $\lambda_i(\sqrt{B^* B})$ ($i = 1, 2, \ldots$) are the eigenvalues of the operator $\sqrt{B^* B}$.

LEMMA 2.1. *In order that $L_D^{-1} \in \mathfrak{S}_p$ ($p > \frac{1}{2}$) it is necessary and sufficient that $A^{-\frac{2p-1}{p}} \in \mathfrak{S}_p$.*

Proof. Assume first that $q(t) \equiv 0$ in expression (1.1). As has been said above, in this case the operator L_D admits separation of variables and its eigenvalues can be expressed by formula (2.2). Since $\nu_k(l_D) = \frac{\pi^2 k^2}{(b-a)^2}$ ($k \in \mathbb{N}$), we have

$$\lambda_{k,m}(L_D) = \frac{\pi^2 k^2}{(b-a)^2} + \mu_m, \quad \mu_m = \lambda_m(A^2).$$

Then the condition $L_D^{-1} \in \mathfrak{S}_p$ is equivalent to the inequality

$$\sum_{k=1}^{\infty}\sum_{m=1}^{\infty}\left(\frac{\pi^2 k^2}{(b-a)^2}+\mu_m\right)^{-p} < \infty. \tag{2.3}$$

Consider the function $\varphi_m(x) = \left(\frac{\pi^2 x^2}{(b-a)^2}+\mu_m\right)^{-p}$, which is decreasing on $[0,\infty)$. It is not difficult to observe that

$$\varphi_m(k+1) \le \int_k^{k+1} \varphi_m(x)\,dx \le \varphi_m(k) \quad (m = 1,2,\ldots;\ k = 0,1,\ldots).$$

Consequently,

$$\sum_{k=1}^{\infty}\left(\frac{\pi^2 k^2}{(b-a)^2}+\mu_m\right)^{-p} < \int_0^{\infty}\left(\frac{\pi^2 x^2}{(b-a)^2}+\mu_m\right)^{-p} dx$$
$$\le \sum_{k=1}^{\infty}\left(\frac{\pi^2 k^2}{(b-a)^2}+\mu_m\right)^{-p} + \mu_m^{-p},$$

and (2.3) is equivalent to the convergence of the series $\sum_{m=1}^{\infty}\int_0^{\infty}\left(\frac{\pi^2 x^2}{(b-a)^2}+\mu_m\right)^{-p} dx$. After the change of variables $\frac{\pi^2 x^2}{(b-a)^2} = s$ we have

$$\int_0^{\infty}\left(\frac{\pi^2 x^2}{(b-a)^2}+\mu_m\right)^{-p} dx = \frac{b-a}{2\pi}\int_0^{\infty}\frac{ds}{\sqrt{s}(s+\mu_m)^p}$$
$$= \frac{b-a}{2\pi}\int_0^{\infty}\frac{s^{\frac{1}{2}-1}\,ds}{(s+\mu_m)^{\frac{1}{2}+(p-\frac{1}{2})}}$$
$$= \frac{b-a}{2\pi}\cdot\frac{\Gamma(p-\frac{1}{2})\Gamma(\frac{1}{2})}{\Gamma(p)}\mu_m^{\frac{1}{2}-p}.$$

Thus, convergence of the series (2.3), i.e. the membership $L_D^{-1} \in \mathfrak{S}_p$, is equivalent to convergence of the series $\sum_{m=1}^{\infty}\left(\mu_m^{\frac{2p-1}{2p}}\right)^{-p}$, hence to $A^{-\frac{2p-1}{p}} \in \mathfrak{S}_p$.

For $q(t) \not\equiv 0$ the validity of the lemma follows from the resolvent formula (4.7) (Chapter 3). □

Let the maximal dissipative (maximal accumulative) extension L_K correspond to condition (1.18) ((1.19)). Taking in Theorem 3.1 (Chapter 3) $K_1 = -E$, on the basis of the lemma proved above we arrive at the following statement.

THEOREM 2.3. *Let $A^{-1} \in \mathfrak{S}_{2p-1}$ $(p \ge 1)$. The resolvent $R_\lambda(L_K)$ $(\lambda \in \rho(L_K))$ of the operator L_K belongs to the ideal $\mathfrak{S}_p$ if and only if $K + E \in \mathfrak{S}_p$.*

COROLLARY 2.2. *If $A^{-1} \in \mathfrak{S}_{2p-1}$ $(p \ge 1)$, then a necessary and sufficient condition for convergence of the series $\sum_{i=1}^{\infty}\lambda_i^{-p}(L_C)$, where L_C is the self-adjoint extension of the minimal operator associated with the boundary condition* (1.20), *is* $\cos C \in \mathfrak{S}_p$.

Specifically, if $A^{-1} \in \mathfrak{S}_1$ $(A^{-\frac{3}{2}} \in \mathfrak{S}_2)$, then the resolvent of the maximal dissipative extension L_K of the operator L_0 is a nuclear operator (a Hilbert–Schmidt operator) if and only if the operator $K + E$ is nuclear (is a Hilbert–Schmidt operator).

COROLLARY 2.3. *If the operators A^{-1} and $K+E$ are nuclear, then the system of eigenvectors and adjoined vectors of the dissipative boundary-value problem* (1.18) *for the expression* (1.1) *is complete in the space $L_2(\mathfrak{H},(a,b))$.*

Proof. Let $z(t) \in L_2(\mathfrak{H},(a,b))$ be arbitrary. Set $-R_\lambda(L_K)z = y$ ($\lambda : \operatorname{Im}\lambda < 0$ is fixed). Evidently, $y \in \mathfrak{D}(L_K)$ and $z = -(L_K - \lambda E)y$. So,

$$\begin{aligned}\operatorname{Im}\big(-R_\lambda(L_K)z, z\big)_{L_2(\mathfrak{H},(a,b))} &= -\operatorname{Im}\big(y, (L_K - \lambda E)y\big)_{L_2(\mathfrak{H},(a,b))} \\ &= -\operatorname{Im}\big(\overline{L_K y, y}\big)_{L_2(\mathfrak{H},(a,b))} + \operatorname{Im}\big(\overline{\lambda}(y,y)_{L_2(\mathfrak{H},(a,b))}\big) \geq 0,\end{aligned}$$

that is $-R_\lambda(L_K)$ is a bounded dissipative operator on $L_2(\mathfrak{H},(a,b))$. Since the operator $K+E$ is nuclear, it follows by Theorem 2.3 that $-R_\lambda(L_K) \in \mathfrak{S}_1$. According to Theorem 6.5, Chapter 1, the set of eigenvectors and adjoined vectors of the operator $-R_\lambda(L_K)$ is complete in $L_2(\mathfrak{H},(a,b))$. It remains to note that the eigenvectors and the adjoined vectors of the operators L_K and $-R_\lambda(L_K)$ are the same. □

2.3. MAXIMAL DISSIPATIVE EXTENSIONS THAT ARE SMOOTH UP TO THE BOUNDARY

Because of Lemma 1.2 the third summand in the representation (1.6) for a vector-function $y(t) \in \mathfrak{D}(L_0^*)$ is continuous into $\mathfrak{H}^{\frac{3}{2}}$ on the closed interval $[a,b]$. The other two summands are infinitely differentiable inside (a,b) into $\mathfrak{H}^\alpha$ ($\alpha > 0$ is arbitrary); on the closed interval $[a,b]$ they are continuous into $\mathfrak{H}^{-\frac{1}{2}}$ only. Therefore a vector-function in the domain of the maximal operator is continuous into $\mathfrak{H}^{\frac{3}{2}}$ inside (a,b) and into $\mathfrak{H}^{-\frac{1}{2}}$ on $[a,b]$. If $y(t) \in \mathfrak{D}(L_0)$, then according to Corollary 1.5 it is continuous into $\mathfrak{H}^{\frac{3}{2}}$ on $[a,b]$. Vector-functions from $\mathfrak{D}(L_D)$ and $\mathfrak{D}(L_N)$ have the same property. Of course, if $y(t) \in \mathfrak{D}(L_D)$ ($y(t) \in \mathfrak{D}(L_N)$), then by substituting its representation (1.6) into the boundary condition $y(a) = y(b) = 0$ ($y'(a) = y'(b) = 0$) we obtain $f_1, f_2 \in \mathfrak{H}^{\frac{3}{2}}$, whence follows the continuity of $y(t)$ into $\mathfrak{H}^{\frac{3}{2}}$ on $[a,b]$.

In this connection it is interesting to describe those maximal dissipative and maximal accumulative extensions L_K of the minimal operator L_0 whose domains consist of vector-functions that are on the interval $[a,b]$ continuous into the space $\mathfrak{H}^\alpha$ ($\alpha > -\frac{1}{2}$) (since $\mathfrak{D}(L_0) \subset \mathfrak{D}(L_K)$ it is clear that $\alpha \leq \frac{3}{2}$).

DEFINITION. We will call a maximal dissipative (maximal accumulative) extension of the minimal operator α-smooth ($-\frac{1}{2} < \alpha \leq \frac{3}{2}$) if any vector-function from its domain is continuous on $[a,b]$ into the space $\mathfrak{H}^\alpha$; $\frac{3}{2}$-smooth extensions are called maximally smooth.

Let

$$\hat{\mathbf{A}}^\alpha = \begin{pmatrix} \hat{A}^\alpha & 0 \\ 0 & \hat{A}^\alpha \end{pmatrix}, \quad \hat{Y} = \{y(a), -y(b)\}, \quad \hat{Y}' = \{y'(a), y'(b)\}.$$

LEMMA 2.2. *A vector-function $y(t) \in \mathfrak{D}(L_0^*)$ belongs to the domain $\mathfrak{D}(L_K)$ of the* *maximal dissipative extension L_K of the minimal operator if and only if the vectors $\hat{Y}, \hat{Y}'$ have a representation*

$$\begin{aligned} \hat{Y} &= -i\hat{\mathbf{A}}^{\frac{1}{2}}(K+E)F, \\ \hat{Y}' &= -\hat{\mathbf{A}}^{-\frac{1}{2}}(K-E)F + i\hat{\mathbf{A}}^{\frac{3}{2}}(K+E)F \quad (F \in \mathfrak{H} \oplus \mathfrak{H}). \end{aligned} \tag{2.4}$$

Proof. We first show that a necessary and sufficient condition for the membership $y(t) \in \mathfrak{D}(L_K)$ is a representation of Y, Y', determined by (1.16), in the form

$$Y = i(K+E)F, \quad Y' = -(K-E)F \quad (F \in \mathfrak{H} \oplus \mathfrak{H}). \tag{2.5}$$

It can be verified immediately that the vectors Y, Y' in (2.5) satisfy the boundary condition (1.18). Conversely, if Y, Y' satisfy (1.18), then setting $F = \frac{1}{2i}(Y + iY')$ we find $KF = \frac{1}{2i}(Y - Y')$, whence the representation (2.5) follows. Expressing now $\hat{Y}, \hat{Y}'$ in terms of Y, Y', from (2.5) we obtain (2.4). □

THEOREM 2.4. *In order that the maximal dissipative extension L_K of the operator L_0 be α-smooth it is necessary and sufficient for the operator $\hat{\mathbf{A}}^{\alpha+\frac{1}{2}}(K+E)$ to be continuous on the space $\mathfrak{H} \oplus \mathfrak{H}$.*

Proof. Let L_K be an α-smooth maximal dissipative extension of the minimal operator, that is, on $[a,b]$ any vector-function $y(t) \in \mathfrak{D}(L_K)$ is continuous into $\mathfrak{H}^\alpha$ $(\frac{1}{2} < \alpha \leq \frac{3}{2})$. Then $\hat{Y} = -i\hat{\mathbf{A}}^{\frac{1}{2}}(K+E)F \in \mathfrak{H}^\alpha \oplus \mathfrak{H}^\alpha$. Since $\hat{\mathbf{A}}^\alpha$ is an isometry of $\mathfrak{H}^\alpha \oplus \mathfrak{H}^\alpha$ onto $\mathfrak{H} \oplus \mathfrak{H}$, the closed operator $\hat{\mathbf{A}}^{\alpha+\frac{1}{2}}(K+E)$ transforms the space $\mathfrak{H} \oplus \mathfrak{H}$ into $\mathfrak{H} \oplus \mathfrak{H}$. By Banach's theorem it is continuous.

Conversely, suppose that the operator $\hat{\mathbf{A}}^{\alpha+\frac{1}{2}}(K+E)$ is continuous on $\mathfrak{H} \oplus \mathfrak{H}$. We know that a vector-function $y(t) \in \mathfrak{D}(L_0^*)$ can be represented in the form of (1.6). Taking into account that $\int_a^b e^{-A|t-s|}A^{-1}h(s)\,ds \in \mathfrak{H}^{\frac{3}{2}}$ for $t \in [a,b]$ (see Lemma 1.2) and because of Lemma 2.2 $y(a), y(b) \in \mathfrak{H}^\alpha$, we can conclude that the f_1, f_2 in representation (1.6) belong to $\mathfrak{H}^\alpha$. So the vector-function $y(t)$ is continuous into $\mathfrak{H}^\alpha$. □

COROLLARY 2.4. *The maximal dissipative extension L_K of the operator L_0 is maximally smooth if and only if the operator $\hat{\mathbf{A}}^2(K+E)$ is continuous on $\mathfrak{H} \oplus \mathfrak{H}$.*

Notice that for the self-adjoint extension L_C defined by condition (1.20) the α-smoothness condition (maximal smoothness condition) is equivalent to the continuity of the operator $\hat{\mathbf{A}}^{\alpha+\frac{1}{2}}\cos C(\hat{\mathbf{A}}^2 \cos C)$ on the space $\mathfrak{H} \oplus \mathfrak{H}$.

COROLLARY 2.5. *If the spectrum of the operator A is discrete, then the spectrum of any α-smooth maximal disspative extension of the minimal operator is also discrete.*

This assertion can be deduced from Theorem 2.2 and the equality

$$K + E = \hat{\mathbf{A}}^{-(\alpha+\frac{1}{2})}(\hat{\mathbf{A}}^{\alpha+\frac{1}{2}}(K+E)).$$

Of course, all results are valid for maximal accumulative extensions of the operator L_0.

2.4. MAXIMAL DISSIPATIVE EXTENSIONS WITH FINITE QUADRATIC FUNCTIONAL

Let $y(t) \in \mathfrak{D}(L_0)$. Integrating by parts $(L_0 y, y)_{L_2(\mathfrak{H},(a,b))}$ and taking into account that $y(a) = y(b) = 0$, we obtain

$$(L_0 y, y)_{L_2(\mathfrak{H},(a,b))} = \int_a^b \left(\|y'(t)\|^2 + \|Ay(t)\|^2 + (q(t)y(t), y(t))\right) dt.$$

The expression appearing at the right-hand side of this equality makes sense also for certain vector-functions not in $\mathfrak{D}(L_0)$. We denote this expression by $D(y)$:

$$D(y) = \int_a^b \left(\|y'(t)\|^2 + \|Ay(t)\|^2 + (q(t)y(t), y(t))\right) dt,$$

and call it the quadratic functional for $l[y]$. The lower semi-bounded functional $D(y)$ is defined not for all $y \in \mathfrak{D}(L_0^*)$. If it is not defined for $y(t)$ we set $D(y) = \infty$.

It is obvious that the notion of quadratic functional introduced in such a way is a generalization of the Dirichlet integral for the Laplace equation.

DEFINITION. We say that L_K is a maximal dissipative (maximal accumulative) extension of L_0 with finite quadratic functional if $D(y) < \infty$ for any vector-function $y(t) \in \mathfrak{D}(L_K)$.

THEOREM 2.5. *In order that the maximal dissipative (maximal accumulative) extension L_K of the minimal operator be an extension with finite quadratic functional it is necessary and sufficient that it be $\frac{1}{2}$-smooth.*

Proof. Assume that the extension L_K is $\frac{1}{2}$-smooth, i.e. any vector-function $y(t) \in \mathfrak{D}(L_K)$ is continuous into $\mathfrak{H}^{\frac{1}{2}}$ on $[a, b]$. Then the vectors f_1, f_2 in the representation (1.6) for $y(t)$ belong to $\mathfrak{H}^{\frac{1}{2}}$. Since by Lemma 1.1 the mappings $f_1 \to e^{-\hat{A}(t-a)} f_1$, $f_2 \to e^{-\hat{A}(b-t)} f_2$, are continuous from $\mathfrak{H}^{-\frac{1}{2}}$ into $L_2(\mathfrak{H}, (a, b))$, the vector-functions

$$y'(t) = -e^{-\hat{A}(t-a)} \hat{A} f_1 + e^{-\hat{A}(b-t)} \hat{A} f_2 + \\ + \tfrac{1}{2}\left(\int_t^b e^{\hat{A}(t-s)} h(s)\, ds - \int_a^t e^{\hat{A}(s-t)} h(s)\, ds\right)$$

and (2.6)

$$(\hat{A}y)(t) = e^{-\hat{A}(t-a)} \hat{A} f_1 + e^{-\hat{A}(b-t)} \hat{A} f_2 + \tfrac{1}{2} \int_a^b e^{-\hat{A}|t-s|} h(s)\, ds$$

belong to $L_2(\mathfrak{H}, (a, b))$. Hence, $D(y) < \infty$.

Conversely, let $y(t) \in \mathfrak{D}(L_0^*)$ and $D(y) < \infty$. Then $y'(t), (\hat{A}y)(t) \in L_2(\mathfrak{H}, (a, b))$. On account of the continuity of the third summands at the right-hand sides of (2.6) on $[a, b]$, we find that the vector-functions $e^{-\hat{A}(t-a)} \hat{A} f_1$ and $e^{-\hat{A}(b-t)} \hat{A} f_2$ belong to $L_2(\mathfrak{H}, (a, b))$. It follows from Corollary 5.3, Chapter 2, that $\hat{A} f_1, \hat{A} f_2 \in \mathfrak{H}^{-\frac{1}{2}}$, i.e.

$f_1, f_2 \in \mathfrak{H}^{\frac{1}{2}}$. Representation (1.6) implies now the continuity of $y(t)$ on $[a,b]$ into the space $\mathfrak{H}^{\frac{1}{2}}$. □

COROLLARY 2.6. *If the spectrum of the operator A is discrete then the spectrum of any maximal dissipative (maximal accumulative) extension with finite quadratic functional is also discrete.*

It can be seen from Theorem 2.5 that L_D is an extension with finite quadratic functional; furthermore, by integrating by parts we find that $(L_D y, y)_{L_2(\mathfrak{H},(a,b))} = D(y)$ $(y \in \mathfrak{D}(L_D))$.

DEFINITION. A maximal dissipative (maximal accumulative) extension L of the operator L_0 is called a D-extension if

$$(Ly, y)_{L_2(\mathfrak{H},(a,b))} = D(y) \quad (\forall y \in \mathfrak{D}(L)).$$

THEOREM 2.6. *The maximal dissipative (maximal accumulative) extension L_K of the minimal operator is D-extension if and only if*

$$-(K^* + E)(K - E) + i(K^* + E)\hat{\mathbf{A}}^2(K + E) = 0$$

(respectively, (2.7)

$$-(K^* + E)(K - E) - i(K^* + E)\hat{\mathbf{A}}^2(K + E) = 0).$$

Proof. We present the proof only for the maximal dissipative extension, because it is similar for the maximal accumulative extension. So, let L_K be a D-extension. By Theorem 2.5 any vector-function $y(t) \in \mathfrak{D}(L_K)$ is continuous on $[a,b]$ into the space $\mathfrak{H}^{\frac{1}{2}}$, hence (see the representation (1.6)) its derivative is continuous on $[a,b]$ into $\mathfrak{H}^{-\frac{1}{2}}$. On the basis of Lemma 1.3 one can integrate by parts the expression $(L_K y, y)_{L_2(\mathfrak{H},(a,b))}$. As a result we have

$$(L_K y, y)_{L_2(\mathfrak{H},(a,b))} = (\hat{Y}', \hat{Y})_{\mathfrak{H}\oplus\mathfrak{H}} + D(y), \tag{2.8}$$

whence $(\hat{Y}', \hat{Y})_{\mathfrak{H}\oplus\mathfrak{H}} = 0$ for an arbitrary vector-function $y(t) \in \mathfrak{D}(L_K)$. Employing Lemma 2.2 we find for any $F \in \mathfrak{H} \oplus \mathfrak{H}$

$$(F, (-(K^* + E)(K - E) + i(K^* + E)\hat{\mathbf{A}}^2(K + E))F)_{\mathfrak{H}\oplus\mathfrak{H}} = 0. \tag{2.9}$$

Note that $(K^*+E)\hat{\mathbf{A}}^2(K+E) = ((K^*+E)\hat{\mathbf{A}})((K^*+E)\hat{\mathbf{A}})^*$. Since the extension L_K is $\frac{1}{2}$-smooth, according to Theorem 2.4 the operator $\hat{\mathbf{A}}(K + E)$, and together with it the operator $(K^* + E)\hat{\mathbf{A}}$, is bounded on $\mathfrak{H} \oplus \mathfrak{H}$. Therefore, $(K^* + E)\hat{\mathbf{A}}^2(K + E)$ is also bounded. The relation (2.7) follows from (2.9).

Conversely, let (2.7) be valid. Then the operator $(K^* + E)\hat{\mathbf{A}}^2(K + E) = ((K^* + E)\hat{\mathbf{A}})((K^* + E)\hat{\mathbf{A}})^*$ is bounded on $\mathfrak{H} \oplus \mathfrak{H}$. By Theorem 2.4 the extension L_K is $\frac{1}{2}$-smooth and equality (2.8) is valid for it. Lemma 2.2 shows the equivalence of (2.7) to the equality $(\hat{Y}', \hat{Y})_{\mathfrak{H}\oplus\mathfrak{H}} = 0$ for any vector-function $y(t)$ in $\mathfrak{D}(L_K)$. Now we obtain from (2.8) that $(L_K y, y)_{L_2(\mathfrak{H},(a,b))} = D(y)$ $(\forall y \in \mathfrak{D}(L_K))$. □

Equality (2.7) implies $E - K^*K = 0$, whence follows the following statement.

COROLLARY 2.7. *Every D-extension is maximal symmetric.*

REMARK 2.1. In the case when a self-adjoint extension of the minimal operator is given by condition (1.20), a necessary and sufficient condition for L_C to be a D-extension is

$$\cos C\hat{\mathbf{A}}^2 \cos C = \cos C \sin C. \tag{2.10}$$

3. The Asymptotics of the Spectrum of the Dirichlet and the Neumann Problem

In this section we study the asymptotic distribution of the eigenvalues of the self-adjoint extensions L_D and L_N associated with the Dirichlet and the Neumann problem for expression (1.1), under the assumption that the spectrum of the operator A $(A \geq E)$ is discrete. Since the operator q defined by $(qy)(t) = q(t)y(t)$ is bounded on $L_2(\mathfrak{H}, (a, b))$, we assume $q(t) \equiv 0$. As in the previous section, transition to the general case is realized using the resolvent formula (4.7) (Chapter 3).

3.1. THE DOMINANT TERM OF THE ASYMPTOTICS

It is convenient to describe the distribution of the eigenvalues of a semi-bounded self-adjoint operator with discrete spectrum by the function

$$N(\lambda; .) = \operatorname{card}\{j : \lambda_j(\cdot) \leq \lambda\},$$

where card S denotes the number of elements of the set S. This function takes values in the set of non-negative integers. It is non-decreasing and tends to ∞ as $\lambda \to \infty$.

As was shown in subsection 2.1, the spectra of the operators L_D and L_N are discrete and their eigenvalues can be found by formula (2.2). The identities

$$N(\lambda; L.) = \sum_{\lambda_j(l.) \leq \lambda} N(\lambda - \lambda_j(l.); A^2) = \sum_{\lambda_j(A^2) \leq \lambda} N(\lambda - \lambda_j(A^2); l.), \tag{3.1}$$

where $L.,l.$ mean either L_D, l_D or L_N, l_N, follow from (2.2). They show that the function $N(\lambda; L.)$ depends on $N(\lambda; A^2)$ linearly and monotonically.

LEMMA 3.1. *The asymptotic equality* $\frac{\sqrt{\lambda}}{N(\lambda; L_D)} = o(1)$ $(\lambda \to \infty)$ *is valid.*

Proof. Indeed,[3]

$$\varliminf_{\lambda\to\infty} (\lambda^{-\frac{1}{2}} N(\lambda; L_D)) \geq \varliminf_{\lambda\to\infty} \sum_{k=1}^{N} \lambda^{-\frac{1}{2}} \left[\frac{b-a}{\pi}(\lambda - \lambda_k(A^2))^{\frac{1}{2}}\right] \geq N\frac{b-a}{\pi}$$

for any natural N. □

---

[3] Here and below in this section, $[\cdot]$ and $\{\cdot\}$ denote the integer and fractional part of a number, respectively.

As $\sigma(l_N) = \sigma(l_d) \cup \{0\}$,

$$N(\lambda; L_N) = N(\lambda; L_D) + N(\lambda; A^2). \tag{3.2}$$

Formula (3.2) reduces the investigation of the spectral asymptotics for L_N to the study of that for L_D.

LEMMA 3.2. *The equality*

$$N(\lambda; L_D) = \tfrac{1}{2}\frac{b-a}{\pi}\sqrt{\lambda}\int_0^1 \frac{N(\lambda t; A^2)}{\sqrt{1-t}}\,\mathrm{d}t - \tau_\lambda N(\lambda; A^2) \tag{3.3}$$

holds, where $\tau_\lambda \in (0,1)$.

Proof. We have from (3.1),

$$\begin{aligned} N(\lambda; L_D) &= \sum_{\lambda_n(A^2)\le\lambda} N(\lambda - \lambda_n(A^2); l_D) = \sum_{\lambda_n(A^2)\le\lambda}\left[\frac{b-a}{\pi}(\lambda-\lambda_n(A^2))^{\frac{1}{2}}\right] \\ &= \sum_{\lambda_n(A^2)\le\lambda}\frac{b-a}{\pi}(\lambda-\lambda_n(A^2))^{\frac{1}{2}} - \sum_{\lambda_n(A^2)\le\lambda}\left\{\frac{b-a}{\pi}(\lambda-\lambda_n(A^2))^{\frac{1}{2}}\right\} \\ &= \frac{b-a}{\pi}\int_0^\lambda (\lambda-\mu)^{\frac{1}{2}}\,\mathrm{d}N(\mu; A^2) - \tau_\lambda N(\lambda; A^2). \end{aligned}$$

After integration by parts and the change of variables $\mu = \lambda t$ we obtain (3.3). □

On the set $L_{\infty,\mathrm{loc}(0,\infty)}$ of locally bounded functions on $[0,\infty)$ we define the transformation

$$\widetilde{\Phi}(\lambda) = \int_0^1 \frac{\Phi(\lambda t)}{\sqrt{1-t}}\,\mathrm{d}t \quad (\forall \Phi \in L_{\infty,\mathrm{loc}(0,\infty)}).$$

THEOREM 3.1. *Suppose that*

$$N(\lambda; A^2) = \Phi(\lambda) + R(\lambda), \quad R(\lambda) = o(1)\Phi(\lambda) \quad (\lambda \to \infty). \tag{3.4}$$

Then the asymptotic equality

$$N(\lambda; L_D) = \frac{b-a}{2\pi}\lambda^{\frac{1}{2}}\widetilde{\Phi}(\lambda) + O(\Phi(\lambda)) + O(\lambda^{\frac{1}{2}})\widetilde{R}(\lambda) \tag{3.5}$$

is valid.

The proof follows immediately from (3.3).

THEOREM 3.2. *If* $\Phi(\lambda)$ *in the decomposition* (3.4) *is such that* $\Phi(\lambda) = o(1)\lambda^{\frac{1}{2}}\widetilde{\Phi}(\lambda)$ $(\lambda \to \infty)$, *then*

$$N(\lambda; L_D) \sim N(\lambda; L_N) \sim \frac{b-a}{2\pi}\lambda^{\frac{1}{2}}\widetilde{\Phi}(\lambda) \quad (\lambda \to \infty). \tag{3.6}$$

Proof. It suffices to show that $\widetilde{R}(\lambda) = o(1)\widetilde{\Phi}(\lambda)$.

The local boundedness of the function $\Phi(\lambda)$ together with the relation $\Phi(\lambda) \to \infty$ as $\lambda \to \infty$ implies its lower semi-boundedness on $(0,\infty)$. By adding, if necessary, a bounded summand to $\Phi(\lambda)$ and taking into account Lemma 3.1 one can assume from the very beginning that $\Phi(\lambda) \ge 1$.

We note also that if $\delta \to 0$, then $\frac{\delta\Phi(\delta t)}{\Phi(t)} \to 0$ uniformly with respect to $t \in (0,\infty)$. In fact, suppose this to be not true. Then there exist $\delta_n \to 0$ and $t_n > 0$ such that $\frac{\delta_n \Phi(t_n \delta_n)}{\Phi(t_n)} \geq \rho > 0$. If the sequence $\{t_n\delta_n\}_{n=1}^{\infty}$ contains an unbounded subsequence $t_{n_k}\delta_{n_k}$, then $\delta_{n_k} N(t_{n_k}\delta_{n_k}; A^2) N^{-1}(t_{n_k}; A^2) \sim \delta_{n_k}\Phi(t_{n_k}\delta_{n_k})\Phi^{-1}(t_{n_k}) \geq \rho$. On the other hand, since $N(t;A^2) \uparrow \infty$, the quantity written from the left does not exceed $\delta_{n_k} \to 0$. So, if such sequences $\{\delta_n\}_{n=1}^{\infty}$ and $\{t_n\}_{n=1}^{\infty}$ exist, then the sequence $\{t_n\delta_n\}_{n=1}^{\infty}$ is necessarily bounded. On account of the local boundedness of $\Phi(t)$, $\delta_n\Phi(t_n\delta_n) = o(1)$, hence $\Phi(t_n) \to 0$ as $n \to \infty$ which contradicts $\Phi(t) \geq 1$.

Setting now $R(t) = \epsilon(t)\Phi(t)$, where $\epsilon(t) = o(1)$ $(t \to \infty)$, we estimate $\widetilde{R}(\lambda)$. For an arbitrary $\delta \in (0,1)$,

$$\int_\delta^1 \frac{\epsilon(t\lambda)\Phi(t\lambda)}{\sqrt{1-t}}\,dt \leq \widetilde{\Phi}(\lambda) \sup_{t\geq\delta\lambda} \epsilon(t),$$

$$\int_0^\delta \frac{\epsilon(t\lambda)\Phi(t\lambda)}{\sqrt{1-t}}\,dt = \delta\int_0^1 \frac{\epsilon(\delta s\lambda)\Phi(\delta s\lambda)}{\sqrt{1-\delta s}}\,ds \leq \widetilde{\Phi}(\lambda) \sup_{\lambda s>0} \frac{\delta\Phi(\delta s\lambda)}{\Phi(s\lambda)}.$$

Choosing $\delta = \delta(\lambda)$ so that $\delta(\lambda) \to 0$ while $\lambda\delta(\lambda) \to \infty$ $(\lambda \to \infty)$, one gets

$$\int_{\delta(\lambda)}^1 \frac{\epsilon(t\lambda)}{\sqrt{1-t}}\Phi(\lambda t)\,dt = o(1)\widetilde{\Phi}(\lambda) \quad (\lambda \to \infty),$$

$$\int_0^{\delta(\lambda)} \frac{\epsilon(t\lambda)}{\sqrt{1-t}}\Phi(\lambda t)\,dt = o(1)\widetilde{\Phi}(\lambda) \quad (\lambda \to \infty).$$

The result desired is obtained by adding these asymptotic equalities. □

REMARK 3.1. It can be seen from the proof of Theorem 3.2 and formula (3.2) that the condition $\Phi(\lambda) = o(\lambda^{\frac{1}{2}})\widetilde{\Phi}(\lambda)$ is necessary and sufficient for $N(\lambda; L_D) \sim N(\lambda; L_N)$. It is not satisfied, for example, for the function $\Phi(\lambda) = e^\lambda$.

From the point of view of applications the case of power asymptotics, when $\Phi(\lambda) = c\lambda^\alpha$ $(\alpha > 0)$, is very important. In this situation

$$\widetilde{\Phi}(\lambda) = c\lambda^\alpha \int_0^1 \frac{t^\alpha}{\sqrt{1-t}}\,dt = cB(\alpha+1;\tfrac{1}{2})\lambda^\alpha.$$

Thus we have the conditions of Theorem 3.2. Therefore the following statement holds.

COROLLARY 3.1. *If* $N(\lambda; A^2) \sim c\lambda^\alpha$, *then*

$$N(\lambda; L_D) \sim N(\lambda; L_N) \sim \frac{b-a}{2\pi} cB(\alpha+1;\tfrac{1}{2})\lambda^{\alpha+\frac{1}{2}}.$$

Because of Lemma 6.2, Chapter 1, one can formulate this assertion in the equivalent form, namely: if $\lambda_n(A^2) \sim c^{-\frac{1}{\alpha}} n^{\frac{1}{\alpha}}$ $(c > 0, \alpha > 0; n \to \infty)$, then

$$\lambda_n(L_D) \sim \lambda_n(L_N) \sim dn^{\frac{1}{\beta}} \quad (n \to \infty), \tag{3.7}$$

where $\beta = \alpha + \frac{1}{2}$, $d = \left(\frac{b-a}{2\pi} cB(\alpha+1;\frac{1}{2})\right)^{-\frac{1}{\beta}}$.

We note one more interesting case.

COROLLARY 3.2. *Let* $N(\lambda; A^2) \sim c\lambda^\alpha m(\lambda)$, *where* $\alpha \geq 0$ *while* $m(\lambda)$ *is a positive function,* $m(\lambda) \uparrow \infty$, *so that* $m(a\lambda) \sim m(\lambda)$ $(\forall a > 0)$. *Then*

$$N(\lambda; L_D) \sim N(\lambda; L_N) \sim \frac{b-a}{2\pi} cB(\alpha+1; \tfrac{1}{2})\lambda^{\alpha+\frac{1}{2}} m(\lambda) \quad (\lambda \to \infty).$$

Proof. If $\Phi(\lambda) = c\lambda^\alpha m(\lambda)$, then

$$\tilde{\Phi}(\lambda) = c\lambda^\alpha \int_0^1 \frac{t^\alpha m(t\lambda)}{\sqrt{1-t}}\, dt \leq cB(\alpha; \tfrac{1}{2})\lambda^\alpha m(\lambda).$$

On the other hand, for any $\delta > 0$,

$$\tilde{\Phi}(\lambda) \geq c\lambda^\alpha \int_\delta^1 \frac{t^\alpha m(t\lambda)}{\sqrt{1-t}}\, dt \geq c\lambda^\alpha m(\delta\lambda) \int_\delta^1 \frac{t^\alpha}{\sqrt{1-t}}\, dt.$$

Because δ is arbitrary,

$$\tilde{\Phi}(\lambda) \geq c\lambda^\alpha m(\delta\lambda)(B(\alpha+1; \tfrac{1}{2}) - \epsilon) \sim c(B(\alpha+1; \tfrac{1}{2}) - \epsilon)\lambda^\alpha m(\lambda),$$

where $\epsilon > 0$ can be chosen arbitrarily small. So the conditions of Theorem 3.2 are satisfied with $\tilde{\Phi}(\lambda) \sim cB(\alpha+1; \frac{1}{2})\lambda^\alpha m(\lambda)$. □

If we have additional information concerning $N(\lambda; A^2)$, then Theorem 3.1 allows us to find not only the dominant term of the asymptotics of the function $N(\lambda; L_D)$, but also lower terms or an estimate of the remainder.

COROLLARY 3.3. *If* $N(\lambda; A^2) = \sum_{k=1}^{N-1} c_k\lambda^{\alpha_k} + O(\lambda^{\alpha_N})$ $(\alpha_1 > \cdots > \alpha_N \geq 0)$, *then*

$$N(\lambda; L_D) = \sum_{k=1}^{N-1} c_k' \lambda^{\alpha_k+\frac{1}{2}} + O(\lambda^{\alpha_1}) + O(\lambda^{\alpha_N+\frac{1}{2}}),$$

$$N(\lambda; L_N) = \sum_{k=1}^{N-1} c_k' \lambda^{\alpha_k+\frac{1}{2}} + O(\lambda^{\alpha_1}) + O(\lambda^{\alpha_N+\frac{1}{2}}),$$

$$(c_k' = c_k \frac{b-a}{2\pi} B(\alpha_k+1; \tfrac{1}{2})).$$

Corollary 3.3 has a more precise formulation.

3.2. LOWER TERMS OF THE ASYMPTOTICS

Under the assumptions of Corollary 3.3 the term $O(\lambda^{\alpha_1})$ has an asymptotical expansion of the form $\mp \sum_{k=1}^{N-1} c_k\lambda^{\alpha_k} + O(\lambda^{\alpha_1-\frac{1}{2}})$, where the sign "−" corresponds to the Dirichlet problem and "+" stands for the Neumann problem.

THEOREM 3.3. *Suppose that the asymptotic equality*

$$N(\lambda; A^2) = \sum_{k=1}^{N} c_k\lambda^{\alpha_k} + \theta(\lambda^{\alpha_N}) \quad (\alpha_1 > \cdots > \alpha_N \geq 0)$$

holds. Then

$$N(\lambda; L_D) = \sum_{k=1}^{N} c'_k \lambda^{\alpha_k+\frac{1}{2}} - \frac{1}{2}\sum_{k=1}^{N} c_k \lambda^{\alpha_k} + \theta(\lambda^{\alpha_N+\frac{1}{2}}) + O(\lambda^{\alpha_1-\frac{1}{2}}),$$

$$N(\lambda; L_N) = \sum_{k=1}^{N} c'_k \lambda^{\alpha_k+\frac{1}{2}} + \frac{1}{2}\sum_{k=1}^{N} c_k \lambda^{\alpha_k} + \theta(\lambda^{\alpha_N+\frac{1}{2}}) + O(\lambda^{\alpha_1-\frac{1}{2}}),$$

where $c'_k = \frac{b-a}{2\pi} c_k B(\alpha_k + 1; \frac{1}{2})$; *the symbol* θ *represents* o *or* O.

To prove this statement we need some auxiliary facts.

Let $\Phi(\lambda)$ be a locally bounded function on $[0,\infty)$. Set

$$N(\lambda; l) * \Phi(\lambda) = \sum_{\lambda_n(l)\le\lambda} \Phi(\lambda - \lambda_n(l)).$$

This "convolution" makes sense and is linear and monotonic with respect to $\Phi(\lambda)$. Now (3.1) becomes

$$N(\lambda; L_D) = N(\lambda; l_D) * N(\lambda; A^2).$$

LEMMA 3.3. *If* $N(\lambda; A^2) = \Phi(\lambda) + \theta(\lambda^\delta)$ $(\delta > 0)$, *then*

$$N(\lambda; L_D) = N(\lambda; l_D) * \Phi(\lambda) + \theta(\lambda^{\delta+\frac{1}{2}}) \quad (\lambda \to \infty).$$

Proof. 1. $\theta = O$. It follows from the assumption of the lemma that there exist constants $c_0 > 0, c_1 > 0$ such that

$$|N(\lambda; A^2) - \Phi(\lambda)| \le c_0\lambda^\delta + c_1 \tag{3.8}$$

for all λ. Formulas (3.1), (3.8) and Corollary 3.1 imply

$$\begin{aligned}|N(\lambda; L_D) - N(\lambda; l_D) * \Phi(\lambda)| &\le \sum_{\lambda_n(l_D)\le\lambda} |N(\lambda - \lambda_n(l_D); A^2) - \\ &\qquad - \Phi(\lambda - \lambda_n(l_D))| \\ &\le \sum_{\lambda_n(l_D)\le\lambda} (c_0(\lambda - \lambda_n(l_D))^\delta + c_1) \\ &= c'_0\lambda^{\delta+\frac{1}{2}} + o(\lambda^{\delta+\frac{1}{2}}) + O(\lambda^{\frac{1}{2}}) = O(\lambda^{\delta+\frac{1}{2}}).\end{aligned}$$

2. $\theta = o$. By assumption, for any $\epsilon > 0$ we can find a constant $c_\epsilon > 0$ such that

$$|N(\lambda; A^2) - \Phi(\lambda)| \le \epsilon\lambda^\delta + c_\epsilon$$

for all λ. By repeating the arguments from the previous case we deduce

$$|N(\lambda; L_D) - N(\lambda; l_D) * \Phi(\lambda)| \le \epsilon'\lambda^{\delta+\frac{1}{2}} + o(\lambda^{\delta+\frac{1}{2}}) + O(\lambda^{\frac{1}{2}}).$$

The result desired is obtained if we observe that $\epsilon' = \frac{b-a}{2\pi} B(1+\delta; \frac{1}{2})\epsilon$ and that ϵ tends to zero. □

The next statement plays a key role in the process of proving Theorem 3.3.

LEMMA 3.4. *Let p, q, α are positive numbers. Denote by $N(\lambda)$ the number of pairs of natural numbers (x, y) which satisfy the inequality $px^2 + qy^{\frac{1}{\alpha}} \leq \lambda$. The relation*

$$N(\lambda) = p^{-\frac{1}{2}} q^{-\alpha} B(\alpha; \tfrac{1}{2}) \lambda^{\alpha+\frac{1}{2}} - \tfrac{1}{2} q^{-\alpha} \lambda^{\alpha} + O(\lambda^{\alpha-\frac{1}{2}}) + O(\lambda^{\frac{1}{2}}) \tag{3.9}$$

holds as $\lambda \to \infty$.

Proof. It is not hard to see that $N(\lambda)$ equals the number of points (x, y) with natural coordinates which are situated inside the closed contour consisting of the curve $px^2 + qy^{\frac{1}{\alpha}} = \lambda$ and the two intervals $[0, m_0]$ and $[0, n_0]$ on the coordinate axes (Figure 1).

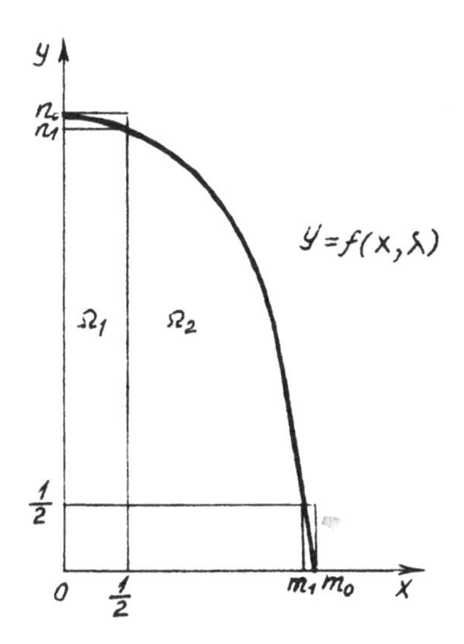

Figure 1.

We consider two cases.

1°. Let $\alpha \in (0, 1]$.

Assume that m_1 is a solution of the equation $q^{-\alpha}(\lambda - px^2)^{\alpha} = \frac{1}{2}$. The straight lines $x = \frac{1}{2}$ and $x = m_1$ divide the domain Ω enveloped by the indicated contour into three parts: Ω_1, Ω_2 and Ω_3 (see Figure 1). Clearly, $N(\lambda) = N_2(\lambda)$ ($N_2(\lambda)$ is the number of points with integer coordinates in the domain Ω_2). It is not difficult to verify that

$$N_2(\lambda) = \sum_{\frac{1}{2} < m \leq m_1} [f(m, \lambda)] = \sum_{\frac{1}{2} < m \leq m_1} f(m, \lambda) - \sum_{\frac{1}{2} < m \leq m_1} \{f(m, \lambda)\},$$

where $f(x, \lambda) = q^{-\alpha}(\lambda - px^2)^{\alpha}$; the summation extends over all integers in the interval $(\frac{1}{2}, m_1)$.

Trivial estimation of the fractional parts yields

$$0 \le \sum_{\frac{1}{2}<m\le m_1} \{f(m,\lambda)\} < \sum_{\frac{1}{2}<m\le m_1} 1 < m_0 = p^{-\frac{1}{2}}\lambda^{\frac{1}{2}}. \tag{3.10}$$

To calculate the sum $\sum_{\frac{1}{2}<m\le m_1} f(m,\lambda)$ we use the Sonin formula (see Sonin [1])

$$\sum_{a<x\le b} f(x) = \int_a^b f(x)\,dx + \rho(b)f(b) - \rho(a)f(a) - \int_a^b \rho(x)f'(x)\,dx,$$

where $f(x) \in C^1[a,b]$, the sum extends over the integers $x \in (a,b]$, and $\rho(x) = [x] - x + \frac{1}{2}$. Since $\rho(\frac{1}{2}) = 0$, $|\rho(m_1)| \le \frac{1}{2}$, $f(m_1,\lambda) \equiv \frac{1}{2}$, we have

$$\sum_{\frac{1}{2}<m\le m_1} f(m,\lambda) = \int_{\frac{1}{2}}^{m_1} f(x,\lambda)\,dx - \int_{\frac{1}{2}}^{m_1} \rho(x)f'_x(x,\lambda)\,dx + O(1)$$

$$= \int_{\frac{1}{2}}^{m_1} f(x,\lambda)\,dx + \int_{\frac{1}{2}}^{m_1} \sigma(x)f''_{xx}(x,\lambda)\,dx -$$

$$- \sigma(m_1)f'_x(m_1,\lambda) + O(1), \tag{3.11}$$

where $\sigma(x) = \int_{\frac{1}{2}}^{x} \rho(t)\,dt$. A direct calculation shows that $f'_x(m_1,\lambda) = \text{const} \cdot \lambda^{\frac{1}{2}}$. Taking into account the inequality $|\sigma(x)| \le \frac{1}{8}$ for all x, we obtain

$$|\sigma(m_1)f'_x(m_1,\lambda)| = O(\lambda^{\frac{1}{2}}).$$

Since in the situation considered $(0 < \alpha \le 1)$ $f''_{xx}(x,\lambda) > 0$ for all $x \in [\frac{1}{2}, m_1]$, the second integral at the right-hand side of (3.11) can be estimated as follows:

$$\left|\int_{\frac{1}{2}}^{m_1} \sigma(x)f''_{xx}(x,\lambda)\,dx\right| \le \frac{1}{8}\int_{\frac{1}{2}}^{m_1} |f''_{xx}(x,\lambda)|\,dx < \frac{1}{8}|f'_x(m_1,\lambda)| = O(\lambda^{\frac{1}{2}}),$$

so $N_2(\lambda) = S_2(\lambda) + O(\lambda^{\frac{1}{2}})$, where $S_i(\lambda)$ is the area of the domain Ω_i $(i = 1,2,3)$.

Since $m_0(\lambda) - m_1(\lambda) = p^{\frac{1}{2}}(\lambda^{\frac{1}{2}} - (\lambda - 2^{-\frac{1}{\alpha}}q)^{\frac{1}{2}}) \to 0$ as $\lambda \to \infty$, we have

$$S_3(\lambda) \le \tfrac{1}{2}(m_0(\lambda) - m_1(\lambda)) = o(1).$$

We now estimate $S_1(\lambda)$. Certainly,

$$\tfrac{1}{2}n_1(\lambda) \le S_1(\lambda) \le \tfrac{1}{2}n_0(\lambda).$$

The relation $n_0(\lambda) - n_1(\lambda) = q^{-\alpha}(\lambda^\alpha - (\lambda - \frac{p}{4})^\alpha) \le \text{const} \cdot \lambda^{\alpha-1}$ implies

$$S_1(\lambda) = \tfrac{1}{2}q^{-\alpha}\lambda^\alpha + O(\lambda^{\alpha-1}). \tag{3.12}$$

So, finally, we can write

$$N_2(\lambda) = S(\lambda) - \tfrac{1}{2}q^{-\alpha}\lambda^\alpha + O(\lambda^{\alpha-\frac{1}{2}}) + O(\lambda^{\frac{1}{2}}) \tag{3.13}$$

($S(\lambda)$ is the area of Ω).

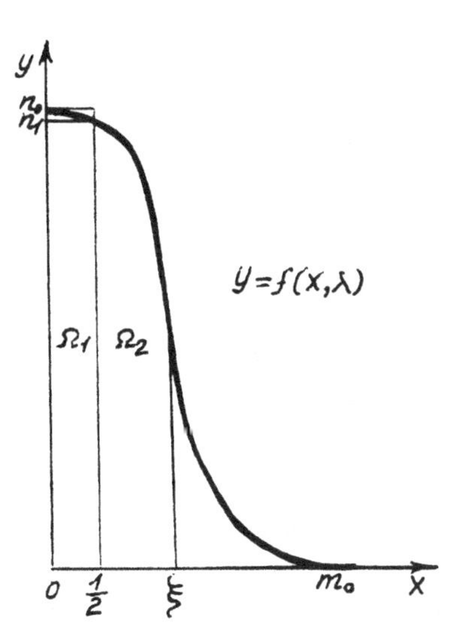

Figure 2.

2°. We now show that formula (3.13) is true also for $\alpha \in (1, \infty)$.

The straight line $x = \frac{1}{2}$ divides the domain considered into two parts: Ω_1, Ω_2 (Figure 2). By repeating the arguments from case 1° (with m_1 replaced by m_0) and taking into account the identity $f(m_0, \lambda) \equiv 0$, we find

$$N(\lambda) = N_2(\lambda) = \int_{\frac{1}{2}}^{m_0} f(x, \lambda)\, dx - \int_{\frac{1}{2}}^{m_0} \rho(x) f'_x(x, \lambda)\, dx + O(\lambda^{\frac{1}{2}}). \tag{3.14}$$

Integration by parts with regard to $\sigma(\frac{1}{2}) = 0$, $f'_x(m_0, \lambda) \equiv 0$ $(\alpha > 1)$ gives

$$\int_{\frac{1}{2}}^{m_0} \rho(x) f'_x(x, \lambda)\, dx = -\int_{\frac{1}{2}}^{m_0} \sigma(x) f''_{xx}(x, \lambda)\, dx. \tag{3.15}$$

But $f''_{xx}(x, \lambda)$ is negative when $\frac{1}{2} < x < \xi(\lambda) = p^{-\frac{1}{2}}(2\alpha - 1)^{-\frac{1}{2}}\lambda^{\frac{1}{2}}$, and positive if $S(\lambda) < x < m_0(\lambda)$, so we get

$$\left| \int_{\frac{1}{2}}^{m_0} \sigma(x) f''_{xx}(x, \lambda)\, dx \right| \leq \tfrac{1}{4} |f'_x(\xi(\lambda), \lambda)| = \text{const} \cdot \lambda^{\alpha - \frac{1}{2}}. \tag{3.16}$$

Comparison of (3.14)-(3.16) yields

$$N_2(\lambda) = S_2(\lambda) + O(\lambda^{\frac{1}{2}}) + O(\lambda^{\alpha - \frac{1}{2}}).$$

Because of relation (3.12), which is valid for $S_1(\lambda)$ just as in the case 1°, we obtain the desired formula (3.13).

It remains only to find $S(\lambda)$. We have

$$S(\lambda) = p^{-\frac{1}{2}} \int_0^{q^{-\alpha}\lambda^\alpha} (\lambda - qy^{\frac{1}{\alpha}})^{\frac{1}{2}}\,dy = p^{-\frac{1}{2}}\lambda^{\frac{1}{2}} \int_0^{q^{-\alpha}\lambda^\alpha} (1 - q\lambda^{-1}y^{\frac{1}{\alpha}})^{\frac{1}{2}}\,dy$$
$$= 2\alpha p^{-\frac{1}{2}} q^{-\alpha}\lambda^{\alpha+\frac{1}{2}} \int_0^{\frac{\pi}{2}} \cos^2 t \sin^{2\alpha-1} t\,dt = 2\alpha p^{-\frac{1}{2}} q^{-\alpha} J \lambda^{\alpha+\frac{1}{2}},$$

where

$$J = \int_0^{\frac{\pi}{2}} \cos^2 t \sin^{2\alpha-1} t\,dt = \tfrac{1}{2}B(\alpha; \tfrac{3}{2}).$$

Substituting this expression for $S(\lambda)$ in (3.13) and employing well-known relations for the beta-function we deduce (3.9). □

Proof of Theorem 3.3. Let the function $N(\lambda; A^2)$ satisfy the asymptotic equality mentioned in the formulation of the theorem. Write it in the form

$$N(\lambda; A^2) = \sum_{k=1}^{N} \operatorname{sign} c_k \left[|c_k|\lambda^{\alpha_k}\right] + \theta(\lambda^{\alpha_N}) = \Phi(\lambda) + \theta(\alpha^N).$$

By the lemmas proved,

$$N(\lambda; L_D) = N(\lambda; l_D) * \Phi(\lambda) + \theta(\lambda^{\alpha_N+\frac{1}{2}})$$
$$= \sum_{k=1}^{N} \operatorname{sign} c_k\, N(\lambda; l_D) * \left[|c_k|\lambda^{\alpha_k}\right] + \theta(\lambda^{\alpha_N+\frac{1}{2}}). \tag{3.17}$$

Each of the functions $\left[|c_k|\lambda^{\alpha_k}\right]$ is the distribution function of eigenvalues for an operator whose eigenvalues are $\lambda_j = |c_k|^{-\frac{1}{\alpha_k}} j^{\frac{1}{\alpha_k}}$ $(j = 1, 2, \ldots)$. By virtue of formula (2.2) and Lemma 3.4, where $p = \frac{\pi^2}{(b-a)^2}$, $q = |c_k|^{-\frac{1}{\alpha_k}}$, $\alpha = \alpha_k$ $(k = 1, \ldots, N)$,

$$N(\lambda; l_D) * \left[|c_k|\lambda^{\alpha_k}\right] = (b-a)\pi^{-1}\alpha_k B(\alpha_k; \tfrac{3}{2})|c_k|\lambda^{\alpha_k+\frac{1}{2}} -$$
$$- \tfrac{1}{2}|c_k|\lambda^{\alpha_k} + O(\lambda^{\alpha_k-\frac{1}{2}}) + O(\lambda^{\frac{1}{2}}). \tag{3.18}$$

Substituting (3.18) into (3.17) we obtain the relation required for $N(\lambda; L_D)$, whence on the basis of (3.2) we arrive at the asymptotic expansion for $N(\lambda; L_N)$. □

4. On the Asymptotics of the Spectra of General Self-Adjoint Problems

In this section we assume that the spectrum of the operator A occurring in expression (1.1) is discrete and that $N(\lambda; A^2) \sim c\lambda^\alpha$ $(c > 0, \alpha > 0)$. Our purpose is to describe the class of self-adjoint extensions of the minimal operator generated by (1.1) that have purely discrete spectrum with dominant term of its asymptotics the same as that of the Dirichlet and Neumann problems. It turns out that maximally smooth extensions, D-extensions and others belong to this class. Moreover, we give a description of extensions whose eigenvalues are arranged more densely than those of the Dirichlet problem.

4.1. A POSITIVE BOUNDARY VALUE SPACE OF THE MINIMAL OPERATOR

Assume that in expression (1.1) $q(t) \equiv 0,\ A \geq E$. The operator L_D is the Friedrichs extension of the minimal operator L_0. In fact, it follows from the construction of this extension that its domain coincides with $\mathfrak{D}(L_0^*) \cap \mathfrak{H}_{L_0}$ (see Theorem 2.1, Chapter 3), where $\mathfrak{H}_{L_0}$ is the completion of $\mathfrak{D}(L_0)$ with respect to the norm

$$\|y\|_{L_0} = (L_0 y, y)^{\frac{1}{2}}_{L_2(\mathfrak{H},(a,b))} = \left(\int_a^b \left(\|y'(t)\|^2 + \|Ay(t)\|^2\right) dt\right)^{\frac{1}{2}}.$$

The inequality $\|y(t)\| \leq c\|y\|_{L_0}$ $(y \in \mathfrak{D}(L_0))$ shows that every vector-function $y(t) \in \mathfrak{H}_{L_0}$ is continuous and vanishes at the ends of the interval, that is, a vector-function $y(t) \in \mathfrak{D}(L_0^*) \cap \mathfrak{H}_{L_0}$ belongs to $\mathfrak{D}(L_D)$. This means that the Friedrichs extension is contained in L_D. Because both of the operators are self-adjoint, they coincide.

LEMMA 4.1. *If the spectrum of the operator A is discrete, then the inequality*

$$n(\lambda; L_D^{-1}P) \leq 2N(\lambda; A^2) \quad (\lambda > 0)$$

is valid, where $n(\lambda;\cdot)$ is the distribution function of s-numbers of an operator (see subsection 6.4, Chapter 1) and P is the orthoprojector from $L_2(\mathfrak{H},(a,b))$ onto $\operatorname{Ker} L_0^*$.

Proof. Denote by $\{e_k\}_{k=1}^{\infty}$ the orthonormal basis in $\mathfrak{H}$ formed by eigenvectors of the operator A^2. Set $L_2(\{e_k\},(a,b)) = L_2(a,b) \otimes \{e_k\}$ $(\{e_k\} = \text{l.s.}\, e_k)$. Then the orthogonal decomposition $L_2(\mathfrak{H},(a,b)) = \oplus_{k=1}^{\infty} L_2(\{e_k\},(a,b))$ is valid. It is not difficult to verify that each of the spaces $L_2(\{e_k\},(a,b))$ is reducing for L_D. This means that $L_2(\{e_k\},(a,b))$, as well as its orthogonal complement, is invariant under L_D and that projection onto it maps $\mathfrak{D}(L_D)$ onto itself. Let $L_D \restriction L_2(\{e_k\},(a,b)) = L_k$. We have $L_D = \oplus_{k=1}^{\infty} L_k$, i.e. if $\mathfrak{D}(L_D) \ni y(t) = \sum_{k=1}^{\infty} y_k(t)$, then $y_k \in \mathfrak{D}(L_k)$ and $L_D y = \sum_{k=1}^{\infty} L_k y_k$. Therefore, $L_D^{-1} = \oplus_{k=1}^{\infty} L_k^{-1}$. The operator L_k^{-1}, acting on $L_2(\{e_k\},(a,b))$, is unitarily equivalent to the operator $\left(l_D + \lambda_k(A^2)E\right)^{-1}$, where $\lambda_k(A^2)$ is the eigenvalue corresponding to the eigenvector e_k, l_D is the self-adjoint extension of the minimal operator l_0 (generated on $L_2(a,b)$ by the expression $-y''$) associated with the boundary condition $y(a) = y(b) = 0$. This implies that

$$\|L_k^{-1}\| = \|(l_D + \lambda_k(A^2)E)^{-1}\| = \left(\frac{\pi^2}{(b-a)^2} + \lambda_k(A^2)\right)^{-1} < \lambda_k^{-1}(A^2).$$

Similarly, the spaces $L_2(\{e_k\},(a,b))$ are reducing for the orthoprojector P. So, $\operatorname{Ker} L_0^* = \oplus_{k=1}^{\infty}\left(\operatorname{Ker}(l_0 + \lambda_k(A^2)E)^* \otimes \{e_k\}\right)$. If we set $P \restriction L_2(\{e_k\},(a,b)) = P_k$, then $P = \oplus_{k=1}^{\infty} P_k$, where the rank of the orthoprojector P_k is equal to $\dim \operatorname{Ker}(l_0 + \lambda_k(A^2)E)^* = 2$. Multiplying the decompositions for L_D^{-1} and P we obtain

$$L_D^{-1}P = \bigoplus_{k=1}^{\infty} L_k^{-1}P_k,$$

whence

$$n(\lambda; L_D P) = \sum_{k=1}^{\infty} n(\lambda; L_k^{-1} P_k) = 2 \sum_{\lambda_k(A^2)\le\lambda} 1 = 2N(\lambda; A^2). \qquad \square$$

According to formula (2.3), Chapter 3,

$$\mathfrak{D}(L_0^*) = \mathfrak{D}(L_D) \dotplus \operatorname{Ker} L_0^*. \tag{4.1}$$

Because of this, $\mathfrak{D}(L_0^*) \ni y = y_1 + y_2$, $\mathfrak{D}(L_D) \ni y_1 = \widetilde{P}y$, $\operatorname{Ker} L_0^* \ni y_2 = y - y_1$, where $\widetilde{P} = L_D^{-1} L_0^*$.

LEMMA 4.2. *Let $F = \{f_1, f_2\} \in \mathfrak{H}\oplus\mathfrak{H}$ be arbitrary. There exists a vector-function $y(t) \in \mathfrak{D}(L_D)$ such that $\{A^{\frac{1}{2}}y'(a), A^{\frac{1}{2}}y'(b)\} = F$.*

Proof. Consider the vector-functions $y_1(t) = (t-a)e^{-A(t-a)}f_1'$ and $y_2(t) = (b-t)e^{-A(b-t)}f_2'$, where $f_i' = A^{-\frac{1}{2}}f_i$ $(i = 1, 2)$. Clearly, $y_1(a) = 0$, $y_1'(a) = f_1'$, $y_2(b) = 0$, $y_2'(b) = f_2'$, the vector-functions $y_i'(t)$ are absolutely continuous into $\mathfrak{H}^{-\frac{1}{2}}$ and $\hat{l}[y_i] \in L_2(\mathfrak{H}, (a,b))$. In view of Theorem 1.1 $y_i(t) \in \mathfrak{D}(L_0^*)$.

Let now a scalar function $\varphi(t) \in W_2^2(a,b)$ be equal to one in a neighbourhood of the point a and vanish in some neighbourhood of b. By Corollary 1.1, the vector-function $y(t) = \varphi(t)y_1(t) + (1-\varphi(t))y_2(t)$ belongs to $\mathfrak{D}(L_0^*)$. Since $y(t)$ vanishes at the ends of the interval $[a,b]$, we have $y \in \mathfrak{D}(L_D)$. It can be verified immediately that $\{A^{\frac{1}{2}}y'(a), A^{\frac{1}{2}}y'(b)\} = F$. $\square$

We denote by Γ_1, Γ_2 the mappings from $\mathfrak{D}(L_0^*)$ into $\mathfrak{H} \oplus \mathfrak{H}$ acting by

$$\Gamma_1 y = \{A^{\frac{1}{2}}y_1'(a), A^{\frac{1}{2}}y_1'(b)\}, \quad \Gamma_2 y = \{A^{-\frac{1}{2}}y(a), -A^{-\frac{1}{2}}y(b)\} \tag{4.2}$$
$$(y = y_1 + y_2, \quad y_1 = L_D^{-1}L_0^* y \in \mathfrak{D}(L_D), \quad y_2 \in \operatorname{Ker} L_0^*).$$

By virtue of Lemma 4.2, Γ_1 is a mapping "onto". Lemma 1.7 shows that Γ_2 is also a mapping "onto". It follows also from (4.2) that $\Gamma_1 \operatorname{Ker} L_0^* = 0$, $\Gamma_2\mathfrak{D}(L_D) = 0$. Therefore, the mapping $\gamma : y \to \{\Gamma_1 y, \Gamma_2 y\}$ maps $\mathfrak{D}(L_0^*)$ onto $(\mathfrak{H} \oplus \mathfrak{H}) \oplus (\mathfrak{H} \oplus \mathfrak{H})$. In addition, because of (1.17), for any $y, z \in \mathfrak{D}(L_0^*)$,

$$\begin{aligned}(L_0^* y, z)_{L_2(\mathfrak{H},(a,b))} &= (L_0^*(y_1 + y_2), z_1 + z_2)_{L_2(\mathfrak{H},(a,b))} \\ &= (L_D y_1, z_1)_{L_2(\mathfrak{H},(a,b))} + (L_0^* y_1, z_2)_{L_2(\mathfrak{H},(a,b))} \\ &= (L_D y_1, z_1)_{L_2(\mathfrak{H},(a,b))} + (Y_1, Z_2')_{\mathfrak{H}\oplus\mathfrak{H}} - (Y_1', Z_2)_{\mathfrak{H}\oplus\mathfrak{H}}.\end{aligned}$$

Since

$$\begin{aligned}Y_1 &= \{-A^{-\frac{1}{2}}y_1(a), A^{-\frac{1}{2}}y_1(b)\} = 0, \\ Y_1' &= \{A^{\frac{1}{2}}(y_1'(a) + Ay_1(a)), A^{\frac{1}{2}}(y_1'(b) - Ay_1(b))\} = \Gamma_1 y, \\ Z_2 &= \{-A^{-\frac{1}{2}}z_2(a), A^{-\frac{1}{2}}z_2(b)\} = \{-A^{-\frac{1}{2}}z(a), A^{-\frac{1}{2}}z(b)\} = -\Gamma_2 z,\end{aligned}$$

we have

$$(L_0^* y, z)_{L_2(\mathfrak{H},(a,b))} = (L_D \widetilde{P}y, \widetilde{P}z)_{L_2(\mathfrak{H},(a,b))} + (\Gamma_1 y, \Gamma_2 z)_{\mathfrak{H}\oplus\mathfrak{H}}.$$

According to what has been said, we can state that the triple $(\mathfrak{H} \oplus \mathfrak{H}, \Gamma_1, \Gamma_2)$ is a positive boundary value space of the minimal operator, corresponding to the decomposition (4.1) in the sense of the definition from Chapter 3, subsection 2.2. Since any positive boundary value space is a boundary value space, there exists (on account of Theorem 1.6, Chapter 3) a one-to-one correspondence between self-adjoint extensions of the operator L_0 and boundary conditions of the form

$$(\cos \widetilde{C})\Gamma_1 y - (\sin \widetilde{C})\Gamma_2 y = 0, \tag{4.3}$$

where $\widetilde{C} = \widetilde{C}^*$, $\|\widetilde{C}\| \leq \frac{\pi}{2}$, $\operatorname{Ker}(\widetilde{C} + \frac{\pi}{2}E) = 0$; Γ_1, Γ_2 are defined by formula (4.2).

Using representation (1.6) we can without difficulty show that

$$\Gamma_1 y = Y' - \Omega Y, \quad \Gamma_2 y = Y, \tag{4.4}$$

where Y, Y' are given by (1.16) while

$$\Omega = 2\begin{pmatrix} A^2 e^{-A(b-a)}(E - e^{-2A(b-a)})^{-1} & 0 \\ 0 & A^2 e^{-A(b-a)}(E - e^{-2A(b-a)})^{-1} \end{pmatrix} \times$$

$$\times \begin{pmatrix} e^{-A(b-a)} & E \\ E & e^{-A(b-a)} \end{pmatrix} \tag{4.5}$$

is a bounded self-adjoint operator on $\mathfrak{H} \oplus \mathfrak{H}$. Starting from Theorem 3.1, Chapter 3, and arguing in exactly the same way as in subsection 2.1, we arrive at the following statement: in order that the spectrum of the self-adjoint extension $L_{\widetilde{C}}$ corresponding to the boundary condition (4.3) be discrete it is necessary and sufficient that the spectrum of the operator A in expression (1.1) be discrete and that $\cos \widetilde{C} \in \mathfrak{S}_\infty$.

In view of Theorem 2.2, Chapter 3, all proper solvable extensions of the operator L_0 are described by boundary conditions of the kind

$$\Gamma_2 y = B\Gamma_1 y, \tag{4.6}$$

where B is a bounded operator on $\mathfrak{H} \oplus \mathfrak{H}$. So, the self-adjoint extension $L_{\widetilde{C}}$ given by equation (4.3) is solvable if and only if the operator $\sin \widetilde{C}$ is boundedly invertible. In fact, if $\sin \widetilde{C}$ has a continuous inverse, then (4.3) can be written in the form of (4.6) with $B = \operatorname{ctg} \widetilde{C}$. Conversely, if a self-adjoint extension is solvable, then it is determined by the boundary condition (4.6). Setting $\cos \widetilde{C} = B(E + B^2)^{-\frac{1}{2}}$, $\sin \widetilde{C} = (E + B^2)^{-\frac{1}{2}}$ one can represent this condition in the form of (4.3) with a continuously invertible operator $\sin \widetilde{C}$.

4.2. ON THE ASYMPTOTICS OF THE SPECTRA OF SELF-ADJOINT PROBLEMS

Let $L_{\widetilde{C}}$ be the self-adjoint extension of the minimal operator L_0 associated with the boundary condition (4.3) and let

$$N(\lambda; L_{\widetilde{C}}) = \operatorname{card}\{j : |\lambda_j(L_{\widetilde{C}})| \leq \lambda\}.$$

THEOREM 4.1. *Suppose that the spectrum of the operator A is discrete and that $N(\lambda; A^2) \sim c\lambda^\alpha$ $(c > 0, \alpha > 0, \lambda \to \infty)$. The spectrum of the extension $L_{\widetilde{C}}$ is discrete and*

$$N(\lambda; L_{\widetilde{C}}) \sim \frac{b-a}{2\pi} B(\alpha+1; \tfrac{1}{2}) c\lambda^{\alpha+\frac{1}{2}} \quad (\lambda \to \infty) \tag{4.7}$$

if and only if $\cos \widetilde{C} \in \mathfrak{S}_\infty$ *and*

$$n(\lambda; \cos \widetilde{C}) = o(\lambda^{\alpha+\frac{1}{2}}) \quad (\lambda \to \infty). \tag{4.8}$$

Proof. Sufficiency. Let $N(\lambda; A^2) \sim c\lambda^\alpha$ $(\lambda \to \infty)$. In accordance with Corollary 3.1,

$$N(\lambda; L_D) \sim \frac{b-a}{2\pi} cB(\alpha+1; \tfrac{1}{2}) \lambda^{\alpha+\frac{1}{2}}.$$

If the asymptotic equality (4.8) holds for the operator $\cos \widetilde{C} \in \mathfrak{S}_\infty$, then, as follows from Theorem 3.2, Chapter 3, and with regard to Lemma 6.2, Chapter 1, the asymptotic distribution of the eigenvalues of the extension $L_{\widetilde{C}}$ is expressed by the relation (4.7).

Necessity. Assume $L_{\widetilde{C}}$ to be a self-adjoint extension of L_0 with discrete spectrum. Let the relation (4.7) be valid for $N(\lambda; L_{\widetilde{C}})$. Then $\cos \widetilde{C} \in \mathfrak{S}_\infty$.

Suppose first that the extension $L_{\widetilde{C}}$ is solvable. As has been noted before, this fact is equivalent to the continuous invertibility of the operator $\sin \widetilde{C}$. Since $L_{\widetilde{C}}^{-1} = PL_{\widetilde{C}}^{-1}P + (E-P)L_{\widetilde{C}}^{-1}P + PL_{\widetilde{C}}^{-1}(E-P) + (E-P)L_{\widetilde{C}}^{-1}(E-P)$, where P is the orthoprojector of $L_2(\mathfrak{H}, (a,b))$ onto $\operatorname{Ker} L_0^*$ and the three last summands do not depend on the choice of the operator $\widetilde{C}$, we have, by Lemma 4.1 and Corollary 3.1,

$$n(\lambda; L_{\widetilde{C}}^{-1}) \sim \frac{b-a}{2\pi} cB(\alpha+1; \tfrac{1}{2}) \lambda^{\alpha+\frac{1}{2}} \sim n(\lambda; PL_{\widetilde{C}}^{-1}P) + n(\lambda; (E-P)L_D^{-1}(E-P)),$$

$$n(\lambda; L_D^{-1}) \sim \frac{b-a}{2\pi} cB(\alpha+1; \tfrac{1}{2}) \lambda^{\alpha+\frac{1}{2}} \sim n(\lambda; (E-P)L_D^{-1}(E-P)),$$

This implies that $n(\lambda; PL_{\widetilde{C}}^{-1}P) = o(\lambda^{\alpha+\frac{1}{2}})$.

Lemma 4.1 also shows that $n(\lambda; L_D^{-1}P) = o(\lambda^{\alpha+\frac{1}{2}})$. Then Corollary 2.1, Chapter 3, implies

$$n(\lambda; \operatorname{ctg} \widetilde{C}) = o(\lambda^{\alpha+\frac{1}{2}}).$$

Since the operator $\sin \widetilde{C}$ is continuously invertible, $n(\lambda; \cos \widetilde{C}) = o(\lambda^{\alpha+\frac{1}{2}})$, i.e. (4.8) holds.

If $\sin \widetilde{C}$ is not continuously invertible, then, on account of $\cos \widetilde{C} \in \mathfrak{S}_\infty$, we find that zero is an isolated eigenvalue of finite multiplicity of the operator $\widetilde{C}$. Denote by P_0 the orthoprojector of $\mathfrak{H} \oplus \mathfrak{H}$ onto the eigensubspace of $\widetilde{C}$ corresponding to the eigenvalue 0. It can be easily seen that $\sin \widetilde{C}_1$, where $\widetilde{C}_1 = \widetilde{C} + \frac{\pi}{2} P_0$, is continuously invertible, while $\cos \widetilde{C}_1 - \cos \widetilde{C}$ is a finite-dimensional operator. Theorem 3.2 (Chapter 3) and Lemma 6.2 (Chapter 1) allow us to conclude that

$n(\lambda; L_{\widetilde{C}_1}) \sim n(\lambda; L_{\widetilde{C}})$. Consequently,

$$n(\lambda; \cos \widetilde{C}) \sim n(\lambda; \cos \widetilde{C}_1) = o(\lambda^{\alpha+\frac{1}{2}}).$$ □

The boundary condition (4.3), describing all self-adjoint extensions of the minimal operator L_0, contains an operator Γ_1 which does not immediately relate a vector-function $y(t)$ from the domain of the extension to the boundary values of $y(t)$ and its derivative. In contrast to (4.3), in the boundary condition (1.20), which also describes all self-adjoint extensions of L_0, the vectors Y, Y' are present; these vectors can clearly be expressed in terms of the values of $y(t)$ and $y'(t)$ at the ends of the interval. This is why the formulation of results concerning the behaviour of the eigenvalue distribution function of a self-adjoint extension with discrete spectrum is more natural in terms of the operator C appearing in (1.20) than in terms of $\widetilde{C}$.

THEOREM 4.2. *Let L_C be the self-adjoint extension of the minimal operator generated by the boundary condition (1.20). Assume that the spectrum of the operator A is discrete and that $N(\lambda; A^2) \sim c\lambda^\alpha$ $(c > 0, \alpha > 0)$. The spectrum of L_C is discrete and (4.7) is valid for $N(\lambda; L_C)$ if and only if $\cos C \in \mathfrak{S}_\infty$ and*

$$n(\lambda; \cos C) = o(\lambda^{\alpha+\frac{1}{2}}). \tag{4.9}$$

Proof. Sufficiency is established in the same way as in Theorem 4.1. We prove necessity.

If the spectrum of the operator L_C is discrete, then by Corollary 2.1 $\cos C \in \mathfrak{S}_\infty$; therefore, the spectrum of C (one can always regard $\|C\| \leq \frac{\pi}{2}$) consists of eigenvalues of finite multiplicity, located on $(-\frac{\pi}{2}, \frac{\pi}{2}]$ and concentrating near $\pm\frac{\pi}{2}$. For $0 < \epsilon < \frac{\pi}{2}$ we set $C_\epsilon = C(E-P_\epsilon)+\frac{\pi}{2}P_\epsilon$, where $P_\epsilon = E_\Delta$, $\Delta = [-\frac{\pi}{2}+\epsilon, \frac{\pi}{2}-\epsilon]$, and E_λ is a resolution of the identity of C. The operator $C - C_\epsilon$, hence $\cos C - \cos C_\epsilon$, is finite dimensional. By virtue of Theorem 3.2, Chapter 3, the spectrum of L_{C_ϵ} is discrete and $N(\lambda; L_{C_\epsilon}) \sim N(\lambda; L_C)$. It is also obvious that the operator C_ϵ is continuously invertible. So, the boundary condition $(\cos C_\epsilon)Y' - (\sin C_\epsilon)Y = 0$ can be rewritten in the form $Y = (\mathrm{ctg} C_\epsilon)Y'$. Taking into account (4.4) we have

$$\mathrm{ctg} C_\epsilon(\Gamma_1 y - \Omega\Gamma_2 y) = -\Gamma_2 y,$$

where Ω is determined by formula (4.5). This implies that

$$(\mathrm{ctg} C_\epsilon)\Gamma_1 y = -(E - (\mathrm{ctg} C_\epsilon)\Omega)\Gamma_2 y.$$

Choosing ϵ so that $\|\mathrm{ctg} C_\epsilon\| < \|\Omega\|^{-1}$, we find

$$\Gamma_2 y = -(E - (\mathrm{ctg} C_\epsilon)\Omega)^{-1}(\mathrm{ctg} C_\epsilon)\Gamma_1 y.$$

By Theorem 4.1,

$$n(\lambda; -(E - (\mathrm{ctg} C_\epsilon)\Omega)^{-1}\mathrm{ctg} C_\epsilon) = o(\lambda^{\alpha+\frac{1}{2}}),$$

hence $n(\lambda; \cos C_\epsilon) = o(\lambda^{\alpha+\frac{1}{2}})$. Since the operator $\cos C - \cos C_\epsilon$ is finite dimensional, $n(\lambda; \cos C) = o(\lambda^{\alpha+\frac{1}{2}})$. □

REMARK 4.1. Because of Lemma 6.2, Chapter 1, we can replace (4.9) in Theorem 4.2 by the condition $|\lambda_n(\cos C)| = o(n^{-\frac{2}{2\alpha+1}})$, which is equivalent to (4.9).

COROLLARY 4.1. *Assume that the spectrum of the operator A is discrete and that $N(\lambda; A^2) \sim c\lambda^\alpha$ $(c > 0, \alpha > 0)$. Let L_C be a β-smooth $(-\frac{1}{2} < \beta \leq \frac{3}{2})$ self-adjoint extension of L_0. If*

$$2\alpha(3 - 2\beta) < 2\beta + 1, \tag{4.10}$$

then

$$N(\lambda; L_C) \sim N(\lambda; L_D) \sim \frac{b-a}{2\pi} cB(\alpha + 1; \tfrac{1}{2})\lambda^{\alpha+\frac{1}{2}}.$$

Proof. According to Theorem 2.4 for the β-smooth self-adjoint extension L_C of L_0 the operator $B = \hat{\mathbf{A}}^{\beta+\frac{1}{2}} \cos C$ is continuous on $\mathfrak{H} \oplus \mathfrak{H}$. This implies (see Chapter 1, subsection 6.4) that

$$n(\lambda; \cos C) \leq n(\|B\|\lambda; \hat{\mathbf{A}}^{-\beta-\frac{1}{2}}) \sim \text{const} \cdot \lambda^{\frac{4\alpha}{2\beta+1}}.$$

If α and β satisfy inequality (4.10), then $n(\lambda; \cos C) = o(\lambda^{\alpha+\frac{1}{2}})$. By Theorem 4.2

$$N(\lambda; L_C) \sim \frac{b-a}{2\pi} cB(\alpha + 1; \tfrac{1}{2})\lambda^{\alpha+\frac{1}{2}}. \qquad \square$$

When $\beta = \frac{3}{2}$, inequality (4.10) is satisfied for arbitrary $\alpha > 0$. So, the following statement is valid.

COROLLARY 4.2. *Under the assumptions of Corollary 4.1 the asymptotic relation $N(\lambda; L_C) \sim N(\lambda; L_D)$ is true for any maximally smooth self-adjoint extension of the minimal operator. In particular, $N(\lambda; L_N) \sim N(\lambda; L_D)$.*

COROLLARY 4.3. *Under the assumptions of Corollary 4.1 with $\alpha < \frac{1}{2}$,*

$$N(\lambda; L) \sim N(\lambda; L_D) \sim \frac{b-a}{2\pi} cB(\alpha + 1; \tfrac{1}{2})\lambda^{\alpha+\frac{1}{2}}$$

for any self-adjoint extension L of the minimal operator with finite quadratic functional.

Proof. Since by Theorem 2.5 a self-adjoint extension of L_0 with finite quadratic functional is $\frac{1}{2}$-smooth, inequality (4.10) is valid for $\alpha < \frac{1}{2}$ $(\beta = \frac{1}{2})$. $\square$

LEMMA 4.3. *A vector-function*

$$y(t) = e^{-\hat{A}(t-a)} f_1 + e^{-\hat{A}(b-t)} f_2 \quad (f_1, f_2 \in \mathfrak{H}^{-\infty}(A) = \bigcup_n \mathfrak{H}^{-n}(A))$$

belongs to the space $W_2^s(\mathfrak{H}, (a, b))$ $(0 < s \in \mathbb{R}^1)$ if and only if $f_1, f_2 \in \mathfrak{H}^{s-\frac{1}{2}}(A)$.

Proof. Suppose first that $s = m$ is an integer.

Since $y^{(m)}(t) = (-1)^m e^{-\hat{A}(t-a)} \hat{A}^m f_1 + e^{-\hat{A}(b-t)} \hat{A}^m f_2 \in L_2(\mathfrak{H}, (a, b))$, according to Corollary 5.3, Chapter 2, $\hat{A}^m f_i \in \mathfrak{H}^{-\frac{1}{2}}(A)$, i.e. $f_i \in \mathfrak{H}^{m-\frac{1}{2}}(A)$.

If s is fractional, then we use an interpolation theorem (see Lions, Magènes [1, p. 41]), setting in it $X = \mathfrak{H}^{m-\frac{1}{2}}(A)$, $Y = \mathfrak{H}^{-\frac{1}{2}}(A)$, $\mathcal{X} = W_2^m(\mathfrak{H}, (a, b))$, $\mathcal{Y} =$

$L_2(\mathfrak{H}, (a,b))$, $\pi = e^{-\hat{A}(t-a)}$ or $\pi = e^{-\hat{A}(b-t)}$, where m, s and θ are related to each other by the relation $(1-\theta)m = s$. □

THEOREM 4.3. *Let the operator A satisfy the conditions of Theorem 4.2. If the inclusion*

$$\mathfrak{D}(L_C) \subset W_2^s(\mathfrak{H}, (a,b)) \quad \left(\frac{4\alpha}{2\alpha+1} < s \leq 2\right)$$

holds for the self-adjoint extension L_C of the minimal operator, then the spectrum of L_C is discrete and $N(\lambda; L_C) \sim N(\lambda; L_D)$.

Proof. Let $y(t) \in \mathfrak{D}(L_C)$. It follows from the inclusion $\frac{1}{2}\int_a^b e^{-\hat{A}|t-\tau|}A^{-1}h(\tau)\,d\tau \in W_2^2(\mathfrak{H}, (a,b))$ $(h \in L_2(\mathfrak{H}, (a,b))$, the representation (1.6) and Lemma 4.3 that $y(t) \in C(\mathfrak{H}^{s-\frac{1}{2}}, [a,b])$ (recall that $\frac{1}{2}\int_a^b e^{-\hat{A}|t-\tau|}A^{-1}h(\tau)\,d\tau \in C(\mathfrak{H}^{\frac{3}{2}}, [a,b])$). Thus, the extension L_C is $(s-\frac{1}{2})$-smooth. But $s-\frac{1}{2}$ satisfies (4.10) only provided $s > \frac{4\alpha}{2\alpha+1}$. By Corollary 4.1, the spectrum of L_C is discrete and $N(\lambda; L_C) \sim N(\lambda; L_D)$. □

THEOREM 4.4. *Assume that the spectrum of the operator A is discrete and that $N(\lambda; A^2) \sim c\lambda^\alpha$ $(c > 0, \alpha > 0)$. Assume that the self-adjoint extension L_C of the minimal operator is a D-extension. Then $N(\lambda; L_C) \sim N(\lambda; L_D)$.*

Proof. Since any D-extension is an extension with finite quadratic functional, the assertion of the theorem for $\alpha < \frac{1}{2}$ is deduced from Corollary 4.3.

Set as before

$$B = \hat{\mathbf{A}} \cos C. \tag{4.11}$$

Taking into account (2.10) we obtain

$$B^*B = \hat{\mathbf{A}}^{-1} B \sin C. \tag{4.12}$$

Since the operator $B \sin C$ is bounded and $|\lambda_n(\hat{\mathbf{A}}^{-1})| \leq \text{const} \cdot n^{-\frac{1}{2\alpha}}$, property (6.8) of s-numbers from Chapter 1 implies

$$\lambda_n(B^*B) \leq \text{const} \cdot n^{-\frac{1}{2\alpha}},$$

hence

$$s_n(B) \leq \text{const} \cdot n^{-\frac{1}{4\alpha}}.$$

Using the properties (6.8) and (6.15) of s-numbers (Chapter 1) we find from (4.12),

$$s_n(B) \leq \text{const} \cdot n^{-\frac{3}{8\alpha}} = \text{const} \cdot n^{-\frac{2^2-1}{2^3\alpha}}.$$

By repeating this argument m times we arrive at the estimate

$$s_n(B) \leq \text{const} \cdot n^{-\frac{2^m-1}{2^{m+1}\alpha}},$$

where the constant depends on m and does not depend on n. This estimate, taking into account equality (4.11) and property (6.15) of s-numbers (Chapter 1), enables us to conclude that

$$|\lambda_n(\cos C)| \leq \text{const} \cdot n^{-\frac{2^{m+1}-1}{2^{m+1}\alpha}}.$$

For a given α it is always possible to choose a number $m = m(\alpha)$ so that $\frac{2^{m+1}-1}{2^{m+1}\alpha} > \frac{2}{2\alpha+1}$. Then $|\lambda(\cos C)| = o(n^{-\frac{2}{2\alpha+1}})$. It remains to apply Remark 4.1. □

In the results discussed above we have selected those self-adjoint extensions of the minimal operator with purely discrete spectrum whose dominant term in the asymptotics of the eigenvalue distribution function is the same as that of the extensions L_D and L_N. Relying on Theorem 3.3 (Chapter 3) it is also possible to describe the extensions for which the dominant term of the asymptotics of their eigenvalue distribution functions increases at infinity faster than that of the Dirichlet problem.

THEOREM 4.5. *Assume the operator A to be as in Theorem 4.2. Let L_C be a self-adjoint extension of L_0 with discrete spectrum. The asymptotic equality*

$$N(\lambda; L_C) = O(\lambda^\beta) \quad (\beta > \alpha + \tfrac{1}{2})$$

holds if and only if

$$n(\lambda; \cos C) = O(\lambda^\beta).$$

Note that all considerations in this section were made under the assumption $q(t) \equiv 0$. In case $q(t) \not\equiv 0$ we must use the resolvent formula (4.7) from Chapter 3.

5. Other Boundary Value Problems

In this section we consider the following boundary value problems for expression (1.1): a) an analogue of the third boundary value problem for the Laplace equation; b) the generalized periodic problem. It is shown that when the spectrum of the operator A is discrete, these problems generate self-adjoint extensions of the minimal operator with purely discrete spectrum. The asymptotics of their eigenvalue distribution functions is almost the same as that in the case of the Dirichlet and Neumann problems.

5.1. THE BOUNDARY VALUE PROBLEM OF THE THIRD KIND

On the set of vector-functions $y(t) \in \mathfrak{D}(L_0^*)$ satisfying the condition

$$\hat{Y}' = B\hat{Y}, \tag{5.1}$$

where $\hat{Y} = \{y(a), -y(b)\}$, $\hat{Y}' = \{y'(a), y'(b)\}$, B is a linear operator on $\mathfrak{H} \oplus \mathfrak{H}$, we define the operator L'_B by $L'_B y = L_0^* y$. It is obvious that L'_B is an extension of the minimal operator. The question arises, under what assumptions on B is the closure of L'_B a self-adjoint operator on $L_2(\mathfrak{H}, (a, b))$? The answer is as follows.

THEOREM 5.1. *In order that $L_B = \overline{L'_B}$ be a maximal dissipative (maximal accumulative, self-adjoint) extension of the minimal operator it is necessary and sufficient that the operator $S = -\hat{\mathbf{A}}^{\frac{1}{2}}(\hat{\mathbf{A}} + B)\hat{\mathbf{A}}^{\frac{1}{2}}$ be maximal accumulative (maximal* *dissipative, self-adjoint) on $\mathfrak{H} \oplus \mathfrak{H}$. If the spectrum of the operator A is discrete, then a necessary and sufficient condition for discreteness of the spectrum of this extension is discreteness of the spectrum of the operator S.*

Recall that $\hat{\mathbf{A}}^\alpha = \begin{pmatrix} \hat{A}^\alpha & 0 \\ 0 & \hat{A}^\alpha \end{pmatrix}$.

Proof. Adding $\hat{\mathbf{A}}\hat{Y}$ to (5.1) and taking into account that $Y = -\hat{\mathbf{A}}^{-\frac{1}{2}}\hat{Y}$ and $Y' = \hat{\mathbf{A}}^{\frac{1}{2}}(\hat{Y}' + \hat{\mathbf{A}}\hat{Y})$, where Y, Y' are determined by formulas (1.16), we obtain the equality

$$Y' - SY = 0 \tag{5.2}$$

which is equivalent to (5.1). It is clear that the set of vector-functions $y(t) \in \mathfrak{D}(L_0^*)$ satisfying (5.1) is the domain of a maximal dissipative (maximal accumulative, self-adjoint) extension of the minimal operator if and only if the pairs $\{Y', Y\}$ associated with those $y(t)$ form a linear relation of the corresponding type. Evidently, for the validity of the last requirement it is necessary and sufficient that the operator S be maximal accumulative (maximal dissipative, self-adjoint). The first assertion of the theorem has been proved.

We prove the second assertion only for the maximal dissipative extensions, for in the other cases the proof is similar. Since (5.2) gives a maximal dissipative linear relation, it can be written (by Theorem 1.4) in the form of (1.18), where the operator K is expressed in terms of S as follows:

$$K = -(S + iE)(S - iE)^{-1}.$$

Because S is maximal accumulative, the operator $(S - iE)^{-1}$ is defined and bounded on the whole space $\mathfrak{H} \oplus \mathfrak{H}$.

Let now the spectrum of the operator A be discrete. On account of Theorem 2.2 the spectrum of the maximal dissipative extension L_K is discrete if and only if the operator $E + K = -2i(S - iE)^{-1}$ is completely continuous on $\mathfrak{H} \oplus \mathfrak{H}$. This completes the proof of the theorem. □

LEMMA 5.1. *If $y(t)$ is an absolutely continuous vector-function on $[a, b]$ such that $y'(t) \in L_2(\mathfrak{H}, (a, b))$, then the inequality*

$$\|y(t)\|^2 \leq c_1 \|y(t)\|^2_{L_2(\mathfrak{H},(a,b))} + c_2 \|y'(t)\|^2_{L_2(\mathfrak{H},(a,b))} \tag{5.3}$$

holds, where $c_1 > 0, c_2 > 0$ and, furthermore, c_2 can be chosen arbitrarily small.

Proof. Denote by $\chi_t(s)$ $(a < t \leq b)$ a continuously differentiable function on $[a, b]$, vanishing at the point a and having the properties: $\chi_t(t) = 1$, $|\chi_t(s)| \leq 1$, $\int_a^b |\chi_t(s)|^2 \, ds = c_2$, where c_2 is a given small number. Set $z(s) = \chi_t(s) y(s)$. Then

$$y(t) = z(t) = \int_a^t z'(s) \, ds = \int_a^t \chi_t'(s) y(s) \, ds + \int_a^t \chi_t(s) y'(s) \, ds.$$

Applying the Cauchy–Buniakowski inequality we get

$$\|y(t)\|^2 \le 2\Big\|\int_a^t \chi_t'(s)y(s)\,ds\Big\|^2 + 2\Big\|\int_a^t \chi_t(s)y'(s)\,ds\Big\|^2$$
$$\le c_1\|y\|^2_{L_2(\mathfrak{H},(a,b))} + c_2\|y'\|^2_{L_2(\mathfrak{H},(a,b))},$$

where

$$c_1 = 2\int_a^b |\chi_t'(s)|^2\,ds, \quad c_2 = 2\int_a^b |\chi_t(s)|^2\,ds.$$

So, the lemma has been proved in case $t \neq a$.

The case $t = a$ is proved in a similar way by taking $\chi_a(s)$ to vanish at b. □

THEOREM 5.2. *Let B be a bounded dissipative (accumulative, self-adjoint) operator on $\mathfrak{H}\oplus\mathfrak{H}$. Then the extension L_B of the operator L_0 generated by the boundary condition (5.1) is maximal dissipative (maximal accumulative, self-adjoint) with finite quadratic functional and lower semi-bounded real part. If the spectrum of A is discrete, the spectrum of the operator L_B is also discrete. If $A^{-1} \in \mathfrak{S}_{2p-1}$ $(p \ge 1)$, then $R_\lambda(L_B) \in \mathfrak{S}_p$. If $N(\lambda; A^2) \sim c\lambda^\alpha$ $(c > 0, \alpha > 0)$ and B is self-adjoint, then $N(\lambda; L_B) \sim N(\lambda; L_D)$.*

Proof. Without loss of generality we may assume $A^2 \ge 2\|B\|^2E$ (this may always be done if we take in (1.1) $A^2 + 2\|B\|^2E$ instead of A^2 and $q(t) - 2\|B\|^2E$ instead of $q(t)$). According to Theorem 5.1, to establish maximal dissipativeness of L_B it is sufficient to show that the operator $S = -\hat{\mathbf{A}}^{\frac12}(B+\hat{\mathbf{A}})\hat{\mathbf{A}}^{\frac12}$ is maximal accumulative.

Of course, the operator S is accumulative. Since $A^2 \ge 2\|B\|^2E$, the operator $\hat{\mathbf{A}} + B$ is invertible, hence S is also invertible, so it is maximal accumulative.

Now we will show that L_B is an extension with finite quadratic functional. To this end it suffices to prove, by Theorems 2.4, 2.5, that the operator $\hat{\mathbf{A}}(K+E)$ is bounded on $\mathfrak{H}\oplus\mathfrak{H}$. Since

$$\begin{aligned} E + K &= -2i(S - iE)^{-1} \\ &= 2i(\hat{\mathbf{A}}^{-1}(\hat{\mathbf{A}}^{-\frac12}B\hat{\mathbf{A}}^{-\frac12} + E)^{-1}\hat{\mathbf{A}}^{-1}(E - iS^{-1})^{-1}) \end{aligned} \tag{5.4}$$

and the operators $(\hat{\mathbf{A}}^{-\frac12}B\hat{\mathbf{A}}^{-\frac12} + E)^{-1}$, $(E - iS^{-1})^{-1}$ are bounded on $\mathfrak{H}\oplus\mathfrak{H}$, the operator $\hat{\mathbf{A}}(K+E)$ is bounded on this space, i.e. L_B is an extension with finite quadratic functional. Therefore

$$(L_By, y)_{L_2(\mathfrak{H},(a,b))} = (\hat{Y}', \hat{Y})_{\mathfrak{H}\oplus\mathfrak{H}} + D(y),$$

where $D(y) = \int_a^b(\|y'(t)\|^2 + \|Ay(t)\|^2 + (q(t)y(t), y(t)))\,dt$,

$$\operatorname{Re}(L_By, y)_{L_2(\mathfrak{H},(a,b))} = \operatorname{Re}(B\hat{Y}, \hat{Y})_{\mathfrak{H}\oplus\mathfrak{H}} + D(y) \ge m(\hat{Y}, \hat{Y})_{\mathfrak{H}\oplus\mathfrak{H}} + D(y) \tag{5.5}$$
$$\Big(m = \inf_{F:\|F\|_{\mathfrak{H}\oplus\mathfrak{H}}=1} \operatorname{Re}(BF, F)_{\mathfrak{H}\oplus\mathfrak{H}}\Big).$$

If $m \geq 0$, then (5.5) implies $\operatorname{Re}\big(L_B y, y\big)_{L_2(\mathfrak{H},(a,b))} \geq 0$, i.e. the real part of the operator L_B is non-negative. When $m < 0$ we use the estimate (5.3). Then

$$\operatorname{Re}\big(L_B y, y\big)_{L_2(\mathfrak{H},(a,b))} \geq D(y) + 2mc_1\|y\|^2_{L_2(\mathfrak{H},(a,b))} + 2mc_2\|y'\|^2_{L_2(\mathfrak{H},(a,b))}.$$

Choosing c_2 small so that $D(y) + 2mc_2\|y'\|^2_{L_2(\mathfrak{H},(a,b))} \geq 0$ we obtain

$$\operatorname{Re}\big(L_B y, y\big)_{L_2(\mathfrak{H},(a,b))} \geq 2mc_1\|y\|^2_{L_2(\mathfrak{H},(a,b))}.$$

Thus, the real part of L_B is lower semi-bounded.

Suppose, next, that the spectrum of the operator A is discrete. It can be seen from formula (5.4) that $K+E \in \mathfrak{S}_\infty$. By Theorem 2.2 the spectrum of the operator L_B is discrete.

If $A^{-1} \in \mathfrak{S}_{2p-1}$ $(p \geq 1)$, then $K+E \in \mathfrak{S}_{p-\frac{1}{2}} \subset \mathfrak{S}_p$ follows from (5.4). According to Theorem 2.3, $R_\lambda(L_B) \in \mathfrak{S}_p$.

Let, finally, $N(\lambda; A^2) \sim c\lambda^\alpha$ $(c > 0, \alpha > 0)$ and the operator B be self-adjoint. Let C correspond to L_B by formula (1.20), $K = e^{-2iC}$. On account of (5.4), $n(\lambda, \cos C) \leq c\lambda^\alpha = o(\lambda^{\alpha+\frac{1}{2}})$. The last assertion of the theorem is deduced from Theorem 4.2. □

This theorem shows that if the operator A^{-1} is nuclear, then $R_\lambda(L_B)$ is nuclear too. According to Corollary 2.3 the system of eigenvectors and adjoined vectors of the operator L_B is complete in $L_2(\mathfrak{H},(a,b))$.

In what follows we suppose, for simplicity, that $q(t) \equiv 0$.

LEMMA 5.2. *Let B_i $(i = 1, 2)$ be bounded self-adjoint linear operators on $\mathfrak{H} \oplus \mathfrak{H}$ with $B_1 \geq B_2$. Then*

$$R_\lambda(L_{B_1}) \leq R_\lambda(L_{B_2})$$

for sufficiently large λ.

Proof. Using formula (4.4) one can rewrite the boundary condition (5.1) (which is equivalent to (5.2)) in the case of expression

$$l[y] = -y'' + (A^2 + k^2E)y \tag{5.6}$$

in the form

$$\Gamma_1 y = (\hat{\mathbf{A}}_k^{\frac{1}{2}}(\hat{\mathbf{A}}_k + B)\hat{\mathbf{A}}_k^{\frac{1}{2}} + \Omega(k))\Gamma_2 y \stackrel{\text{def}}{=} \Lambda(B,k)\Gamma_2 y,$$

where

$$\hat{\mathbf{A}}_k^2 = \begin{pmatrix} \hat{A}^2 + k^2E & 0 \\ 0 & \hat{A}^2 + k^2E \end{pmatrix}.$$

It follows from the form of $\Omega(k)$ (see formula (4.5)) that $\|\Omega(k)\| = o(1)$ as $|k| \to \infty$. Since B is bounded on $\mathfrak{H} \oplus \mathfrak{H}$, the operator $\Lambda(B,k)$, defined on the set $\mathfrak{D}(\Lambda(B,k)) = \mathfrak{D}(\hat{\mathbf{A}}_k^2) = \mathfrak{D}(A^2) \oplus \mathfrak{D}(A^2)$, has a bounded inverse for sufficiently large $|k|$. If $B_2 \leq B_1$, then $\Lambda^{-1}(B_2,k) \geq \Lambda^{-1}(B_1,k)$. By virtue of Lemma 3.4 (Chapter 3) $\big(L_{B_1}^{(k)}\big)^{-1} \leq \big(L_{B_2}^{(k)}\big)^{-1}$, where $L_{B_i}^{(k)}$ denotes the extension of the minimal operator generated by expression (5.6) corresponding to the boundary condition (5.1) with B replaced by B_i. Since $L_{B_i}^{(k)} = L_{B_i} + k^2 E$, the proof of the lemma is complete. □

This lemma and Corollary 6.1, Chapter 1, imply the following assertion.

COROLLARY 5.1. *Under the conditions of Lemma* 5.2, $\lambda_n(L_{B_1}) \geq \lambda_n(L_{B_2})$.

LEMMA 5.3. *Let S be a lower semi-bounded self-adjoint operator on a Hilbert space $\mathfrak{H}$ with discrete spectrum. Suppose also that V is a bounded self-adjoint operator on $\mathfrak{H}$. Then the spectrum of the operator $S+V$ is discrete and*

$$\lambda_k(S+V) \in [\lambda_k(S) - \|V\|, \lambda_k(S) + \|V\|] \quad (k \in \mathbb{N}).$$

Proof. The assertions of the lemma follow from the two-sided inequality $S - \|V\|E \leq S + V \leq S + \|V\|E$. □

COROLLARY 5.2. *Under the conditions of Lemma* 5.3 *the asymptotic equality*

$$N(\lambda; S) = \sum_{k=1}^{N} a_k \lambda^{\alpha_k} + \theta(\lambda^{\alpha_N}) \quad (\alpha_1 > \cdots > \alpha_N \geq 0)$$

implies

$$N(\lambda; S+V) = \sum_{k=1}^{N} a_k \lambda^{\alpha_k} + \theta(\lambda^{\alpha_N}) + O(\lambda^{\alpha_1 - 1}).$$

THEOREM 5.3. *If B is a bounded self-adjoint operator on $\mathfrak{H} \oplus \mathfrak{H}$, then the relation*

$$|\lambda_n(L_B) - \lambda_n(L_N)| = O(1) \tag{5.7}$$

is valid, where the eigenvalues $\lambda_n(\cdot)$ of the extensions L_B $(B \neq 0)$ and L_N are indexed in increasing order.

Proof. We first prove the assertion of the theorem in the special case $B = \eta E$ $(\eta \in \mathbb{R}^1)$. It is not difficult to see that the operator $L_{\eta E}$ admits separation of variables:

$$L_{\eta E} = \overline{l_\eta \otimes E + E \otimes A^2},$$

where l_η is the self-adjoint extension of the minimal operator, generated on the space $L_2(a,b)$ by the differential expression $-y''$ and the boundary conditions $y'(a) - \eta y(a) = 0$; $y'(b) + \eta y(b) = 0$. So (see Chapter 1, subsection 4.7) the spectrum of the operator $L_{\eta E}$ coincides with the arithmetical sum of the spectra of l_η and A^2.

It is known (see, for example, Levitan, Sargsyan [1, p. 22]) that

$$\lambda_n(l_\eta) = \frac{\pi^2}{(b-a)^2}(n-1)^2 + \frac{4\pi}{(b-a)^2}\eta + O(\tfrac{1}{n}) \quad (n \to \infty).$$

Therefore, the operator l_η is representable in the form

$$l_\eta = l'_\eta + l''_\eta$$

where the self-adjoint operators l'_η and l''_η on $L_2(a,b)$ have the same orthonormal system of eigenfunctions as that of l_η. Moreover,

$$\lambda_n(l'_\eta) = \frac{\pi^2}{(b-a)^2}(n-1)^2 = \lambda_n(l_N),$$

while the operator l''_η is bounded. Thus $L_{\eta E} = \overline{l'_\eta \otimes E + E \otimes A^2} + l''_\eta \otimes E$, furthermore, the eigenvalues of the operators $\overline{l' \otimes E + E \otimes A^2}$ and L_N coincide, while the operator $l''_\eta \otimes E$ is bounded on $L_2(\mathfrak{H},(a,b))$. According to Lemma 5.3 the relation (5.7) holds in this special case.

Let now B be an arbitrary bounded self-adjoint operator on $\mathfrak{H} \oplus \mathfrak{H}$. Since $-\|B\|E \leq B \leq \|B\|E$, from Corollary 5.1 we conclude that

$$\lambda_n(L_{-\|B\|E}) \leq \lambda_n(L_B) \leq \lambda_n(L_{\|B\|E}),$$

whence

$$\lambda_n(L_N) - c \leq \lambda_n(L_B) \leq \lambda_n(L_N) + c$$

with some $c > 0$. This completes the proof. □

Comparison of Theorems 5.3, 3.3 and Corollary 5.2 leads to the following result.

THEOREM 5.4. *Assume that the distribution function of the eigenvalues of the operator A^2 satisfies the condition*

$$N(\lambda; A^2) = \sum_{k=1}^{N} c_k \lambda^{\alpha_k} + \theta(\lambda^{\alpha_N}) \quad (\alpha_1 > \alpha_2 \cdots > \alpha_N > 0),$$

while the operator B in the boundary condition (5.1) is bounded and self-adjoint. Then

$$N(\lambda; L_B) = \sum_{k=1}^{N} c'_k \lambda^{\alpha_k + \frac{1}{2}} + \tfrac{1}{2}\sum_{k=1}^{N} c_k \lambda^{\alpha_k} + \theta(\lambda^{\alpha_N + \frac{1}{2}}) + \theta(\lambda^{\alpha_1 - \frac{1}{2}}),$$

where $c'_k = \frac{b-a}{2\pi} c_k B(\alpha_k + 1; \frac{1}{2})$.

REMARK 5.1. Theorems 5.3 and 5.4 are also true if $q(t) \not\equiv 0$. This follows from Lemma 5.3 and the resolvent formula (4.7) (Chapter 3).

5.2. THE PERIODIC BOUNDARY VALUE PROBLEM

We consider for expression (1.1) a problem of the form

$$y'(a) - Uy'(b) = 0, \quad y(a) - Uy(b) = 0, \tag{5.8}$$

where U is a unitary operator on $\mathfrak{H}$, commuting with A (i.e. with its resolution of the identity). On the set of vector-functions $y(t) \in \mathfrak{D}(L_0^*)$ satisfying the conditions (5.8) we define the operator L'_U by $L'_U y = L_0^* y$. Its closure in the space $L_2(\mathfrak{H},(a,b))$, denoted by L_U, is clearly an extension of the minimal operator L_0.

THEOREM 5.5. *The operator L_U is a lower semi-bounded, maximally smooth, self-adjoint extension of the minimal operator. Under the assumption that the spectrum of the operator A is discrete the spectrum of the extension L_U is discrete. If $A^{-1} \in \mathfrak{S}_{2p-1}$ $(p \geq 1)$, we have $R_\lambda(L_U) \in \mathfrak{S}_p$ $(\operatorname{Im}\lambda \neq 0)$. If $N(\lambda; A^2) \sim c\lambda^\alpha$ $(c > 0, \alpha > 0)$, then $N(\lambda; L_U) \sim N(\lambda; L_D)$.*

Proof. After elementary manipulations we find that (5.8) is equivalent to

$$\begin{aligned} &y_a' - U y_b' - A^2 y_a - U A^2 y_b = 0, \\ &U^* y_a - y_b = 0, \end{aligned} \tag{5.9}$$

where, as before

$$\begin{aligned} &y_a = \hat{A}^{-\frac{1}{2}} y(a), \quad y_b = \hat{A}^{-\frac{1}{2}} y(b), \\ &y_a' = A^{\frac{1}{2}}(y'(a) + \hat{A}y(a)), \quad y_b' = A^{\frac{1}{2}}(y'(b) - \hat{A}y(b)). \end{aligned}$$

We write (5.9) in matrix form:

$$\begin{pmatrix} E & -U \\ 0 & 0 \end{pmatrix} \begin{pmatrix} y_a' \\ y_b' \end{pmatrix} + \begin{pmatrix} A^2 & -UA^2 \\ -U^* & -E \end{pmatrix} \begin{pmatrix} -y_a \\ y_b \end{pmatrix} = 0.$$

Multiplying this equality from the left by the matrix inverse to $\begin{pmatrix} A^2 & -UA^2 \\ -U^* & -E \end{pmatrix}$ we obtain

$$\frac{1}{2} \begin{pmatrix} A^{-2} & 0 \\ 0 & A^{-2} \end{pmatrix} \begin{pmatrix} -E & U \\ U^* & -E \end{pmatrix} Y' - Y = 0$$

$(Y = \{-y_a, y_b\},\ Y' = \{y_a', y_b'\})$. Set

$$\cos C = \hat{\mathbf{A}}^{-2} B (E + \hat{\mathbf{A}}^{-4} B^2)^{-\frac{1}{2}}, \quad \sin C = (E + \hat{\mathbf{A}}^{-4} B^2)^{-\frac{1}{2}}, \qquad B = \begin{pmatrix} -E & -U \\ U^* & -E \end{pmatrix}. \tag{5.10}$$

Then the conditions (5.8) turn into

$$(\cos C) Y' - (\sin C) Y = 0. \tag{5.11}$$

It is clear now that the extension L_U coincides with the extension L_C given by condition (5.11), where the operator C $(\|C\| \leq \frac{\pi}{2})$ is uniquely found from (5.10) in terms of the operator U. Since $\hat{\mathbf{A}}^{-4} B^2$ is positive, the operator $(E + \hat{\mathbf{A}}^{-4} B^2)^{-\frac{1}{2}}$ is bounded. According to Corollary 2.4, boundedness of $\hat{\mathbf{A}}^2 \cos C$ implies maximal smoothness of the extension L_U. If the spectrum of A is discrete, then the spectrum of L_U is discrete (see Corollary 2.5). Under the condition $A^{-1} \in \mathfrak{S}_{2p-1}$ $(p \geq 1)$, $\cos C \in \mathfrak{S}_{p-\frac{1}{2}} \subset \mathfrak{S}_p$, then by Theorem 2.3 we have $R_\lambda(L_U) \in \mathfrak{S}_p$. The last assertion of the theorem is a consequence of Corollary 4.2.

Further, since L_U is a maximally smooth extension of the minimal operator, it is an extension with finite quadratic functional. By integration by parts we verify that $(L_U y, y)_{L_2(\mathfrak{H},(a,b))} = D(y) > m\|y\|^2_{L_2(\mathfrak{H},(a,b))}$ $(m = \text{const})$. Thus, the operator L_U is lower semi-bounded. □

6. The Case of a Variable Operator-Valued Coefficient

We consider the differential expression

$$(l[y](t) = -y''(t) + A(t)y(t) + q(t)y(t) \quad (t \in [a,b],\ -\infty < a < b < \infty), \tag{6.1}$$

where the $A(t)$ are self-adjoint operators on $\mathfrak{H}$ for which: 1) the domain $\mathfrak{D} = \mathfrak{D}(A(t))$ does not depend on t; 2) $A(t) \geq mE$ ($m \in \mathbb{R}^1$); 3) the vector-function $A(t)f$ (any $f \in \mathfrak{D}$) is strongly continuously differentiable on $[a,b]$ into $\mathfrak{H}$; the conditions on $q(t)$ are the same as in section 1.

In exactly the same way as in the case of a constant operator $A(t) = A^2$ we define the operators L_0 (minimal) and L_0^* (maximal) generated on the space $L_2(\mathfrak{H},(a,b))$ by the expression (6.1).

In this section we study the domains of the minimal and maximal operators and then describe in terms of boundary conditions all maximal dissipative and maximal accumulative, in particular self-adjoint, extensions of L_0. A separate place among them is occupied by the extensions with discrete spectrum and the extensions whose domain consists of vector-functions of higher smoothness (they are also selected by boundary conditions).

We will mainly use that very scheme of research which was applied in sections 1-3 for the expression with a constant operator coefficient. In the situation considered the role of $e^{-A|t-s|}$ is played by the Green function $G(t,s)$ of one of the boundary value problems for the expression (6.1); for the sake of simplicity we choose the Neumann problem $y'(a) = y'(b) = 0$.

It is convenient for us to regard $a = 0$. Of course, this restriction does not lead to a loss of generality, because the general case reduces to this special one by means of a linear change of variables.

6.1. CONSTRUCTION OF THE GREEN FUNCTION FOR THE NEUMANN PROBLEM

Assume that $A(t) \geq E$. Next to the expression (6.1) we consider the differential expression

$$(l_k[y])(t) = -y''(t) + (A(t) + k^2E)y(t) = -y''(t) + A_k(t)y(t), \tag{6.2}$$

where k is a positive number. Construct the operator-function $G(t,\tau)$ satisfying for any fixed $\tau \in [0,b]$ the equation

$$l_k[G(t,\tau)] = 0 \quad (t \neq \tau) \tag{6.3}$$

and the Neumann boundary condition

$$\left.\frac{\partial G(t,\tau)}{\partial t}\right|_{t=0} = \left.\frac{\partial G(t,\tau)}{\partial t}\right|_{t=b} = 0. \tag{6.4}$$

It can be immediately verified that the operator-function

$$F(t,\tau) = \tfrac{1}{2}A_k^{-\frac{1}{2}}(\tau)\big(V_\tau(|t-\tau|) + (E - V_\tau(2b))^{-1}\times$$
$$\times\,(V_\tau(t+\tau) + V_\tau(2b-t+\tau) + V_\tau(2b+t-\tau) + V_\tau(2b-t+\tau))\big),$$

where $V_\tau(t) = e^{-t\sqrt{A_k(\tau)}}$ $(t, \tau \in [0, b])$, satisfies the equation

$$\frac{\partial^2 F(t,\tau)}{\partial t^2} = A_k(\tau) F(t,\tau) \quad (t \neq \tau) \tag{6.5}$$

and the conditions (6.4). We will look for the function $G(t,\tau)$ in the form

$$G(t,\tau) = F(t,\tau) + \int_0^b F(t,s)\Phi(s,\tau)\, ds. \tag{6.6}$$

Substitution of expression (6.6) into (6.3) taking into account (6.5) shows that $\Phi(t,\tau)$ has to satisfy a Fredholm equation of the second kind:

$$\Phi(t,\tau) = K(t,\tau) + \int_0^b K(t,s)\Phi(s,\tau)\, ds, \tag{6.7}$$

with kernel $K(t,\tau) = (A(\tau) - A(t))F(t,\tau)$. We prove that equation (6.7) is solvable for k large enough. To this end we estimate the operator-function $K(t,\tau)$, using the fact that under the restrictions imposed on $A(t)$ every power $A^\alpha(t)$ $(0 \leq \alpha \leq 1)$ of $A(t)$ has domain $\mathfrak{D}_\alpha$ independent of t, $A^\alpha(t)$ is strongly continuously differentiable on $\mathfrak{D}_\alpha$, while the function $\|A^\alpha(t)A^{-\alpha}(s)\|$ is bounded on $[0,b] \times [0,b]$ uniformly with respect to $\alpha \in [-1, 1]$, and, in addition,

$$\|A^{-\alpha}(s)(A(t) - A(\tau))A^{-(1-\alpha)}(s)\| \leq M_\alpha |t - \tau| \tag{6.8}$$
$$(0 \leq s, t, \tau \leq b,\ \alpha \in [0, 1])$$

(see, for instance, S. Krein [1, Chapter II, §1]). It follows from the form of the operator-function $F(t,\tau)$ that the operators $A^{\frac{1}{2}}(\tau)V_\tau(-|t-\tau|)F(t,\tau)$ are bounded jointly in t, τ and k. This implies, taking into account (6.8),

$$\begin{aligned} \|K(t,\tau)\| &= \|(A(\tau) - A(t))A^{-1}(\tau)A(\tau)F(t,\tau)\| \\ &\leq M_0|t-\tau|\|A^{\frac{1}{2}}(\tau)V_\tau(-|t-\tau|)F(t,\tau)A^{\frac{1}{2}}(\tau)V_\tau(|t-\tau|)\| \\ &\leq M\|A^{\frac{1}{2}}(\tau)V_\tau(|t-\tau|)\|\,|t-\tau|. \end{aligned}$$

Due to the obvious inequality $2\sqrt{A(\tau) + k^2 E} \geq \sqrt{A(\tau)} + kE$ we next get that

$$\|A^{\frac{1}{2}}(\tau)V_\tau(|t-\tau|)\| \leq \left\|\sqrt{A(\tau)}\, e^{-\frac{\sqrt{A(\tau)}+kE}{2}|t-\tau|}\right\|.$$

In accordance with the spectral theory of self-adjoint operators (see Chapter 1, subsection 5.3) the last norm coincides with the maximum of the function $\lambda \exp(-\frac{\lambda+k}{2}|t-\tau|)$, where λ belongs to the spectrum of the operator $\sqrt{A(\tau)}$. Direct computation shows that this maximum does not exceed $\frac{2}{e|t-\tau|}e^{-\frac{k|t-\tau|}{2}}$. So, finally we have

$$\|K(t,\tau)\| \leq c \exp(-k\frac{|t-\tau|}{2})$$

(c does not depend on k). Hence, for sufficiently large k,

$$\int_0^b \|K(t,s)\|\, ds < 1. \tag{6.9}$$

This implies that if k is large enough, then one can solve equation (6.7) by the method of successive approximation. The solution is looked for in the form

$$\Phi(t,s) = \sum_{n=0}^{\infty} K^{(n)}(t,s), \tag{6.10}$$

where

$$K^{(0)}(t,s) = K(t,s), \quad K^{(n+1)}(t,s) = \int_0^b K^{(n)}(t,x)K(x,s)\,dx. \tag{6.11}$$

The proof of the fact that the formulas (6.10) and (6.11) give really a solution of equation (6.7) is not essentially different from proving solvability for the scalar Fredholm integral equation of the second kind by means of successive approximations (see, for example, Sobolev [2]). On account of (6.10), (6.11) and the estimate for $\|K(t,s)\|$ obtained above, we deduce that the operator-functions $\Phi(t,s)$, $F(t,s)$ and $G(t,s)$ are uniformly bounded on $[0,b]\times[0,b]$ in the operator norm. Moreover, the relations (6.8) allow us to establish that the operator-functions $A^{-\alpha}(\tau)K(t,s)A^{\alpha}(\tau)$, $A^{-\alpha}(\tau)\Phi(t,s)A^{\alpha}(\tau)$, $A^{-\alpha}(\tau)F(t,s)A^{\alpha}(\tau)$ and $A^{-\alpha}(\tau)G(t,s)A^{\alpha}(\tau)$ are uniformly bounded not only in t and s but also in τ (here $0\le\alpha\le 1$).

It can be directly verified that the operator-function $G(t,s)$ just constructed has the property that for any $\mathfrak{D}$-valued vector-function $h(s)$ such that $A(0)h(s)$ is strongly continuous into $\mathfrak{H}$, the vector-function

$$y(t) = \int_0^b G(t,s)h(s)\,ds \tag{6.12}$$

is a $\mathfrak{D}$-valued solution of the equation $(l_k[y])(t) = h(t)$ that is twice strongly continuously differentiable into $\mathfrak{H}$ and satisfies the Neumann boundary condition $y'(0)=y'(b)=0$. We will call $G(t,s)$ the Green function for the Neumann problem for (6.2).

In what follows the number k is chosen so that equation (6.7) is solvable and, hence, the Green function for the Neumann problem for (6.2) can be constructed.

Starting from the operator $\sqrt{A(t_0)}$ we construct, as has been done in Chapter 2, subsection 2.1, the Hilbert scale of spaces $\mathfrak{H}^{\alpha} = \mathfrak{H}^{\alpha}(\sqrt{A(t_0)})$. Since the operators $A^{\frac{\alpha}{2}}(t)A^{-\frac{\alpha}{2}}(s)$ are uniformly bounded with respect to $t,s\in[0,b]$ and $\alpha\in[-2,2]$, the spaces $\mathfrak{H}^{\alpha}(\sqrt{A(t_0)})$ coincide, as sets, for different t_0 and the same α, and their norms are equivalent. For the sake of being specific we set $t_0=0$, and we denote by $\hat{A}(t)$ the extension of $A(t)$ in accordance with the definition from Chapter 2, subsection 3.1.

The uniform boundedness of the operator-functions $A^{-\alpha}(\tau)K(t,s)A^{\alpha}(\tau)$, $A^{-\alpha}(\tau)\Phi(t,s)A^{\alpha}(\tau)$, $A^{-\alpha}(\tau)F(t,s)A^{\alpha}(\tau)$ and $A^{-\alpha}(\tau)G(t,s)A^{\alpha}(\tau)$, which was proved above, means that for any $t,s,\tau\in[0,b]$ each operator $K(t,s)$, $\Phi(t,s)$, $F(t,s)$, $G(t,s)$ has an extension to an operator $\hat{K}(t,s)$, $\hat{\Phi}(t,s)$, $\hat{F}(t,s)$, $\hat{G}(t,s)$, respectively, mapping any $\mathfrak{H}^{-\alpha}$ into itself; furthermore, these operator-functions are bounded uniformly in t,s relative to the operator norm in any $\mathfrak{H}^{-\alpha}$ $(0\le\alpha\le 2)$.

LEMMA 6.1. *The linear operators $f \to \hat{F}(t,s)f$ and $f \to \hat{G}(t,s)f$ are continuous from $\mathfrak{H}^{-\frac{1}{2}}$ into $L_2(\mathfrak{H},(0,b))$ for any fixed $s \in [0,b]$ and their norms are bounded by a constant not depending on s. A similar statement is true for the operator $f \to \hat{F}(t,s)f$ when t is fixed.*

Proof. The assertion of the lemma for the operators $f \to \hat{F}(t,s)$ when t or s is fixed follows from the fact that all operators $\sqrt{A(s)}\, V_s(-|t-s|)F(t,s)$ are bounded uniformly in t,s and from the arguments used in the proof of Lemma 1.1 for the vector-function $A^{-\frac{1}{2}}(s)V_s(|t-s|)f$. As for the operator $f \to \hat{G}(t,s)f$ when s is fixed, by virtue of (6.6) and the above reasoning it remains to estimate the integral

$$\int_0^b \Big\| \int_0^b \hat{F}(t,\tau)\hat{\Phi}(\tau,s)f\, d\tau \Big\|^2 dt \le b^2 \int_0^b \int_0^b \|\hat{F}(t,\tau)\hat{\Phi}(\tau,s)f\|^2 d\tau\, dt$$
$$= b^2 \int_0^b \Big(\int_0^b \|\hat{F}(t,\tau)\hat{\Phi}(\tau,s)f\|^2 dt \Big) d\tau.$$

But the construction of $\Phi(t,s)$ itself leads to the estimate $\|\hat{\Phi}(\tau,s)f\|_{\mathfrak{H}^{-\frac{1}{2}}} \le$ $\le c\|f\|_{\mathfrak{H}^{-\frac{1}{2}}}$, uniform in τ, s. This, together with the already proved part of the lemma, yields the desired result. □

REMARK 6.1. If, in addition to the conditions 1)–3), the operators $A(t)$ also satisfy the condition

4) there exist neighbourhoods U_0 and U_b of the points 0 and b, respectively, such that the inequality

$$\|(A(t)-A(0))A^{-1}(0)\| \le L_1 t^2 \quad (\|(A(t)-A(b))A^{-1}(b)\| \le L_2(b-t)^2)$$

holds for all $t \in U_0$ $(t \in U_b)$,

then one can show that the operator $f \to \hat{\Phi}(t,0)f$ $(f \to \hat{\Phi}(t,b)f)$ is continuous from $\mathfrak{H}^{-\frac{1}{2}}$ into $L_2(\mathfrak{H},(0,b))$.

Indeed, from the construction of the operator-function $\Phi(t,s)$ it follows that we first have to prove the assertion for the operators $f \to \hat{K}(t,0)f$ and $f \to \hat{K}(t,b)f$. In turn, it is not difficult to see that

$$\int_0^b \|\hat{K}(t,0)f\|^2 dt \le \int_0^b \|(A(t)-A(0))A^{-1}(0)\|^2 \|\hat{A}(0)\hat{F}(t,0)f\|^2 dt$$
$$\le c\int_0^b t^4 \|\hat{A}^{\frac{1}{2}}(0)\hat{V}_0(t)f\|^2 dt.$$

After fourfold integration by parts the problem reduces to estimating the integral considered in Lemma 1.1. The result required follows from this. In a similar way we can consider the case $f \to \hat{K}(t,b)f$. Next, taking into account that the operator-function $K(t,s)$ is uniformly bounded in t,s and using (6,10), (6,11), we can prove the statement for the operators $f \to \hat{\Phi}(t,0)f$ and $f \to \hat{\Phi}(t,b)f$.

If condition 4) is not satisfied, then we can only state that the operators

$f \to \hat{\Phi}(t,0)f$ and $f \to \hat{\Phi}(t,b)f$ are continuous from $\mathfrak{H}^{\beta-\frac{1}{2}}$ $(-\frac{3}{2} \leq \beta \leq 2)$ into $L_2(\mathfrak{H},(0,b))$.

6.2. THE DOMAIN OF THE MAXIMAL OPERATOR

As in subsection 6.1 we may assume $A(t) \geq E$, $q(t) \equiv 0$. Next to the operators L_0 and L_0^* we consider the operator L_N, constructed as follows: on the set $\mathfrak{D}'_N$ of vector-functions $y(t) = \sum_{k=1}^{n} \varphi_k(t) f_k$ ($f_k \in \mathfrak{D}$, $\varphi_k \in W_2^2(0,b)$, $\varphi'_k(0) = \varphi'_k(b) = 0$) we define the operator L'_N by $L'_N y = l[y]$; this operator is symmetric, by formula (1.2), which is valid for $y(t), z(t) \in \mathfrak{D}(L'_N)$; its closure is denoted by L_N.

Let $W_2^2(\mathfrak{H}; A(t); (0,b))$ be the set of vector-functions $y(t)$ in $L_2(\mathfrak{H},(0,b))$ with

$$|||y||| = \left(\int_0^b (\|y(t)\|^2 + \|y''(t)\|^2 + \|A(t)y(t)\|^2)\,dt\right)^{\frac{1}{2}} < \infty,$$

where the derivative $y''(t)$ is understood in the strong sense. Since the functions $\|A^{-1}(t)A(0)\|$, $\|A^{-1}(0)A(t)\|$ are bounded on $[0,b]$, this norm is equivalent to any of the norms

$$|||y|||' = \left(\int_0^b (\|y(t)\|^2 + \|y''(t)\|^2 + \|A(0)y(t)\|^2)\,dt\right)^{\frac{1}{2}},$$

$$|||y|||'' = \left(\int_0^b (\|y(t)\|^2 + \|y''(t)\|^2 + \|(A(0)+B)y(t)\|^2)\,dt\right)^{\frac{1}{2}}$$

where B ($\|B\| < 1$) is a bounded operator on $\mathfrak{H}$.

LEMMA 6.2. *The operator L_N coincides with the closure of its restriction to the set $\mathfrak{D}''_N$ of vector-functions in $W_2^2(\mathfrak{H}; A(t); (0,b))$ whose derivative vanishes at the points 0 and b.*

Proof. Since $\mathfrak{D}''_N \supset \mathfrak{D}'_N$ we only need to prove that for any vector-function $y(t) \in \mathfrak{D}''_N$ there exists a sequence of elements $y_n(t) \in \mathfrak{D}'_N$ converging to $y(t)$ in the sense of the graph of the operator L_N (it is sufficient to show that $y_n(t)$ converges to $y(t)$ in the norm $|||\cdot|||''$).

Select an operator B with $\|B\| < 1$ and such that the operator $A(0)+B$ has an orthonormal system $\{f_k\}_{k=1}^{\infty}$ consisting of eigenvectors that is complete in $\mathfrak{H}$ (this is possible in view of von Neumann's theorem; see, for example, Ahiezer, Glazman [1, section 94, Chapter 7]). Set $y_n(t) = \sum_{k=1}^{n} (y(t), f_k) f_k \in \mathfrak{D}'_N$. The sequence of vector-functions $y_n(t)$ possesses all the required properties. □

LEMMA 6.3. *The operator L_N is self-adjoint.*

Proof. It suffices to prove the self-adjointness of the operator $L_N + k^2E$ for a certain positive k. This operator is symmetric. Since $L_N + k^2E \geq (1+k^2)E$, the set $\mathfrak{R}(L_N + k^2E)$ is closed. So, it is sufficient to verify that $\mathfrak{R}(L_N + k^2E)$ is dense in $L_2(\mathfrak{H},(0,b))$. But, all $\mathfrak{D}$-valued vector-functions $h(s)$ such that $A(0)h(s)$ is strongly continuous into $\mathfrak{H}$ belong to $\mathfrak{R}(L_N + k^2E)$. This can be seen from the fact that vector-functions of the form (6.12) with such an $h(s)$ belong to $\mathfrak{D}''_N \subset \mathfrak{D}(L_N)$, and

$(L_N + k^2 E)y = h$. The denseness of this set of functions $h(s)$ in $L_2(\mathfrak{H}, (0, b))$ leads to the desired result. □

COROLLARY 6.1. *The Green function of the Neumann problem has the property* $G^*(t,s) = G(s,t)$.

Proof. As we can observe from the proof of Lemma 6.3, the inverse of $L_N + k^2 E$ coincides with the integral operator $h(s) \to \int_0^b G(t,s)h(s)\,ds$, which is bounded, in view of the uniform boundedness of $\|G(t,s)\|$, and self-adjoint, being the inverse of a self-adjoint operator. The relation required follows from here. □

LEMMA 6.4. *Let* $h(s) \in L_2(\mathfrak{H}, (0, b))$. *Then the vector-functions* $\int_0^b G(t,s)h(s)\,ds$ *and* $\int_0^b F(t,s)h(s)\,ds$ *are strongly continuous on* $[0, b]$ *into the space* $\mathfrak{H}^{\frac{3}{2}}$.

Proof. As in subsection 1.1, the proof is a consequence of the fact that integral operators in question are adjoint to the operators appearing in Lemma 6.1. □

REMARK 6.2. Analogously to the case of a constant coefficient we can show that the vector-functions mentioned in Lemma 6.4 are strongly continuously differentiable (in the strong sense) on $[0, b]$ into the space $\mathfrak{H}^{\frac{1}{2}}$.

DEFINITION. Consider the equation

$$(l_k[y])(t) = h(t), \quad t \in [0, b], \tag{6.13}$$

where $h(t)$ is a weakly measurable vector-function taking values in $\mathfrak{H}$. A $\mathfrak{H}$-valued vector-function $y(t)$ that is twice weakly differentiable into the space $\mathfrak{H}^{-2}$ and satisfies the equation

$$(\hat{l}_k[y])(t) = -y''(t) + \hat{A}_k(t)y(t) = h(t) \quad (t \in [0, b]) \tag{6.14}$$

is called a weak solution of (6.13).

LEMMA 6.5. *The domain* $\mathfrak{D}(L_0^*)$ *of the maximal operator* L_0^* *generated by expression* (6.1) *consists precisely of the vector-functions* $y(t)$ *that can be represented in the form*

$$y(t) = -\hat{G}(t,0)f_1 + \hat{G}(t,b)f_2 + \int_0^b G(t,s)h(s)\,ds, \tag{6.15}$$

where $f_1 = -y'(0) \in \mathfrak{H}^{-\frac{1}{2}}$, $f_2 = y'(b) \in \mathfrak{H}^{-\frac{1}{2}}$, $h = (L_0^* + k^2 E)y \in L_2(\mathfrak{H}, (0, b))$.

Proof. It follows from Lemmas 6.1, 6.4 and the method of constructing $G(t,s)$ that any vector-function of the kind (6.15) is a weak solution of equation (6.13), and belongs to $L_2(\mathfrak{H}, (0, b))$. The fact that this function belongs to $\mathfrak{D}(L_0^*)$ and that $(L_0^* + k^2 E)y = h$ is proved in exactly the same way as in lemma 1.4.

Conversely, let $y(t) \in \mathfrak{D}(L_0^*)$, $h = (L_0^* + k^2 E)y \in L_2(\mathfrak{H}, (0, b))$. Then $((L_0 + k^2 E)z, y)_{L_2(\mathfrak{H},(0,b))} = (z, h)_{L_2(\mathfrak{H},(0,b))}$ for an arbitrary $z(t) \in \mathfrak{D}(L_0)$. Substituting

$z(t) = \varphi(t)f$, where $f \in \mathfrak{D}$, $\varphi(t) \in \overset{\circ}{W}{}_2^2(0,b)$, we arrive at the equality

$$\int_0^b \varphi''(t)(f, y(t))\,dt = \int_0^b \varphi(t)(f, \hat{A}_k(t)y(t) - h(t))\,dt.$$

Set

$$w(t) = \int_0^t (t-s)(\hat{A}_k(s)y(s) - h(s))\,ds.$$

The $\mathfrak{H}^{-2}$-valued vector-function $w(t)$ has an absolutely continuous derivative into $\mathfrak{H}^{-2}$, while its second derivative, equal to $\hat{A}_k(t)y(t) - h(t)$, belongs to $L_2(\mathfrak{H}^{-2}, (0,b))$. This means that $\int_0^b \varphi''(t)f, y_1(t))\,dt = 0$ for any $\varphi(t) \in \overset{\circ}{W}{}_2^2(0,b)$, where $y_1(t) = y(t) - w(t) - g_1 - g_2 t$ $(\forall g_1, g_2 \in \mathfrak{H}^{-2})$. Choosing g_1, g_2 so that the vector-function $\psi(t) = \int_0^t (t-s)y_1(s)\,ds$ satisfies the conditions $\psi(b) = \psi'(b) = 0$ (these conditions determine g_1 and g_2 uniquely) and setting $\varphi(t) = (f, \psi(t))$, we obtain $\int_0^b |(y_1(t), f)|^2\,dt = 0$ for any $f \in \mathfrak{D}$. Therefore, $y(t)$ coincides almost everywhere with the vector-function $w(t) + g_1 + g_2 t$, hence it has a strong derivative into $\mathfrak{H}^{-2}$, $y'(t)$ is absolutely continuous into $\mathfrak{H}^{-2}$ and $y''(t) \in L_2(\mathfrak{H}, (0,b))$. This and Theorem 7.2 (Chapter 1) imply $y'(0) = f_1 \in \mathfrak{H}^{-\frac{3}{2}}$, $y'(b) = f_2 \in \mathfrak{H}^{-\frac{3}{2}}$.

Using such vectors f_1, f_2 we construct the vector-function $\tilde{y}(t)$ as in (6.15), and have to show that $u(t) = y(t) - \tilde{y}(t) \equiv 0$. In fact, $u(t) \in \mathfrak{D}(L_0^*)$, $u'(0) = u'(b) = 0$, so $u(t) \in \mathfrak{D}(L_N^* + k^2E) = \mathfrak{D}(L_N + k^2E)$ and $(L_N + k^2E)u = 0$. Since $L_N + k^2E \geq (1+k^2)E$, we have $u(t) \equiv 0$. □

It follows from Lemma 6.5 that Theorem 1.1 and Corollaries 1.1, 1.2 can be literally transfered to the maximal operator generated on $L_2(\mathfrak{H}, (0,b))$ by expression (6.1), because all the arguments used in the process of proving these statements remain valid.

6.3. REGULARIZED BOUNDARY VALUES AND THE GREEN FORMULA

There are three summands in the representation (6.15) for a vector-function $y(t)$ in the domain of the maximal operator. As was proved in subsection 6.2, the third of them is strongly continuous into $\mathfrak{H}^{\frac{3}{2}}$ and strongly continuously differentiable on $[0,b]$ into $\mathfrak{H}^{\frac{1}{2}}$. Next we consider the vector-function $\hat{G}(t,0)f_1$ $(f_1 \in \mathfrak{H}^{-\frac{3}{2}})$. As can be seen from (6.6), it also consists of two terms: $\hat{F}(t,0)f_1$ and $\int_0^b F(t,s)\Phi(s,0)f_1\,ds$. The first of them is infinitely differentiable on $(0,b]$ into $\mathfrak{H}$, its boundary value at 0 belongs to $\mathfrak{H}^{-\frac{1}{2}}$, while the boundary value of its derivative belongs to $\mathfrak{H}^{-\frac{3}{2}}$. As for the second summand, we can only say that because of Lemma 6.1 and Remark 6.1 it is strongly continuous into $\mathfrak{H}^{\frac{1}{2}}$ and strongly continuously differentiable into $\mathfrak{H}^{-\frac{1}{2}}$ on $[0,b]$. So, we know now that the vector-function $\hat{G}(t,0)f_1$ is strongly continuous into $\mathfrak{H}^{\frac{1}{2}}$ and strongly continuously differentiable into $\mathfrak{H}^{-\frac{1}{2}}$ on $(0,b]$; its boundary value at 0 belongs to $\mathfrak{H}^{-\frac{1}{2}}$ while the boundary value of its derivative belongs to $\mathfrak{H}^{-\frac{3}{2}}$.

Further, whatever the positive number ϵ, $\epsilon < b$ would be, the vector-function $\hat{G}(t,0)f_1$ belongs to the domain of the maximal operator $L^*_{0,\epsilon}$ generated by expression (6.1) on the interval $[\epsilon, b]$, and $L^*_{0,\epsilon}(\hat{G}(t,0)f_1) = 0$, $\hat{G}'(b,0)f_1 = 0$. So, if $t \in [\epsilon, b]$, then $\hat{G}(t,0)f_1 = \hat{G}_\epsilon(t,\epsilon)f_\epsilon$, where $G_\epsilon(t,s)$ is the Green function of the Neumann problem for expression (6.1), considered on $[\epsilon, b]$, and $f_\epsilon = \hat{G}'_t(\epsilon,0)f_1 \in \mathfrak{H}^{-\frac{1}{2}}$. Applying once more the arguments used above, which are based on Lemma 6.1 and Remark 6.1, to the vector-function under discussion we find that, since $\epsilon \in (0,b)$ is arbitrary, the vector-function $\hat{G}(t,0)f_1$ is strongly continuous on $(0,b]$ into $\mathfrak{H}^{\frac{3}{2}}$ and strongly continuously differentiable into $\mathfrak{H}^{\frac{1}{2}}$. An analogous assertion is valid for the vector-function $\hat{G}(t,b)f_2$ $(f_2 \in \mathfrak{H}^{-\frac{3}{2}})$ on $[0,b)$, hence for the vector-function $y(t) \in \mathfrak{D}(L^*_0)$ on $(0,b)$. The boundary values $y(0)$ and $y(b)$ belong to $\mathfrak{H}^{-\frac{1}{2}}$ while $y'(0)$ and $y'(b)$ belong to $\mathfrak{H}^{-\frac{3}{2}}$.

Assume that f_0, f_b are arbitrary vectors in $\mathfrak{H}^{-\frac{1}{2}}$, and f'_0, f'_b are vectors in $\mathfrak{H}^{-\frac{3}{2}}$. The question is: are there any vector-functions $y(t), z(t) \in \mathfrak{D}(L^*_0)$ for which

$$y(0) = f_0, \quad y'(0) = f'_0, \quad z(b) = f_b, \quad z'(b) = f'_b. \tag{6.16}$$

The following lemma gives the answer.

LEMMA 6.6. *Let* $f_0, f_b \in \mathfrak{H}^{-\frac{1}{2}}$; $f'_0, f'_b \in \mathfrak{H}^{-\frac{3}{2}}$. *There exist vector-functions* $y(t), z(t) \in \mathfrak{D}(L^*_0)$ *satisfying* (6.16) *if and only if*

$$f_0 + \hat{G}(0,0)f'_0 \in \mathfrak{H}^{\frac{3}{2}}, \quad f_b - \hat{G}(b,b)f'_b \in \mathfrak{H}^{\frac{3}{2}}. \tag{6.17}$$

Proof. We prove this lemma only for the point 0. For the point b the argument is analogous. If we assume that there exists a vector-function $y(t) \in \mathfrak{D}(L^*_0)$ satisfying the two first equalities in (6.16), then the proof of the first condition in (6.17) is similar as in Lemma 1.5; the only difference is that representation (6.15) for $y(t) \in \mathfrak{D}(L^*_0)$ must be used instead of (1.6).

Conversely, let $f_0 \in \mathfrak{H}^{-\frac{1}{2}}$, $f'_0 \in \mathfrak{H}^{-\frac{3}{2}}$ be such that $f_0 + \hat{G}(0,0)f'_0 \in \mathfrak{H}^{\frac{3}{2}}$. We will look for a vector-function $y(t) \in \mathfrak{D}(L^*_0)$ satisfying the conditions $y(0) = f_0$, $y'(0) = f'_0$ in the form $y(t) = -\hat{G}(t,0)f'_0 + \int_0^b G(t,s)h(s)\,ds$. We only have to properly select $h(s) \in L_2(\mathfrak{H},(0,b))$. As can be seen from this representation of $y(t)$, to prove the lemma it suffices to show that it is possible to find, for any $f \in \mathfrak{H}^{\frac{3}{2}}$, a vector-function $h(s) \in L_2(\mathfrak{H},(0,b))$ such that $f = \int_0^b G(0,s)h(s)\,ds$ (in other words, for any $f \in \mathfrak{H}^{\frac{3}{2}}$ to find a vector-function $x(t) \in \mathfrak{D}(L_N)$ for which $x(0) = f$).

On account of Theorem 7.2 (Chapter 1) we can guarantee the existence of a vector-function $u(t) \in W^2_2(\mathfrak{H}; A(t); (0,b))$ satisfying the conditions $u(0) = f$, $u'(0) = 0$. Evidently, $u(t) \in \mathfrak{D}(L^*_0)$. Set $x(t) = \varphi(t)u(t)$, where $\varphi(t)$ is a scalar twice continuously differentiable function which equals 1 in a neighbourhood of zero and vanishes in a neighbourhood of the point b. It is not difficult to verify that the vector-function $x(t)$ is the one desired. □

COROLLARY 6.2. *Let the operators* $A(t)$ *have the property* 4) (*see Remark* 6.1) *and let* $f_0, f_b \in \mathfrak{H}^{-\frac{1}{2}}$; $f_0', f_b' \in \mathfrak{H}^{-\frac{3}{2}}$. *In order that vector-functions* $y(t), z(t) \in \mathfrak{D}(L_0^*)$ *satisfying* (6.16) *exist it is necessary and sufficient that*

$$\sqrt{\hat{A}(0)}\, f_0 + f_0' \in \mathfrak{H}^{\frac{1}{2}}, \quad \sqrt{\hat{A}(b)}\, f_b - f_b' \in \mathfrak{H}^{\frac{1}{2}}. \tag{6.18}$$

This follows from the fact that under condition 4) the vector $\hat{G}(0,0)f_0'$ $(\hat{G}(b,b)f_b')$ can be represented as the sum of $\hat{A}_k^{-\frac{1}{2}}(0)f_0'$ $(\hat{A}_k^{-\frac{1}{2}}(b)f_b')$ and a vector from $\mathfrak{H}^{\frac{3}{2}}$ (see Remark 6.1 and Lemma 6.4). In turn, $\hat{A}_k^{-\frac{1}{2}}(0)f_0' = \hat{A}^{-\frac{1}{2}}(0)f_0' + g_0$, where $g_0 = (\hat{A}_k^{-\frac{1}{2}}(0) - \hat{A}^{-\frac{1}{2}}(0))f_0' \in \mathfrak{H}^{\frac{3}{2}}$. Analogously, $\hat{A}_k^{-\frac{1}{2}}(b)f_b' = \hat{A}^{-\frac{1}{2}}(b)f_b' + g_b$, where $g_b = (\hat{A}_k^{-\frac{1}{2}}(b) - \hat{A}^{-\frac{1}{2}}(b))f_b' \in \mathfrak{H}^{\frac{3}{2}}$. □

We show that if condition 4) does not hold, then Corollary 6.2 is not true, in general, that is, there exists a vector-function $y(t) \in \mathfrak{D}(L_0^*)$ for which $y'(0) + \sqrt{\hat{A}(0)}\, y(0) \notin \mathfrak{H}^{\frac{1}{2}}$. We give an example. Let $A(t) = (1+t^2)A$, where $A \geq E$ is a self-adjoint operator. For any $f \in \mathfrak{H}^{-\frac{1}{2}}$ we set $y(t) = \exp((\frac{1}{2}E - \sqrt{\frac{1}{4}E + A}\,)\ln(t+1))f \in \mathfrak{D}(L_0^*)$. Then $y'(0) + \sqrt{\hat{A}(0)}\, y(0) = \frac{1}{2}f + \sqrt{\hat{A}}\, f - \sqrt{\frac{1}{4}E + \hat{A}}\, f \notin \mathfrak{H}^{\frac{1}{2}}$.

THEOREM 6.1. *For any* $y(t), z(t) \in \mathfrak{D}(L_0^*)$ *formula* (1.12) *with* $a = 0$ *is valid where*

$$\begin{aligned} \overline{y}_0 &= y(0) + \hat{G}(0,0)y'(0), & \overline{y}_0' &= y'(0), \\ \overline{y}_b &= y(b) - \hat{G}(b,b)y'(b), & \overline{y}_b' &= y'(b). \end{aligned} \tag{6.19}$$

Proof. It follows from Corollary 6.1 that the operators $G(0,0)$, $G(b,b)$ are self-adjoint on $\mathfrak{H}$. Therefore, formula (1.12) holds for any pair of vector-functions from $\mathfrak{D}(L')$ ($\mathfrak{D}(L')$ is defined in a similar way as in subsection 1.1). As was noted before, the operator L_0^* is the closure of its restriction to $\mathfrak{D}(L')$. By passing in the indicated formula to the limit with respect to the norm of the graph of the operator L_0^* we have established the validity of (1.12) for arbitrary $y(t), z(t) \in \mathfrak{D}(L_0^*)$. □

COROLLARY 6.3. *If the operators* $A(t)$ *satisfy condition* 4) *then for any* $y(t)$, $z(t) \in \mathfrak{D}(L_0^*)$ *formula* (1.12) *with* $a = 0$,

$$\begin{aligned} &\overline{y}_0 = y(0), \quad \overline{y}_0' = y'(0) + \sqrt{\hat{A}(0)}\, y(0), \\ &\overline{y}_b = y(b), \quad \overline{y}_b' = y'(b) - \sqrt{\hat{A}(b)}\, y(b) \end{aligned} \tag{6.20}$$

holds.

To obtain the proof it is necessary to combine the arguments used in the proof of Corollary 6.2 with those of Theorem 6.1.

Note that all the assertions of subsection 1.3 can be completely transfered to the case considered. To do this we must replace in the corresponding reasoning the representation (1.6) by (6.15) (in this connection it is necessary to properly reformulate Corollary 1.4).

6.4. DESCRIPTION OF MAXIMAL DISSIPATIVE AND ACCUMULATIVE EXTENSIONS OF THE MINIMAL OPERATOR. SPECIAL CLASSES OF EXTENSIONS

Let $y(t) \in \mathfrak{D}(L_0^*)$. The vectors $\overline{y}_0 = y(0) + \hat{G}(0,0)y'(0)$, $\overline{y}_b = y(b) - \hat{G}(b,b)y'(b)$ belong to $\mathfrak{H}^{\frac{3}{4}}$, while $\overline{y}_0' = y'(0) \in \mathfrak{H}^{-\frac{1}{4}}$, $\overline{y}_b' = y'(b) \in \mathfrak{H}^{-\frac{1}{4}}$. Consequently, the vectors

$$Y = \{-y_0, y_b\}, \quad Y' = \{y_0', y_b'\}, \tag{6.21}$$

where $y_0 = A^{\frac{3}{4}}(0)\overline{y}_0$, $y_b = A^{\frac{3}{4}}(b)\overline{y}_b$, $y_0' = \hat{A}^{-\frac{3}{4}}(0)\overline{y}_0'$, $y_b' = \hat{A}^{-\frac{3}{4}}(b)\overline{y}_b'$, are regularized values of $y(t)$ and its derivative on the boundary, belong to the space $\mathfrak{H} \oplus \mathfrak{H}$. By arguments similar to those in subsection 1.4, it can be proved that the triple $(\mathfrak{H} \oplus \mathfrak{H}, \Gamma_1, \Gamma_2)$, where $\Gamma_1 y = Y$, $\Gamma_2 y = Y'$, is a boundary value space in the sense of Chapter 3, subsection 1.4 of the symmetric operator L_0. After this, Theorem 1.4 and Remark 1.1 can be automatically extended to the case of a variable coefficient A. The only difference is that the vectors Y and Y' appearing in these statements must be constructed according to (6.21), and not (1.16). In particular, the operator L_N is associated with $K = -E$ (or, what is the same, $C = 0$).

Corollaries 6.2 and 6.3 show that when the operators $A(t)$ satisfy an additional condition 4) the above-mentioned statements remain valid if the vectors Y and Y' are defined by (6.21), where $y_0 = \hat{A}^{-\frac{1}{4}}(0)y(0)$, $y_b = \hat{A}^{-\frac{1}{4}}(b)y(b)$, $y_0' = \hat{A}^{\frac{1}{4}}(0)(y'(0) + \sqrt{\hat{A}(0)}\, y(0))$, $y_b' = \hat{A}^{\frac{1}{4}}(b)(y'(b) - \sqrt{\hat{A}(b)}\, y(b))$, i.e. just as in the case of a constant operator, discussed in §1.4.

Having described the maximal dissipative and maximal accumulative extensions of the operator L_0 in terms of boundary values, we can use the general results of Chapter 3, subsection 3.2 (on the comparison by resolvents of extensions) and, by choosing the operator L_N as standard, obtain results concerning a description of various specific classes of extensions close to L_N in their properties. Since the path of investigation in this situation is the same, in principle, as in the case of a constant operator coefficient $A(t) \equiv A^2$ (see sections 2, 3) we cite only the assertions whose formulation or proof is different from the corresponding ones in the sections indicated.

On the basis of results from Chapter 1, subsection 3.6 (about properties of integral operators with operator kernels) the following assertion is obtained.

LEMMA 6.7. *If all operators $A^{-1}(t)$ $(t \in [0,b])$ are completely continuous on $\mathfrak{H}$, then the resolvent of the operator L_N is a completely continuous operator on $L_2(\mathfrak{H}, (0,b))$.*

Proof. It follows from the fact that the Green function of the Neumann problem for expression (6.1) is bounded, uniformly in t, s, and takes values in the set of completely continuous operators on $\mathfrak{H}$. □

Theorem 3.1 (Chapter 3) with $K_1 = -E$ and the statement just proved lead to the following result.

THEOREM 6.2. *Let $A^{-1}(t) \in \mathfrak{S}_\infty$ in $[\mathfrak{H}]$ $(t \in [0, b])$. In order that $R_\lambda(L_K) \in \mathfrak{S}_\infty$ in $[L_2(\mathfrak{H}, (0, b))]$ it is necessary and sufficient that $E + K \in \mathfrak{S}_\infty$ in $[\mathfrak{H} \oplus \mathfrak{H}]$. Under the assumption that $R_\lambda(L_N) \in \mathfrak{S}_p$ $(p \geq 1)$ in $[L_2(\mathfrak{H}, (0, b))]$, we have $R_\lambda(L_K) \in \mathfrak{S}_p$ in $[L_2(\mathfrak{H}, (0, b))]$ if and only if $E + K \in \mathfrak{S}_p$ in $[\mathfrak{H} \oplus \mathfrak{H}]$.*

As in subsection 2.2, this theorem implies the following assertion.

COROLLARY 6.4. *If the operators $R_\lambda(L_N)$ and $E + K$ are nuclear, then the system of eigenvectors and adjoined vectors of the operator L_K is complete in the space $L_2(\mathfrak{H}, (0, b))$.*

As in subsection 2.3 we can define an α-smooth maximal dissipative (maximal accumulative) extension of the minimal operator $(-\frac{1}{2} \leq \alpha \leq \frac{3}{2})$. We introduce the following notations:

$$\hat{Y} = \{y(0), y(b)\}, \quad \hat{Y}' = \{-y'(0), y'(b)\},$$
$$\hat{\mathbf{A}}^\alpha = \begin{pmatrix} \hat{A}^{\frac{\alpha}{2}}(0) & 0 \\ 0 & \hat{A}^{\frac{\alpha}{2}}(b) \end{pmatrix}, \quad \mathbf{G} = \begin{pmatrix} \hat{G}(0,0) & 0 \\ 0 & \hat{G}(b,b) \end{pmatrix}. \tag{6.22}$$

The following lemma is an analogue of Lemma 2.2.

LEMMA 6.8. *A vector-function $y(t) \in \mathfrak{D}(L_0^*)$ belongs to the domain $\mathfrak{D}(L_K)$ of the maximal dissipative extension L_K of the operator L_0 if and only if the vectors $\hat{Y}, \hat{Y}'$ have a representation*

$$\hat{Y} = i\hat{\mathbf{A}}^{-\frac{3}{2}}(K - E)F + \mathbf{G}\hat{\mathbf{A}}^{\frac{3}{2}}(K + E)F,$$
$$\hat{Y}' = \hat{\mathbf{A}}^{\frac{3}{2}}(K + E)F \quad (F \in \mathfrak{H} \oplus \mathfrak{H}).$$

This lemma is rather different from Lemma 2.2 in form. The way to prove it is the same if we use the representation (6.15) and formula (6.21) instead of (1.6).

It follows from this lemma that Theorem 2.4 and Corollaries 2.4, 2.5 can be extended to the case under discussion.

The definitions of a maximal dissipative (maximal accumulative) extension with finite quadratic functional and a D-extension are analogous to the corresponding ones in subsection 2.4. Theorems 2.5, 2.6, Corollaries 2.6, 2.7 and Remark 2.1 are repeated word-for-word if we take the relations

$$(K^* - E)(K + E) + i(K^* + E)\hat{\mathbf{A}}^{\frac{3}{2}}\hat{\mathbf{G}}\hat{\mathbf{A}}^{\frac{3}{2}}(K + E) = 0,$$
$$(K^* - E)(K + E) - i(K^* + E)\hat{\mathbf{A}}^{\frac{3}{2}}\hat{\mathbf{G}}\hat{\mathbf{A}}^{\frac{3}{2}}(K + E) = 0$$

and

$$(\cos C)\hat{\mathbf{A}}^{\frac{3}{2}}\hat{\mathbf{G}}\hat{\mathbf{A}}^{\frac{3}{2}} \cos C = \cos C \sin C$$

instead of (2.7) and (2.10).

We give without proof (the proof differs in a non-essential way from the case $A(t) \equiv A^2$) a few results concerning the asymptotic behaviour of eigenvalues of self-adjoint extensions of the operator L_0. The spectrum of the operators $A(t)$ $(0 \leq t \leq b)$ is assumed to be discrete.

THEOREM 6.3. *Let $N(\lambda; L_N) \sim d\lambda^\delta$ $(d, \delta > 0)$. In order that the self-adjoint extension L_C of the minimal operator has a purely discrete spectrum satisfying $N(\lambda; L_C) \sim N(\lambda; L_N)$ it is sufficient that $\cos C \in \mathfrak{S}_\infty$ and $n(\lambda; \cos C) \sim o(\lambda^\delta)$.*

COROLLARY 6.5. *Let $N(\lambda; L_N) \sim d\lambda^{\alpha+\frac{1}{2}}$ $(\alpha > 0)$, $N(\lambda; A(0)) \sim a_1\lambda^\alpha$, $N(\lambda; A(b)) \sim a_2\lambda^\alpha$ $(a_1, a_2 > 0,\ \alpha < \frac{1}{2})$. Then any self-adjoint extension L_C with finite quadratic functional has the property $N(\lambda; L_C) \sim d\lambda^{\alpha+\frac{1}{2}}$. This relation with $\alpha > 0$ is true for any D-extension.*

THEOREM 6.4. *Suppose that $N(\lambda; L_N) \sim d\lambda^{\alpha+\frac{1}{2}}$ $(d > 0, \alpha > 0)$, $N(\lambda; A(0)) \sim a_1\lambda^\alpha$, $N(\lambda; A(b)) \sim a_2\lambda^\alpha$, $(a_1, a_2 > 0)$. The self-adjoint extension L_C of the minimal operator with discrete spectrum satisfies the relation $N(\lambda; L_C) = O(\lambda^\beta)$ $(\beta > \alpha + \frac{1}{2})$ if and only if $n(\lambda; \cos C) = O(\lambda^\beta)$.*

Consider the operator L'_B generated by the expression (6.1) on the set of vector-functions $y(t) \in \mathfrak{D}(L_0^*)$ for which

$$\hat{Y}' + B\hat{Y} = 0, \tag{6.23}$$

where $\hat{Y}, \hat{Y}'$ are defined by formulas (6.22) and B is a continuous operator on $\mathfrak{H} \oplus \mathfrak{H}$. Clearly, $L'_B \supseteq L_0$.

THEOREM 6.5. *If the operator B is dissipative (accumulative, self-adjoint), then the closure L_B of L'_B is a 1-smooth maximal dissipative (maximal accumulative, self-adjoint) extension with finite quadratic functional and lower semi-bounded real part. Provided that the spectrum of $A(t)$ $(0 \le t \le b)$ is discrete, the resolvent $R_\lambda(L_B)$ is completely continuous. If $N(\lambda; L_N) \sim d\lambda^{\alpha+\frac{1}{2}}$ $(d > 0, \alpha > 0)$, $N(\lambda; A(0)) \sim a_1\lambda^\alpha$, $N(\lambda; A(b)) \sim a_2\lambda^\alpha$ $(a_1, a_2 > 0)$, then $N(\lambda; L_B) \sim d\lambda^{\alpha+\frac{1}{2}}$.*

In order to prove this theorem we must express the vectors $\hat{Y}$ and $\hat{Y}'$ in terms of the vectors Y, Y' occurring in (6.21) and apply to the boundary value problem obtained in this way the general results formulated above. The lower semi-boundedness of the real part of L_B is proved just as in the case of a constant coefficient.

Note, finally, that if the operators $A(t)$ satisfy the additional condition 4), then by using the regularized boundary values (6.20) we can eliminate $\hat{\mathrm{G}}$ from all relations, and give these relations a form resembling those in subsection 2.3.

7. Applications to Partial Differential Equations

The purpose of this section is to illustrate the results obtained in this chapter in the case of certain classes of partial differential equations. The examples given show that if we want to describe all self-adjoint extensions of the minimal operator generated by a partial differential equation, then we have to use not only the values of functions and their normal derivatives on the boundary of a domain, which usually appear in the classical problems of mathematical physics, but also the boundary values of their tangential derivatives. As will be seen, even for the Laplace equation in a rectangle the general self-adjoint problems are essentially different in their spectral properties from the classical problems (for instance, from the Dirichlet problem). In particular, self-adjoint problems with spectrum filling the whole real axis are possible in this case. It will be also shown that the operator methods developed in this chapter enable us to study certain boundary value problems for non-elliptic equations, which have not yet been studied thoroughly. Moreover, they allow us to obtain new results even for the classical problems for the Laplace equation in tube domains.

7.1 THE LAPLACE EQUATION IN A STRIP

Let $\mathfrak{H} = L_2(\mathbb{R}^1)$ and let A^2 be the closure on $L_2(\mathbb{R}^1)$ of the operator defined by the differential expression $-\frac{d^2}{dx^2}$ on the set of all infinitely differentiable functions with compact support. It is a positive self-adjoint operator (see, for example, Naimark [2, Chapter VII]). In this case the minimal operator L_0 generated on the space $L_2(\mathfrak{H}, (0,b))$ by the expression $l[y] = -y'' + A^2 y$ coincides with the minimal operator generated on $L_2(\Omega) = L_2(\mathbb{R}^1 \times [0,b])$ by the Laplace expression

$$-\Delta u = -\left(\frac{\partial^2 u}{\partial t^2} + \frac{\partial^2 u}{\partial x^2}\right).$$

Under Fourier transformation the operator A^α becomes multiplication by λ^α. So the space $\mathfrak{H}^j(A)$ coincides with the Sobolev space $W_2^j(\mathbb{R}^1)$.

Since A^2 does not have a bounded inverse, we use for the construction of regularized boundary values the operator $\widetilde{A}^2 = A^2 + E$ instead of A^2. Differentiation with respect to t in the situation considered means differentiation along the normal to the boundary of the domain, while the action of the operator $\widetilde{A}$ coincides, roughly speaking, with differentiation in the tangential direction (because $\mathfrak{D}(\widetilde{A})$ coincides with the domain of the operator of differentiation with respect to x). Hence, the expressions $-y'(b) + \widetilde{A}y(b)$ and $y'(0) + \widetilde{A}y(0)$ represent the sum of the tangential and normal (along the interior normal) derivatives on the boundary of the strip Ω.

According to Theorem 1.4, all maximal dissipative extensions of the minimal operator generated by the expression $-\Delta$ in Ω are described by boundary conditions of the form

$$(K - E)\widetilde{\mathbf{A}}^{-\frac{1}{2}}U + i(K + E)\widetilde{\mathbf{A}}^{\frac{1}{2}}U' = 0, \tag{7.1}$$

where

$$\widetilde{\mathbf{A}}^\alpha = \begin{pmatrix} \widetilde{A}^\alpha & 0 \\ 0 & \widetilde{A}^\alpha \end{pmatrix}, \quad U = \begin{pmatrix} -u(x,0) \\ u(x,b) \end{pmatrix}, \quad U' = \begin{pmatrix} \frac{\partial u(x,0)}{\partial \nu} + \frac{\partial u(x,0)}{\partial \tau} \\ \frac{\partial u(x,b)}{\partial \nu} - \frac{\partial u(x,b)}{\partial \tau} \end{pmatrix}, \tag{7.2}$$

$\widetilde{A} = \frac{\partial}{\partial \tau}$ is differentiation in the tangential direction, and K is a contraction on $L_2(\mathbb{R}^1) \oplus L_2(\mathbb{R}^1)$.

Thus, in contrast to the classical problems of mathematical physics, in the description of maximal dissipative (among them self-adjoint) extensions, in this situation the boundary values of tangential derivatives as well as the boundary values of the functions and their normal derivatives are involved.

In this case the quadratic functional is the well-known Dirichlet integral for the Laplace expression.

7.2. SELF-ADJOINT BOUNDARY VALUE PROBLEMS FOR THE LAPLACE EQUATION IN A TUBE DOMAIN

Let $\Xi = \Omega \times [0, b]$ be a cylinder in $\mathbb{R}^{n+1}$ whose base is a bounded domain Ω in $\mathbb{R}^n$ with piecewise smooth boundary $\partial\Omega$. We take for A^2 the closure on $L_2(\Omega)$ of the operator generated by the expression

$$-\Delta_n = -\Big(\frac{\partial^2}{\partial x_1^2} + \frac{\partial^2}{\partial x_2^2} + \cdots + \frac{\partial^2}{\partial x_n^2}\Big),$$

on the set of infinitely differentiable functions on $\Omega \cup \partial\Omega$ that vanish on $\partial\Omega$. As is well-known (see, for example, Birman, Solomyak [1]), the operator A^2 is self-adjoint and positive definite. Its spectrum is discrete and

$$N(\lambda; A^2) \sim c_1 \lambda^{\frac{n}{2}} \quad (\lambda \to \infty). \tag{7.3}$$

For certain special domains Ω,

$$N(\lambda; A^2) \sim c_1 \lambda^{\frac{n+1}{2}} - c_2 \lambda^{\frac{n-1}{2}} + o(\lambda^{\frac{n-1}{2}}) \tag{7.4}$$

(here $c_1 = \frac{v_n}{(2\pi)^n} \operatorname{mes} \Omega$, $c_2 = \frac{v_{n-1}}{4(2\pi)^{n-1}} \operatorname{mes} \partial\Omega$, v_m is the volume of the unit ball in $\mathbb{R}^m$).

Using the expression

$$l[u] = -\frac{\partial^2 u}{\partial t^2} - \Delta_n u \tag{7.5}$$

we construct on the space $L_2(\Xi)$ the minimal operator L_0 as the closure of the operator $L_0' : u \to l[u]$ on the set of functions $u \in C^\infty(\overline{\Xi})$ satisfying the conditions

$$\begin{aligned} &u\big|_{\partial\Omega \times [0,b]} = 0, \\ &u(x,0) = \frac{\partial u}{\partial t}(x,0) = u(x,b) = \frac{\partial u}{\partial t}(x,b) = 0. \end{aligned} \tag{7.6}$$

The maximal operator L_0^* is the adjoint of L_0. The operators L_0 and L_0^* coincide with the minimal and maximal operators generated on $L_2(L_2(\Omega), (0, b))$ by the operator differential expression $-y'' + A^2 y$.

The deficiency numbers of the operator L_0 are infinite. So, it has self-adjoint extensions on $L_2(\Xi)$. By Theorem 1.4, any such extension is given by condition (7.1) where K is a unitary operator on $L_2(\Omega) \oplus L_2(\Omega)$ and U, U' are defined by (7.2).

It follows from Theorem 2.1, that there is a self-adjoint boundary value problem of the kind (7.1), (7.6) for the expression $-\Delta_{n+1}$ whose spectrum fills the whole real axis. Problems with purely discrete spectrum are associated with the operators

K from the boundary condition (7.1) for which $K+E \in \mathfrak{S}_\infty$ (see Theorem 2.2). In particular, the problems corresponding to the extensions L_D and L_N are contained in this class. The operator L_D is generated by the boundary conditions

$$u(x,0) = u(x,b) = u(x,t)\big|_{t\in[0,b],x\in\partial\Omega} = 0,$$

and L_N by the conditions

$$u_t'(x,0) = u_t'(x,b) = u(x,t)\big|_{t\in[0,b],x\in\partial\Omega} = 0.$$

Corollary 3.1 shows that

$$N(\lambda; L_D) \sim N(\lambda; L_N) \sim \frac{b-1}{2\pi} c_1 B\left(\tfrac{n+2}{2}; \tfrac{1}{2}\right) \lambda^{\frac{n+1}{2}}.$$

If the domain Ω is such that the asymptotic equality (7.4) holds, then Theorem 3.3 implies the relation

$$N(\lambda; L_D) \sim \tilde{c}_1 \lambda^{\frac{n+1}{2}} + \tilde{c}_2 \lambda^{\frac{n}{2}} + o(\lambda^{\frac{n}{2}}),$$
$$N(\lambda; L_N) \sim \tilde{c}_1 \lambda^{\frac{n+1}{2}} + \tilde{c}_2' \lambda^{\frac{n}{2}} + o(\lambda^{\frac{n}{2}}),$$

where the constants $\tilde{c}_1, \tilde{c}_2, \tilde{c}_2'$ can be explicitly expressed in terms of $\operatorname{mes}\Omega$.

Now we consider boundary condidions of the form

$$\frac{\partial u}{\partial \nu}(x,b) = B_1 u(x,b), \quad \frac{\partial u}{\partial \nu}(x,0) = B_2 u(x,0), \tag{7.7}$$

where B_1, B_2 are bounded self-adjoint operators on $L_2(\Omega)$. It follows from Theorem 5.2 that the problem (7.6), (7.7) for the expression (7.5) generates a lower semi-bounded self-adjoint operator L_B on the space $L_2(\Xi)$ with discrete spectrum and finite quadratic functional (the Dirichlet integral). In accordance with Theorem 5.3 the asymptotic relation $\lambda_m(L_B) = \lambda_m(L_N) + O(1)$ holds for the eigenvalues of this problem. Theorem 5.5 shows that the problem

$$\begin{aligned} &u(x,0) = e^{i\alpha_1} u(x,b), \\ &\frac{\partial u(x,0)}{\partial \nu} = e^{i\alpha_2} \frac{\partial u(x,b)}{\partial \nu}, \\ &u\big|_{\partial\Omega\times(0,b)} = 0 \qquad (\alpha_k \in \mathbb{R}^1 (k=1,2)) \end{aligned}$$

for the expression (7.5) generates on $L_2(\Xi)$ a self-adjoint operator $\widetilde{L}$ with discrete spectrum, furthermore

$$N(\lambda; \widetilde{L}) \sim \frac{b-a}{2\pi} c_1 \lambda^{\frac{n+1}{2}}.$$

It should be noted that if it is known in advance that the functions in the domain of a self-adjoint extension L of the minimal operator have a certain smoothness, $\mathfrak{D}(L) \subset W_2^s(\Xi)$ for example, then by virtue of Theorem 4.3 it is possible to state that the spectrum of the extension L is discrete and $N(\lambda; L) \sim N(\lambda; L_D)$ when $\frac{2n}{n+1} < s \leq 2$.

We also stress that the class of self-adjoint boundary value problems for the Laplace operator in a cylinder can be considerably enlarged if one takes for A^2 the operator generated by the Laplace expression in Ω and a self-adjoint condition on $\partial\Omega$ different from the Diriclet condition (for example, the Neumann condition or third boundary value problem). It is possible to obtain in this way a wide class of self-adjoint boundary problems which can be studied by the method indicated.

The scheme for studying spectral properties of boundary value problems for the Laplace expression in a cylinder presented above admits a generalization to partial differential equations of the form

$$(l[u])(x,t) = -\frac{\partial^2 u(x,t)}{\partial t^2} + \sum_{|\alpha|\le 2m} a_\alpha(x,t) D_x^\alpha u(x,t) + q(x,t)u(x,t) \tag{7.8}$$
$$(x \in \Omega,\ t \in [0,b]),$$

where $\alpha = \{\alpha_1, \ldots, \alpha_n\}$ is a vector of integers; $D_x^\alpha = D_{x_1}^{\alpha_1} \cdots D_{x_n}^{\alpha_n}$, $|\alpha| = \alpha_1 + \cdots + \alpha_n$; $D_{x_j}^{\alpha_i} = \frac{\partial^{\alpha_i}}{\partial x_j^{\alpha_i}}$; $a_\alpha(x,t)$ is a function that is twice continuously differentiable with respect to $t \in [0,b]$ and $|\alpha|$ times differentiable with respect to $x \in \Omega \cup \partial\Omega$; and $q(x,t)$ is a real-valued function, measurable and bounded on Ξ. In addition, the expression

$$M = \sum_{|\alpha|\le 2m} a_\alpha(x,t) D_x^\alpha$$

must be formally self-adjoint and strongly elliptic (see, for instance, Berezansky [4, Chapters III, VII]). The Dirichlet condition

$$\left.\frac{\partial^j u}{\partial \nu_j}\right|_{\partial\Omega\times[0,b]} = 0 \quad (j = 0, 1, \ldots, m-1) \tag{7.9}$$

(ν is the interior normal) or another self-adjoint condition is given on the side surface of the cylinder.

The minimal operator is constructed from the expression (7.8) and conditions (7.9). Its self-adjoint extensions are described by boundary conditions of the form (7.1) on the bases of the cylinder.

If the domain Ω is bounded and its boundary is piecewise smooth, then the operator A^2 generated on $L_2(\Omega)$ by the expression M and the Dirichlet boundary condition on $\partial\Omega$ is positive self-adjoint with discrete spectrum. The eigenvalue distribution function $N(\lambda; A^2) \sim c\lambda^{\frac{n}{2m}}$, where the constant c can be explicitly expressed in terms of the measure of Ω (see Birman, Solomyak [1]). Just as for the Laplace expression it is possible in this case to consider variations of the problem (7.8), (7.9), (7.1), depending on the choice of a unitary operator K, and investigate their spectral properties. For example, the spectrum of the problem (7.8), (7.9), (7.7) is discrete and its eigenvalues satisfy the asymptotic relation

$$\lambda_k \sim c'^{-\frac{2m}{M+n}} k^{\frac{2m}{m+n}},$$

where $c' = \frac{b}{2\pi} c B(\frac{n}{2m} + 1; \frac{1}{2})$.

Concluding this section we would like to note that the expression (7.8) is not elliptic if $m > 1$.

CHAPTER 5

Boundary Values of Solutions of Differential Equations in a Banach Space

1. Semigroups of Operators and their Generators

This section is devoted to a brief presentation (mainly without proofs) of some results in the theory of semigroups needed to investigate solutions of operator differential equations in a Banach space.

1.1. CONTINUOUS SEMIGROUPS

Let $\mathfrak{F}$ be a locally convex Hausdorff space. A one-parameter family $U(t)$ $(t \geq 0)$ of continuous linear operators from $\mathfrak{F}$ into $\mathfrak{F}$ is called a semigroup of continuous linear operators on $\mathfrak{F}$ if

(i) $U(0) = E$;
(ii) $U(t+s) = U(t)U(s)$ for every $t, s > 0$ (the semigroup property).

A semigroup $U(t)$ of continuous linear operators is called strongly continuous at the point $t_0 \geq 0$ if

$$\lim_{t \to t_0} (U(t)x - U(t_0)x) = 0 \quad (\forall x \in \mathfrak{F}).$$

It is not difficult to show that the strong continuity of $U(t)$ at the point t_0 implies its strong continuity at any $t \geq t_0$. In the sequel we consider only semigroups which are strongly continuous for all $t > 0$. If a semigroup is strongly continuous at the point 0, then it is said to be of class C_0.

A semigroup of class C_0 (a C_0-semigroup) is called equi-continuous if for any continuous semi-norm $p(x)$ on $\mathfrak{F}$ there exists a continuous semi-norm $q(x)$ on $\mathfrak{F}$ such that $p(U(t)x) \leq q(x)$ $(\forall t \geq 0,\ \forall x \in \mathfrak{F})$.

We denote by $\mathfrak{D}(A_0)$ the set of all elements $x \in \mathfrak{F}$ for which

$$\lim_{t \to 0} \frac{x - U(t)x}{t} \stackrel{\text{def}}{=} A_0 x \tag{1.1}$$

exists.

The linear operator A_0 defined on $\mathfrak{D}(A_0)$ by formula (1.1) is the infinitesimal operator of the semigroup $U(t)$. If A_0 is closeable, then $\overline{A}_0$ is called the generating operator or simply the generator of $U(t)$. The fact that A is the generator of a semigroup $U(t)$ is written as $U(t) = e^{-At}$.

As a rule we will deal with semigroups of bounded linear operators on Banach spaces. For any C_0-semigroup of bounded linear operators on a Banach space $\mathfrak{B}$ the limit

$$\lim_{t\to\infty} \frac{\ln \|U(t)\|}{t} = \omega_0 < \infty \tag{1.2}$$

exists; it is called the type of this semigroup. If A_0 is the infinitesimal operator of a semigroup $U(t)$, then, as can be seen from (1.1), the vector-function $U(t)x$ $(x \in \mathfrak{D}(A_0))$ satisfies the differential equation

$$\frac{dy(t)}{dt} + A_0 y(t) = 0, \quad t \geq 0.$$

If the generating operator $A = \overline{A}_0$ exists, then $U(t)x \in \mathfrak{D}(A)$ $(\forall x \in \mathfrak{D}(A))$ and

$$\frac{dU(t)x}{dt} = -AU(t)x = -U(t)Ax \quad (t > 0).$$

Let $U(t)$ be a C_0-semigroup of bounded linear operators on $\mathfrak{B}$ and A_0 be its infinitesimal operator. Consider also the operator

$$\widetilde{A}_0 x = \operatorname*{w\text{-}lim}_{t\to 0} \frac{x - U(t)x}{t}$$

(w-lim stands for the weak limit), defined on the set of all x for which this limit exists. Since strong convergence implies weak convergence, $\widetilde{A}_0$ is an extension of A_0. It turns out that this extension is not proper. Actually, $\widetilde{A}_0 = A_0$. Moreover, the following theorem is valid.

THEOREM 1.1 (Yosida). *Let $U(t)$ be a semigroup of bounded linear operators on a Banach space $\mathfrak{B}$. If the operator-function $U(t)$ is weakly continuous at zero, then $U(t)$ is a C_0-semigroup.*

For a C_0-semigroup on $\mathfrak{B}$ the operator $A_0 = A$ is closed and $\overline{\mathfrak{D}(A)} = \mathfrak{B}$. The resolvent set of the operator $-A$ contains the half-plane $\operatorname{Re}\lambda > \omega_0$ and

$$\begin{aligned} &(A + \lambda E)^{-1}x = \int_0^\infty e^{-\lambda t} U(t)x \, dt, \\ &\lim_{\operatorname{Re}\lambda\to\infty} \lambda(A + \lambda E)^{-1}x = x \quad (x \in \mathfrak{B}) \end{aligned} \tag{1.3}$$

for all λ in this half-plane.

If $U(t)$ and $V(t)$ are C_0-semigroups of bounded linear operators with generators A and B, respectively, and if $A = B$, then $U(t) = V(t)$ $(t \geq 0)$.

The following assertion allows us to answer the very important question: when does an operator A generate a C_0-semigroup?

THEOREM 1.2 (Hille–Phillips–Yosida). *An (unbounded) linear operator A on a Banach space $\mathfrak{B}$ is the generator of a C_0-semigroup if and only if:*

(i) *A is closed and $\overline{\mathfrak{D}(A)} = \mathfrak{B}$;*

(ii) *There exist real numbers $M > 0$ and ω such that for every real $\lambda > \omega$,*

$$\lambda \in \rho(-A) \quad \text{and} \quad \|(A + \lambda E)^{-n}\| \leq \frac{M}{(\lambda - \omega)^n} \quad (n = 1, 2, \ldots).$$

If $U(t)$ is of class C_0, then for any $\lambda > \omega > \omega_0$, $\widetilde{U}(t) = e^{-\lambda t}U(t)$ is also a C_0-semigroup, and its generator is $\widetilde{A} = A + \lambda E$. Moreover, it follows from (1.2) that

$$\|\widetilde{U}(t)\| \leq ce^{-\lambda t}e^{\omega t} \leq c.$$

A semigroup satisfying the condition

$$\|U(t)\| \leq c < \infty \quad (c = \text{const})$$

is called equi-bounded, or simply bounded. For such a semigroup $\omega_0 \leq 0$, so the resolvent of the operator $-A$ exists for any λ with $\operatorname{Re}\lambda > 0$. In addition, in this case

$$\|(A + \lambda E)^{-n}\| \leq \frac{c}{\lambda^n} \quad (\lambda > 0).$$

If a semigroup $U(t)$ is bounded, then it is possible to introduce in $\mathfrak{B}$ an equivalent norm, for example

$$\|x\|_1 = \sup_{0 \leq t < \infty} \|U(t)x\|,$$

so that the operator $U(t)$ has (new) norm not exceeding one. A semigroup with the property $\|U(t)\| \leq 1$ ($\forall t \geq 0$) is called a contraction semigroup. Theorem 1.2 enables us to conclude that a densely defined closed operator on $\mathfrak{B}$ generates a contraction C_0-semigroup if and only if

$$(0, \infty) \subset \rho(-A) \quad \text{and} \quad \|(A + \lambda E)^{-1}\| \leq \frac{1}{\lambda} \quad (\lambda > 0).$$

Let now $\mathfrak{B} = \mathfrak{H}$ be a Hilbert space. A closed linear operator A on $\mathfrak{H}$ with $\overline{\mathfrak{D}(A)} = \mathfrak{H}$ is called accretive if

$$\operatorname{Re}(Ax, x) \geq 0 \quad (\forall x \in \mathfrak{D}(A)).$$

Suppose that A is the generator of a contraction C_0-semigroup $U(t)$ and $x \in \mathfrak{D}(A)$. Then the function $h(t) = (U(t)x, U(t)x)$ is differentiable and $h'(t) = -(U(t)Ax, U(t)x) - (U(t)x, U(t)Ax)$. But, $h(t) \leq h(0)$, hence $h'(0) = -2\operatorname{Re}(Ax, x) \leq$ ≤ 0. Thus, the operator A is accretive. Inverting this argument, it is possible to show that if an accretive operator generates a C_0-semigroup, then this semigroup has to be a contraction C_0-semigroup. Taking into account that if $U(t)$ is a contraction C_0-semigroup, then the C_0-semigroup $U^*(t)$ generated by the operator A^* is a contraction semigroup, we can establish that the operator A^* is also accretive.

Of great importance is the question: when does an accretive operator generate a contraction C_0-semigroup of operators on $\mathfrak{H}$. The answer to this question is provided by the following theorem.

THEOREM 1.3 (Phillips). *Let A be an accretive operator on $\mathfrak{H}$ and suppose that the set $\mathfrak{R}(A+E)$ of values of the operator $A+E$ coincides with the whole space. Then A is the generator of a contraction C_0-semigroup.*

Generally speaking, A being accretive need not imply that A^* is accretive. Nevertheless, the following theorem holds.

THEOREM 1.4 (Lyance). *Let A be a densely defined closed linear operator on $\mathfrak{H}$. Suppose that both operators A and A^* are accretive. Then A and A^* generate contraction C_0-semigroups on $\mathfrak{H}$.*

Let us return to a Banach space. Let A be the generator of a C_0-semigroup $U(t)$ of bounded linear operators on $\mathfrak{B}$. If $\mathfrak{B}$ is reflexive and $U(t)$ is bounded, then the decomposition

$$\mathfrak{B} = \operatorname{Ker} A \dotplus \overline{\mathfrak{R}(A)} \tag{1.4}$$

holds. We denote by P the projector onto the subspace $\operatorname{Ker} A$ in this decomposition. The following statement is true.

THEOREM 1.5. *If $U(t)$ is a bounded C_0-semigroup of linear operators on a reflexive Banach space $\mathfrak{B}$, then for any $x \in \mathfrak{B}$,*

$$\lim_{t\to\infty} \tfrac{1}{t} \int_0^t U(s)x\, ds = Px. \tag{1.5}$$

Proof. Let $\mathfrak{B}_0'$ be the set of elements $x \in \mathfrak{B}$ for which $\lim_{t\to\infty} \frac{1}{t}\int_0^t U(s)x\, ds$ exists and let $\mathfrak{B}_0 = \overline{\mathfrak{B}_0'}$. We define operators B_t by $B_t x = \frac{1}{t}\int_0^t U(s)x\, ds$, $x \in \mathfrak{B}_0'$. Since the semigroup $U(t)$ is bounded, the family B_t is equi-bounded: $\|B_t\| \le c$. According to the Banach–Steinhaus theorem, this family converges strongly on the set $\mathfrak{B}_0$. It is obvious that $\operatorname{Ker} A \subset \mathfrak{B}_0$ and $B_t x = x$ for any $x \in \operatorname{Ker} A$, any $t > 0$. If $x \in \mathfrak{R}(A)$, i.e. $x = Ay$ $(y \in \mathfrak{D}(A))$, then $B_t x = \frac{1}{t}\int_0^t U(s)Ay\, ds = -\frac{1}{t}\int_0^t U'(s)y\, ds = -\frac{U(t)-U(0)}{t} y \to 0$ as $t \to \infty$. Therefore, $\overline{\mathfrak{R}(A)} \subset \mathfrak{B}_0$ and $\lim_{t\to\infty} B_t x = 0$ when $x \in \overline{\mathfrak{R}(A)})$. This implies that

$$\lim_{t\to\infty} B_t x = \lim_{t\to\infty} B_t(Px + (E-P)x) = Px \quad (\forall x \in \mathfrak{B}). \qquad \square$$

1.2. HOLOMORPHIC SEMIGROUPS

Let $\mathfrak{B}$ be a Banach space. A semigroup $U(t)$ of bounded linear operators on it is called (strongly) differentiable if for any $x \in \mathfrak{B}$ the function $U(t)x$ is strongly differentiable with respect to t for all $t > 0$. The following theorem characterizes differentiability of semigroups.

THEOREM 1.6. *Assume that the semigroup $U(t)$ is strongly differentiable and A_0 is its infinitesimal operator. Then*

$$U(t)x \in \mathfrak{B}^\infty(A) \quad (\forall x \in \mathfrak{B},\ \forall t > 0),$$

the function $U(t)x$ is infinitely strongly differentiable and

$$\frac{d^n U(t)x}{dt^n} = (-1)^n A^n U(t)x.$$

Let $\theta \in (0, \frac{\pi}{2}]$. A semigroup $U(t)$ of bounded linear operators on the space $\mathfrak{B}$ that is strongly continuous on $(0, \infty)$ is called holomorphic (analytic) with angle θ if it is defined for all z from the sector $S_\theta = \{z : |\arg z| < \theta\}$ and

(i) $U(z_1 + z_2) = U(z_1)U(z_2) \quad (\forall z_1, z_2 \in S_\theta)$;
(ii) for any $x \in \mathfrak{B}$ the function $U(z)x$ is analytic in the sector S_θ;
(iii) for each $x \in \mathfrak{B}$

$$\|U(z)x - x\| \to 0 \quad (z \to 0 \text{ in any closed subsector of } S_\theta).$$

If in addition to these properties the family $U(z)$ is equi-bounded in the sector S_{θ_1} for an arbitrary $\theta_1 < \theta$, then the semigroup $U(t)$ is called a bounded holomorphic semigroup with angle θ. We formulate certain defining properties for such semigroups.

THEOREM 1.7. *A densely defined closed operator A on a Banach space $\mathfrak{B}$ is the generator of a bounded holomorphic semigroup $U(t) = e^{-At}$ with angle $\theta \leq \frac{\pi}{2}$ if and only if*

$$\sigma(A) \subset \overline{S_{\frac{\pi}{2}-\theta}} = \{z : |\arg z| \leq \tfrac{\pi}{2} - \theta\}$$

and if for any $\theta_1 < \theta$, any $z \in \mathbb{C}^1 \backslash S_{\frac{\pi}{2}-\theta_1}$

$$\|(A + zE)^{-1}\| \leq \frac{M_1}{\operatorname{dist}(z, \overline{S_{\frac{\pi}{2}-\theta_1}})}, \tag{1.6}$$

where M_1 depends only on θ_1.

THEOREM 1.8. *Let $U(t)$ be a C_0-semigroup of bounded linear operators on a Banach space $\mathfrak{B}$ and A be its generator: $U(t) = e^{-At}$. The semigroup $U(t)$ is bounded holomorphic with angle θ if and only if it is differentiable and there exists a positive constant c such that*

$$\|A^n e^{-At}\| \leq c^n n^n t^{-n} \tag{1.7}$$

for any $t > 0$, any $n \in \mathbb{N}$ $(\theta = \operatorname{arctg}(ce^{-1}))$.

Note that condition (1.7) is equivalent to the estimate

$$\|Ae^{-At}\| \leq ct^{-1} \quad (\forall t > 0).$$

If $U(t)$ is a holomorphic semigroup with angle θ, then for any $\theta_1 < \theta$ and any $x \in \mathfrak{B}$ the set $\{\|U(z)x\|\}$ is bounded in

$$\widetilde{R}_{1,\theta_1} = \{z : |\arg z| \leq \theta_1,\ |z| \leq 1\};$$

so according to the uniform boundedness principle the function $\|U(z)\|$ is bounded on $\widetilde{R}_{1,\theta_1}$. The semigroup property permits us to conclude that

$$\|U(z)\| \le Me^{\widetilde{\omega}|z|},$$

where $M, \widetilde{\omega} > 0$ are certain constants and $z \in \overline{S}_{\theta_1}$ is arbitrary. Hence, there exists $\omega > 0$ such that $e^{-\omega z}U(z)$ is a bounded holomorphic semigroup with angle θ_1. This is valid for all $\theta_1 < \theta$; however, in general ω depends on θ_1. The following theorem is a consequence of Theorems 1.7, 1.8.

THEOREM 1.9. *A densely defined closed operator A on a Banach space $\mathfrak{B}$ generates a holomorphic semigroup with angle θ if and only if for any $\theta_1 < \theta$ there exist constants $M, \omega > 0$ such that $\lambda \notin \overline{S_{\frac{\pi}{2}-\theta_1}}$ implies $\lambda - \omega \in \rho(A)$ and also*

$$\|(A - (\lambda - \omega)E)^{-1}\| \le \frac{M}{\operatorname{dist}(\lambda, \overline{S_{\frac{\pi}{2}-\theta_1}})}.$$

Furthermore e^{-At} is infinitely differentiable $(t > 0)$ in the uniform operator topology and there exist constants $M_0, \omega_0 > 0$ such that

$$\|A^m e^{-At}\| \le M_0 m^m e^{\omega_0 t} t^{-m}.$$

The following theorem provides a sufficient condition for stability of the analyticity property of a semigroup under perturbations of its generator.

THEOREM 1.10. *Assume that A is the generator of a holomorphic semigroup with angle θ on a Banach space $\mathfrak{B}$. Let B be a linear operator for which*

a) $\mathfrak{D}(B) \supset \mathfrak{D}(A)$;

b) *there exists $b > 0$ such that*

$$\|Bx\| \le a\|Ax\| + b\|x\| \quad (\forall x \in \mathfrak{D}(A))$$

for all $a > 0$. Then $A + B$ generates a holomorphic semigroup on $\mathfrak{D}(A)$ with angle θ.

Like the case of self-adjoint operators on a Hilbert space, for generators of contraction semigroups a theorem concerning their essential domains is valid.

THEOREM 1.11. *Let A be the generator of a contraction C_0-semigroup on a Banach space $\mathfrak{B}$, let $\mathfrak{D} \subset \mathfrak{D}(A)$ be a dense set that is invariant with respect to the operators e^{-At} $(\forall t > 0)$. Then $\mathfrak{D}$ is an essential domain of the operator A i.e. $\overline{A \restriction \mathfrak{D}} = A$.*

2. Functions of the Generator of a Contraction C_0-semigroup

As can be seen from Chapter 2, the investigation of differential equations in a Hilbert space conducted there was based on an operational calculus for self-adjoint operators. The main purpose of this section is to discuss some new results concerning an operational calculus for the generators of continuous semigroups on a Banach space. It should be noted that the class of functions for which it is constructed is considerably narrower than in the case discussed before. It is the class of so-called absolutely concave functions, whose properties we also briefly dwell on.

2.1. ABSOLUTELY CONCAVE FUNCTIONS

DEFINITION. A function of the form

$$\varphi(x) = a + \int_0^\infty \frac{x}{xs+1}\, d\sigma(s), \quad x \geq 0, \quad a = \varphi(0_+) \geq 0, \tag{2.1}$$

where σ is a non-negative measure on $[0,\infty)$ satisfying

$$\int_0^\infty \frac{d\sigma(s)}{1+s} < \infty,$$

is called absolutely concave.

It can be seen from the definition that an absolutely concave function is continuous at zero and has an analytic continuation to the domain $\mathbb{C}^1 \backslash (-\infty, 0]$. Moreover, it maps the right half-plane $\operatorname{Re} z \geq 0$ into intself, and in this half-plane $|\arg \varphi(z)| \leq |\arg z|$ (we assume $-\pi < \arg z < \pi$).

Of course, if $\varphi(x)$ is absolutely concave, then $\psi(z) = \varphi(-\frac{1}{z})$ can be represented as

$$\psi(z) = a + \int_0^\infty \frac{d\sigma(s)}{s-z}. \tag{2.2}$$

Conversely, if $\psi(z)$ is a function of the kind (2.2), then $\varphi(x) = \psi(-\frac{1}{x})$ is absolutely concave.

The class S of functions of the form of (2.2) was introduced and studied by M. Krein. A function $\psi(z)$ belongs to this class if and only if:

(i) $\psi(z)$ is holomorphic in the upper half-plane and $\operatorname{Im} \psi(z) \geq 0$ for $\operatorname{Im} z > 0$;
(ii) $\psi(z)$ is continuous and positive on the negative semi-axis $(-\infty, 0)$.

This implies that a function $\varphi(x)$ is absolutely concave if and only if it is continuous and positive on $(0,\infty)$ and has an analytic continuation into the upper half-plane $\operatorname{Im} z > 0$ such that $\operatorname{Im} \varphi(z) \geq 0$.

We will be interested only in those absolutely concave functions $\varphi(x)$ that satisfy the conditions

$$\lim_{x\to\infty} \frac{\varphi(x)}{x} = 0, \qquad \varphi(0) = 0.$$

We denote by S_0 the set of such functions. It is more convenient to use another integral representation for a function in S_0, namely

$$\varphi(x) = \int_0^\infty \frac{x}{x+s}\, d\sigma(s). \tag{2.3}$$

Here σ is a non-negative measure for which $\int_0^\infty \frac{d\sigma(s)}{1+s} < \infty$. This representation can be derived from (2.1) by replacing $d\sigma(s)$ by $s\,d\sigma(s)$ and subsequently s by s^{-1}. The part of the function corresponding to $\sigma\{0\}x$ disappears under these manipulations.

We cite a few properties of functions in S_0 which will be used later on.

THEOREM 2.1. *If $\varphi(x) \in S_0$, then*

$$\varphi^n(x) = \int_0^\infty \frac{x^n}{(x+s)^n}\, d\sigma_n(s), \quad \int_0^\infty \frac{d\sigma_n(s)}{(1+s)^n} < \infty,$$

where the measure σ_n is non-negative.

The proof of this theorem can be found in Hirschman, Widder [1, p. 194].

LEMMA 2.1. *For any $\varphi(x) \in S_0$, $\mu > 0$ and any natural n,*

$$\int_0^\infty e^{-\mu\varphi(s)}\, d\sigma_n(s) \le \frac{2^n n!}{\mu^n}. \tag{2.4}$$

Proof. Since $\varphi(x)$ $(x > 0)$ is positive and increases monotonically, there the inverse of $\varphi(x)$ exists; it is the function $g(x)$ defined by $g(\varphi(x)) = x$, and it has the same properties. Substituting $g(x)$ for x in the equality

$$e^{-\mu\varphi(x)}\varphi^n(x) = \int_0^\infty \frac{x^n e^{-\mu\varphi(x)}}{(x+s)^n}\, d\sigma_n(s)$$

we obtain

$$e^{-\mu x}x^n = \int_0^\infty \left(\frac{g(x)}{g(x)+s}\right)^n e^{-\mu x}\, d\sigma_n(s).$$

Integrating both sides of this equality with respect to x we deduce that

$$\begin{aligned} n!\mu^{-(n+1)} &= \int_0^\infty \left(\int_0^\infty \left(\frac{g(x)}{g(x)+s}\right)^n e^{-\mu x}\, dx\right) d\sigma_n(s) \\ &\ge \int_0^\infty \left(\int_{\varphi(s)}^\infty \left(\frac{g(x)}{g(x)+s}\right)^n e^{-\mu x}\, dx\right) d\sigma_n(s) \\ &\ge \frac{1}{2^n\mu}\int_0^\infty e^{-\mu\varphi(s)}\, d\sigma_n(s), \end{aligned}$$

whence (2.4) follows. □

DEFINITION. An infinitely differentiable function $f(x)$ $(x > 0)$ is called absolutely monotone if

$$(-1)^n f^{(n)}(x) \ge 0 \quad (\forall n \ge 0).$$

THEOREM 2.2 (Bernstein). *A function $f(x)$ $(x \ge 0)$ has a representation*

$$f(x) = \int_0^\infty e^{-x\lambda}\, d\sigma(\lambda) \tag{2.5}$$

with a finite non-negative measure σ if and only if $f(x)$ is absolutely monotone on $(0,\infty)$ and $f(0) = f(0_+) < \infty$. The measure is determined uniquely.

The proof of this theorem can be found in, for example, Berezansky [4, section 4, Chapter 8].

THEOREM 2.3. *If $\varphi \in S_0$, then*

$$e^{-t\varphi(x)} = \int_0^\infty e^{-x\lambda}\, d\mu_t(\lambda) \quad (t \geq 0), \tag{2.6}$$

where μ_t is a convolution family of probability measures on $[0, \infty)$.

Recall that a family of measures μ_t with $\int_0^\infty d\mu_t(\lambda) = 1$ is called a convolution family if $\mu_0 = \delta_0$ (δ_0 is the Dirac δ-function), $\mu_{t_1} * \mu_{t_2} = \mu_{t_1+t_2}$ ($t_1, t_2 \geq 0$, $*$ is a symbol of convolution), and $\mu_t \to \delta_0$ weakly as $t \to 0_+$.

Proof. We verify that the function $e^{-t\varphi(x)}$ is absolutely monotone with respect to x. It follows from the representation (2.3) for $\varphi(x)$ that

$$\varphi^{(n)}(x) = (-1)^{n-1} n! \int_0^\infty \frac{s}{(x+s)^{n+1}}\, d\sigma(s)$$

for any natural n. Therefore, $(-1)^{n-1}\varphi^{(n)}(x) \geq 0$. By induction it follows that

$$\left(e^{f(x)}\right)^{(n)} = P_n(f^{(n)}, \ldots, f')e^{f(x)},$$

where P_n is the sum of products of the form of $a\left(f^{(n)}\right)^{k_n} \cdots \left(f'\right)^{k_1}$, $a \geq 0$, $nk_n + \cdots + 2k_2 + k_1 = n$. This formula and our knowledge of the sign of $\varphi^{(n)}(x)$ imply the inequalities for $\left(e^{-t\varphi(x)}\right)^{(n)}$. On the basis of Theorem 2.2 we arrive at the representation (2.6).

The fact that $\mu_0 = \delta_0$ and $\int_0^\infty d\mu_t(\lambda) = 1$ is obvious. The uniqueness of the measure σ in (2.5) implies that $\mu_{t_1} * \mu_{t_2} = \mu_{t_1+t_2}$ ($t_1, t_2 \geq 0$). It remains to show that $\mu_t \to \delta_0$ weakly as $t \to 0_+$. Since the function $e^{-t\varphi(z)}$ is analytic in the half-plane $\operatorname{Re} z > 0$ and continuous for $\operatorname{Re} z \geq 0$, and since the function $f(x) = \int_0^\infty e^{-x\lambda}\, d\mu_t(\lambda)$ ($t \geq 0$ is arbitrary but fixed) has the same properties, the equality $e^{-t\varphi(z)} = \int_0^\infty e^{-z\lambda}\, d\mu_t(\lambda)$ is true for any z with $\operatorname{Re} z \geq 0$. This equality leads to the relation

$$\int_0^\infty e^{ix\lambda}\, d\mu_t(\lambda) = e^{-t\varphi(ix)} \to 1 \quad \text{as} \quad t \to 0_+.$$

Now the weak convergence $\mu_t \to \delta_0$ ($t \to 0_+$) is a direct consequence of the continuity theorem for the characteristic function (see, for example, Billingsley [1, section 1.7]). □

Let $\varphi \in S_0$. Set

$$k(t) = \max_{x \geq 0}(\varphi(x) - tx) \quad (t > 0).$$

The maximum exists, in view of the continuity of $\varphi(x) - tx$ on $[0, \infty)$ and the fact that $\varphi(x) - tx \to -\infty$ as $x \to \infty$ ($t > 0$). The function $k(t)$ does not increase. Since $\varphi(x) - tx$ vanishes at $x = 0$, the function $k(t)$ is non-negative. It is strongly positive at those t for which there exists an $x_t > 0$ such that $\varphi'(x_t) = t$. This follows from the monotone decrease of the function $\varphi'(x) - t$. Note that also

$$ak\left(\tfrac{t}{a}\right) = \max_{x \geq 0}(a\varphi(x) - tx) \quad (\forall\, a > 0). \tag{2.7}$$

LEMMA 2.2. *Let $\varphi \in S_0$. Then for any $s > a > 0$,*

$$\int_0^\infty e^{ak(\frac{t}{a})}\, d\mu_s(t) \leq c\left(1 + \sqrt{\frac{a}{s-a}}\right)^2 \quad (c = \text{const}). \tag{2.8}$$

Proof. Assume that $s = (1+\epsilon)a$. As before we denote by g the function inverse to φ. By virtue of (2.6),

$$e^{-(1+\epsilon)a\varphi(x)} = \int_0^\infty e^{-\lambda x}\, d\mu_{(1+\epsilon)a}(\lambda),$$

whence

$$e^{-\epsilon x} = \int_0^\infty e^{x - \lambda g(\frac{x}{a})}\, d\mu_{(1+\epsilon)a}(\lambda). \tag{2.9}$$

Suppose first that $k(\lambda) \to 0$ as $\lambda \to \infty$. Take $b > 1$. Since $k(\lambda)$ does not increase, we can find a number $\lambda_0 > 0$ such that

$$ak\left(\tfrac{\lambda}{a}\right) \leq \ln b$$

for all $\lambda \geq \lambda_0$. Then

$$\int_{\lambda_0}^\infty e^{ak(\frac{\lambda}{a})}\, d\mu_{(1+\epsilon)a}(\lambda) \leq e^{\ln b}\int_0^\infty d\mu_{(1+\epsilon)a} = b.$$

We integrate (2.9) with respect to x and obtain

$$\epsilon^{-1} = \int_0^\infty \left(\int_0^\infty \exp\left(x - \lambda g\left(\tfrac{x}{a}\right)\right) dx\right) d\mu_{(1+\epsilon)a}(\lambda).$$

This implies that

$$\epsilon^{-1} \geq \int_0^{\lambda_0} \left(\int_0^\infty \exp\left(x - \lambda g\left(\tfrac{x}{a}\right)\right) dx\right) d\mu_{(1+\epsilon)a}(\lambda).$$

If $\lambda \leq \lambda_0$, then

$$\ln b \leq ak\left(\tfrac{\lambda}{a}\right) = x_\lambda - \lambda g\left(\tfrac{x_\lambda}{a}\right),$$

where x_λ is the point at which the function (2.7) attains its maximum. Therefore $x_\lambda \geq \ln b$. Next, because g is monotone, we have for $x \in [0, x_\lambda]$,

$$x - \lambda g\left(\tfrac{x}{a}\right) \geq x - \lambda g\left(\tfrac{x_\lambda}{a}\right) = x_\lambda - \lambda g\left(\tfrac{x_\lambda}{a}\right) - (x_\lambda - x) = ak\left(\tfrac{\lambda}{a}\right) - (x_\lambda - x).$$

Hence,

$$\int_0^\infty \exp\left(x - \lambda g\left(\tfrac{x}{a}\right)\right) dx \geq \int_0^{x_\lambda} e^{ak(\frac{\lambda}{a})} e^{-(x_\lambda - x)}\, dx$$

$$= e^{ak(\frac{\lambda}{a})}\int_0^{x_\lambda} e^{-x}\, dx \geq e^{ak(\frac{\lambda}{a})}\int_0^{\ln b} e^{-x}\, dx = e^{ak(\frac{\lambda}{a})}\left(1 - \tfrac{1}{b}\right).$$

So,

$$\int_0^{\lambda_0} e^{ak(\frac{\lambda}{a})}\, d\mu_{(1+\epsilon)a}(\lambda) \le \frac{b}{\epsilon(b-1)}.$$

Finally,

$$\int_0^{\infty} e^{ak(\frac{\lambda}{a})}\, d\mu_{(1+\epsilon)a}(\lambda) \le b + \frac{b}{\epsilon(b-1)}.$$

Setting $b = 1 + \frac{1}{\sqrt{\epsilon}}$ we obtain (2.8) with $c = 1$.

If $k(\lambda) \to \alpha > 0$ $(\lambda \to \infty)$, then the inequality (2.8) holds where $c = e^{\alpha}$. □

We denote by S_p the class of functions $\varphi \in S_0$ for which $x\varphi'(x)$ is monotone and

$$x\varphi'(x) \uparrow \infty \quad \text{as} \quad x \to \infty.$$

Clearly, for such functions $\lim_{x\to\infty} \frac{\varphi(x)}{\ln x} = \infty$.

If $\varphi(x) \in S_p$, then the numbers

$$m_n(\varphi) = \max_{x\ge 0}(x^n e^{-\varphi(x)}) = x_n^n e^{-\varphi(x_n)} \quad (n = 0, 1, \ldots)$$

$(x_n\varphi'(x_n) = n)$ are well-defined. This is obvious if we observe that $\lim\limits_{x\to 0_+}(x\varphi'(x)) = 0$. Indeed, since the function $x\varphi'(x)$ increases, we have $t\varphi'(t) \ge c = \lim\limits_{x\to 0_+}(x\varphi'(x))$. Hence, $\varphi(1) - \varphi(\epsilon) = \int_\epsilon^1 \varphi'(t)\, dt \ge c \ln \frac{1}{\epsilon}$ for any $\epsilon > 0$. If $\epsilon \in (0,1)$, then $\varphi(1) \ge \varphi(\epsilon) + c\ln\frac{1}{\epsilon}$. By letting ϵ tend to 0 we obtain $c = 0$.

LEMMA 2.3. *Let $\varphi \in S_p$. Then the sequence $m_n(\varphi)$ has the following properties:*

(a) *if $\alpha \ge 1$, then $\alpha^n m_n(\varphi) \le m_n(\alpha^{-1}\varphi)$;*

(b) $\exists c > 0 : m_{n-1}(\varphi) \le c m_n(\varphi)$ *for any $n \in \mathbb{N}$;*

(c) *There exists a $c \ge 0$ such that*

$$\inf_{0\le k\le n} \frac{m_k(\varphi)}{x^k} \le c \exp\left(-\frac{\varphi(x)}{2}\right)$$

for any $n \ge 0$ and any $x \in [0, x_n]$ $(x_n\varphi'(x_n) = n)$.

Proof. (a) Let $\alpha \ge 1$. Then

$$\varphi(\alpha x) = \int_0^\infty \frac{\alpha x}{\alpha x + s}\, d\sigma(s) \le \alpha \int_0^\infty \frac{x}{x+s}\, d\sigma(s) = \alpha\varphi(x),$$

whence

$$\begin{aligned}\alpha^n m_n(\varphi) &= \alpha^n \max_{x\ge 0}(x^n e^{-\varphi(x)}) = \max_{x\ge 0}((\alpha x)^n e^{-\varphi(x)}) \le \max_{x\ge 0}((\alpha x)^n e^{-\frac{1}{\alpha}\varphi(\alpha x)}) \\ &= \max_{x\ge 0}(x^n e^{-\frac{1}{\alpha}\varphi(x)}) = m_n(\alpha^{-1}\varphi).\end{aligned}$$

(b) Let $x_n\varphi'(x_n) = n$. Since $x\varphi'(x)$ monotonically increases from zero to infinity, $\lim_{n\to\infty} x_n = \infty$. Therefore, for a certain k, $x_{k-1} \ge 1$. For all $n > k$ we have

$x_{n-1} > 1$. Then

$$m_{n-1}(\varphi) = x_{n-1}^{n-1} e^{-\varphi(x_{n-1})} \leq x_{n-1}^{n} e^{-\varphi(x_{n-1})} \leq m_n(\varphi) \quad (n \geq k).$$

Setting $c_1 = \max_{1 \leq n \leq k-1} \frac{m_{n-1}}{m_n}$ we obtain $m_{n-1} \leq c m_n$ for all natural n, where $c = \max\{c_1, 1\}$.

(c) In (b) we have found a natural number k such that $x_k \geq 1$. For $x \in [0, x_k]$ we have

$$\inf_{0 \leq k \leq n} \frac{m_k(\varphi)}{x^k} \leq m_0(\varphi) = 1 \leq e^{\frac{1}{2}\varphi(x_k)} e^{-\frac{1}{2}\varphi(x)} = c_1 e^{-\frac{1}{2}\varphi(x)} \quad (n = 0, 1, \ldots).$$

Let $n > k$, $u \in [x_k, x_n]$ and $r = [u\varphi'(u)]$ ($[\cdot]$ is the integer part of a number). Then $k \leq r \leq n$ and $u \geq 1$. Since $\max_{x \geq 0}(x^{u\varphi'(u)} e^{-\varphi(x)}) = u^{u\varphi'(u)} e^{-\varphi(u)}$, the inequalities

$$x^r e^{-\varphi(x)} \leq x^{u\varphi'(u)} e^{-\varphi(x)} \leq u^{u\varphi'(u)} e^{-\varphi(u)} \leq u^{r+1} e^{-\varphi(u)} \quad (x \geq 1)$$

hold. Therefore $m_r(\varphi) \leq u^{r+1} e^{-\varphi(u)}$ and

$$\inf_{0 \leq k \leq n} \frac{m_k(\varphi)}{u^k} \leq \frac{m_r(\varphi)}{u^r} \leq u e^{-\varphi(u)}.$$

Put $c_2 = \max_{u \geq 0}(u e^{-\frac{1}{2}\varphi(u)})$. Then $u u^{-\varphi(u)} \leq c_2 e^{-\frac{1}{2}\varphi(u)}$. Combining both cases we obtain

$$\inf_{0 \leq k \leq n} \frac{m_k(\varphi)}{x^k} \leq \max\{c_1, c_2\} e^{-\frac{1}{2}\varphi(x)} \quad (x \in [0, x_n]). \qquad \square$$

We present some examples of absolutely concave functions occurring rather frequently.

(a) $\ln(x+1) = \int_1^\infty \frac{x}{x+\lambda} \cdot \frac{d\lambda}{\lambda}$,

$$k(t) = \begin{cases} \ln \frac{1}{t} - 1 + t & \text{if } 0 \leq t \leq 1, \\ 0 & \text{if } t > 1; \end{cases}$$

b) $x^\alpha = \frac{\sin \pi\alpha}{\pi} \int_0^\infty \frac{x}{x+\lambda} \lambda^{\alpha-1} \, d\lambda \quad (0 < \alpha < 1)$,

$$k(t) = (1-\alpha)\alpha^{\frac{\alpha}{1-\alpha}} t^{-\frac{\alpha}{1-\alpha}}, \quad m_n(x^\alpha) = \left(\frac{n}{\alpha e}\right)^{\frac{n}{\alpha}}.$$

(c) $\varphi_1(x) = \int_e^\infty \frac{x}{x+\lambda} \cdot \frac{d\lambda}{\lambda \ln \lambda}, \quad \ln\ln(x+e) - 1 < \varphi_1(x) \leq \ln\ln(x+e)$;

for small t the function $k(t)$ behaves as $\ln\ln \frac{1}{t}$;

(d) $\varphi_2(x) = \int_1^\infty \frac{x}{x+\lambda} \cdot \frac{\ln \lambda}{\lambda} \, d\lambda, \quad \frac{1}{2}\ln^2(x+1) \leq \varphi_2(x) < \frac{1}{2}\ln^2(x+1) + \frac{\pi^2}{6}$.

The function $k(t)$ behaves as $\ln^2 \frac{1}{t}$ if t is small enough. For large n,

$$\left(2^n e^{\frac{\pi^2}{6}}\right)^{-1} e^{\frac{n^2}{2}} \leq m_n(\varphi_2) \leq e^{\frac{n^2}{2}}.$$

The functions (a)–(d) belong to the class S_0, while (b) and (d) belong to the class S_p.

2.2. ABSOLUTELY CONCAVE FUNCTIONS OF THE GENERATOR OF A CONTRACTION C_0-SEMIGROUP

Let A be the generator of a contraction C_0-semigroup on a Banach space $\mathfrak{B}$. According to Theorem 1.2 this is equivalent to the estimate

$$\|(A+\lambda E)^{-1}\| \le \tfrac{1}{\lambda} \quad (\lambda > 0).$$

Suppose that a function $\varphi(x)$ belongs to S_0, that is, $\varphi(x) = \int_0^\infty \frac{x}{x+\lambda}\,d\sigma(\lambda)$ with a non-negative measure σ for which $\int_0^\infty \frac{d\sigma(s)}{1+s} < \infty$. We define the operator $\varphi(A)$ by

$$\varphi(A)f = \int_0^\infty A(A+\lambda E)^{-1}f\,d\sigma(\lambda) \tag{2.10}$$

for $f \in \mathfrak{D}(A)$. The integral converges because

$$\Big\|\int_0^\infty A(A+\lambda E)^{-1}f\,d\sigma(\lambda)\Big\| \le \int_0^1 \|E-\lambda(A+\lambda E)^{-1}\|\,d\sigma(\lambda)\|f\| + \\ + \int_1^\infty \lambda^{-1}\,d\sigma(\lambda)\|Af\| \le 2\int_0^1 d\sigma(\lambda)\|f\| + \int_1^\infty \lambda^{-1}\,d\sigma(\lambda)\|Af\| < \infty.$$

This operator is closeable. In fact, let $f_n \to 0$ ($f_n \in \mathfrak{D}(A)$), and let $\epsilon > 0$ be arbitrary. Set $A_\epsilon = A + \epsilon E$. Then

$$\|A_\epsilon^{-1}\varphi(A)f_n\| \le \int_0^\infty \|A_\epsilon^{-1}A(A+\lambda E)^{-1}f_n\|\,d\sigma(\lambda)$$

$$\le \int_0^1 \|A_\epsilon^{-1}\|\|A(A+\lambda E)^{-1}\|\,d\sigma(\lambda)\|f_n\| + \int_1^\infty \|A_\epsilon^{-1}A\|\|(A+\lambda E)^{-1}\|\,d\sigma(\lambda)\|f_n\|$$

$$\le 2\epsilon^{-1}\int_0^1 d\sigma(\lambda)\|f_n\| + 2\int_1^\infty \tfrac{1}{\lambda}\,d\sigma(\lambda)\|f_n\| \to 0 \quad (n\to\infty).$$

This implies that the operator $A_\epsilon^{-1}\varphi(A)$ is continuous. If we now assume that $\varphi(A)f_n \to g$ as $n \to \infty$, then $A_\epsilon^{-1}\varphi(A)f_n \to A_\epsilon^{-1}g = 0$, whence $g = 0$. □

By $\varphi(A)$ we will mean the closure of its restriction to $\mathfrak{D}(A)$.

For $f \in \mathfrak{D}(A^n)$ the formula

$$\varphi^n(A)f = \int_0^\infty A^n(A+\lambda E)^{-n}\,d\sigma_n(\lambda) \tag{2.11}$$

holds, where the measure σ_n is determined by the equation $\varphi^n(x) = \int_0^\infty x^n(x+\lambda)^{-n}\,d\sigma_n(\lambda)$. A detailed proof is given in Hirsh [1], Pustylnik [1].

Fractional powers of the operator A,

$$A^\alpha f = \frac{\sin\pi\alpha}{\pi}\int_0^\infty \lambda^{\alpha-1}A(A+\lambda E)^{-1}f\,d\lambda \quad (0<\alpha<1;\ f \in \mathfrak{D}(A)), \tag{2.12}$$

and its logarithm,

$$\ln(A+E)f = \int_1^\infty A(A+\lambda E)^{-1}f\frac{d\lambda}{\lambda}, \tag{2.13}$$

are the most important and useful for our purpose.

The operator A_ϵ generates a contraction C_0-semigroup. Therefore $\varphi(A_\epsilon)$ makes sense.

LEMMA 2.4. *If $\varphi \in S_0$, then the inequality*

$$\|\varphi(A)f - \varphi(A_\epsilon)f\| \leq \varphi(\epsilon)\|f\| \tag{2.14}$$

is valid for any $f \in \mathfrak{D}(A)$.

Proof. By (2.10),

$$\varphi(A)f - \varphi(A_\epsilon)f = \int_0^\infty (A(A+\lambda E)^{-1} - A_\epsilon(A+(\lambda+\epsilon)E)^{-1})f\,d\sigma(\lambda).$$

By the Hilbert resolvent formula

$$\begin{aligned} A(A+\lambda E)^{-1} - A_\epsilon(A+(\lambda+\epsilon)E)^{-1} &= A((A+\lambda E)^{-1} - (A+(\lambda+\epsilon)E)^{-1} - \\ &\quad - \epsilon(A+(\lambda+\epsilon)E)^{-1} \\ &= \epsilon A(A+\lambda E)^{-1}(A+(\lambda+\epsilon)E)^{-1} - \\ &\quad - \epsilon(A+(\lambda+\epsilon)E)^{-1} \\ &= \epsilon(A+(\lambda+\epsilon)E)^{-1}(A(A+\lambda E)^{-1} - E) \\ &= -\epsilon\lambda(A+(\lambda+\epsilon)E)^{-1}(A+\lambda E)^{-1}, \end{aligned}$$

whence

$$\|\varphi(A)f - \varphi(A_\epsilon)f\| \leq \int_0^\infty \frac{\epsilon}{\lambda+\epsilon}\,d\sigma(\lambda)\|f\| = \varphi(\epsilon)\|f\|.$$

□

It follows from this lemma that $\varphi(A)f = \varphi(A_\epsilon)f + Bf$ $(f \in \mathfrak{D}(A))$, where the operator B is bounded on $\mathfrak{D}(A)$, hence on all of $\mathfrak{B}$. Therefore, $\mathfrak{D}(\varphi(A_\epsilon))$ does not depend on ϵ and coincides with $\mathfrak{D}(\varphi(A))$. Moreover, the inequality (2.14) implies that $\varphi(A_\epsilon)f \to \varphi(A)f$ $(\epsilon \to 0)$ in the norm of $\mathfrak{B}$ for any $f \in \mathfrak{D}(\varphi(A))$.

THEOREM 2.4. *Let A be the generator of a contraction C_0-semigroup. Suppose that $\varphi \in S_0$. Then $\varphi(A)$ generates a contraction semigroup of class C_0, and the formula*

$$e^{-\varphi(A)t} = \int_0^\infty e^{-A\lambda}\,d\mu_t(\lambda) \quad (\forall t \geq 0),$$

is valid, where μ_t, with $\int_0^\infty d\mu_t(\lambda) = 1$, is a non-negative measure on $[0,\infty)$.

Proof. Define the operator $V(t)$ by

$$V(t) = \int_0^\infty e^{-A\lambda}\,d\mu_t(\lambda),$$

where μ_t is the convolution family of probability measures occurring in representation (2.6). It is not difficult to see that $V(t)$ is a contraction semigroup. The weak convergence of μ_t to δ_0 as $t \to 0_+$ implies that for any $f \in \mathfrak{B}$ and $l \in \mathfrak{B}'$,

$$l(V(t)f) = \int_0^\infty l(e^{-A\lambda}f)\,d\mu_t(\lambda) \to l(f) \quad \text{as} \quad t \to 0_+.$$

Thus, the semigroup $V(t)$ is continuous at zero. Let B be its generator: $V(t) = e^{-Bt}$. We show now that B coincides with $\varphi(A)$.

Because of (2.6) we have

$$e^{-t(\varphi(x)+1)} = \int_0^\infty e^{-x\lambda} e^{-t}\,\mathrm{d}\mu_t(\lambda) = \int_0^\infty e^{-x\lambda}\,\mathrm{d}\nu_t(\lambda), \tag{2.15}$$

where $\int_0^\infty \mathrm{d}\nu_t(\lambda) = e^{-t}$. The measures ν_t are positive linear functionals l_t on the space $C_0\,[0,\infty)$ of continuous functions on $[0,\infty)$ that tend to 0 as $t \to \infty$. A norm in it is the sup-norm. If $f \in C_0\,[0,\infty)$, then $l_t(f) = \int_0^\infty f(\lambda)\,\mathrm{d}\nu_t(\lambda)$. Obviously $|l_t(f)| \le e^{-t}\|f\|_{C_0\,[0,\infty)}$ and the function $l_t(f)$ (f is fixed) is measurable with respect to t. Let $l(f) = \int_0^\infty l_t(f)\,\mathrm{d}t$. This is a positive continuous linear functional on $C_0\,[0,\infty)$. According to the Riesz theorem there exists a finite positive measure ν on $[0,\infty)$ such that $l(f) = \int_0^\infty f(\lambda)\,\mathrm{d}\nu(\lambda)$ ($\forall\, f \in C_0\,[0,\infty)$), i.e.

$$\int_0^\infty \Big(\int_0^\infty f(\lambda)\,\mathrm{d}\nu_t(\lambda)\Big)\,\mathrm{d}t = \int_0^\infty f(\lambda)\,\mathrm{d}\nu(\lambda).$$

Now it follows from (2.15) that

$$\int_0^\infty e^{-x\lambda}\,\mathrm{d}\nu(\lambda) = \int_0^\infty e^{-t(\varphi(x)+1)}\,\mathrm{d}t = \big(\varphi(x)+1\big)^{-1} \quad (x > 0). \tag{2.16}$$

On the other hand,

$$\begin{aligned}\varphi(x) &= \int_0^\infty \frac{x}{x+\lambda}\,\mathrm{d}\sigma(\lambda) = \int_0^\infty x\Big(\int_0^\infty e^{-t(x+\lambda)}\,\mathrm{d}t\Big)\,\mathrm{d}\sigma(\lambda)\\ &= \int_0^\infty xe^{-tx}h(t)\,\mathrm{d}t, \quad h(t) = \int_0^\infty e^{-t\lambda}\,\mathrm{d}\sigma(\lambda),\end{aligned}$$

whence

$$\begin{aligned}\varphi(x)+1 &= \int_0^\infty xe^{-tx}h(t)\,\mathrm{d}t + x\int_0^\infty e^{-tx}\,\mathrm{d}t = \int_0^\infty xe^{-tx}(h(t)+1)\,\mathrm{d}t\\ &= \int_0^\infty xe^{-tx}\,\mathrm{d}\rho(t), \quad \mathrm{d}\rho(t) = (h(t)+1)\,\mathrm{d}t.\end{aligned} \tag{2.17}$$

There exists a simple connection between the measures ρ and ν. By virtue of (2.16), (2.17),

$$1 = \int_0^\infty e^{-x\lambda}\,\mathrm{d}\nu(\lambda)\int_0^\infty xe^{-tx}\,\mathrm{d}\rho(t) = \int_0^\infty xe^{-tx}\,\mathrm{d}(\rho*\nu)(t) = \int_0^\infty xe^{-tx}\,\mathrm{d}t.$$

By the uniqueness theorem for the Laplace transform, $\rho*\nu$ is the Lebesgue measure.

As before, set $A_\epsilon = A + \epsilon E$ ($\epsilon > 0$). The semigroup $e^{-A_\epsilon t}$ has the property $\|e^{-A_\epsilon t}\| \le e^{-\epsilon t} \to 0$ as $t \to \infty$. Denoting by B_ϵ the generator of the semigroup $\int_0^\infty e^{-A_\epsilon\lambda}\,\mathrm{d}\mu_t(\lambda)$ and taking into account formula (1.3), we find that for any $f \in \mathfrak{B}'$ and $x \in \mathfrak{B}$,

$$f((B_\epsilon + E)^{-1}x) = f\Big(\int_0^\infty e^{-t}\Big(\int_0^\infty e^{-A_\epsilon\lambda}\,\mathrm{d}\mu_t(\lambda)x\Big)\,\mathrm{d}t\Big)$$

$$= \int_0^\infty \left(\int_0^\infty f(e^{-A_\epsilon \lambda} x) \, d\nu_t(\lambda) \right) dt = \int_0^\infty l_t(f(e^{-A_\epsilon \lambda} x)) \, dt$$
$$= l(f(e^{-A_\epsilon \lambda} x)) = \int_0^\infty f(e^{-A_\epsilon \lambda} x) \, d\nu(\lambda)$$
$$= f\left(\int_0^\infty e^{-A_\epsilon \lambda} \, d\nu(\lambda) x \right).$$

So we have

$$\int_0^\infty e^{-A_\epsilon \lambda} \, d\nu(\lambda) = (B_\epsilon + E)^{-1}.$$

Acting in the way we did for (2.17), we find that

$$(\varphi(A_\epsilon) + E)x = \int_0^\infty A_\epsilon e^{-A_\epsilon t} x \, d\rho(t),$$

whence

$$(\varphi(A_\epsilon) + E)(B_\epsilon + E)^{-1} x = \int_0^\infty \int_0^\infty A_\epsilon e^{-A_\epsilon (t+\lambda)} x \, d\nu(\lambda) \, d\rho(t)$$
$$= \int_0^\infty A_\epsilon e^{-A_\epsilon t} x \, dt = \int_0^\infty (-e^{-A_\epsilon t} x)' \, dt = x,$$

because $\|e^{-A_\epsilon t}\| \to 0$ $(t \to \infty)$. Consequently, $\varphi(A_\epsilon) = B_\epsilon$ on $\mathfrak{D}(A)$. But $e^{-B_\epsilon t} x = \int_0^\infty e^{-A_\epsilon \lambda} x \, d\mu_t(\lambda) = \lim S_\pi x$, where S_π are the corresponding integral sums. Since A_ϵ is closed, $e^{-B_\epsilon t} \mathfrak{D}(A) \subset \mathfrak{D}(A)$. Hence $\overline{B_\epsilon \upharpoonright \mathfrak{D}(A)} = B_\epsilon$ (see Theorem 1.11). Therefore, $B_\epsilon = \varphi(A_\epsilon)$ and $\varphi(A_\epsilon)$ is the generator of a contraction semigroup of class C_0. As can be seen from Lemma 2.4, $\varphi(A)$ differs from $\varphi(A_\epsilon)$ by a bounded operator B with $\|B\| \leq \varphi(\epsilon)$. This results in the fact that $\varphi(A)$ generates a semigroup of class C_0. Thus, there exists a $\lambda > 0$ with $-\lambda \in \rho(\varphi(A))$. For this λ,

$$\|(B + \lambda E)^{-1} - (\varphi(A) + \lambda E)^{-1}\| \leq \|(B + \lambda E)^{-1} - (\varphi(A_\epsilon) + \lambda E)^{-1}\| +$$
$$+ \|(\varphi(A_\epsilon) + \lambda E)^{-1} - (\varphi(A) + \lambda E)^{-1}\|$$
$$\leq \|(B + \lambda E)^{-1} - (\varphi(A_\epsilon) + \lambda E)^{-1}\| +$$
$$+ \|(\varphi(A_\epsilon) + \lambda E)^{-1}\| \|(\varphi(A) + \lambda E)^{-1}\| \|\varphi(A_\epsilon) - \varphi(A)\|$$
$$\leq o(1) + \lambda^{-1} \|(\varphi(A) + \lambda E)^{-1}\| \varphi(\epsilon) = o(1),$$

because

$$\|e^{-Bt} - e^{-\varphi(A_\epsilon)t}\| \leq \int_0^\infty \|e^{-A\lambda} - e^{-A_\epsilon \lambda}\| \, d\mu_t(\lambda)$$
$$\leq \int_0^\infty (1 - e^{-\epsilon \lambda}) \, d\mu_t(\lambda) \to 0 \quad (\epsilon \to 0)$$

and

$$\|(B+\lambda E)^{-1} - (\varphi(A_\epsilon)+\lambda E)^{-1}\| \leq \int_0^\infty e^{-\lambda t} \|e^{-Bt} - e^{-\varphi(A_\epsilon)t}\| \, dt \to 0 \quad (\epsilon \to 0)$$

(the validity of passing to the limit under the integral signs is ensured by Lebesgue's theorem).

So, $(B+\lambda E)^{-1}=(\varphi(A)+\lambda E)^{-1}$, whence $B=\varphi(A)$. The proof is complete. □

Note that the definition of logarithm of the generator of a contraction C_0-semigroup by formula (2.13) is completely justified, since

$$(1+x)^{-t}=\frac{1}{\Gamma(t)}\int_0^\infty u^{t-1}e^{-u}e^{-ux}\,du$$

implies

$$e^{-t\ln(E+A)}=\frac{1}{\Gamma(t)}\int_0^\infty u^{t-1}e^{-u}e^{-Au}\,du.$$

It is not difficult to verify that the last integral equals $(E+A)^{-t}$ (see, for example, S. Krein [1, subsection 7, §5, Chapter 1]).

Suppose now that the semigroup e^{-At} is bounded holomorphic (with some angle θ). Condition (1.7),

$$\|A^n e^{-At}\|\le c^n n^n t^{-n}\quad(t>0,\ n\in\mathbb{N},\ c=\mathrm{const}),$$

is necessary and sufficient for e^{-At} to possess the indicated properties. If $\varphi\in S_0$, then on account of (2.11), $\mathfrak{D}(A^n)\subset\mathfrak{D}(\varphi^n(A))$. Therefore, $\mathfrak{R}(e^{-At})\subset\mathfrak{D}(\varphi^n(A))$ for all $t>0$.

LEMMA 2.5. *If A is the generator of a bounded holomorphic semigroup and $\varphi\in S_0$, then for any $t>0$ and any $n\in\mathbb{N}$ there exists a $d>0$ such that*

$$\|\varphi^n(A)e^{-At}\|\le d^n\max_{x\ge 0}(\varphi^n(x)e^{-tx}).$$

Proof. We suppose e^{-At} to be contractions. Since

$$\varphi^n(A)e^{-At}=\int_0^\infty A^n e^{-At}(A+sE)^{-n}\,d\sigma_n(s),$$

we have

$$\begin{aligned}\|\varphi^n(A)e^{-At}\|&\le\int_0^{\frac{cn}{2t}}\|A(A+sE)^{-1}\|^n\,d\sigma_n(s)+\\&\quad+\int_{\frac{cn}{2t}}^\infty\|A^n e^{-At}\|\|(A+sE)^{-1}\|^n\,d\sigma_n(s)\\&\le\int_0^{\frac{cn}{2t}}(1+s\|(A+sE)^{-1}\|)^n\,d\sigma_n(s)+\left(\tfrac{cn}{t}\right)^n\int_{\frac{cn}{2t}}^\infty\frac{1}{s^n}\,d\sigma_n(s)\\&\le 2^n\int_0^{\frac{cn}{2t}}d\sigma_n(s)+\left(\tfrac{cn}{t}\right)^n\int_{\frac{cn}{2t}}^\infty\frac{1}{s^n}\,d\sigma_n(s)\\&\le 3^n\int_0^\infty\left(\tfrac{cn}{t}\right)^n\left(\tfrac{cn}{t}+s\right)^{-n}d\sigma_n(s)=3^n\varphi^n\left(\tfrac{cn}{t}\right)\\&\le(3e^c)^n\max_{x\ge 0}(\varphi^n(x)e^{-tx})\end{aligned}$$

(c is the number appearing in the estimate (1.7). □

THEOREM 2.5. *If A is the generator of a bounded holomorphic semigroup and $\varphi \in S_0$, then $\varphi(A)$ also generates a bounded holomorphic semigroup.*

Proof. It is sufficient to show that the estimate (1.7) is valid for $\varphi(A)$, that is, there exist constants $p, q > 0$ such that

$$\|\varphi^n(A)e^{-\varphi(A)t}\| \leq pq^n n! t^{-n} \quad (t > 0,\ n \in \mathbb{N}).$$

On the basis of Theorem 2.4, Lemmas 2.2 and 2.5, and also (2.7), we conclude that

$$\begin{aligned}
\left\|\tfrac{1}{n!}\varphi^n(A)e^{-\varphi(A)t}\right\| &\leq \frac{1}{n!}\int_0^\infty \|\varphi^n(A)e^{-A\lambda}\|\, d\mu_t(\lambda) \\
&\leq \frac{(2d)^n}{n!t^n}\int_0^\infty \max_{x\geq 0}\left(\frac{1}{2^n}t^n\varphi^n(x)e^{-\lambda x}\right) d\mu_t(\lambda) \\
&\leq \frac{(2d)^n}{t^n}\int_0^\infty \max_{x\geq 0}\exp\left(\tfrac{t\varphi(x)}{2} - \lambda x\right) d\mu_t(\lambda) \\
&= \frac{(2d)^n}{t^n}\int_0^\infty e^{\frac{1}{2}tk\left(\frac{2\lambda}{t}\right)}\, d\mu_t(\lambda) \\
&= c\frac{(2d)^n}{t^n}\left(1+\sqrt{\tfrac{t}{2t-t}}\right)^2 = 2c\frac{(2d)^n}{t^n}.
\end{aligned}$$

□

3. Boundary Values at Zero of Solutions of a First-order Differential Equation in a Banach Space

In this section we investigate the behaviour in a neighbourhood of zero of a solution of equation (4.1) from Chapter 2, considered on $(0, \infty)$ and when A is the generator of a holomorphic semigroup on a Banach space.

3.1. A REPRESENTATION OF A SOLUTION

Let $\mathfrak{B}$ be a Banach space. Let A be the generator of a semigroup of bounded linear operators on $\mathfrak{B}$ that is strongly continuous on $(0, \infty)$. Consider the equation

$$y'(t) + Ay(t) = 0, \quad t > 0. \tag{3.1}$$

Recall that a function $y(t) : (0, \infty) \to \mathfrak{D}(A)$ is called a solution inside the interval $(0, \infty)$ of (3.1) if it is strongly continuously differentiable and satisfies (3.1). In particular, it follows from this definition that $e^{-At}f$ is a solution inside $(0, \infty)$ of (3.1) for any $f \in \mathfrak{D}(A)$.

LEMMA 3.1. *If $y(t)$ is a solution inside $(0, \infty)$ of (3.1), then $e^{-At}y(s) = y(t + s)$ for any $t > 0$ and any $s > 0$.*

Proof. Set $f(t) = e^{-At}y(s - t)$ $t \in (0, s), s > 0)$. Since $y(s - t) \in \mathfrak{D}(A)$, the vector-function $f(t)$ is strongly continuously differentiable on $(0, s)$ and

$$f'(t) = -Ae^{-At}y(s - t) + e^{-At}Ay(s - t) = 0.$$

Hence, $e^{-At}y(s-t) = h(s)$, where the vector-function $h(s)$ is continuous on $(0,\infty)$. This implies that $e^{-At}y(s) = h(s+t)$ for all $s,t > 0$. Because $y(s) \in \mathfrak{D}(A)$ we find that $e^{-At}y(s)$ is strongly continuously differentiable at zero; by letting t tend to zero we obtain $y(s)=h(s)$. Thus, $e^{-At}y(s)=y(s+t)$ and the proof is complete. □

In the sequel we will assume that e^{-At} satisfies the following conditions:

a) e^{-At} $(t > 0)$ is strongly continuously differentiable;

b) $\operatorname{Ker} e^{-At_0} = \{0\}$ for some $t_0 > 0$ (this is equivalent to $\operatorname{Ker} e^{-At} = \{0\}$ for any $t > 0$). We also assume, for simplicity, that e^{-At} is a contraction semigroup. This restriction is not essential but considerably simplifies the exposition. It should be stressed here that neither denseness of $\mathfrak{D}(A)$ in $\mathfrak{B}$ nor strong continuity of the semigroup e^{-At} at zero is required. These facts are equivalent when $\sup_{0\le t\le 1}\|e^{-At}\| < \infty$.

Note also that holomorphic semigroups satisfy the requirements a) and b).

Now we will introduce the spaces necessary to describe solutions inside $(0,\infty)$ of equation (3.1).

We denote by $\mathfrak{B}_{-t}(A)$ the Banach space obtained by completion of $\mathfrak{B}$ with respect to the norm

$$\|f\|_{\mathfrak{B}_{-t}} = \|e^{-At}f\| \quad (f \in \mathfrak{B}).$$

Sometimes, when it is clear what operator is kept in mind, the symbol A will be omitted, as before. It follows from the inequality

$$\|f\|_{\mathfrak{B}_{-t}} \le \|f\|$$

that convergence in $\mathfrak{B}$ implies convergence in $\mathfrak{B}_{-t}$ to the same element.

LEMMA 3.2. *If $t > s > 0$, then we have topological (i.e. dense and continuous) imbeddings*

$$\mathfrak{B} \subseteq \mathfrak{B}_{-s} \subseteq \mathfrak{B}_{-t}.$$

Proof. Since e^{-At} is a contraction semigroup we have for $f \in \mathfrak{B}$,

$$\|f\|_{\mathfrak{B}_{-t}} = \|e^{-At}f\| = \|e^{-A(t-s)}e^{-As}f\| \le \|e^{-As}f\| = \|f\|_{\mathfrak{B}_{-s}}. \tag{3.2}$$

The norms $\|\cdot\|_{\mathfrak{B}_{-t}}$ and $\|\cdot\|_{\mathfrak{B}_{-s}}$ are compatible. To prove this we have to show that if a sequence $f_n \in \mathfrak{B}$ converges to zero in one of the norms and is fundamental in the other, then f_n converges to zero in both norms. It is sufficient to check that the relations $\|f_n\|_{\mathfrak{B}_{-t}} \to 0$ $(n\to\infty)$ and $\|f_n - f_m\|_{\mathfrak{B}_{-s}} \to 0$ $(m,n\to\infty)$ imply $\|f_n\|_{\mathfrak{B}_{-s}} \to 0$ $(n\to\infty)$. Since the sequence $\{f_n\}_{n=1}^{\infty}$ is fundamental in $\mathfrak{B}_{-s}$, the sequence $\{e^{-As}f_n\}_{n=1}^{\infty}$ is fundamental in $\mathfrak{B}$. So, $e^{-As}f_n \overset{\mathfrak{B}}{\to} y$ $(n\to\infty)$ for a certain $y \in \mathfrak{B}$. Since $\|f_n\|_{\mathfrak{B}_{-t}} = \|e^{-At}f_n\| \to 0$ $(n\to\infty)$, we have $\|e^{-A(t-s)}e^{-As}f_n\| \to 0$ $(n\to\infty)$, i.e. $e^{-A(t-s)}y = 0$. Property b) of the semigroup e^{-At} yields $y=0$.

The imbedding $\mathfrak{B}_{-s} \subseteq \mathfrak{B}_{-t}$ follows from the compatibility of the norms considered and inequality (3.2) for them. Since $\mathfrak{B}_{-s}$ contains the set $\mathfrak{B}$, which is dense in $\mathfrak{B}_{-t}$, this imbedding is dense. The denseness of $\mathfrak{B}$ in $\mathfrak{B}_{-s}$ preserves the relation (3.2) between the norms for $f \in \mathfrak{B}_{-s}$. Therefore, the imbedding operator is continuous. □

We construct the family of operators $S(t)$ as follows: for every $t > 0$, $S(t)$ is the extension by continuity of e^{-At} to $\mathfrak{B}_{-t}$. This definition is correct because

$$\|e^{-At}f\|_{\mathfrak{B}_{-t}} = \|e^{-2At}f\| \leq \|e^{-At}f\| = \|f\|_{\mathfrak{B}_{-t}} \quad (f \in \mathfrak{B})$$

and $\mathfrak{B}$ is dense in $\mathfrak{B}_{-t}$.

LEMMA 3.3 *The space $\mathfrak{B}_{-t}(A)$ is isometrically isomorphic to $\overline{\mathfrak{R}(e^{-At})}$ (the norm is induced by that of $\mathfrak{B}$). Moreover, this isomorphism is realized by the operator $S(t)$:*

$$S(t)\mathfrak{B}_{-t}(A) = \overline{\mathfrak{R}(e^{-At})} \quad \text{and} \quad \|S(t)f\| = \|f\|_{\mathfrak{B}_{-t}}. \tag{3.3}$$

Proof. Let $x \in \mathfrak{B}_{-t}$. Then there exists a sequence $x_n \in \mathfrak{B}$ such that $x_n \xrightarrow{\mathfrak{B}_{-t}} x$. Therefore $e^{-At}x_n = S(t)x_n \xrightarrow{\mathfrak{B}_{-t}} S(t)x$. Since the sequence $e^{-At}x_n \in \mathfrak{R}(e^{-At})$ is fundamental in $\mathfrak{B}$, it converges in $\mathfrak{B}$, hence in $\mathfrak{B}_{-t}$, to a certain element $y \in \overline{\mathfrak{R}(e^{-At})}$. By virtue of the uniqueness of a limit, $S(t)x = y \in \overline{\mathfrak{R}(e^{-At})}$. Thus, the inclusion $S(t)\mathfrak{B}_{-t} \subseteq \overline{\mathfrak{R}(e^{-At})}$ has been proved.

Conversely, let $x \in \overline{\mathfrak{R}(e^{-At})}$, $e^{-At}y_n \xrightarrow{\mathfrak{B}} x$ $(y_n \in \mathfrak{B})$. Then $\|y_n - y_m\|_{\mathfrak{B}_{-t}} = \|e^{-At}(y_n - y_m)\| \to 0$ $(n, m \to \infty)$. Consequently, the sequence y_n converges in $\mathfrak{B}_{-t}$ to a certain element $y \in \mathfrak{B}_{-t}$. Thus, $y_n \to y$ in $\mathfrak{B}_{-t}$ and $S(t)y_n \to x$ in $\mathfrak{B}$, hence in $\mathfrak{B}_{-t}$. This implies that $S(t)y = x$, so the converse inclusion $\overline{\mathfrak{R}(e^{-At})} \subseteq S(t)\mathfrak{B}_{-t}$ holds. It remains to observe that the equality $\|S(t)f\| = \|f\|_{\mathfrak{B}_{-t}}$ holds for elements f from the set $\mathfrak{B}$, which is dense in $\mathfrak{B}_{-t}$. □

COROLLARY 3.1. *The operator $S(t) : \mathfrak{B}_{-t} \to \mathfrak{B}$ is bounded.*

Assume

$$\mathfrak{B}_{\{-\}}(A) = \operatorname*{limpr}_{t\to 0} \mathfrak{B}_{-t}(A), \quad \mathfrak{B}_{(-)}(A) = \operatorname*{limind}_{t\to\infty} \mathfrak{B}_{-t}(A).$$

Certainly, $\mathfrak{B}_{\{-\}}(A)$ $(\mathfrak{B}_{(-)}(A))$ is a locally convex space; it may be represented as the projective (inductive) limit of the countable "decreasing" ("increasing") family of Banach spaces $\mathfrak{B}_{-\frac{1}{n}}(A)$ as $n \to \infty$ $(n \to 0)$.

THEOREM 3.1. *The family of operators $S(t)$ forms an equi-continuous semigroup of class C_0 in $\mathfrak{B}_{\{-\}}(A)$.*

Proof. a) The semigroup property.

Let $t_1 > t_2 > 0$. Then $S(t_1)S(t_2)$ and $S(t_1 + t_2)$ are defined on $\mathfrak{B}_{-t_2}$. Furthermore, the equality $S(t_1 + t_2) = S(t_1)S(t_2)$ holds on the set $\mathfrak{B}$, which is dense in $\mathfrak{B}_{-t_2}$. Since the operators $S(t_1)$, $S(t_2)$ and $S(t_1 + t_2)$ are continuous on $\mathfrak{B}_{-t_2}$, the semigroup property holds on $\mathfrak{B}_{-t_2}$, hence on $\mathfrak{B}_{\{-\}}$.

b) Continuity at zero.

It is necessary to prove that $\|S(t)f - f\|_{\mathfrak{B}_{-s}} \to 0$ as $t \to 0$ for any $f \in \mathfrak{B}_{\{-\}}$, any $s > 0$. As $S(\frac{s}{2})f \in \mathfrak{B}$ and $e^{-\frac{1}{2}As}S(\frac{s}{2})f \in \mathfrak{D}(A)$, it follows from Lemma 3.3 that

$$\|S(t)f - f\|_{\mathfrak{B}_{-s}} = \|S(t+s)f - S(s)f\| = \|(e^{-At} - E)e^{-\frac{1}{2}As}S(\tfrac{s}{2})f\| \to 0 \quad (t \to 0).$$

c) Equi-continuity.

Since any continuous semi-norm on $\mathfrak{B}_{\{-\}}$ is majorized by a certain norm $\|\cdot\|_{\mathfrak{B}_{-s}}$, the equi-continuity of the semigroup $S(t)$ follows from the inequality

$$\|S(t)f\|_{\mathfrak{B}_{-s}} = \|e^{-At}S(s)f\| \le \|S(s)f\| = \|f\|_{\mathfrak{B}_{-s}}.$$ □

COROLLARY 3.2. *For all* $t, s > 0$,

$$S(t+s) = e^{-As}S(t).$$

LEMMA 3.4. *Any solution inside* $(0, \infty)$ *of* (3.1) *has, as* $t \to 0$, *a limit in the space* $\mathfrak{B}_{\{-\}}(A)$.

Proof. Let $y(t)$ be a solution inside $(0, \infty)$ of (3.1). Since $y(t)$ is continuous on $(0, \infty)$, we have according to Lemma 3.1,

$$\|y(t) - y(s)\|_{\mathfrak{B}_{-p}} = \|e^{-Ap}(y(t) - y(s))\| = \|y(t+p) - y(s+p)\| \to 0 \quad (t, s \to 0)$$

for any $p > 0$. Therefore, $y(t)$ tends in $\mathfrak{B}_{\{-\}}$ to a certain element $y_0 \in \mathfrak{B}_{\{-\}}$: $y(t) \overset{\mathfrak{B}_{\{-\}}}{\longrightarrow} y_0$ as $t \to 0$. □

THEOREM 3.2. *A vector-function* $y(t)$ *is a solution inside* $(0, \infty)$ *of equation* (3.1) *if and only if it can be represented in the form*

$$y(t) = S(t)y_0, \quad y_0 \in \mathfrak{B}_{\{-\}}(A). \tag{3.4}$$

Proof. Let $y_0 \in \mathfrak{B}_{\{-\}}$. Then for any $t > 0$ and any sufficiently small $\epsilon > 0$ we have $S(t)y_0 = S(t-\epsilon)e^{-\frac{1}{2}A\epsilon}S(\frac{\epsilon}{2})y_0$. Since $e^{-\frac{1}{2}A\epsilon}S(\frac{\epsilon}{2})y_0 \in \mathfrak{D}(A)$, the vector-function $S(t)y_0$ is a solution inside (ϵ, ∞) of (3.1). Because $\epsilon > 0$ is arbitrary, $S(t)y_0$ is a solution inside $(0, \infty)$ of (3.1).

Conversely, assume that $y(t)$ is a solution inside $(0, \infty)$ of (3.1). By Lemma 3.4 there exists an element $y_0 \in \mathfrak{B}_{\{-\}}$ which is the limit of $y(t)$ as $t \to 0$. Set $x(t) = y(t) - S(t)y_0$. The function $x(t)$ is a solution inside $(0, \infty)$ of (3.1) and $x(t) \overset{\mathfrak{B}_{\{-\}}}{\longrightarrow} 0$ as $t \to 0$. Therefore $x(t) \to 0$ as $t \to 0$ in all spaces $\mathfrak{B}_{-s}$ $(s > 0)$, whence

$$\|x(t)\|_{\mathfrak{B}_{-s}} = \|S(s)x(t)\| = \|x(t+s)\| \to 0 \quad (t \to 0).$$

Consequently, $x(s) \equiv 0$ on $(0, \infty)$ and (3.4) is valid. □

A statement analogous to the Fatou theorem holds for solutions inside $(0,\infty)$ of equation (3.1). Namely, the following theorem is valid.

THEOREM 3.3. *Let $\mathfrak{B}$ be the dual of a Banach space $\mathfrak{B}^1$, and let the operator e^{-At_0} be the adjoint of a certain operator on $\mathfrak{B}^1$. If $y(t)$ is a solution inside $(0,\infty)$ of (3.1) and $y(t) = S(t)y_0$ $(y_0 \in \mathfrak{B}_{\{-\}})$, then*

$$(y_0 \in \mathfrak{B}) \Longleftrightarrow \left(\sup_{0<t\le 1} \|S(t)y_0\| < \infty\right). \tag{3.5}$$

Proof. Since $S(t_0)$ is an isometric isomorphism of $\mathfrak{B}_{-t_0}(A)$ onto $\overline{\mathfrak{R}(e^{-At_0})}$ (Lemma 3.3), convergence of a generalized sequence x_α to x with respect to a certain topology in S_{-t_0} is equivalent to convergence of $S(t_0)x_\alpha$ to $S(t_0)x$ with respect to the corresponding topology in $\overline{\mathfrak{R}(e^{-At_0})}$. Let $\sigma(\mathfrak{B},\mathfrak{B}^1)$ be the weak topology in $\mathfrak{B}$, as in a dual space. It is clear that $\sigma(\mathfrak{B},\mathfrak{B}^1)$ is weaker than the norm-topology and, of course, $\mathfrak{B}$ equipped with $\sigma(\mathfrak{B},\mathfrak{B}^1)$ is a Hausdorff space (therefore, uniqueness of a limit holds). The topology τ on $\mathfrak{B}_{-t_0}$ corresponding to $\sigma(\mathfrak{B},\mathfrak{B}^1)$ is also weaker than the norm-topology.

Set $b = \sup_{0<t\le 1}\|S(t)y_0\|$. According to Theorem 7.1 (Banach–Alaoglu) the ball in $\mathfrak{B}$ of radius b with centre at zero is compact in the weak topology $\sigma(\mathfrak{B},\mathfrak{B}^1)$. Therefore, there exists a generalized subsequence $\{t_\alpha\}_{\alpha\in\Lambda}$ of $\{t\}_{0<t\le 1}$ such that $S(t_\alpha)y_0 \overset{\sigma(\mathfrak{B},\mathfrak{B}^1)}{\longrightarrow} g$, where $g \in \mathfrak{B}$, $\|g\| \le b$. Thus, for any $l \in \mathfrak{B}^1$,

$$\begin{aligned}(S(t_0)S(t_\alpha)y_0)(l) &= (e^{-At_0}S(t_\alpha)y_0)(l) = S(t_\alpha)y_0(Cl) \to g(Cl)\\ &= (e^{-At_0}g)(l) = (S(t_0)g)(l)\end{aligned}$$

(here $e^{-At_0} = C^*$), i.e. $S(t_0)S(t_\alpha)y_0 \to S(t_0)g$ in the space $\overline{\mathfrak{R}(e^{-At_0})}$ with the $\sigma(\mathfrak{B},\mathfrak{B}^1)$-topology. This implies the convergence $S(t_\alpha)y_0 \to g$ in the space $\mathfrak{B}_{-t_0}$ with the τ-topology. Since $\left\|S(t_\alpha)y_0 - y_0\right\|_{\mathfrak{B}_{-t_0}} \to 0$, we have $y_0 = g \in \mathfrak{B}$.

The converse assertion follows from the fact that if $y_0 \in \mathfrak{B}$, then $\|S(t)y_0\| = \|e^{-At}y_0\| \le \|y_0\|$. □

COROLLARY 3.3. *If $\mathfrak{B}$ is reflexive, then (3.5) is true.*

REMARK 3.1. Note, that $y_0 \in \mathfrak{B}$ does not imply continuity of the vector-function $S(t)y_0 = e^{-At}y_0$ at the point 0 in this space.

COROLLARY 3.4. *If in addition to the conditions of Theorem 3.3 e^{-At} is a contraction C_0-semigroup on $\mathfrak{B}$, then continuity at zero of a solution inside $(0,\infty)$ of (3.1) is equivalent to boundedness of this solution in a neighbourhood of zero in the norm of $\mathfrak{B}$.*

3.2. SOME SPACES OF SMOOTH VECTORS AND AN ANALOGUE OF THE PALEY–WIENER THEOREM

Let

$$\mathfrak{B}_t(A) = \mathfrak{R}(e^{-At}), \quad \|f\|_{\mathfrak{B}_t} = \|e^{At}f\| \quad (t > 0)$$

(e^{At} is the inverse of e^{-At}). Since $\mathfrak{R}(e^{-At}) = \mathfrak{D}(e^{At})$ and e^{At} is closed, $\mathfrak{B}_t$ is a Banach space with respect to the norm $\|\cdot\|_{\mathfrak{B}_t}$ of the graph of e^{At}.

LEMMA 3.5. *If* $\overline{\mathfrak{R}(e^{-At})} = \mathfrak{B}$ $(\forall t > 0)$ *and* $t > s > 0$, *then the imbedding* $\mathfrak{B}_t \subset \mathfrak{B}_s$ *is topological.*

Proof. The condition $\overline{\mathfrak{R}(e^{-At})} = \mathfrak{B}$ $(\forall t > 0)$ implies that the imbedding $\mathfrak{B}_t \subset \mathfrak{B}_s$ is dense. Indeed, let $x = e^{-As}y$ $(y \in \mathfrak{B})$ be an arbitrary element from $\mathfrak{B}_s$. Since $\overline{\mathfrak{R}(e^{-A(t-s)})} = \mathfrak{B}$ there exists a sequence $z_n \in \mathfrak{B}$ such that $\|e^{-A(t-s)}z_n - y\| \to 0$ as $n \to \infty$. But

$$\|e^{-A(t-s)}z_n - y\| = \|e^{As}(e^{-At}z_n - e^{-As}y\| = \|e^{As}(e^{-At}z_n - x)\|.$$

Hence, the sequence $e^{-At}z_n \in \mathfrak{B}_t$ approximates x in the norm of $\mathfrak{B}_s$.

Next, if $x \in \mathfrak{B}_{t+s}$, then the inequality $\|x\|_s \le \|x\|_t$ holds, which is obtained in the following way. For any $y \in \mathfrak{B}$ we have $\|e^{-At}y\| \le \|e^{-As}y\|$. Setting $y = e^{A(t+s)}x$ we find that $\|e^{-At}e^{A(s+t)}x\| \le \|e^{-As}e^{A(s+t)}x\|$, whence $\|e^{As}x\| \le \|e^{At}x\|$, i.e. $\|x\|_{\mathfrak{B}_s} \le \|x\|_{\mathfrak{B}_t}$. As $\overline{\mathfrak{B}_{t+s}} = \mathfrak{B}_t$ and $\overline{\mathfrak{B}_{t+s}} = \mathfrak{B}_s$, the latter inequality is valid for any $x \in \mathfrak{B}_t$. Thus the imbedding $\mathfrak{B}_t \subset \mathfrak{B}_s$ is continuous.

So we have

$$\mathfrak{B}_t \subset \mathfrak{B}_s, \quad \|x\|_{\mathfrak{B}_s} \le \|x\|_{\mathfrak{B}_t} \quad (t > s > 0;\ x \in \mathfrak{B}_t)$$

and the imbedding is dense and continuous.

The norms $\|\cdot\|_{\mathfrak{B}_t}$ and $\|\cdot\|_{\mathfrak{B}_s}$ are compatible. Indeed, let $\|x_n\|_{\mathfrak{B}_s} \to 0$ $(n \to \infty)$ and $\|x_n - x_m\|_{\mathfrak{B}_t} \to 0$ $(n, m \to \infty)$. Then the sequence $\{e^{At}x_n\}_{n=1}^{\infty}$ is fundamental in $\mathfrak{B}$. Therefore, there exists $y \in \mathfrak{B}$ such that $e^{At}x_n = e^{A(t-s)}e^{As}x_n \xrightarrow{\mathfrak{B}} y$. Moreover, $e^{As}x_n \to 0$. Since the operator $e^{A(t-s)}$ is closed, $y = 0$. □

THEOREM 3.4. *If* A *is the generator of a holomorphic semigroup, then the set* $\mathfrak{A}_c(A)$ *of its entire vectors is dense in* $\mathfrak{B}$.

Proof. We first assume that e^{-At} is a bounded holomorphic semigroup. In this case $\bigcap_{t>0} \mathfrak{B}_t(A) \subseteq \mathfrak{A}_c(A)$. In fact, suppose that $f \in \bigcap_{t>0} \mathfrak{R}(e^{-At})$. Then for any $t > 0$, $f = e^{-At}\varphi(t)$, $\varphi(t) \in \mathfrak{B}$. Under the conditions imposed on the semigroup e^{-At}, we have $f \in \mathfrak{B}^\infty(A)$ and (1.7) holds, i.e. $\|A^n e^{-At}\| \le c^n n^n t^{-n}$ for some $c > 0$. Therefore,

$$\|A^n f\| = \|A^n e^{-At}\varphi(t)\| \le c^n n^n t^{-n}\|\varphi(t)\|.$$

Let $\alpha > 0$ be arbitrary. Selecting t_0 so that $ect_0^{-1} < \alpha$ and taking into account the inequality $(ne^{-1})^n < n!$, we find that

$$\|A^n f\| \le c^n n^n t_0^{-n} \|\varphi(t_0)\| \le \alpha^n n! \|f\|_{\mathfrak{B}_{t_0}},$$

i.e. $f \in \mathfrak{A}_c(A)$. The last inequality shows that

$$\|f\|_{C_\alpha\langle n!\rangle} \le \|f\|_{\mathfrak{B}_{t_0}}. \tag{3.6}$$

So it is sufficient to prove that $\overline{\bigcap_{t>0} \mathfrak{B}_t(A)} = \mathfrak{B}$.

To this end we use Lemma 1.1, Chapter 2. For the spaces $\mathfrak{B}_n$ appearing in this lemma we take the spaces $\mathfrak{B}_t(A)$ with $t = 0, 1, 2, \ldots$. Each of them is dense in $\mathfrak{B}$. For if this was not true then there would exist a functional $l \in \mathfrak{B}'$ such that $f(s,x) = l(e^{-As}x) = 0$ for any $x \in \mathfrak{B}$ and any $s \ge t$. But the function $f(s,x)$ is holomorphic for $s > 0$ (x is arbitrary but fixed) and continuous at zero. Hence, $f(0,x) = l(x) = 0$, i.e. $l = 0$. By virtue of Lemma 3.5, the imbeddings $\mathfrak{B}_{n+1}(A) \subset \mathfrak{B}_n(A)$ are topological. Then $\bigcap_{t>0} \mathfrak{B}_t(A) = \bigcap_{n\in\mathbb{N}} \mathfrak{B}_n(A)$ is dense in $\mathfrak{B}$.

Suppose now that the semigroup e^{-At} is holomorphic with angle θ but does not possess the boundedness property. Then (see subsection 1.2) there exists a constant $\omega > 0$ such that $e^{-\omega z}e^{-Az}$ is a bounded holomorphic semigroup with angle $\theta_1 < \theta$ whose generator is $A + \omega E$. We show that $\mathfrak{A}_c(A + \omega E) = \mathfrak{A}_c(A)$.

Let $f \in \mathfrak{A}_c(A)$, i.e. $\|A^n f\| \le c\alpha^n n!$ for any $\alpha > 0$. Then

$$\begin{aligned}
\|(A+\omega E)^n f\| &= \left\| \sum_{k=0}^{n} C_n^k \omega^k A^{n-k} f \right\| \le \sum_{k=0}^{n} C_n^k \omega^k \|A^{n-k} f\| \\
&\le c \sum_{k=0}^{n} \frac{n! \omega^k \alpha^{n-k}(n-k)!}{k!(n-k)!} = c\alpha^n n! \sum_{k=0}^{n} \frac{(\omega\alpha^{-1})^k}{k!} \\
&\le c e^{\omega\alpha^{-1}} \alpha^n n! = c_\alpha \alpha^n n!.
\end{aligned}$$

Hence $f \in \mathfrak{A}_c(A+\omega E)$. If we put $B = A + \omega E$, then $A = B - \omega E$; so, the inclusion $\mathfrak{A}_c(A+\omega E) \subset \mathfrak{A}_c(A)$ is proved analogously. □

THEOREM 3.5. *If A is the generator of a holomorphic semigroup, then the equality*

$$\mathfrak{A}_c(A) = \bigcap_{t>0} \mathfrak{B}_t(A) \tag{3.7}$$

is valid, and

$$e^{At} f = \sum_{n=0}^{\infty} \frac{A^n f}{n!} t^n$$

for any entire vector f of A.

Proof. The inclusion $\bigcap_{t>0} \mathfrak{R}(e^{-At}) \subseteq \mathfrak{A}_c(A)$ has been proved in the previous theorem. As was also shown above, in order to prove the converse inclusion it suffices to show it under the assumption that A generates a bounded holomorphic semigroup.

So, let $f \in \mathfrak{A}_c$. Take $t > 0$, $\alpha = (4t)^{-1}$. By assumption there exists a positive constant $c_0 > 0$ such that

$$\|A^n f\| \le c_0 \alpha^n n! = c_0 n! 4^{-n} t^{-n}.$$

Set $g = \sum_{n=0}^{\infty} \frac{A^n f}{n!} t^n$ (the series converges because $f \in \mathfrak{A}_c$). If we prove that $e^{-At} g = f$, then the membership $f \in \mathfrak{R}(e^{-At})$, hence the equality (3.7), will be established. So we prove that $e^{-At} g = f$.

Let $g_n = \sum_{p=0}^{n-1} \frac{A^p f}{p!} t^p$. The vector g_n is infinitely differentiable for the operator A ($g_n \in \mathfrak{B}^\infty(A)$) and $g_n \to g$ in $\mathfrak{B}$. Therefore $e^{-At} g_n \to e^{-At} g$ ($n \to \infty$). We show that $e^{-At} g_n \to f$ ($n \to \infty$). The desired result follows from this. Using the formula

$$e^{-At} x = \sum_{k=0}^{n-1} \frac{(-1)^k A^k x}{k!} t^k + \frac{1}{(n-1)!} \int_0^t (t-s)^{n-1} e^{-As} A^n x \, ds \quad (x \in \mathfrak{D}(A^n))$$

(see Hille–Phillips [1, subsection 11.6]) we obtain

$$e^{-At} g_n = \sum_{k=0}^{n-1} \frac{(-1)^k t^k}{k!} \sum_{p=0}^{n-1} \frac{A^{k+p}}{p!} t^p + \frac{1}{(n-1)!} \int_0^t (t-s)^{n-1} e^{-As} A^n g_n \, ds.$$

But

$$\begin{aligned}
\sum_{k=0}^{n-1} \frac{(-1)^k t^k}{k!} \sum_{p=0}^{n-1} \frac{A^{p+k} f}{p!} t^p &= \sum_{k=0}^{n-1} \sum_{p=0}^{n-1} (-1)^k C_{k+p}^k \frac{A^{p+k} f}{(p+k)!} t^{p+k} \\
&= \sum_{k=0}^{2n-2} \left(\sum_{m=0}^{\min\{k,n-1\}} (-1)^m C_k^m \right) \frac{A^k f}{k!} t^k \\
&= \sum_{k=0}^{n-1} \left(\sum_{m=0}^{k} (-1)^m C_k^m \right) \frac{A^k f}{k!} t^k + \\
&\quad + \sum_{k=n}^{2n-2} \left(\sum_{m=0}^{n-1} (-1)^m C_k^m \right) \frac{A^k f}{k!} t^k \\
&= f + \sum_{k=n}^{2n-2} \left(\sum_{m=0}^{n-1} (-1)^m C_k^m \right) \frac{A^k f}{k!} t^k = f + f_n.
\end{aligned}$$

Thus,

$$e^{-At} g_n - f = f_n + p_n, \tag{3.8}$$

where

$$p_n(t) = \frac{1}{(n-1)!} \int_0^t (t-s)^{n-1} e^{-As} A^n g_n \, ds.$$

We separately estimate each summand at the right-hand side of (3.8). We have

$$\|f_n\| \le \sum_{k=n}^{2n-2} \sum_{m=0}^{n-1} C_k^m \frac{\|A^k f\|}{k!} t^k \le \sum_{k=n}^{2n-2} \Big(\sum_{m=0}^{k} C_k^m\Big) \frac{\|A^k f\|}{k!} t^k$$

$$\le c_0 \sum_{k=n}^{2n-2} \frac{2^k k! t^k}{4^k t^k k!} = c_0 \sum_{k=n}^{2n-2} \frac{1}{2^k} < \frac{c_0}{2^{n-1}}.$$

Further,

$$\|p_n(t)\| = \Big\| \frac{1}{(n-1)!} \int_0^t (t-s)^{n-1} e^{-As} A^n g_n \, ds \Big\|$$

$$\le \frac{1}{(n-1)!} \int_0^t (t-s)^{n-1} \|A^n g_n\| \, ds.$$

Since

$$\|A^n g_n\| \le \sum_{k=0}^{n-1} \frac{\|A^{k+n} f\|}{k!} t^k \le \frac{c_0 n!}{4^n t^n} \sum_{k=0}^{n-1} \frac{(n+k)!}{n! k!} \frac{1}{4^k}$$

$$= \frac{c_0 n!}{4^n t^n} \sum_{k=0}^{n-1} C_{n+k}^k \frac{1}{4^k} \le \frac{c_0 n!}{4^n t^n} \sum_{k=0}^{n-1} \frac{2^{k+n}}{4^k} = \frac{2 c_0 n!}{(2t)^n},$$

we get

$$\|p_n(t)\| \le \frac{2 c_0 n!}{(n-1)!(2t)^n} \int_0^t (t-s)^{n-1} \, ds = \frac{c_0}{2^{n-1}}.$$

The equality (3.8) enables us to conclude that

$$\|e^{-At} g_n - f\| \le \frac{c_0}{2^{n-1}} + \frac{c_0}{2^{n-1}} = \frac{c_0}{2^{n-2}} \to 0 \quad (n \to \infty).$$

So, $f = e^{-At} g$. We also obtain

$$\|f\|_{\mathfrak{B}_t} = \|e^{At} f\| = \|g\| = \Big\| \sum_{n=0}^{\infty} \frac{A^n f}{n!} t^n \Big\| \tag{3.9}$$

$$\le \sup_n \frac{\|A^n f\|}{n! \alpha^n} \sum_{n=0}^{\infty} (\alpha t)^n = \sum_{n=0}^{\infty} \frac{1}{4^n} \|f\|_{C_\alpha\langle n! \rangle} = \tfrac{4}{3} \|f\|_{C_\alpha\langle n! \rangle}. \qquad \square$$

REMARK 3.2. It can be seen from the proof of Theorems 3.4 and 3.5 that the equality

$$\mathfrak{A}(A) = \bigcup_{t>0} \mathfrak{B}_t(A)$$

is valid, where $\mathfrak{A}(A)$ is the set of analytic vectors of the operator A.

Put, by definition,

$$\mathfrak{B}_{(+)}(A) = \operatorname*{limpr}_{t\to\infty} \mathfrak{B}_t(A), \quad \mathfrak{B}_{\{+\}}(A) = \operatorname*{limind}_{t\to 0} \mathfrak{B}_t(A).$$

The estimates (3.6) and (3.9) prove the topological equalities

$$\mathfrak{B}_{(+)}(A) = \mathfrak{A}_c(A), \quad \mathfrak{B}_{\{+\}}(A) = \mathfrak{A}(A).$$

REMARK 3.3. It has been established in the proof of Theorem 3.5 that if $f \in C_\alpha\langle n!\rangle(A)$, i.e. $\|A^n f\| \le c\alpha^n n!$ ($\forall\, n \in \mathbb{N}$), then $f \in \mathfrak{B}_{\frac{1}{4\alpha}}(A) = \mathfrak{R}(e^{-\frac{1}{4\alpha}A})$.

LEMMA 3.6. *Let A be the generator of a bounded holomorphic semigroup and $\varphi \in S_0$. Then there exists a number $r > 0$ such that*

$$\mathfrak{R}(e^{-At}) \subseteq \mathfrak{R}(e^{-\varphi(A)\mu}),$$
$$\|e^{\varphi(A)\mu}e^{-At}\| \le \tfrac{4}{3}e^{r\mu k(\frac{t}{r\mu})}$$

for any $t>0$ and any $\mu>0$. Here $e^{\varphi(A)\mu} = (e^{-\varphi(A)\mu})^{-1}$, $k(s) = \sup_{x\ge 0}(\varphi(x) - sx)$

Proof. According to Lemma 2.5 there exists a positive constant $d > 0$ independent of t such that $\|\varphi^n(A)e^{-At}\| \le d^n \max_{x\ge 0}(\varphi^n(x)e^{-tx})$. This implies that

$$\begin{aligned}\frac{\mu^n}{n!}\|\varphi^n(A)e^{-At}\| &\le \frac{1}{4^n}\max_{x\ge 0}\left(\frac{(4d\mu\varphi(x))^n}{n!}e^{-tx}\right)\\ &\le \frac{1}{4^n}\max_{x\ge 0} e^{4d\mu\varphi(x)-tx} = \frac{1}{4^n}e^{4d\mu k(\frac{t}{4d\mu})}.\end{aligned}$$

Then Theorem 3.5 shows that $\mathfrak{R}(e^{-At}) \subseteq \mathfrak{R}(e^{-\varphi(A)\mu})$ and

$$\|e^{\varphi(A)\mu}e^{-At}\| \le e^{r\mu k(\frac{t}{r\mu})}\sum_{n=0}^{\infty}\frac{1}{4^n} = \tfrac{4}{3}e^{r\mu k(\frac{t}{r\mu})},$$

where $r = 4d$ does not depend on t and μ (and φ). □

Theorem 2.5 states that the semigroup $e^{-\varphi(A)t}$ ($\varphi \in S_0$) is bounded holomorphic if e^{-At} has those properties. Starting from $e^{-\varphi(A)t}$ we introduce the spaces

$$\mathfrak{B}_{\{-\}}(\varphi(A)) = \operatorname*{limpr}_{t\to 0} \mathfrak{B}_{-t}(\varphi(A)), \quad \mathfrak{B}_{(-)}(\varphi(A)) = \operatorname*{limind}_{t\to\infty} \mathfrak{B}_{-t}(\varphi(A)),$$

where $\mathfrak{B}_{-t}(\varphi(A))$ is the completion of $\mathfrak{B}$ with respect to the norm $\|f\|_{\mathfrak{B}_{-t}(\varphi(A))} = \|e^{-\varphi(A)t}f\|$.

THEOREM 3.6. *Let A be the generator of a bounded holomorphic semigroup and $\varphi \in S_0$. The following topological imbeddings hold:*

$$\begin{aligned}\mathfrak{B}_{(+)}(A) &\subseteq \mathfrak{B}_{\{+\}}(A) \subseteq \mathfrak{B}_{(+)}(\varphi(A)) \subseteq \mathfrak{B}_{\{+\}}(\varphi(A)) \subseteq \mathfrak{B}\\ &\subseteq \mathfrak{B}_{\{-\}}(\varphi(A)) \subseteq \mathfrak{B}_{(-)}(\varphi(A)) \subseteq \mathfrak{B}_{\{-\}}(A) \subseteq \mathfrak{B}_{(-)}(A).\end{aligned}$$

Proof. Only the imbeddings $\mathfrak{B}_{\{+\}}(A) \subseteq \mathfrak{B}_{(+)}(\varphi(A))$ and $\mathfrak{B}_{(-)}(\varphi(A)) \subseteq \mathfrak{B}_{\{-\}}(A)$ need to be checked. However, they are consequences of the boundedness of the operators $e^{\varphi(A)\mu}e^{-At}$ ($\forall\,\mu > 0, \forall\, t > 0$), which was proved in Lemma 3.6. The proof is the same as in Lemmas 3.2, 3.5. The continuity of the imbeddings is also shown as in those lemmas.

Let us pass to denseness. This is obvious for the spaces containing $\mathfrak{B}$, because $\mathfrak{B}$ is dense in all of them. It remains to prove the denseness of the imbeddings $\mathfrak{B}_{(+)}(A) \subseteq \mathfrak{B}_{\{+\}}(A) \subseteq \mathfrak{B}_{(+)}(\varphi(A)) \subseteq \mathfrak{B}_{\{+\}}(\varphi(A)) \subseteq \mathfrak{B}$. The denseness of all these imbeddings except the second one follows immediately from Theorems 3.4 and 3.5. We will prove, for instance, that $\mathfrak{B}_{(+)}(A)$ is dense in $\mathfrak{B}_{\{+\}}(A)$. It is necessary to show that for any element of the kind $e^{-As}f$ ($f \in \mathfrak{B}$, $s > 0$) there exists a sequence $f_n \in \bigcap_{t>0} \mathfrak{R}(e^{-At})$ such that $\|e^{-As}f - f_n\|_{\mathfrak{B}_s} \to 0$ ($n \to \infty$), i.e. such that $\|f - e^{As}f_n\| \to 0$ ($n \to \infty$). Since $e^{As}f_n \in \bigcap_{t\geq 0} \mathfrak{R}(e^{-At})$, the existence of such a sequence follows from the denseness of $\mathfrak{B}_{(+)}(A)$ in $\mathfrak{B}$.

Now we want to prove that $\mathfrak{B}_{\{+\}}(A)$ is dense in $\mathfrak{B}_{(+)}(\varphi(A))$. For this it is sufficient to show that $\mathfrak{B}_1(A) = \mathfrak{R}(e^{-A})$ is dense in $\mathfrak{B}_t(\varphi(A))$ ($t \to 0$ is arbitrary), i.e. that for any $f \in \mathfrak{B}$ there exists a sequence $f_n \in \mathfrak{B}$ such that $\|e^{\varphi(A)t}e^{-A}f_n - f\| \to 0$ ($n \to \infty$). In other words, it is necessary to prove that the set $\mathfrak{R}(e^{\varphi(A)t}e^{-A})$ is dense in $\mathfrak{B}$. Assume that this is not true. Then there would exist an $l \in \mathfrak{B}'$, $l \neq 0$, which vanishes on the set $e^{\varphi(A)t}e^{-A}\mathfrak{B}$, i.e. $l(e^{\varphi(A)t}e^{-A}f) = 0$ for any $f \in \mathfrak{B}$. By the estimate

$$\|\varphi^n(A)e^{-A}\| \leq c(\mu)(4\mu)^{-n}n!$$

(for any $\mu > 0$), obtained in the proof of Lemma 3.6, we conclude that $e^{-A}f$ ($\forall f \in \mathfrak{B}$) is an entire vector for the operator $\varphi(A)$. Therefore the function $e^{z\varphi(A)}e^{-A}f$, hence $l(e^{\varphi(A)z}e^{-A}f)$, is entire. By the semigroup property it equals 0 on (t, ∞), whence $l(e^{\varphi(A)z}e^{-A}f) \equiv 0$. Setting $z = 0$ we get $l(e^{-A}f) = 0$. Since $\mathfrak{R}(e^{-A})$ is dense in $\mathfrak{B}$, we have $l = 0$. The proof is complete. □

LEMMA 3.7. *Let $\varphi \in S_p$, let A be the generator of a bounded holomorphic semigroup, let $f \in \mathfrak{B}^\infty(A)$, and let*

$$\exists \mu > 0, \exists c > 0 : \|A^n f\| \leq c m_n(\mu\varphi) \quad (n = 0, 1, \ldots),$$

where

$$m_n(\mu\varphi) = \max_{x\geq 0}(x^n e^{-\mu\varphi(x)}).$$

Then

$$\exists c_0 > 0 : \|\varphi^n(A)f\| \leq c_0 \frac{8^n n!}{\mu^n}. \tag{3.10}$$

Proof. According to (2.11),

$$\|\varphi^n(A)f\| = \left\| \int_0^\infty A^n(A + sE)^{-n} f \, d\sigma_n(s) \right\|$$
$$\leq \int_0^{s_n} \|A^n(A + sE)^{-n} f\| \, d\sigma_n(s) + \int_{s_n}^\infty \|A^n(A + sE)^{-n} f\| \, d\sigma_n(s). \tag{3.11}$$

Here $\mu s_n\varphi'(s_n) = n$. By Lemmas 2.3 (assertion c)) and 2.1,

$$\int_0^{s_n} \|A^n(A+sE)^{-n}f\|\,d\sigma_n(s)$$

$$\le \int_0^{s_n} \|A^{k_1}(A+sE)^{-k_1}f\|\|A^{n-k_1}(A+sE)^{-(n-k_1)}\|\,d\sigma_n(s)$$

$$\le 2^{n-k_1}\int_0^{s_n} \inf_{0\le k\le n}\|A^k(A+sE)^{-k}f\|\,d\sigma_n(s)$$

$$\le c\cdot 2^n\int_0^{s_n} \inf_{0\le k\le n}\frac{m_k(\mu\varphi)}{s^k}\,d\sigma_n(s)$$

$$\le c_1\cdot 2^n\int_0^{s_n} e^{-\frac{1}{2}\mu\varphi(s)}\,d\sigma_n(s)$$

$$\le c_1\cdot 2^n\int_0^{\infty} e^{-\frac{1}{2}\mu\varphi(s)}\,d\sigma_n(s) \le c_2\frac{8^n n!}{\mu^n}$$

(here $\|A^{k_1}(A+sE)^{-k_1}\| = \min_{0\le k\le n}\|A^k(A+sE)^{-k}\|$). Further,

$$\int_{s_n}^{\infty} \|A^n(A+sE)^{-n}f\|\,d\sigma_n(s) \le m_n(\mu\varphi)\int_{s_n}^{\infty}\frac{d\sigma_n(s)}{s^n}$$

$$\le \frac{2^n m_n(\mu\varphi)}{s_n^n}\int_{s_n}^{\infty}\frac{s_n^n\,d\sigma_n(s)}{(s+s_n)^n}$$

$$\le \frac{2^n m_n(\mu\varphi)}{s_n^n}\varphi^n(s_n) = \frac{2^n\varphi^n(s_n)}{s_n^n}s_n^n e^{-\mu\varphi(s_n)}$$

$$= \frac{2^n(\mu\varphi(s_n))^n}{\mu^n n!}e^{-\mu\varphi(s_n)}n!$$

$$\le \frac{2^n e^{-\mu\varphi(s_n)}e^{\mu\varphi(s_n)}}{\mu^n}n! = \frac{2^n n!}{\mu^n}.$$

Substituting the estimates obtained in (3.11) leads to

$$\|\varphi^n(A)f\| \le c_0\left(\tfrac{8}{\mu}\right)^n n! \quad (c_0 = c_2+1).$$

□

LEMMA 3.8. *Let A be the generator of a bounded holomorphic semigroup, $\varphi\in S_0$. Then for any $t\ge 0$ there exist $d>0$ and $c_t>0$ such that*

$$\|A^n e^{-\varphi(A)t}\| \le c_t m_n(dt\varphi).$$

(In particular, $\bigcup_{t\ge 0}\mathfrak{R}(e^{-\varphi(A)t}) = \mathfrak{A}(\varphi(A)) \subset \mathfrak{B}^\infty(A)$).

Proof. On the basis of (1.7),

$$\|A^n e^{-\varphi(A)t}\| \le \int_0^\infty \|A^n e^{-As}\|\,d\mu_t(s) \le c^n n^n\int_0^\infty\frac{d\mu_t(s)}{s^n}.$$

We estimate the last integral. Integrate the equality

$$\frac{1}{(n-1)!}x^{n-1}e^{-t\varphi(x)} = \frac{1}{(n-1)!}\int_0^\infty x^{n-1}e^{-sx}\,d\mu_t(s)$$

with respect to $x \in [0, \infty)$. Because of the positivity of the integrands we can interchange the order of integration at the right-hand side. After this we obtain

$$\int_0^\infty \frac{d\mu_t(s)}{s^n} = \frac{1}{(n-1)!} \int_0^\infty x^{n-1} e^{-t\varphi(x)}\, dx$$
$$\le \frac{1}{(n-1)!} m_{n-1}(\tfrac{1}{2}t\varphi) \int_0^\infty e^{-\frac{1}{2}t\varphi(x)}\, dx.$$

The latter integral converges as $\lim_{x\to\infty} \frac{\varphi(x)}{\ln x} = \infty$. Taking into account assertion b) of Lemma 2.3 we find that

$$\int_0^\infty \frac{d\mu_t(s)}{s^n} \le \frac{d_t}{(n-1)!} m_n(\tfrac{1}{2}t\varphi).$$

Therefore

$$\|A^n e^{-\varphi(A)t}\| \le \frac{d_t c^n n^n}{(n-1)!} m_n(\tfrac{1}{2}t\varphi) \le \frac{d_t c^n e^n n!}{(n-1)!} m_n(\tfrac{1}{2}t\varphi) \le c_t (6c)^n m_n(\tfrac{1}{2}t\varphi).$$

By virtue of Lemma 2.3 (assertion a)),

$$\|A^n e^{-\varphi(A)t}\| \le c_t m_n(dt\varphi),$$

where $d = \frac{1}{12}c^{-1}$. □

These lemmas help us to characterize the spaces $\mathfrak{B}_{(+)}(\varphi(A))$ and $\mathfrak{B}_{\{+\}}(\varphi(A))$ in terms of the operator A. Namely, the following theorem, which is an obvious consequence of Lemmas 3.7, 3.8, Theorem 3.5 and Remark 3.2, is valid.

THEOREM 3.7. *Let e^{-At} be a bounded holomorphic semigroup and $\varphi \in S_p \cup \{x\}$. Then $f \in \mathfrak{B}_{\{+\}}(\varphi(A))$ ($f \in \mathfrak{B}_{(+)}(\varphi(A))$ if and only if $f \in \mathfrak{B}^\infty(A)$ and if for some (any) $\mu > 0$ there exists a $c > 0$ ($c = c(\mu) > 0$) such that*

$$\|A^n f\| \le c m_n(\mu\varphi) \quad (\forall n \in \mathbb{N}).$$

COROLLARY 3.5. *If A is the generator of a bounded holomorphic semigroup, then the equivalences*

$$(f \in \mathfrak{B}_{\{+\}}(A^\alpha)\ (f \in \mathfrak{B}_{(+)}(A^\alpha)),\ \alpha \in (0,1]) \iff$$
$$\iff (f \in \mathfrak{B}^\infty(A) \text{ and } \exists \mu > 0\ (\forall \mu > 0)$$
$$\exists c > 0\ (c = c(\mu) > 0) : \|A^n f\| \le c^n n^{\frac{n}{\alpha}},\ \forall n \in \mathbb{N})$$

hold.

LEMMA 3.9. *Let $\varphi_1, \varphi_2 \in S_0$ and $\sup_{x\ge 0}(4d\varphi_1(x) - \varphi_2(x)) = M < \infty$ ($d = 3e^c$, c is the constant appearing in (1.7)). If A is the generator of a bounded holomorphic semigroup, then*

$$\mathfrak{R}(e^{-\varphi_2(A)(1+\epsilon)}) \subseteq \mathfrak{R}(e^{-\varphi_1(A)})$$

($\epsilon > 0$ is arbitrary). Hence, the operator $e^{\varphi_1(A)}e^{-\varphi_2(A)(1+\epsilon)}$ is bounded. Its norm can be bounded by

$$\|e^{\varphi_1(A)}e^{-\varphi_2(A)(1+\epsilon)}\| \leq \tfrac{4}{3}e^{M+\alpha}\left(1+\frac{1}{\sqrt{\epsilon}}\right)^2, \tag{3.12}$$

where $\alpha = \lim_{t\to\infty} k(t)$, $k(t) = \max_{x\geq 0}(\varphi_2(x) - tx)$.

Proof. By virtue of Theorem 2.4 we have

$$e^{-\varphi_2(A)(1+\epsilon)} = \int_0^\infty e^{-A\lambda}\,d\mu_{1+\epsilon}(\lambda).$$

Applying Lemma 2.5 to $f \in \mathfrak{D}(\varphi_1^n(A))$ we obtain

$$\frac{1}{n!}\|\varphi_1^n(A)e^{-\varphi_2(A)(1+\epsilon)}f\| \leq \frac{\|f\|}{4^n}\int_0^\infty \max_{x\geq 0} e^{4d\varphi_1(x)-\lambda x}\,d\mu_{1+\epsilon}(\lambda).$$

Since the operator $\varphi_1^n(A)$ is closed, this estimate is valid for any $f \in \mathfrak{B}$. According to the assumption, $4d\varphi_1(x) \leq \varphi_2(x) + M$. Therefore in view of Lemma 2.2,

$$\begin{aligned}\frac{1}{n!}\|\varphi_1^n(A)e^{-\varphi_2(A)(1+\epsilon)}\| &\leq \frac{e^M}{4^n}\int_0^\infty \max_{x\geq 0} e^{\varphi_2(x)-\lambda x}\,d\mu_{1+\epsilon}(\lambda)\\ &\leq \frac{e^{M+\alpha}}{4^n}\left(1+\frac{1}{\sqrt{\epsilon}}\right)^2.\end{aligned} \tag{3.13}$$

Taking into account Remark 3.3 we find that $\mathfrak{R}(e^{-\varphi_2(A)(1+\epsilon)}) \subseteq \mathfrak{R}(e^{-\varphi_1(A)})$. Inequality (3.12) follows from the estimates (3.13). □

COROLLARY 3.6. *If A is the generator of a bounded holomorphic semigroup, $\varphi, \psi \in S_0$ and $\lim_{x\to\infty}\frac{\psi(x)}{\varphi(x)} = a > 0$, then the following topological imbeddings hold:*

$$\mathfrak{B}_{\{+\}}(\varphi(A)) \supseteq \mathfrak{B}_{\{+\}}(\psi(A)), \quad \mathfrak{B}_{(+)}(\varphi(A)) \supseteq \mathfrak{B}_{(+)}(\psi(A)),$$
$$\mathfrak{B}_{\{-\}}(\varphi(A)) \subseteq \mathfrak{B}_{\{-\}}(\psi(A)), \quad \mathfrak{B}_{(-)}(\varphi(A)) \subseteq \mathfrak{B}_{(-)}(\psi(A)).$$

If $\lim_{x\to\infty}\frac{\psi(x)}{\varphi(x)} = \infty$, then

$$\mathfrak{B}_{\{+\}}(\psi(A)) \subseteq \mathfrak{B}_{(+)}(\varphi(A)), \quad \mathfrak{B}_{(-)}(\varphi(A)) \subseteq \mathfrak{B}_{\{-\}}(\psi(A)).$$

It follows from these assertions that if $\varphi \in S_p$, then

$$\mathfrak{B}_{\{+\}}(\varphi(A)) \subseteq \mathfrak{B}^\infty(A), \quad \mathfrak{B}^{-\infty}(A) \stackrel{\text{def}}{=} \mathfrak{B}_{\{-\}}(\ln(A+E)) \subseteq \mathfrak{B}_{\{-\}}(\varphi(A)).$$

3.3. BOUNDARY VALUE PROPERTIES OF SOLUTIONS AT ZERO

Let us revert to the equation (3.1). The following theorem is the main result of this subsection. It establishes dependence between the rate of growth of a solution $y(t) = S(t)y_0$ ($y_0 \in \mathfrak{B}_{\{-\}}(A)$) inside $(0,\infty)$ of the equation considered as t approaches the point 0 and "smoothness" of its boundary value y_0, where "smoothness" is understood as belonging to a certain space of type $\mathfrak{B}_{\{-\}}(\varphi(A))$ or $\mathfrak{B}_{(-)}(\varphi(A))$.

THEOREM 3.8. *Let e^{-At} be a bounded holomorphic semigroup on a Banach space $\mathfrak{B}$ which is the dual of a Banach space $\mathfrak{B}^1$. Suppose also that there exists a $t_0 > 0$ such that e^{-At_0} is the adjoint of a certain operator on $\mathfrak{B}^1$. For $\varphi \in S_0$ the following assertions for a solution $y(t) = S(t)y_0$ inside $(0, \infty)$ of (3.1) are equivalent:*

a) $y_0 \in \mathfrak{B}_{(-)}(\varphi(A))$ $\quad (y_0 \in \mathfrak{B}_{\{-\}}(\varphi(A)))$;

b) *there exist $a > 0, c > 0$ (for any $a > 0$ there exists a $c = c(a) > 0$) such that*

$$\|y(t)\| \le c\exp(ak(\tfrac{t}{a})) \quad (\forall t > 0),$$

where $k(t) = \max_{x \ge 0}(\varphi(x) - tx)$. Moreover, $y(t) \to y_0$ $(t \to 0)$ in the topology of the space $\mathfrak{B}_{(-)}(\varphi(A))$ $(\mathfrak{B}_{\{-\}}(\varphi(A)))$.

Proof. The inequality

$$\|e^{\varphi(A)a}e^{-At}\| \le \tfrac{4}{3}\exp\left(rak\left(\frac{t}{ra}\right)\right) \tag{3.14}$$

(r does not depend on t and φ) proved in Lemma 3.6 is the central moment in proving the implication a) $\Longrightarrow$ b). The operator $e^{\varphi(A)a}$ is extended to an isometric operator from $\mathfrak{B}$ onto $\mathfrak{B}_{-a}(\varphi(A))$ (the proof is the same as that of Lemma 3.3). We denote this extension by P_a.

The operator $S(t)$ restricted to $\mathfrak{B}_{-a}(\varphi(A))$ is continuous as an operator on $\mathfrak{B}_{-a}(\varphi(A))$ because $\|e^{-\varphi(A)a}e^{-At}f\| \le \|e^{-\varphi(A)a}f\|$ for any $f \in \mathfrak{B}$. The operator P_a^{-1} is also continuous on $\mathfrak{B}_{-a}(\varphi(A))$. This follows from the inequality $\|e^{-\varphi(A)a}f\|_{\mathfrak{B}_{-a}(\varphi(A))} \le \|f\|_{\mathfrak{B}_{-a}(\varphi(A))}$ $(\forall f \in \mathfrak{B})$. Therefore, commutability of $S(t)$ and P_a^{-1} on $\mathfrak{B}$, a set dense in $\mathfrak{B}_{-a}(\varphi(A))$, implies their commutability on $\mathfrak{B}_{-a}(\varphi(A))$. Hence, for any $f \in \mathfrak{B}$, we have $S(t)P_a f = P_a S(t) f = e^{-\varphi(A)a}e^{-At}f$. Taking into account that for any $y_0 \in \mathfrak{B}_{-a}(\varphi(A))$ there exists a vector $f \in \mathfrak{B}$ such that $y_0 = P_a f$ and using the inequality (3.14) mentioned above we arrive at the assertion required.

Proof of the implication b) $\Longrightarrow$ a). Suppose $y(t) = S(t)y_0$ $(y_0 \in \mathfrak{B}_{\{-\}}(A))$ satisfies the estimate $\|y(t)\| \le c\exp(ak(\frac{t}{a}))$ $(t \to 0)$. According to Theorem 2.4,

$$e^{-2\varphi(A)a} = \int_0^\infty e^{-A\lambda}\, d\mu_{2a}(\lambda).$$

Using (2.8) this implies

$$\begin{aligned}\|e^{-2\varphi(A)a}y(t)\| &= \|e^{-2\varphi(A)a}S(t)y_0\| \le \int_0^\infty \|e^{-A\lambda}S(t)y_0\|\, d\mu_{2a}(\lambda)\\ &= \int_0^\infty \|e^{-At}S(\lambda)y_0\|\, d\mu_{2a}(\lambda) \le c\int_0^\infty e^{ak(\frac{\lambda}{a})}\, d\mu_{2a}(\lambda)\\ &\le 4ce^{\alpha} = c_1 \quad (\alpha = \lim_{t\to\infty} k(t)).\end{aligned}$$

Thus, $\|S(t)y_0\|_{\mathfrak{B}_{-2a}(\varphi(A))} \le c_1$ for all $t \ge 0$. By Lemma 3.3, $\mathfrak{B}_{-2a}(\varphi(A))$ is isometrically isomorphic to $\mathfrak{B}$. So there exist $g \in \mathfrak{B}_{-2a}(\varphi(A))$ and a generalized sequence t_α such that $S(t_\alpha)y_0 \to g$ in the weak topology of $\mathfrak{B}_{-2a}(\varphi(A))$, regarded as a dual space. This implies the convergence $P_{2a}^{-1}S(t_\alpha)y_0 \to P_{2a}^{-1}g$ in the weak topology $\sigma(\mathfrak{B}, \mathfrak{B}^1)$ of $\mathfrak{B}$. In exactly the same way as in Theorem 3.3 we arrive at the conclusion that $S(t_0)P_{2a}^{-1}S(t_\alpha)y_0 \to S(t_0)P_{2a}^{-1}g$ in the weak topology $\sigma(\mathfrak{B}, \mathfrak{B}^1)$ of $\mathfrak{B}$. Then $P_{2a}^{-1}S(t_\alpha)y_0 \to P_{2a}^{-1}g$ in the weak topology of $\mathfrak{B}_{-t_0}(A)$, regarded as a dual space. But $S(t_\alpha)y_0 \to y_0$ with respect to the norm of $\mathfrak{B}_{-t_0}(A)$ while, on account of the inequality $\|P_{2a}^{-1}f\|_{\mathfrak{B}_{-t_0}(A)} = \|e^{-2\varphi(A)a}f\|_{\mathfrak{B}_{-t_0}(A)} \le \|f\|_{\mathfrak{B}_{-t_0}(A)}$ $(f \in \mathfrak{B})$, the operator P_{2a}^{-1} is bounded on $\mathfrak{B}_{-t_0}(A)$. Consequently, $P_{2a}^{-1}y_0 = P_{2a}^{-1}g$, whence $y_0 = g \in \mathfrak{B}_{-2a}(\varphi(A))$. □

COROLLARY 3.7. *Let e^{-At} be a bounded holomorphic semigroup on a reflexive Banach space $\mathfrak{B}$ and $\varphi \in S_0$. If $y(t) = S(t)y_0$ $(y_0 \in \mathfrak{B}_{\{-\}}(A))$, i.e. $y(t)$ is a solution inside $(0,\infty)$ of (3.1), then the following equivalences hold:*

$$(y_0 \in \mathfrak{B}_{(-)}(\varphi(A))) \iff (\exists\, \mu > 0,\ \exists\, c > 0 : \|y(t)\| \le ce^{\mu k(\frac{t}{\mu})})$$
$$(y_0 \in \mathfrak{B}_{\{-\}}(\varphi(A))) \iff (\forall\, \mu > 0\ \exists\, c = c(\mu) > 0 : \|y(t)\| \le ce^{\mu k(\frac{t}{\mu})}).$$

COROLLARY 3.8. *Let $y(t) = S(t)y_0$ be a solution inside $(0,\infty)$ of (3.1). Under the assumptions of Theorem 3.8 the relations*

(a) $$(y_0 \in \mathfrak{B}^{-\infty}(A)) \iff (\exists\, n > 0 : \sup_{t>0}(t^n\|y(t)\|) < \infty);$$

(b) $$(y_0 \in \mathfrak{B}_{(-)}(A^\alpha)\ (0 < \alpha < 1)) \iff$$
$$\iff (\exists\, a > 0,\ \exists\, c > 0 : \|y(t)\| \le c\exp(at^{-\beta}),\ \beta = \frac{\alpha}{1-\alpha});$$
$$(y_0 \in \mathfrak{B}_{\{-\}}(A^\alpha)\ (0 < \alpha < 1)) \iff$$
$$\iff (\forall\, a > 0\ \exists\, c = c(a) > 0 : \|y(t)\| \le c\exp(at^{-\beta}),\ \beta = \frac{\alpha}{1-\alpha}),$$

are valid.

The absolutely concave functions $\varphi_1(x)$ and $\varphi_2(x)$ mentioned in the examples c) and d) (subsection 2.1) allow us to describe the vectors $y_0 \in \mathfrak{B}_{\{-\}}(A)$, which are the boundary values at 0 of the solutions inside $(0,\infty)$ of (3.1) whose growth in a neighbourhood of 0 is at most $\ln^a \frac{1}{t}$ or $t^{a \ln t}$.

3.4. NON-TRIVIALITY OF THE SPACES CONSTRUCTED

We elucidate when the spaces $\mathfrak{B}_t$ and $\mathfrak{B}_{-t}$ introduced in the previous subsections do not coincide with $\mathfrak{B}$ and with each other. Let A_1 and A_2 be contractions with zero kernels on $\mathfrak{B}$ and let $\|A_1x\| \le \|A_2x\|$ $(\forall\, x \in \mathfrak{B})$. Denote by $\mathfrak{B}^1$ and $\mathfrak{B}^2$ the completions of $\mathfrak{B}$ with respect to the norms $\|A_1x\|$ and $\|A_2x\|$. Since $\|x\|$, $\|A_1x\|$ and $\|A_2x\|$ are compatible, we have $\mathfrak{B} \subseteq \mathfrak{B}^2 \subseteq \mathfrak{B}^1$. If there exists a constant $c > 0$ such that

$$\|A_1x\| \ge c\|A_2x\| \quad (\forall\, x \in \mathfrak{B}),$$

then $\mathfrak{B}^1 = \mathfrak{B}^2$.

The converse is also true. If $\mathfrak{B}^1 = \mathfrak{B}^2$, then the imbedding operator from $\mathfrak{B}^2$ into $\mathfrak{B}^1$, being continuous and injective, is continuously invertible, i.e. $\|A_1 x\| \geq c\|A_2 x\|$ $(\forall\, x \in \mathfrak{B})$ for some $c > 0$.

Analogously, for the Banach spaces $\mathfrak{R}(A_i)$ with norms $\|A_i^{-1}x\|$ $(i = 1,2)$ the inclusion $\mathfrak{R}(A_1) \subseteq \mathfrak{R}(A_2)$ holds; equality holds if and only if the operator $A_1 A_2^{-1}$ is bounded on $\mathfrak{R}(A_2)$.

We apply these facts to the spaces $\mathfrak{B}_t(A)$ and $\mathfrak{B}_{-t}(A)$ constructed using a semigroup e^{-At} of contraction operators on $\mathfrak{B}$. In this connection we have to keep in mind that in this case the operators e^{At}, the inverses of e^{-At}, are either all bounded or all unbounded (see Hille–Phillips [1, 16.2]). This implies that either $\mathfrak{B}_{-t}(A) = \mathfrak{B}$ for all $t > 0$ or that all $\mathfrak{B}_{-t}(A)$ are not equal to $\mathfrak{B}$ and differ from each other. The same is valid for $\mathfrak{B}_t(A)$.

If the imbedding $\mathfrak{B}_t(A) \subset \mathfrak{B}$ is strict, then according to Banach's result cited in subsection 7.1, Chapter 1, $\mathfrak{B}_t(A) = \mathfrak{R}(e^{-At})$ is a set of the first category. Since $\mathfrak{B}_{\{+\}}(A) = \bigcup_{n=1}^{\infty} \mathfrak{B}_{\frac{1}{n}}(A)$, this is also a set of the first category and hence $\mathfrak{B}_{\{+\}}(A) \neq \mathfrak{B}$.

Assume that $\mathfrak{B} \neq \mathfrak{B}_{-t}(A)$. Since in this case all spaces $\mathfrak{B}_{-t}(A)$ participating in constructing the countably-normed space $\mathfrak{B}_{\{-\}}(A) = \operatorname{limpr}_{n\to\infty} \mathfrak{B}_{-\frac{1}{n}}(A)$ are different, by virtue of Lemma 7.1, Chapter 1, there exists an element $x \in \mathfrak{B}_{\{-\}}(A)$ such that $\|x\|_{\mathfrak{B}_{-\frac{1}{n}}} > n$. Since $\|x\|_{\mathfrak{B}_{-\frac{1}{n}}} \leq \|x\|$ we have $\mathfrak{B}_{\{-\}}(A) \neq \mathfrak{B}$. Combining this and Theorem 3.2 we arrive at the following result.

THEOREM 3.9. *Let e^{-At} be a differentiable $(t > 0)$ contraction semigroup of operators with zero kernels. Then any solution inside $(0,\infty)$ of (3.1) can be represented in the form*

$$y(t) = e^{-At} y_0 \quad (y_0 \in \mathfrak{B})$$

if and only if

$$\|e^{-At_0} x\| \geq c\|x\| \quad (\forall\, x \in \mathfrak{B})$$

for some $c, t_0 > 0$.

COROLLARY 3.9. *If e^{-At} is a differentiable $(t > 0)$ C_0-semigroup, then any solution inside $(0,\infty)$ of (3.1) is continuous at zero if and only if the operator e^{-At_0} (some $t_0 > 0$) is lower semi-bounded.*

It is not difficult to describe the holomorphic semigroups possessing the property mentioned in this corollary. If A is bounded, then clearly e^{-At} has a bounded inverse for any $t > 0$. Conversely, if $\|e^{-At_0} f\| \geq c\|f\|$ $(f \in \mathfrak{B})$ for some $t_0 > 0$, then $c\|Af\| \leq \|e^{-At_0} Af\| \leq c_1\|f\|$, that is, the operator A is bounded.

If an absolutely concave function $\varphi(x)$ is not bounded (which is equivalent to $\int_0^\infty d\sigma(s) = \infty$), then $\varphi(A)$ and A are bounded simultaneously (see Hirsh [1]). Therefore, if the generator A of a holomorphic semigroup is unbounded and $\varphi \in S_0$

is unbounded, then the imbeddings

$$\mathfrak{B}_{(+)}(\varphi(A)) \subset \mathfrak{B}_{\{+\}}(\varphi(A)) \subset \mathfrak{B} \subset \mathfrak{B}_{\{-\}}(\varphi(A)) \subset \mathfrak{B}_{(-)}(\varphi(A))$$

are strict.

Let now $\psi \in S_0 \cup \{x\}$ and $\lim_{x\to\infty} \frac{\psi(x)}{\varphi(x)} = \infty$. In view of Corollary 3.6, $\mathfrak{B}_{\{+\}}(\psi(A)) \subset \mathfrak{B}_{(+)}(\varphi(A))$ and $\mathfrak{B}_{(-)}(\varphi(A)) \subset \mathfrak{B}_{\{-\}}(\psi(A))$. Set $p(x) = \sqrt{\varphi(x)\psi(x)}$. Certainly, $\lim_{x\to\infty} \frac{p(x)}{\varphi(x)} = \infty$ and $\lim_{x\to\infty} \frac{\psi(x)}{p(x)} = \infty$. Furthermore, the function $p(x)$ is absolutely concave. Indeed, $p(x) \geq 0$ and it is continuous on $[0,\infty)$. The function $p^2(x)$ has an analytic continuation into the upper half-plane $\operatorname{Im} z > 0$. Since $\operatorname{Im} \varphi(z) > 0$ and $\operatorname{Im} \psi(z) > 0$, the function $p^2(z)$ does not take non-negative values. By selecting the branch of $\sqrt{z}$ which is analytic in $\mathbb{C}^1\backslash[0,\infty)$, we conclude that $p(x)$ can be analytically continued into the half-plane $\operatorname{Im} z > 0$, where $\operatorname{Im} p(z) > 0$. On the basis of Corollary 3.6,

$$\mathfrak{B}_{\{+\}}(\psi(A)) \subset \mathfrak{B}_{(+)}(p(A)) \subset \mathfrak{B}_{\{+\}}(p(A)) \subset \mathfrak{B}_{(+)}(\varphi(A)),$$
$$\mathfrak{B}_{(-)}(\varphi(A)) \subset \mathfrak{B}_{\{-\}}(p(A)) \subset \mathfrak{B}_{(-)}(p(A)) \subset \mathfrak{B}_{\{-\}}(\psi(A)).$$

Because of the unboundedness of $p(x)$, $\mathfrak{B}_{(\pm)}(p(A)) \neq \mathfrak{B}_{\{\pm\}}(p(A))$, hence the imbeddings $\mathfrak{B}_{\{+\}}(\psi(A)) \subset \mathfrak{B}_{(+)}(\varphi(A))$ and $\mathfrak{B}_{(-)}(\varphi(A)) \subset \mathfrak{B}_{\{-\}}(\psi(A))$ are strict.

The results of the foregoing discussion are collected in the following theorem.

THEOREM 3.10. *Suppose an unbounded operator A generates a holomorphic semigroup e^{-At} on $\mathfrak{B}$. If a function $\varphi \in S_0$ is unbounded, then all the imbeddings*

$$\mathfrak{B}_{(+)}(A) \subset \mathfrak{B}_{\{+\}}(A) \subset \mathfrak{B}_{(+)}(\varphi(A)) \subset \mathfrak{B}_{\{+\}}(\varphi(A)) \subset \mathfrak{B}$$
$$\subset \mathfrak{B}_{\{-\}}(\varphi(A)) \subset \mathfrak{B}_{(-)}(\varphi(A)) \subset \mathfrak{B}_{\{-\}}(A)$$

are strict. If $\psi \in S_0$ and $\lim \frac{\psi(x)}{\varphi(x)} = \infty$, then the imbeddings $\mathfrak{B}_{\{+\}}(\psi(A)) \subset \mathfrak{B}_{(+)}(\varphi(A))$ and $\mathfrak{B}_{(-)}(\varphi(A)) \subset \mathfrak{B}_{\{-\}}(\psi(A))$ are strict too.

COROLLARY 3.10. *If $\varphi \in S_p$, then the imbedding $\mathfrak{B}_{\{+\}}(\varphi(A)) \subset \mathfrak{B}^{\infty}(A)$ is strict.*

Theorem 3.10 and the corollary just formulated show that in the case of an unbounded A and holomorphic e^{-At} the elements of $\mathfrak{B}^{\infty}(A)$ and $\mathfrak{B}_{\{-\}}(A)$ behave rather differently. There exists, for example, an $f \in \mathfrak{B}^{\infty}(A)$ such that $\|A^n f\|$ increases (as $n \to \infty$) arbitrary rapidly, as well as an $f \in \mathfrak{B}_{\{-\}}(A)$ with arbitrary growth of $\|S(t)f\|$ (as $t \to 0$).

4. First-Order Equations of Parabolic Type in the Case of Degeneration

In this section we consider the equation

$$a(t)y'(t) + Ay(t) = 0, \quad t \in (0,T], \ T < \infty, \tag{4.1}$$

where A is the generator of a semigroup of bounded linear operators on a Banach space $\mathfrak{B}$ that is strongly continuous on $(0,\infty)$, and $a(t)$ is a scalar positive continuous function on $(0,T]$. We investigate the behaviour of its solutions in a neighbourhood of zero in dependence on the order of degeneration of the equation at the point 0. In passing solutions of the abstract inversely parabolic equation are studied.

4.1. THE CASE OF WEAK DEGENERATION

A solution on $(0,T]$ of equation (4.1) is a function $y(t) : (0,T] \to \mathfrak{D}(A)$ that is strongly differentiable on $(0,T]$ and satisfies (4.1) there.

LEMMA 4.1. *Let $y_1(t)$ and $y_2(t)$ be solutions on $(0,T]$ of equation (4.1). If $y_1(t_0) = y_2(t_0)$ for some $t_0 \in (0,T]$, then $y_1(t) \equiv y_2(t)$ on $[t_0,T]$.*

Proof. Because of the linearity of the equation it is sufficient to prove that if a solution $y(t)$ on $(0,T]$ vanishes at the point t_0, then $y(t) \equiv 0$ on $[t_0,T]$. Fix $t \in [t_0,T]$. On the interval $[t_0,T]$ we consider the function

$$z(s) = e^{-A\int_s^t a^{-1}(\xi)\,d\xi} y(s).$$

Since $y(s) \in \mathfrak{D}(A)$, the vector-function $z(s)$ is strongly continuously differentiable on $[t_0,t]$ and

$$z'(s) = a^{-1}(s)Ae^{-A\int_s^t a^{-1}(\xi)\,d\xi} y(s) - a^{-1}(s)e^{-A\int_s^t a^{-1}(\xi)\,d\xi} Ay(s) = 0.$$

Hence, $z(s)$ is constant on $[t_0,t]$. Since $z(t_0) = 0$, $z(s) \equiv 0$ on $[t_0,t]$. Taking into account that $z(t) = y(t)$ and that $t \in [t_0,T]$ is arbitrary we obtain $y(t) \equiv 0$ on $[t_0,T]$. □

REMARK 4.1. As can be seen from the proof above, if $y_1(t), y_2(t)$ are solutions on $[t_0,T]$ of (4.1) and $y_1(t_0) = y_2(t_0)$, then $y_1(t) \equiv y_2(t)$.

THEOREM 4.1. *Every solution on $(0,T]$ of (4.1) has on any interval $[t_0,T]$ ($t_0 > 0$) a representation*

$$y(t) = e^{-A\int_{t_0}^t a^{-1}(\xi)\,d\xi} g \quad (g \in \mathfrak{D}(A),\ t \in [t_0,T]),$$

where $g = y(t_0)$.

The proof easily follows from Remark 4.1. In fact, if $y(t)$ is a solution on $[t_0,T]$ of (4.1), then the vector-function $z(t) = e^{-A\int_{t_0}^t a^{-1}(\xi)\,d\xi} y(t_0)$ is, on the same interval, also a solution of this equation, and $z(t_0) = y(t_0)$. □

Depending on the character of degeneration of the equation at zero we distinguish two cases:

a) $\int_0^T a^{-1}(t)\,dt < \infty$ (the case of weak degeneration)

b) $\int_0^T a^{-1}(t)\,dt = \infty$ (the case of strong degeneration).

In the case a) we assume that the semigroup e^{-At} is bounded in a neighbourhood of zero and strongly continuously differentiable on $(0,\infty)$, and that $\operatorname{Ker} e^{-At} = \{0\}$ $(t > 0)$. Holomorphic semigroups as well as more general semigroups generated by the abstract parabolic equation (see S. Krein [1, subsection 5, §3, Chapter 1]), in particular, satisfy these requirements. The change of variables $\tau(t) = \int_0^t a^{-1}(\xi)\,d\xi$ turns the equation (4.1) into (3.1), i.e.

$$z'(\tau) + Az(\tau) = 0, \quad \tau \in (0,b),\ b = \int_0^T a^{-1}(\xi)\,d\xi.$$

By Lemma 4.1, a solution on $(0,b)$ of this equation can be uniquely continued, by $z(nT + \eta) = e^{-AnT}z(\eta)$, to a solution inside $(0,\infty)$. According to Theorem 3.2 a vector-function $y(t)$ is a solution on $(0,T]$ of equation (4.1) if and only if

$$y(t) = S\Big(\int_0^t a^{-1}(\xi)\,d\xi\Big)y_0 \quad (y_0 \in \mathfrak{B}_{\{-\}}(A)). \tag{4.2}$$

The solution (4.2) tends to y_0 as $t \to 0$ in the topology of the space $\mathfrak{B}_{\{-\}}(A)$. If $\mathfrak{B}$ is the dual of a Banach space $\mathfrak{B}^1$, while the operator e^{-At_0} is, for some $t_0 > 0$, the adjoint of an operator on $\mathfrak{B}^1$, then the equivalence

$$(y_0 \in \mathfrak{B}) \iff \big(\sup_{0<t\le T} \|y(t)\| < \infty\big)$$

holds.

Thus, $\mathfrak{B}_{\{-\}}(A)$ is the maximal space of initial data for formulating the initial value (Cauchy) problem $\big(y(t) \xrightarrow{\mathfrak{B}_{\{-\}}(A)} y_0\big)$ for equation (4.1) in the case of weak degeneration; the space $\mathfrak{B}$ (if it is reflexive) can be interpreted as the set of initial data of all bounded solutions. We would like to note here that if $A \ge 0$ is a self-adjoint operator on a Hilbert space $\mathfrak{H}$, then $\mathfrak{H}_{\{-\}}(A)$ coincides with the set $\mathfrak{A}'(A)$ of coanalytic vectors of A.

4.2. THE CASE OF STRONG DEGENERATION

We now consider the case b), assuming holomorphy of e^{-At}. The change of variables $\tau(t) = \int_t^T a^{-1}(\xi)\,d\xi$ turns (4.1) into the so-called abstract inversely parabolic equation, i.e. an equation

$$y'(t) - Ay(t) = 0 \quad (t \in [0,\infty)). \tag{4.3}$$

This equation will be the subject of investigation in the present subsection. By a solution on $[0,\infty)$ of it we mean, as usual, a continuously differentiable function $y(t) : [0,\infty) \to \mathfrak{D}(A)$ satisfying (4.3).

LEMMA 4.2. *Let A in equation (4.3) be the generator of a holomorphic semigroup on $\mathfrak{B}$. If two solutions $y_1(t), y_2(t)$ of this equation on $[0,\infty)$ coincide in a certain point $t_0 \in (0,\infty)$, then $y_1(t) = y_2(t)$ for all $t \in [0,t_0]$.*

Proof. It is sufficient to show that if a solution $y(t)$ on $[0,\infty)$ of equation (4.3) vanishes at some $t_0 > 0$, then $y(t) \equiv 0$ on $[0, t_0]$.

It is obvious that $z(t) = y(t_0 - t)$ $(t \in [0, t_0])$ is a solution of the initial value problem

$$z'(t) + Az(t) = 0, \quad t \in [0, t_0],$$
$$z(0) = 0.$$

By Remark 4.1, $z(t) \equiv 0$ $(t \in [0, t_0])$, whence $y(t) \equiv 0$ on $[0, t_0]$. □

The general form of solutions on $[0,\infty)$ of (4.3) is established in the following theorem.

THEOREM 4.2. *Let A be the generator of a holomorphic semigroup e^{-At} on a Banach space $\mathfrak{B}$. In order that a vector-function $y(t)$ be a solution on $[0,\infty)$ of equation (4.3) it is necessary and sufficient that*

$$y(t) = e^{At} f = \sum_{n=0}^{\infty} \frac{A^n f}{n!} t^n \quad (f \in \mathfrak{A}_c(A)) \tag{4.4}$$

(e^{At} is the inverse of e^{-At}; as was shown in Theorem 3.5 it is defined on $\mathfrak{A}_c(A)$ and its action on an $f \in \mathfrak{A}_c(A)$ is given by formula (4.4)).

Proof. It is not difficult to verify that a vector-function of the kind (4.4) is a solution on $[0,\infty)$ of (4.3).

Conversely, let $y(t)$ be any solution of the same equation on $[0,\infty)$. The vector-function $z(t) = e^{-A(t_0-t)}y(t_0)$ $(0 < t \le t_0)$ satisfies (4.3) and $z(t_0) = y(t_0)$. By virtue of Lemma 4.2, $y(t) = e^{-A(t_0-t)}y(t_0)$ on $[0, t_0]$. We fix t and let t_0 vary in the last equality. Then $y(t) = e^{-A\tau}y(t_0)$, where $\tau = t_0 - t > 0$ is arbitrary. Consequently, $y(t) \in \bigcap_{\tau>0} \mathfrak{R}(e^{-A\tau})$. By Theorem 3.5, $y(t) \in \mathfrak{A}_c(A)$. In particular, $y(0) \in \mathfrak{A}_c(A)$. But

$$y(t) = e^{-A(t_0-t)}y(t_0) = e^{At}e^{-At_0}y(t_0).$$

Taking into account that $e^{-At_0}y(t_0) = y(0)$ and setting $y(0) = f$, on the basis of the same Theorem 3.5 we obtain the representation (4.4). □

REMARK 4.2. It can be seen from the representation (4.4) that $y(t)$ $(t \ge 0$ arbitrary) is an entire vector of the operator A, and that the vector-function y is entire.

The following statement is a kind of uniqueness theorem.

THEOREM 4.3. *Let A be the generator of a bounded holomorphic semigroup with angle $\varphi \le \frac{\pi}{2}$. If a solution $y(t)$ on $[0,\infty)$ of (4.3) satisfies the estimate*

$$\|y(t)\| \le c_a \exp(at^\beta), \quad \beta < \frac{\pi}{2(\pi - \varphi)},$$

with certain constants $a > 0$, $c_a > 0$, then $y(t) \equiv g \in \operatorname{Ker} A$; in case $\operatorname{Ker} A = \{0\}$, $y(t) \equiv 0$.

Proof. As was noted in the proof of Theorem 4.2, a solution $y(t)$ on the interval $[0, t_0]$ ($\forall t_0 > 0$) of (4.3) can be represented in the form $y(t) = e^{-A(t_0-t)}y(t_0)$. Then $y^{(n)}(t) = A^n e^{-A(t_0-t)}y(t_0)$, whence, on account of (1.7),

$$\|y^{(n)}(t)\| = \|A^n e^{-A(t_0-t)}y(t_0)\| \leq c^n n^n \|y(t_0)\|(t_0-t)^{-n}.$$

Setting $t = 0$ in the representation (4.4) we find that $y^{(n)}(0) = A^n f$, hence $\|A^n f\| \leq c_a c^n n^n \exp(at_0^\beta)t_0^{-n}$. Taking $t_0 = n^{\frac{1}{\beta}}$ we obtain

$$\|A^n f\| \leq c_a c^n n^n e^{an} n^{-\frac{n}{\beta}} = c_a b^n n^{-\frac{n(1-\beta)}{\beta}}.$$

Applying to the operator $B = tA$ the, easily verified, identity

$$(B+E)^{-1}f = \sum_{k=0}^{n-1}(-1)^k B^k f + (-1)^n B^n (B+E)^{-1}f,$$

which is valid provided $-1 \in \rho(B)$, and observing that $\|t^n A^n (tA+E)^{-1}f\| \leq \text{const} \cdot \left(btn^{-\frac{1-\beta}{\beta}}\right)^n$, we conclude that

$$g(t) = (tA+E)^{-1}f = \sum_{k=0}^{\infty}(-1)^k t^k A^k f.$$

The latter series converges for all complex t. So

$$g(z) = (zA+E)^{-1}f$$

is an entire function. Its order is $\rho = \overline{\lim_{r\to+\infty}} \ln\ln M(r)$ where $M(r) = \max_{|z|=r}\|g(z)\|$. It is wellknown in complex analysis (see, for example, Markushevich [1, §1, Chapter 7]) that

$$\rho = \overline{\lim_{n\to\infty}} \frac{\ln n}{\ln\left(\sqrt[n]{\|A^n f\|}\right)^{-1}},$$

whence

$$\rho \leq \overline{\lim_{n\to\infty}} \frac{\ln n}{\ln\left(c_a^{-\frac{1}{n}} b^{-1} n^{\frac{1-\beta}{\beta}}\right)} \leq \frac{\beta}{1-\beta} < \frac{\pi}{\pi - 2\varphi}.$$

If we put $\varphi = \frac{\pi}{2} - \alpha$, then $\rho < \frac{\pi}{2\alpha}$.

Since A is the generator of a bounded holomorphic semigroup with angle φ, the resolvent of A is defined and holomorphic in the sector $S = \{z : \alpha < |\arg z| \leq \pi\}$. For z from this sector $\|(A+zE)^{-1}\| \leq \frac{M}{|z|}$. Therefore,

$$\|g(z)\| = \frac{1}{|z|}\|(A + \tfrac{1}{z}E)^{-1}\| \leq M \quad (z \in S). \tag{4.5}$$

In particular, $g(z)$ is bounded on the rays $\arg z = \pm(\alpha+\epsilon)$. Now take the sector $\{z : |\arg z| \leq \alpha+\epsilon\}$ of angle $\gamma\pi$, where $\gamma = \frac{2(\alpha+\epsilon)}{\pi}$ ($\epsilon > 0$) is selected so that $\rho < \frac{\pi}{2(\alpha+\epsilon)}$. This is possible because $\rho < \frac{\pi}{2\alpha}$. Since for the order ρ of the function $g(z)$ we have

$\rho < \frac{\pi}{2(\alpha+\epsilon)} = \frac{1}{\gamma}$, and $g(z)$ is bounded on the sides of the chosen angle, according to the Phragmén–Lindelöf theorem (see Markushevich [1]) $\|g(z)\| \le M_1 < \infty$ inside the angle. Taking into account (4.5) we deduce that $g(z)$ is bounded in the whole complex plane. By virtue of Liouville's theorem $g(z) \equiv g$.

Thus, $(tA + E)^{-1} f = g \in \mathfrak{B}$, i.e. $f = tAg + g$, which is possible only if $Ag = 0$. So, $f = g \in \operatorname{Ker} A$ and $y(t) \equiv g$. □

THEOREM 4.4 *If a normal operator A on a Hilbert space $\mathfrak{H}$ generates a bounded holomorphic semigroup with angle φ and $y(t)$ is a solution on $[0,\infty)$ of equation* (4.3), *then the estimate*

$$\|y(t)\| \le c\exp(\epsilon t) \quad (\forall\, \epsilon > 0)$$

implies $y(t) \equiv g \in \operatorname{Ker} A$.

Proof. The representation (4.4) and the principal spectral theorem for normal operators yield

$$e^{-2\epsilon t}\|y(t)\|^2 = \int_{|\arg\lambda| \le \varphi} e^{2(\operatorname{Re}\lambda - \epsilon)t}\, d(E_\lambda f, f) < c^2.$$

For all λ with $\operatorname{Re}\lambda > \epsilon$ the integrand tends to infinity as $t \to +\infty$. By Fatou's Theorem $(E_\Delta f, f) = 0$ if $\Delta \subset \{\lambda : \operatorname{Re}\lambda > \epsilon\}$. Since $\epsilon > 0$ is arbitrary, $(E_\Delta f, f) = 0$ when $0 \notin \Delta$. Hence $f \in \operatorname{Ker} A$ and $y(t) = e^{At} f \equiv f$. □

The element f appearing in the representation (4.4) of a solution $y(t)$ on $[0,\infty)$ may be characterized more exactly if we know the behaviour of $y(t)$ in a neighbourhood of infinity. The next theorem gives such a characterization.

THEOREM 4.5. *Let A be the generator of a bounded holomorphic semigroup, and let $y(t) = e^{At} f$ $(f \in \mathfrak{A}_c(A))$, i.e. $y(t)$ be a solution on $[0,\infty)$ of* (4.3). *Then the relation*

$$\begin{aligned} &(f \in \mathfrak{G}_{(\beta)}(A)\ (f \in \mathfrak{G}_{\{\beta\}}(A)),\ 0 \le \beta < 1) \Longleftrightarrow \\ &\Longleftrightarrow (\forall\, a > 0\ (\exists\, a > 0),\ \exists\, c > 0\ (\exists\, c = c(a) > 0) : \|y(t)\| \le c\exp(at^\alpha)), \quad (4.6) \end{aligned}$$

is valid, where α and β are related by $\beta = \frac{\alpha-1}{\alpha}$.

Proof. Suppose that for any $a > 0$ there exists a $c_a > 0$ such that $\|y(t)\| \le c_a e^{at^\alpha}$, $\alpha \ge 1$. In the proof of Theorem 4.3 it was established that

$$\|A^n f\| = \|y^{(n)}(0)\| \le c^n n^n t_0^{-n} \|y(t_0)\| \quad (\forall\, t_0 > 0).$$

Using the estimate (4.6) with $t_0 = \left(\frac{n}{a}\right)^{\frac{1}{\alpha}}$ we obtain

$$\|A^n f\| \le c_a c^n n^n t_0^{-n} \exp(a t_0^\alpha) = c_a (c e a^{\frac{1}{\alpha}})^n n^{\frac{n(\alpha-1)}{\alpha}}.$$

Since a is arbitrary, $f \in \mathfrak{G}_{(\beta)}(A)$ with $\beta = \frac{\alpha-1}{\alpha}$.

Conversely, let $f \in \mathfrak{G}_{(\beta)}(A)$ $(0 \le \beta < 1)$. Because of (4.4),

$$\|y(t)\| = \Big\| \sum_{n=0}^{\infty} \frac{A^n f}{n!} t^n \Big\| \le c \sum_{n=0}^{\infty} \frac{a^n n^{n\beta}}{n!} t^n \quad (\forall\, a > 0).$$

Consider the entire function $\sum_{n=0}^{\infty} \frac{a^n n^{n\beta}}{n!} z^n$. Its order is

$$\rho = \overline{\lim_{n\to\infty}} \frac{\ln n}{\ln\Big(\sqrt[n]{\frac{a^n n^{n\beta}}{n!}}\Big)^{-1}} = \overline{\lim_{n\to\infty}} \frac{\ln n}{\ln n^{1-\beta}} = \frac{1}{1-\beta}.$$

Its type is $\sigma = \mu^\rho (e\rho)^{-1}$, where $\mu = \overline{\lim_{n\to\infty}} \Big(n^{\frac{1}{\rho}} \sqrt[n]{\frac{a^n n^{n\beta}}{n!}} \Big) = \lim_{n\to\infty} \frac{n^{1-\beta} a n^\beta}{n} = a$, i.e. $\sigma = e^{-1}(1-\beta)a^{\frac{1}{1-\beta}}$. Therefore

$$\|y(t)\| \le c_a \exp(a t^\alpha) \quad \Big(\alpha = \frac{1}{1-\beta} \ge 1,\ \forall\, a > 0\Big).$$

The proof for $\mathfrak{G}_{\{\beta\}}(A)$ is analogous. □

We now revert to the case of strong degeneration of equation (4.1). To this end it is necessary to return to the initial variable in the arguments above. After that we arrive at the following conclusion: Let A be the generator of a holomorphic semigroup with angle $\varphi \le \frac{\pi}{2}$. Under the assumption $\int_0^T a^{-1}(t)\, dt = \infty$, a vector-function $y(t)$ is a solution on $(0,T]$ of (4.1) if and only if

$$y(t) = e^{A\int_t^T a^{-1}(\xi)\, d\xi} f = \sum_{n=0}^{\infty} \frac{A^n f}{n!} \Big(\int_t^T a^{-1}(\xi)\, d\xi \Big)^n \quad (f \in \mathfrak{A}_c(A)). \tag{4.7}$$

At a fixed $t \in (0,T)$ we have $y(t) \in \mathfrak{A}_c(A)$, while the vector-function $y(t)$ is entire. If $\|y(t)\| \le c_a \exp\big(a\big(\int_t^T a^{-1}(\xi)\, d\xi\big)^\beta\big)$, where $\beta < \frac{\pi}{2(\pi-\varphi)}$, $c_a > 0$, $a > 0$ are constants, then $y(t) \equiv g \in \operatorname{Ker} A$. Analogously, Theorem 4.5 can be reformulated (characterizing in detail the element f in the representation (4.6) of a solution, in dependence on the behaviour of this solution near zero).

5. The Behaviour of Solutions of First-Order Operator Differential Equations at Infinity

We consider equation (3.1),

$$y'(t) + Ay(t) = 0, \quad t \in (0, \infty),$$

where A is the generator of a continuous semigroup on $(0,\infty)$ of contractions on a Banach space $\mathfrak{B}$. So far we were interested in studying boundary values of its solutions inside $(0,\infty)$ as t approaches the left end of the interval $(0,\infty)$. Here we investigate their behaviour as $t \to \infty$. Moreover, we study the asymptotic behaviour $(t \to \infty)$ of solutions of the corresponding inhomogeneous equation, in dependence on the asymptotics at infinity of its right-hand side.

5.1. EXISTENCE OF THE CESÀRO LIMIT

It is not difficult to observe that in the general case it is not necessary that a solution of the equation considered has a finite limit as $t \to \infty$ in the space $\mathfrak{B}$. The following example confirms this. Let A be a self-adjoint operator on a Hilbert space $\mathfrak{H}$. Then the semigroup e^{-iAt} generated by iA has no limit at $+\infty$. However, under additional restrictions on the initial space and initial semigroup it is possible for a solution to have a limit in a certain generalized sense.

THEOREM 5.1. *Let $\mathfrak{B}$ be a reflexive Banach space. Suppose that A generates a bounded C_0-semigroup on $\mathfrak{B}$. If $y(t)$ is a solution inside $(0,\infty)$ of equation (3.1), then $\lim_{t\to\infty} \frac{1}{t-t_0}\int_{t_0}^t y(\tau)\,d\tau$ exists, and*

$$\lim_{t\to\infty} \frac{1}{t-t_0}\int_{t_0}^t y(\tau)\,d\tau = \lim_{\lambda\to 0} \lambda(A+\lambda E)^{-1} y(t_0) = Py(t_0),$$

where P is the projector of $\mathfrak{B}$ onto Ker A *in the decomposition*

$$\mathfrak{B} = \overline{\mathfrak{R}(A)} \dotplus \operatorname{Ker} A.$$

The limit does not depend on the choice of the point $t_0 > 0$.

Proof. As was shown in Lemma 3.1, the solution $y(t)$ can be represented in the form $y(t) = e^{-A(t-t_0)}y(t_0)$ on the interval $[t_0,\infty)$ (any $t_0 > 0$). Then

$$\frac{1}{t-t_0}\int_{t_0}^t y(\tau)\,d\tau = \frac{1}{t-t_0}\int_{t_0}^t e^{-A(\tau-t_0)}y(t_0)\,d\tau$$
$$= \frac{1}{t-t_0}\int_0^{t-t_0} e^{-As}y(t_0)\,ds = \frac{1}{\tau}\int_0^{\tau} e^{-As}y(t_0)\,ds.$$

According to Theorem 1.5,

$$\lim_{t\to\infty} \frac{1}{t-t_0}\int_{t_0}^t y(s)\,ds = Py(t_0). \tag{5.1}$$

It remains to observe that for sufficiently large t,

$$\frac{1}{t-t_1}\int_{t_1}^t e^{-A(s-t_1)}y(t_1)\,ds = \frac{1}{t-t_1}\int_{t_1}^{t_0} e^{-A(s-t_1)}y(t_1)\,ds+$$
$$+\frac{t-t_0}{t-t_1}\cdot\frac{1}{t-t_0}\int_{t_0}^t e^{-A(s-t_0)}e^{-A(t_0-t_1)}y(t_1)\,ds$$
$$= \frac{1}{t-t_1}\int_{t_1}^{t_0} e^{-A(s-t_1)}y(t_1)\,ds+$$
$$+\frac{t-t_0}{t-t_1}\cdot\frac{1}{t-t_0}\int_{t_0}^t e^{-A(s-t_0)}y(t_0)\,ds$$

(for instance, if $0 < t_1 < t_0$), which implies the independence of the limit in (5.1) of the choice of the point t_0. □

DEFINITION. We will say that a $\mathfrak{B}$-valued vector-function $f(t)$ that is continuous on $(0,\infty)$ has (as $t\to\infty$) Cesàro asymptotics $t^\alpha f_0$ ($\alpha\ge 0$, $f_0\in\mathfrak{B}$) of order $\beta\ge 1$ if

$$\lim_{t\to\infty}\frac{\Gamma(\alpha+\beta+1)(C,\beta,f(t))}{\Gamma(\alpha+1)\Gamma(\beta+1)t^\alpha}=f_0, \tag{5.2}$$

where Γ is the Euler gamma-function and

$$(C,\beta,f(t))=\frac{\beta}{(t-t_0)^\beta}\int_{t_0}^t (t-s)^{\beta-1}f(s)\,ds \quad (t_0>0).$$

It should be noted that the limit in (5.2) does not depend on the choice of the point t_0 in the definition of $(C,\beta,f(t))$. This follows from the fact that for $t_0'>t_0$ and sufficiently large t,

$$\frac{1}{(t-t_0)^\beta}\int_{t_0}^t (t-s)^{\beta-1}f(s)\,ds-\frac{1}{(t-t_0')^\beta}\int_{t_0'}^t (t-s)^{\beta-1}f(s)\,ds$$
$$=\frac{1}{(t-t_0')^\beta}\int_{t_0}^{t_0'}(t-s)^{\beta-1}f(s)\,ds+\frac{1}{(t-t_0)^\beta}\left(1-\frac{(t-t_0)^\beta}{(t-t_0')^\beta}\right)\int_{t_0}^t(t-s)^{\beta-1}f(s)\,ds$$

and

$$\left\|\frac{1}{(t-t_0')^\beta}\int_{t_0}^{t_0'}(t-s)^{\beta-1}f(s)\,ds\right\|\le c\frac{(t-t_0)^{\beta-1}}{(t-t_0')^\beta}\le\frac{c_1}{t-t_0}\to 0 \quad (t\to\infty).$$

The proof is analogous if $0<t_0'<t_0$.

When $f(t)$ is continuous on $[0,\infty)$,

$$(C,\beta,f(t))=\beta t^{-\beta}\int_0^t (t-s)^{\beta-1}f(s)\,ds$$

is the usual (C,β)-mean of the function f.

The afore-mentioned definition of Cesàro asymptotics is correct in the sense that if $f(t)$ has ordinary aymptotics $t^\alpha f_0$ $(t\to\infty)$, i.e.

$$\lim_{t\to\infty}\frac{f(t)}{t^\alpha}=f_0 \quad (f_0\in\mathfrak{B}), \tag{5.3}$$

then $f(t)$ has Cesàro asymptotics $t^\alpha f_0$ of arbitrary order $\beta\ge 1$. Indeed, suppose (5.3) holds. Then

$$\lim_{t\to\infty}\frac{\Gamma(\alpha+\beta+1)(C,\beta,f(t))}{\Gamma(\alpha+1)\Gamma(\beta+1)t^\alpha}=\lim_{t\to\infty}\frac{\Gamma(\alpha+\beta+1)\beta(t-t_0)^{-\beta}}{\Gamma(\alpha+1)\Gamma(\beta+1)t^\alpha}\int_{t_0}^t(t-s)^{\beta-1}f(s)\,ds$$
$$=\lim_{t\to\infty}\frac{\Gamma(\alpha+\beta+1)\beta}{\Gamma(\alpha+1)\Gamma(\beta+1)(t-t_0)^\beta t^\alpha}\left(\int_{t_0}^t(t-s)^{\beta-1}s^\alpha f_0\,ds+\int_{t_0}^t(t-s)^{\beta-1}\delta(s)s^\alpha\,ds\right)$$
$$=\lim_{t\to\infty}\frac{\Gamma(\alpha+\beta+1)\beta}{\Gamma(\alpha+1)\Gamma(\beta+1)t^{\alpha+\beta}}\int_{t_0}^t(t-s)^{\beta-1}s^\alpha\,ds\ f_0+\frac{\Gamma(\alpha+\beta+1)\beta}{\Gamma(\alpha+1)\Gamma(\beta+1)}\lim_{t\to\infty}I(t)$$
$$=f_0+\frac{\Gamma(\alpha+\beta+1)}{\Gamma(\alpha+1)\Gamma(\beta+1)}\lim_{t\to\infty}I(t),$$

where

$$I(t) = \frac{1}{(t-t_0)^\beta t^\alpha} \int_{t_0}^{t} (t-s)^{\beta-1}\delta(s)s^\alpha\, ds \quad (\delta(s) \to 0 \text{ as } s \to \infty).$$

We estimate $\|I(t)\|$ for sufficiently large t. We have

$$\|I(t)\| \le \frac{1}{(t-t_0)^\beta t^\alpha}\left(\left\|\int_{t_0}^{T} (t-s)^{\beta-1}\delta(s)s^\alpha\, ds\right\| + \left\|\int_{T}^{t} (t-s)^{\beta-1}\delta(s)s^\alpha\, ds\right\|\right)$$

(T is selected in such a way that $\|\delta(s)\| < \epsilon$ for a given arbitrarily small $\epsilon > 0$). Since $\delta(s)s^\alpha$ is continuous, one obtains

$$\|I(t)\| \le c\frac{(t-t_0)^\beta - (t-T)^\beta}{(t-t_0)^\beta t^\alpha} + \epsilon\frac{t^\alpha (t-T)^\beta}{\beta(t-t_0)^\beta t^\alpha},$$

whence $\|I(t)\| \to 0$ $(t \to \infty)$. Consequently,

$$\lim_{t\to\infty} \frac{\Gamma(\alpha+\beta+1)(C,\beta,f(t))}{\Gamma(\alpha+1)\Gamma(\beta+1)t^\alpha} = f_0.$$

Obviously, Theorem 5.1 states the existence (under assumptions on $\mathfrak{B}$ and A) of the first Cesàro limit (as $t \to \infty$) for any solution inside $(0,\infty)$ of the above-mentioned homogeneous equation.

We now consider the inhomogeneous equation

$$y'(t) + Ay(t) = f(t) \quad (t > 0). \tag{5.4}$$

We want to study the asymptotic behaviour at infinity of its solutions inside $(0,\infty)$, in dependence on that of the function $f(t)$ at the right-hand side of (5.4). This function is assumed to be continuous on $[0,\infty)$. We also assume that $f(t)$ satisfies one of the following conditions.

1°. The values of $f(t)$ belong to $\mathfrak{D}(A)$ and the function $Af(t)$ $(t \ge 0)$ is continuous into $\mathfrak{B}$.

2°. The function $f(t)$ is continuously differentiable into $\mathfrak{B}$.

The following theorem holds.

THEOREM 5.2. *Let A be the generator of a bounded C_0-semigroup on a reflexive Banach space $\mathfrak{B}$. If a continuous vector-function $f(t)$ on $[0,\infty)$ satisfying one of the conditions 1° or 2° has Cesàro asymptotics $t^\alpha f_0$ $(t \to \infty)$ of order $\beta \ge 1$, then the solution $y(t)$ inside $(0,\infty)$ of (5.4) for which $y(t_0) = 0$ (with a certain $t_0 > 0$) has Cesàro asymptotics $\frac{t^{\alpha+1}}{\alpha+1}Pf_0$ of the same order β (the operator P is defined in Theorem 5.1).*

Proof. It is not difficult to verify that the solution inside $(0,\infty)$ of (5.4) with $y(t_0) = 0$ can be represented on $[t_0,\infty)$ by

$$y(t) = \int_{t_0}^{t} e^{-A(t-s)} f(s)\, ds. \tag{5.5}$$

By the condition of the theorem,

$$\lim_{t\to\infty} \frac{\Gamma(\alpha+\beta+1)\beta}{\Gamma(\alpha+1)\Gamma(\beta+1)(t-t_0)^\beta t^\alpha} \int_{t_0}^t (t-s)^{\beta-1} f(s)\, ds = f_0.$$

It is necessary to prove that

$$g(t) = \frac{\Gamma(\alpha+\beta+2)}{\Gamma(\alpha+1)\Gamma(\beta)(t-t_0)^\beta t^{\alpha+1}} \int_{t_0}^t (t-s)^{\beta-1} y(s)\, ds \to Pf_0 \quad (t\to\infty).$$

For a continuous function $x(t) : (0,\infty) \to \mathfrak{B}$ we denote by $x_\beta(t)$ its primitive of order β, i.e.

$$x_\beta(t) = \frac{1}{\Gamma(\beta)} \int_{t_0}^t (t-s)^{\beta-1} x(s)\, ds.$$

Then

$$g(t) = \frac{\Gamma(\alpha+\beta+2)}{\Gamma(\alpha+1)(t-t_0)^\beta t^{\alpha+1}} y_\beta(t).$$

We use the fact that $y_\beta(t)$ is the solution of equation (5.4) with $f(t) = f_\beta(t)$ for which $y_\beta(t_0) = 0$. In fact, if $\beta = 1$, then $y_1(t) = \int_{t_0}^t y(s)\, ds$ and

$$y_1'(t) + Ay_1(t) = y(t) + \int_{t_0}^t (-y'(s) + f(s))\, ds = f_1(t).$$

For $\beta > 1$,

$$\begin{aligned} y_\beta'(t) + Ay_\beta(t) &= \frac{\beta-1}{\Gamma(\beta)} \int_{t_0}^t (t-s)^{\beta-2} y(s)\, ds + \\ &+ \frac{1}{\Gamma(\beta)} \int_{t_0}^t (t-s)^{\beta-1} (-y'(s) + f(s))\, ds \\ &= \frac{\beta-1}{\Gamma(\beta)} \int_{t_0}^t (t-s)^{\beta-2} y(s)\, ds - \\ &- \frac{\beta-1}{\Gamma(\beta)} \int_{t_0}^t (t-s)^{\beta-2} y(s)\, ds + f_\beta(t) = f_\beta(t). \end{aligned}$$

Representing $y_\beta(t)$ on $[t_0, \infty)$ in the form (5.5) we obtain

$$\begin{aligned} g(t) &= \frac{\Gamma(\alpha+\beta+2)}{\Gamma(\alpha+1)(t-t_0)^\beta t^{\alpha+1}} \int_{t_0}^t e^{-A(t-s)} f_\beta(s)\, ds \\ &= \frac{\Gamma(\alpha+\beta+2)}{\Gamma(\beta)\Gamma(\alpha+1)(t-t_0)^\beta t^{\alpha+1}} \int_{t_0}^t e^{-A(t-s)} \int_{t_0}^s (s-\tau)^{\beta-1} f(\tau)\, d\tau\, ds \\ &= \frac{\Gamma(\alpha+\beta+2)}{\beta\Gamma(\beta)\Gamma(\alpha+1)(t-t_0)^\beta t^{\alpha+1}} \int_{t_0}^t e^{-A(t-s)} (s-t_0)^\beta (C,\beta,f(s))\, ds \\ &= \frac{(\alpha+\beta+1)}{(t-t_0)^\beta t^{\alpha+1}} \int_{t_0}^t e^{-A(t-s)} (s-t_0)^{\alpha+\beta} \frac{\Gamma(\alpha+\beta+1)(C,\beta,f(s))}{\Gamma(\alpha+1)\Gamma(\beta+1)(s-t_0)^\alpha}\, ds \\ &= \frac{(\alpha+\beta+1)}{(t-t_0)^\beta t^{\alpha+1}} \int_{t_0}^t e^{-A(t-s)} (s-t_0)^{\alpha+\beta} (f_0 + \delta(s))\, ds = I_1(t) + I_2(t), \end{aligned}$$

where $\delta(s) \to 0$ as $s \to \infty$.

We estimate $I_1(t)$ and $I_2(t)$ for t large. To this end we select, for an arbitrarily small $\epsilon > 0$, a T so that $\|\delta(s)\| < \epsilon$ for $s > T$. Taking into account the boundedness of e^{-At} we find that

$$\|I_2(t)\| \le \frac{c(\alpha+\beta+1)T^{\alpha+\beta+1}}{(t-t_0)^\beta t^{\alpha+1}} + \epsilon\frac{t^{\alpha+\beta+1}}{(t-t_0)^\beta t^{\alpha+1}},$$

whence $\|I_2(t)\| \to 0$ as $t \to \infty$. Next,

$$I_1(t) = \frac{\alpha+\beta+1}{(t-t_0)^\beta t^{\alpha+1}} \int_0^{t-t_0} e^{-A\tau}(t-t_0-\tau)^{\alpha+\beta} f_0 \, d\tau.$$

Since $\lim_{t\to\infty} \frac{1}{t}\int_0^t e^{-A\tau} f_0 \, d\tau = Pf_0$, by Theorem 18.2.1 from Hille–Phillips [1], $\gamma t^{-\gamma}\int_0^t (t-\tau)^{\gamma-1} e^{-A\tau} f_0 \, d\tau \to Pf_0$ for an arbitrary $\gamma > 1$. Therefore $I_1(t) \to Pf_0$ as $t \to \infty$. □

REMARK 5.1. Taking into account that by Theorem 5.1 any solution inside $(0,\infty)$ of the homogeneous equation corresponding to (5.4) has Cesàro asymptotics $Py(t_0)$ of order $\beta = 1$, on the basis of the theorem just proved we conclude that under the assumptions (on $\mathfrak{B}$, A, and $f(t)$) of this theorem:

$$(C,\beta,y(t)) \sim Py(t_0) + \frac{\Gamma(\alpha+1)\Gamma(\beta+1)}{\Gamma(\alpha+\beta+2)} t^{\alpha+1} Pf_0 \quad (t \to \infty), \ \beta \ge 1,$$

for any solution on $(0,\infty)$ of (5.4).

5.2. ABEL ASYMPTOTICS

Let $\gamma(t) : (0,\infty) \to (0,\infty)$ be a continuous function, $\lim_{t\to\infty}\gamma(t) = \infty$ and

$$\forall\,\delta > 0 \quad \exists\, c = c(\delta) > 0 : |\gamma(t)| < ce^{\delta t}. \tag{5.6}$$

Denote by $\hat{\gamma}(\lambda; t_0)$ its transform

$$\hat{\gamma}(\lambda; t_0) = \int_{t_0}^{\infty} e^{-\lambda t}\gamma(t)\, dt, \tag{5.7}$$

which exists for λ with $\operatorname{Re}\lambda > 0$ by virtue of (5.6). It is obvious that $\hat{\gamma}(\lambda; t_0) > 0$, $\hat{\gamma}(\lambda; t_0) \to \infty$ as $\lambda \to 0$.

DEFINITION. We say that a continuous vector-function $f(t)$ on $(0,\infty)$ has Abel asymptotics $\gamma(t) f_0$ $(t \to \infty)$ if

$$\lim_{\lambda\to 0} \frac{\hat{f}(\lambda; t_0)}{\hat{\gamma}(\lambda; t_0)} = f_0 \quad (f_0 \in \mathfrak{B})$$

(here $\hat{f}(\lambda; t_0)$ is the transform (5.7) of $f(t)$). This definition is correct because the implication

$$\left(\lim_{t\to\infty} \frac{f(t)}{\gamma(t)} = f_0\right) \Longrightarrow \left(\lim_{\lambda\to 0} \frac{\hat{f}(\lambda; t_0)}{\hat{\gamma}(\lambda; t_0)} = f_0\right) \tag{5.8}$$

holds. In fact,

$$\lim_{\lambda\to 0}\frac{\hat f(\lambda;t_0)}{\hat\gamma(\lambda;t_0)}=\lim_{\lambda\to 0}\frac{1}{\hat\gamma(\lambda;t_0)}\int_{t_0}^{\infty}(\gamma(t)f_0+\gamma(t)\alpha(t))e^{-\lambda t}\,dt$$
$$=f_0+\lim_{\lambda\to 0}\frac{1}{\hat\gamma(\lambda;t_0)}\int_{t_0}^{\infty}\gamma(t)\alpha(t)e^{-\lambda t}\,dt,$$

where $\alpha(t)\to 0$ as $t\to\infty$. If we select $T>0$ so that $\|\alpha(t)\|<\epsilon$ for $t>T$ ($\epsilon>0$ is arbitrarily small), then the inequalities

$$\left\|\frac{1}{\hat\gamma(\lambda;t_0)}\int_{t_0}^{\infty}\gamma(t)\alpha(t)e^{-\lambda t}\,dt\right\|\le\frac{1}{\hat\gamma(\lambda;t_0)}\int_{t_0}^{T}\gamma(t)\|\alpha(t)\|e^{-\lambda t}\,dt+$$
$$+\frac{\epsilon}{\hat\gamma(\lambda;t_0)}\int_{T}^{\infty}\gamma(t)e^{-\lambda t}\,dt$$
$$\le\frac{c}{\hat\gamma(\lambda;t_0)}+\epsilon$$

hold, whence (5.8) follows.

Note that the definition of Abel asymptotics does not depend on the choice of the point $t_0>0$, since (for $t_0'<t_0$, for example)

$$\lim_{\lambda\to 0}\frac{\int_{t_0'}^{\infty}e^{-\lambda t}f(t)\,dt}{\int_{t_0'}^{\infty}e^{-\lambda t}\gamma(t)\,dt}=\lim_{\lambda\to 0}\frac{\lambda\left(\int_{t_0'}^{t_0}e^{-\lambda t}f(t)\,dt+\int_{t_0}^{\infty}e^{-\lambda t}f(t)\,dt\right)}{\lambda\left(\int_{t_0'}^{t_0}e^{-\lambda t}\gamma(t)\,dt+\int_{t_0}^{\infty}e^{-\lambda t}\gamma(t)\,dt\right)}$$

and

$$\left\|\lambda\int_{t_0'}^{t_0}e^{-\lambda t}f(t)\,dt\right\|\to 0,\quad\left\|\lambda\int_{t_0'}^{t_0}e^{-\lambda t}\gamma(t)\,dt\right\|\to 0\quad(\lambda\to 0).$$

If $f(t)$ and $\gamma(t)$ are continuous on $[0,\infty)$, then

$$\lim_{\lambda\to 0}\frac{\hat f(\lambda;t_0)}{\hat\gamma(\lambda;t_0)}=\lim_{\lambda\to 0}\frac{\hat f(\lambda)}{\hat\gamma(\lambda)},$$

where $\hat f(\lambda)$ is the Laplace transform of f.

THEOREM 5.3. *Let A be the generator of a bounded C_0-semigroup on a reflexive Banach space $\mathfrak{B}$. Let $f(t)$ be a continuous vector-function on $(0,\infty)$ satisfying in this interval one of the conditions 1° or 2° from the previous subsection. If $f(t)$ has Abel asymptotics $\gamma(t)f_0$ as $t\to\infty$, then the solution $y(t)$ inside $(0,\infty)$ of (5.4) with $y(t_0)=0$ (for a certain $t_0>0$) has Abel asymptotics $\beta(t)Pf_0$, where $\beta(t)=\int_{t_0}^{t}\gamma(s)\,ds$.*

Proof. Let $y(t)$ be the solution inside $(0,\infty)$ of (5.4) satisfying the condition $y(t_0)=0$. Then $\lambda\hat y(\lambda;t_0)+A\hat y(\lambda;t_0)=\hat f(\lambda;t_0)$, i.e.

$$\hat y(\lambda;t_0)=(A+\lambda E)^{-1}\hat f(\lambda;t_0).$$

So,

$$\lim_{\lambda\to 0}\frac{\hat{y}(\lambda;t_0)}{\hat{\beta}(\lambda;t_0)} = \lim_{\lambda\to 0}\frac{(A+\lambda E)^{-1}\hat{f}(\lambda;t_0)}{\int_{t_0}^{\infty} e^{-\lambda t}\left(\int_{t_0}^{t}\gamma(s)\,ds\right)dt}$$

$$= \lim_{\lambda\to 0}\frac{(A+\lambda E)^{-1}\hat{f}(\lambda;t_0)}{\lambda^{-1}\hat{\gamma}(\lambda;t_0)} = Pf_0. \qquad \square$$

REMARK 5.2. If a continuous vector-function $f(t)$ on $[0,\infty)$ has Cesàro asymptotics $t^\alpha f_0$ ($\alpha \geq 0$, $f_0 \in \mathfrak{B}$) of arbitrary order $\beta \geq 1$ as $t \to \infty$, then its Abel asymptotics is the same. Indeed,

$$\frac{\int_0^\infty e^{-\lambda t} f(t)\,dt}{\int_0^\infty e^{-\lambda t} t^\alpha\,dt} = \frac{\lambda^{\alpha+1}\int_0^\infty e^{-\lambda t} f(t)\,dt}{\Gamma(\alpha+1)} = \frac{\lambda^{\alpha+\beta+1}\int_0^\infty e^{-\lambda t} t^\beta (C,\beta,f(t))\,dt}{\Gamma(\alpha+1)\Gamma(\beta+1)}$$

$$= \lambda^{\alpha+\beta+1}\int_0^\infty e^{-\lambda t} t^\beta \left(\frac{t^\alpha}{\Gamma(\alpha+\beta+1)} f_0 + \frac{\delta(t)t^\alpha}{\Gamma(\alpha+\beta+1)}\right) dt$$

$$= \frac{\lambda^{\alpha+\beta+1}}{\Gamma(\alpha+\beta+1)}\int_0^\infty e^{-\lambda t} t^{\alpha+\beta}\,dt f_0 + \mathcal{I}(\lambda) = f_0 + \mathcal{I}(\lambda),$$

where $\delta(t) \to 0$ as $t \to \infty$. Choosing T so that $\|\delta(t)\| < \epsilon$ ($\epsilon > 0$ is arbitrary) and writing $\mathcal{I}(\lambda)$ as $\int_0^T + \int_T^\infty$ one can deduce that $\|\mathcal{I}(\lambda)\| \to 0$ as $\lambda \to 0$ and obtain the desired result.

This implies that if $f(t)$ has Cesàro asymptotics $t^\alpha f_0$, then the particular solution of the inhomogeneous equation (5.4) with initial datum $y(0) = 0$ has power Abel asymptotics which coincides with the Cesàro asymptotics, namely $\frac{t^{\alpha+1}}{\alpha+1} P f_0$.

REMARK 5.3. Taking into account that for a C_0-semigroup e^{-At}, we have $\int_0^\infty e^{\lambda s} e^{-As}\,ds = -\lambda(A+\lambda E)^{-1}$, on the basis of Theorems 5.1 and 5.3 we conclude that under the conditions of Theorem 5.3,

$$(\mathcal{A}, y(t)) \stackrel{\text{def}}{=} \lambda\hat{y}(\lambda;t_0) \sim Py(t_0) + \hat{\gamma}(\lambda;t_0)Pf_0 \quad (\lambda \to 0,\ \forall t_0 > 0).$$

for any solution $y(t)$ inside $(0,\infty)$ of (5.4).

5.3. ASYMPTOTIC AND EXPONENTIAL STABILITY

We consider equation (3.1) on $[0,\infty)$, i.e.

$$y'(t) + Ay(t) = 0, \quad t \in [0,\infty),$$

where A is the generator of a semigroup e^{-At} of bounded linear operators on $\mathfrak{B}$ of class C_0. It is well-known that any solution on $[0,\infty)$ of this equation can be represented as

$$y(t) = e^{-At} f \quad (f = y(0) \in \mathfrak{D}(A)).$$

DEFINITION. We will call a vector-function of the form

$$y(t) = e^{-At}f, \quad f \in \mathfrak{B},$$

a weak solution on $[0,\infty)$ of equation (3.1).

It is possible to show (see Pazy [2]) that this definition is equivalent to the following: a vector-function $y(t)$ is a weak solution on $[0,\infty)$ of equation (3.1) if and only if it is continuous on $[0,\infty)$ and if for every functional $f \in \mathfrak{D}(A^*)$ the function $f(y(t))$ is absolutely continuous on $[0,\infty)$ and

$$\frac{\mathrm{d}}{\mathrm{d}t} f(y(t)) + A^* f(y(t)) = 0 \quad (\forall t \geq 0).$$

DEFINITION. We will say that the equation (3.1), considered on $[0,\infty)$, is exponentially stable if there exists a number $\omega_0 > 0$ such that

$$\|y(t)\| \leq c_y e^{-\omega_0 t} \quad (c_y = \text{const})$$

for any weak solution $y(t)$, and asymptotically stable if

$$\|y(t)\| \to 0 \quad \text{as} \quad t \to \infty.$$

THEOREM 5.4. *Let $\gamma(t) \to 0$ $(t \to \infty)$ be a continuous and positive function on $[0,\infty)$. Suppose also that the operator A generates a C_0-semigroup on a Banach space $\mathfrak{B}$. If for any weak solution $y(t)$ on $[0,\infty)$ of (3.1) there exists a constant $c_y > 0$ such that*

$$\|y(t)\| \leq c_y \gamma(t), \tag{5.9}$$

then this equation is exponentially stable.

Proof. Denote by $C(\mathfrak{B},[0,\infty),\gamma(t))$ the Banach space of all continuous $\mathfrak{B}$-valued vector-functions on $[0,\infty)$ for which $\sup_{t\geq 0}(\gamma^{-1}(t)\|y(t)\|) < \infty$, with the norm

$$\|y\|_{C(\mathfrak{B},[0,\infty),\gamma(t))} = \sup_{t\geq 0} \frac{\|y(t)\|}{\gamma(t)}.$$

Define a linear mapping as follows:

$$\mathfrak{B} \ni f \to e^{-At} f \in C(\mathfrak{B},[0,\infty),\gamma(t)). \tag{5.10}$$

By (5.9) this definition is correct. We prove that the mapping (5.10) is closed.

Let $f_n \to 0$ $(n \to \infty)$ and $e^{-At} f_n \to y(t)$ in $C(\mathfrak{B},[0,\infty),\gamma(t))$. Since the semigroup e^{-At} is strongly continuous, $e^{-At} f_n \to 0$, uniformly on each compact set. Therefore $y(t) = 0$ on an arbitrary compact set, i.e. $y(t) \equiv 0$. Since the mapping (5.10) is defined on the whole space $\mathfrak{B}$, according to the closed graph theorem it is continuous, that is,

$$\|e^{-At} f\| \leq c\gamma(t)\|f\| \quad (c = \text{const}).$$

We now choose $T > 0$ so that $c\gamma(t) < \frac{1}{2}$ as $t \geq T$. Then $\|e^{-At}\| < \frac{1}{2}$, and if $t = nT + s$ $(0 \leq s < T)$ we have

$$\|e^{-At}\| \leq \|e^{-AnT}\| \|e^{-As}\| \leq M \|e^{-At}\|^n < M \cdot 2^{-n} \leq M' e^{-\omega_0 t},$$

where $M' = 2M$, $\omega_0 = T^{-1} \ln 2$. The exponential stability of equation (3.1) follows from this. □

REMARK 5.4. The proof produced above shows that the operator A need not generate a semigroup of class C_0. It is sufficient for e^{-At} to be bounded in a neighbourhood of zero.

COROLLARY 5.1. *Let A be the generator of a C_0-semigroup of bounded linear operators on $\mathfrak{B}$. Equation (3.1) is exponentially stable on $[0, \infty)$ if and only if*

$$\int_0^\infty \|y(t)\|^p \, dt < \infty \tag{5.11}$$

for some p, $1 \leq p < \infty$, and any weak solution $y(t)$.

Proof. We first show that condition (5.11) implies that the semigroup e^{-At} is uniformly bounded. Since e^{-At} is a semigroup of class C_0, there exist $M_1 \geq 1$ and $\omega \geq 0$ such that $\|e^{-At}\| \leq M_1 e^{\omega t}$. If $\omega = 0$ then the uniform boundedness of the semigroup is obvious.

Suppose that $\omega > 0$. It then follows from (5.11) that $e^{-At} f \to 0$ as $t \to \infty$ for any $f \in \mathfrak{B}$. Otherwise we could find a number $\delta > 0$, an element $f \in \mathfrak{B}$ and a sequence $t_j \to \infty$ such that $\|e^{-At_j} f\| \geq \delta$. Without loss of generality we may assume that $t_{j+1} - t_j > \omega^{-1}$. Set $\Delta_j = [t_j - \omega^{-1}, t_j]$ (then the length $m(\Delta_j)$ of the interval Δ_j equals $\omega^{-1} > 0$ and the intervals Δ_j do not overlap). For $t \in \Delta_j$, $t = t_j - \omega^{-1} + \eta\omega^{-1}$ $(0 < \eta < 1)$ we would have

$$\begin{aligned}\delta \leq \|e^{-At_j} f\| &= \|e^{-A(t_j - \omega^{-1} + \eta\omega^{-1} + \omega^{-1}(1-\eta))} f\| = \|e^{-At} e^{-A\omega^{-1}(1-\eta)} f\| \\ &\leq \|e^{-A\omega^{-1}(1-\eta)}\| \|e^{-At} f\| \leq M_1 e^{1-\eta} \|e^{-At} f\| \leq M_1 e \|e^{-At} f\|,\end{aligned}$$

whence $\|e^{-At} f\| \geq \delta (M_1 e)^{-1}$ and therefore

$$\begin{aligned}\int_0^\infty \|y(t)\|^p \, dt &= \int_0^\infty \|e^{-At} f\|^p \, dt \geq \sum_{j=1}^\infty \int_{\Delta_j} \|e^{-At} f\|^p \, dt \\ &\geq \left(\frac{\delta}{M_1 e}\right)^p \sum_{j=1}^\infty m(\Delta_j) = \infty,\end{aligned}$$

which contradicts (5.11). Hence, $e^{-At} f \to 0$ $(t \to \infty)$ for each $f \in \mathfrak{B}$. The uniform boundedness principle yields the inequality $\|e^{-At}\| \leq M$. We may assume that $M = 1$, i.e. that e^{-At} is a contraction semigroup. If this is not the case, we introduce in $\mathfrak{B}$ the new norm $\|f\|_1 = \sup_{t \geq 0} \|e^{-At} f\|$, which is equivalent to the initial one by virtue of the inequalities $\|f\| \leq \|f\|_1 \leq M \|f\|$. In this norm e^{-At} is a contraction semigroup and satisfies (5.11). Then the function

$\|e^{-At}f\|^p$ ($f \in \mathfrak{B}$) does not increase. Since it is integrable on the semi-axis $(0,\infty)$, $c_f > \int_{\frac{t}{2}}^{t} \|e^{-A\tau}f\|^p \, d\tau \geq \frac{t}{2}\|e^{-At}f\|^p$ ($c_f = \text{const}$), whence the estimate

$$\|e^{-At}f\|^p \leq \frac{c'_f}{1+t} \quad (c'_f = \text{const})$$

follows. Consequently,

$$\|y(t)\| = \|e^{-At}f\| \leq \frac{c''_f}{(1+t)^{\frac{1}{p}}}.$$

Thus, for any weak solution $y(t)$ on $[0,\infty)$ of (3.1) the estimate (5.9) with $\gamma(t) = (1+t)^{-\frac{1}{p}}$ is true. In view of Theorem 5.4 the considered equation is exponentially stable.

The converse is obvious. □

Theorem 5.4 shows that if equation (3.1) is asymptotically, but not exponentially, stable, then there exist weak solutions which tend to zero at infinity arbitrary slowly. An important example demonstrating this fact is the case when A is a positive self-adjoint operator on a Hilbert space $\mathfrak{H}$ with continuous spectrum $\sigma_c(A)$ containing zero. In this case any weak solution $y(t) = e^{-At}f$ ($f \in \mathfrak{H}$) of equation (3.1) tends to zero as $t \to \infty$; however, not necessary exponentially. It is interesting that the rate of its approach to zero is determined by the smoothness of the element $f = y(0)$ relative to the operator A^{-1}, which exists because $0 \in \sigma_c(A)$ and A is positive and self-adjoint. We will dwell on this question in some detail.

Let a function $\gamma(t) \in C[0,\infty)$ be non-negative, $\gamma(t) \to \infty$ as $t \to \infty$ and $\forall \epsilon > 0 \quad \exists c_\epsilon > 0 : \gamma(t) \leq c_\epsilon \exp(\epsilon t)$. Set

$$\tilde{\gamma}(\lambda) = \int_0^\infty e^{-2\lambda t}\gamma(t)\, dt, \quad G(\lambda) = \left(\tilde{\gamma}\left(\tfrac{1}{\lambda}\right) + 1\right)^{\frac{1}{2}} \quad (\lambda \geq 0). \tag{5.12}$$

Obviously, $G(\lambda) \in C[0,\infty)$ and $G(\lambda) \to \infty$ as $\lambda \to \infty$.

We denote by Y_γ the set of all weak solutions $y(t)$ on $[0,\infty)$ of (3.1) for which $\int_0^\infty \|y(t)\|^2 \gamma(t)\, dt < \infty$. Y_γ is a Hilbert space with respect to the inner product

$$(y, z)_\gamma = \int_0^\infty (y(t), z(t))\gamma(t)\, dt + (y(0), z(0)).$$

It is clear that Y_γ contains all vector-functions of the form $y(t) = e^{-At}E_\Delta h$ ($\forall h \in \mathfrak{H}$), where Δ is an arbitrary interval in $(0,\infty)$ located at a positive distance from zero.

LEMMA 5.1. *Let $A \geq 0$ be a self-adjoint operator on a Hilbert space $\mathfrak{H}$, $0 \in \sigma_c(A)$. A weak solution $y(t)$ on $[0,\infty)$ of equation (3.1) belongs to the space Y_γ if and only if $y(0) \in \mathfrak{H}_G(A^{-1})$, where the function $G(\lambda)$ can be expressed in terms of $\gamma(t)$ by formula (5.12). Moreover, the mapping $Y_\gamma \ni y(t) \to y(0) \in \mathfrak{H}_G(A^{-1})$ is an isometric isomorphism.*

We would like to recall here that $\mathfrak{H}_G(A^{-1}) = \mathfrak{D}(G(A^{-1}))$, with as norm that of the graph of the operator $G(A^{-1})$.

Proof. Let $y(t) = e^{-At}y_0$ ($y_0 = y(0)$) be a weak solution on $[0,\infty)$ of (3.1). By virtue of the principal spectral theorem for self-adjoint operators,

$$\begin{aligned}\int_0^\infty \|y(t)\|^2\gamma(t)\,dt &= \int_0^\infty \|e^{-At}y(0)\|^2\gamma(t)\,dt \\ &= \int_0^\infty \Big(\int_0^\infty e^{-2\lambda t}\,d(E_\lambda y(0), y(0)\Big)\gamma(t)\,dt \\ &= \int_0^\infty \tilde{\gamma}(\lambda)\,d(E_\lambda y(0), y(0)) = \int_0^\infty \tilde{\gamma}(\tfrac{1}{\mu})\,d(F_\mu y(0), y(0)),\end{aligned}$$

where $F_\mu = E - E_{\frac{1}{\mu}}$ is the resolution of the identity of the operator A^{-1}. Therefore,

$$\begin{aligned}\|y\|_\gamma^2 &= \int_0^\infty \|y(t)\|^2\gamma(t)\,dt + \|y(0)\|^2 \\ &= \int_0^\infty G^2(\lambda)\,d(F_\lambda y(0), y(0)) = \|y(0)\|^2_{\mathfrak{H}_G(A^{-1})}\end{aligned}$$

which proves the lemma. □

The following assertions are direct consequences of this lemma.

THEOREM 5.5. *Let $A \geq 0$ be a self-adjoint operator on $\mathfrak{H}$ with $0 \in \sigma_c(A)$. For a weak solution $y(t)$ on $[0,\infty)$ of equation (3.1) to possess the property*

$$\forall\alpha > 0 \quad \exists c = c(\alpha) > 0 : \|y(t)\| \leq c(1+t)^{-\alpha}$$

it is necessary and sufficient that $y(0) \in \mathfrak{H}^\infty(A^{-1})$.

Proof. Put $\gamma(t) = t^\alpha$ ($\alpha > 0$). It is not difficult to calculate that in this case $G^2(\lambda) = \Gamma(\alpha+1)2^{-(\alpha+1)}\lambda^{\alpha+1} + 1$. It follows from Lemma 5.1 that

$$\mathfrak{H}^{\frac{\alpha+1}{2}}(A^{-1}) = \mathfrak{D}(A^{-\frac{\alpha+1}{2}}) = \{f : \int_0^\infty \|e^{-At}f\|^2t^\alpha\,dt < \infty\}.$$

We show that $f \in \bigcap_{\alpha>0}\mathfrak{H}^\alpha(A^{-1}) = \mathfrak{H}^\infty(A^{-1})$ if and only if for any $\alpha > 0$ there exists $c = c(\alpha) > 0$ such that

$$\|e^{-At}f\| \leq ct^{-\alpha} \quad (t > 0). \tag{5.13}$$

It is obvious that the inequality (5.13) for any $\alpha > 0$ implies $f \in \bigcap_{\alpha>0}\mathfrak{H}^\alpha(A^{-1})$. Conversely, assume that $f \in \bigcap_{\alpha>0}\mathfrak{H}^\alpha(A^{-1})$. Since $s^{2\alpha-1}\|e^{-As}f\|^2 \geq$ $\geq (\frac{t}{2})^{2\alpha-1}\|e^{-At}f\|^2$ for an arbitrary $\alpha > \frac{1}{2}$ and $s \in [\frac{t}{2}, t]$ we have

$$\left(\frac{t}{2}\right)^{2\alpha}\|e^{-At}f\|^2 \leq \int_{\frac{t}{2}}^t s^{2\alpha-1}\|e^{-As}f\|^2\,ds \leq \int_0^\infty s^{2\alpha-1}\|e^{-As}f\|^2\,ds < c'_\alpha < \infty.$$

This implies that $\|e^{-At}f\| \leq c_\alpha t^{-\alpha}$, where $c_\alpha = 2^\alpha(c'_\alpha)^{\frac{1}{2}}$. □

Thus, the space $\mathfrak{H}^\infty(A^{-1})$ of infinitely differentiable vectors of the operator A^{-1} coincides with the set of initial data of those weak solutions on $[0,\infty)$ of equation (3.1) which decrease at the infinity faster than any power. It turns out that initial values $y(0)$ ensuring that $y(t)$ decreases at infinity faster than those in Theorem 5.5 are smoother than vectors from $\mathfrak{H}^\infty(A^{-1})$. Namely, the following assertion is valid.

THEOREM 5.6. *Under the assumptions of Theorem* 5.5 *on the operator A a weak solution $y(t) = e^{-At}y_0$ ($y_0 = y(0)$) posseses the property*

$$\exists c > 0 \quad \exists \alpha > 0 : \|y(t)\| \le c\exp(-\alpha t^p), \quad 0 < p < 1$$

if and only if $y(0) \in \mathfrak{G}_{\{\beta\}}(A^{-1})$ where $\beta = \frac{1-p}{p}$.

Proof. Set $\gamma_\alpha(t) = \exp(\alpha t^p)$ ($\alpha > 0$, $0 < p < 1$). The change of variables $t = z\lambda^{-\frac{1}{1-p}}$ yields

$$\tilde{\gamma}_\alpha(\lambda) = \int_0^\infty \exp(-2\lambda t + \alpha t^p)\,dt = \lambda^{-\frac{1}{1-p}} \int_0^\infty \exp(\lambda^{-\frac{p}{1-p}}(\alpha z^p - 2z))\,dz.$$

We denote by $z_{0,\alpha}$ the point at which the function $h_\alpha(z) = \alpha z^p - 2z$ takes a maximum, i.e. $z_{0,\alpha} = \left(\frac{\alpha p}{2}\right)^{\frac{1}{1-p}}$. Since $h''_\alpha(z) \le 0$ for all $z \in [0,\infty)$, the function $h_\alpha(z)$ is convex. Therefore $h_\alpha(z) \ge h_\alpha(z_{0,\alpha})z(z_{0,\alpha})^{-1}$ ($z \in [0, z_{0,\alpha}]$). This implies that

$$\begin{aligned}
\tilde{\gamma}(\lambda) &= \lambda^{-\frac{1}{1-p}} \int_0^\infty \exp(\lambda^{-\frac{p}{1-p}} h_\alpha(z))\,dz \ge \lambda^{-\frac{1}{1-p}} \int_0^{z_{0,\alpha}} \exp(\lambda^{-\frac{p}{1-p}} h_\alpha(z))\,dz \\
&\ge \lambda^{-\frac{1}{1-p}} \int_0^{z_{0,\alpha}} \exp(\lambda^{-\frac{p}{1-p}} h_\alpha(z_{0,\alpha}) z z_{0,\alpha}^{-1})\,dz \\
&= \lambda^{-\frac{1}{1-p}} z_{0,\alpha} (\exp(\lambda^{-\frac{p}{1-p}} h_\alpha(z_{0,\alpha})) - 1)\lambda^{\frac{p}{1-p}} h_\alpha^{-1}(z_{0,\alpha}) \\
&= \lambda^{-1} z_{0,\alpha} h_\alpha^{-1}(z_{0,\alpha})(\exp(\lambda^{-\frac{p}{1-p}} h_\alpha(z_{0,\alpha})) - 1).
\end{aligned}$$

Due to this estimate, there exists for any ϵ, $0 < \epsilon < 1$, a constant $c_{\alpha,\epsilon}$ such that

$$\tilde{\gamma}(\lambda) \ge c_{\alpha,\epsilon} \exp(\lambda^{-\frac{p}{1-p}} h_\alpha(z_{0,\alpha})(1-\epsilon)) \tag{5.14}$$

if $\lambda \in [0, b]$, where b is any finite number.

Let $z_{1,\alpha} = \left(\frac{\alpha}{2}\right)^{\frac{1}{1-p}}$, $h_\alpha(z_{1,\alpha}) = 0$. For all $z \in [0, z_{1,\alpha}]$ we have $h_\alpha(z) \le h_\alpha(z_{0,\alpha})$, and for $z \in [z_{1,\alpha}, \infty)$, $h_\alpha(z) \le h'_\alpha(z_{1,\alpha})(z - z_{1,\alpha})$, $h'_\alpha(z_{1,\alpha}) < 0$. Therefore

$$\begin{aligned}
\tilde{\gamma}_\alpha(\lambda) &= \lambda^{-\frac{1}{1-p}} \int_0^\infty \exp(\lambda^{-\frac{p}{1-p}} h_\alpha(z))\,dz \\
&\le \lambda^{-\frac{1}{1-p}} \int_0^{z_{1,\alpha}} \exp(\lambda^{-\frac{p}{1-p}} h_\alpha(z_{0,\alpha}))\,dz + \\
&\quad + \lambda^{-\frac{1}{1-p}} \int_{z_{1,\alpha}}^\infty \exp(\lambda^{-\frac{p}{1-p}} h'_\alpha(z_{1,\alpha})(z - z_{1,\alpha}))\,dz \\
&= \lambda^{-\frac{1}{1-p}} z_{1,\alpha} \exp(\lambda^{-\frac{p}{1-p}} h_\alpha(z_{0,\alpha})) - (\lambda h'_\alpha(z_{1,\alpha}))^{-1},
\end{aligned}$$

whence one can conclude that for any $\alpha > 0$, $\epsilon > 0$ there exists a constant $K_{\alpha,\epsilon} > 0$ such that

$$\tilde{\gamma}_\alpha(\lambda) \le K_{\alpha,\epsilon} \exp(\lambda^{-\frac{p}{1-p}} h_\alpha(z_{0,\alpha})(1+\epsilon)). \tag{5.15}$$

Taking into account that the functions $\tilde{\gamma}(\lambda)$ and $\exp(\lambda^{-\frac{p}{1-p}} h_\alpha(z_{0,\alpha})(1\pm\epsilon))-1$ tend to zero as $\lambda \to \infty$, by (5.14) and (5.15) we arrive at the conclusion that for any $\alpha > 0$ and any $\epsilon \in (0,1)$ there exist constants $K^1_{\alpha,\epsilon}$ and $K^2_{\alpha,\epsilon}$ such that

$$K^1_{\alpha,\epsilon} \exp(\lambda^{-\frac{p}{1-p}} h_\alpha(z_{0,\alpha})(1-\epsilon)) \le \tilde{\gamma}(\lambda)+1 \le K^2_{\alpha,\epsilon} \exp(\lambda^{-\frac{p}{1-p}} h_\alpha(z_{0,\alpha})(1+\epsilon)).$$

This estimate and the fact that $h_\alpha(z_{0,\alpha}) \to 0$ $(\alpha \to 0)$ enable us to state that

$$\mathfrak{G}_{\{\beta\}}(A^{-1}) = \operatorname*{limind}_{\alpha\to 0} \mathfrak{H}_{\exp(\alpha\lambda^{\frac{1}{\beta}})}(A^{-1}) = \operatorname*{limind}_{\alpha\to 0} \mathfrak{H}_{G_\alpha}(A^{-1}),$$

where $G_\alpha(\lambda) = (\tilde{\gamma}_\alpha(\lambda^{-1}) + 1)^{\frac{1}{2}}$, $\beta = \frac{p}{1-p}$.

According to Lemma 5.1 the equivalence

$$(y(0) \in \mathfrak{H}_{G_\alpha}(A^{-1})) \Longleftrightarrow \left(\int_0^\infty \|y(t)\|^2 \exp(\alpha t^p)\, dt < \infty\right)$$

holds. Therefore

$$(y(0) \in \mathfrak{G}_{\{\beta\}}(A^{-1})) \Longleftrightarrow \left(\exists \alpha > 0\ \exists c > 0 : \int_0^\infty \|y(t)\|^2 \exp(\alpha t^p)\, dt \le c\right) \quad (5.16)$$

The latter is equivalent to the pointwise estimate

$$\|y(t)\| \le c \exp(-\alpha' t^p). \quad (5.17)$$

Indeed, if (5.17) is valid for $y(t)$, then an immediate check verifies that the right-hand side of (5.16) is valid. Conversely, let (5.16) be valid for a weak solution $y(t) = e^{-At}y_0$ $(y_0 = y(0))$ on $[0,\infty)$ of (3.1). Since

$$\|e^{-As}y_0\|^2 \exp(\alpha s^p) \ge \|e^{-At}y_0\|^2 \exp(\alpha(\tfrac{t}{2})^p)$$

for $s \in [\frac{t}{2}, t]$, we obtain the inequalities

$$\begin{aligned}\tfrac{t}{2}\|e^{-At}y_0\|^2 \exp(\alpha(\tfrac{t}{2})^p) &\le \int_{\frac{t}{2}}^t \|e^{-As}y_0\|^2 \exp(\alpha s^p)\, ds \\ &\le \int_0^\infty \|e^{-As}y_0\|^2 \exp(\alpha s^p)\, ds \le c,\end{aligned}$$

i.e.

$$\|e^{-At}y_0\| \le (2ct^{-1})^{\frac{1}{2}} \exp\left(-\frac{\alpha t^p}{2^{p+1}}\right).$$

This completes the proof. □

COROLLARY 5.2. *If $A = A^* \ge 0$ and $0 \in \sigma_c(A)$, then*

$$(f \in \mathfrak{A}(A^{-1})) \Longleftrightarrow (\|e^{-At}f\| \le c \exp(-\alpha t^{\frac{1}{2}})).$$

EXAMPLE. Let A be the closure in the space $\mathfrak{H} = L_2(\mathbb{R}^1)$ of the operator $f \to -\frac{d^2 f}{dx^2}$, given on the set of infinitely differentiable functions with compact support. Then equation (3.1) becomes

$$\frac{\partial u(x,t)}{\partial t} = \frac{\partial^2 u(x,t)}{\partial x^2} \quad (t \ge 0,\ x \in \mathbb{R}^1). \quad (5.18)$$

It follows from Theorems 5.5, 5.6 that the relation $\int_{-\infty}^{\infty} |u(x,t)|^2 \, dx = o(t^{-k})$ as $t \to \infty$ ($\forall\, k \in \mathbb{N}$), where $u(x,t)$ is the solution of the Cauchy problem $u(x,0) = f(x) \in L_2(\mathbb{R}^1)$ for the heat equation (5.18) on the half-space $t \geq 0$, is possible if and only if the integrals $\int_{-\infty}^{\infty} |\tilde{f}(\lambda)|^2 \lambda^{-2n} \, d\lambda$ exist for any natural n ($\tilde{f}(\lambda)$ is the Fourier transform of f). Moreover,

$$\int_{-\infty}^{\infty} |u(x,t)|^2 \, dx \leq c \exp(-\alpha t^p) \quad (\alpha > 0,\ 0 < p < 1,\ c = \text{const})$$

if and only if

$$\exists\, \alpha' > 0 : \sup_n \frac{\int_{-\infty}^{\infty} |\tilde{f}(\lambda)|^2 \lambda^{-4n} \, d\lambda}{(\alpha')^n n^{n\beta}} < \infty \quad \left(\beta = \frac{1-p}{p}\right).$$

6. Applications

6.1. THE HEAT EQUATION

We consider the equation

$$D_t u(t,x) = \Delta_x u(t,x), \quad \{t,x\} \in (0,\infty) \times \mathbb{R}^n. \tag{6.1}$$

By a solution of it in the half-space $t > 0$ we understand a function $u(t,x)$: $(0,\infty) \times \mathbb{R}^n \to \mathbb{R}^1$, continuously differentiable with respect to $t \in (0,\infty)$ and twice continuously differentiable with respect to the variables $x_1, \ldots, x_n$ for all $x = \{x_1, \ldots, x_n\} \in \mathbb{R}^n$, satisfying (6.1). Formally we may give $u(x,t)$ as

$$u(t,x) = \frac{1}{(4\pi t)^{\frac{n}{2}}} \int_{\mathbb{R}^n} \exp\left(-\frac{|x-y|^2}{4t}\right) f(y) \, dy, \quad f(x) = u(0,x). \tag{6.2}$$

We are interested in the existence of boundary values ($t \to 0$) of solutions $u(t,x)$ in function spaces of L_p-type with weight, and in this connection we are interested in giving a meaning to formula (6.2). As weights we select positive functions $q(x)$ such that

$$q(x) \leq \exp(a|x|^2), \quad \forall\, a > 0$$

and

$$\forall t > 0\ \exists\, c_t > 0 : \frac{1}{(4\pi t)^{\frac{n}{2}}} \int_{\mathbb{R}^1} \exp\left(-\frac{|x-y|^2}{4t}\right) q(y) \, dy \leq c_t q(x) \ (\forall\, x \in \mathbb{R}^n). \tag{6.3}$$

A simple sufficient condition for (6.3) to be valid is the estimate $q(x+y) \leq c(y) q(x)$ where $c(y) \in L_1(\mathbb{R}^n, e^{-a|y|^2} dy)$ for any $a > 0$. The functions $P(|x|)$ (where P is a polynomial without positive roots), $\exp(b|x|^\alpha)$ ($0 \leq \alpha \leq 1$, $b \in \mathbb{R}^1$), e^{bx} and others may be used as examples of functions with these properties.

We now consider the spaces $L_p(\mathbb{R}^n, q(x)\, dx)$ ($1 \leq p < \infty$). Set $L_\infty(\mathbb{R}^n, q(x)) = \{f : \exists\, c > 0 : |f(x)| \leq c q(x)$ almost everywhere$\}$ with norm which is equal to the least constant c in the last inequality. The space $L_\infty(\mathbb{R}^n, q(x))$ is dual to

$L_1(\mathbb{R}^n, q(x)\,dx)$ relative to $\int_{\mathbb{R}^n} f(x)g(x)\,dx$. It is not difficult to show that the family of bounded linear operators

$$(W(t)f)(x) = \frac{1}{(4\pi t)^{\frac{n}{2}}} \int_{\mathbb{R}^n} \exp\Big(-\frac{|x-y|^2}{4t}\Big) f(y)\,dy$$

is a semigroup on each of the spaces mentioned above. Indeed, inequality (6.3) implies that $\|W(t)\|_{[L_1(\mathbb{R}^n,q(x)\,dx)]} \le c_t$. Since $\|W(t)\|_{[L_\infty(\mathbb{R}^n,\,dx)]} \le 1$, it follows from the Riesz–Thorin interpolation theorem (see Reed, Simon [2, IX.4]) that $\|W(t)\|_{[L_p(\mathbb{R}^n,q(x)\,dx)]} \le c_t^{p^{-1}}$. For $L_\infty(\mathbb{R}^n, q(x))$ we also obtain from (6.3) that $\|W(t)\|_{[L_\infty(\mathbb{R}^n,q(x))]} \le c_t$.

We will next prove that the semigroup $W(t)$ is infinitely differentiable on $(0,\infty)$ into all spaces $L_p(\mathbb{R}^n, q(x)\,dx)$. Since $u \le 2e^{\frac{u}{2}}$, we have

$$\|W'(t)f\|_{L_1(\mathbb{R}^n,q(x)\,dx)} \le \frac{1}{(4\pi t)^{\frac{n}{2}}} \int_{\mathbb{R}^n}\int_{\mathbb{R}^n} \Big(\frac{n}{2t} + \frac{|x-y|^2}{4t^2}\Big) \exp\Big(-\frac{|x-y|^2}{4t}\Big) \times$$

$$\times\, q(x)|f(x)|\,dx\,dy$$

$$\le \frac{1}{(4\pi t)^{\frac{n}{2}}} \cdot \frac{n}{t} \int_{\mathbb{R}^n}\int_{\mathbb{R}^n} \Big(\tfrac{1}{2} + \frac{|x-y|^2}{4t}\Big) \exp\Big(-\frac{|x-y|^2}{4t}\Big) q(x)|f(y)|\,dx\,dy$$

$$\le 2(4\pi t)^{-\frac{n}{2}} \frac{n}{t} \int_{\mathbb{R}^n}\int_{\mathbb{R}^n} \exp\Big(\tfrac{1}{4} + \frac{|x-y|^2}{8t}\Big) \exp\Big(-\frac{|x-y|^2}{4t}\Big) q(x)|f(y)|\,dx\,dy$$

$$\le \frac{2}{(4\pi t)^{\frac{n}{2}}} e^{\frac{1}{4}} \frac{n}{t} \int_{\mathbb{R}^n} \Big(\int_{\mathbb{R}^n} \exp\Big(-\frac{|x-y|^2}{8t}\Big) q(x)\,dx\Big) |f(y)|\,dy$$

$$\le 2e^{\frac{1}{4}} \cdot 2^{\frac{n}{2}} c_{2t} \frac{n}{t} \int_{\mathbb{R}^n} |f(y)| q(y)\,dy = 2e^{\frac{1}{4}} c_{2t} \cdot 2^{\frac{n}{2}} \cdot \frac{n}{t} \|f\|_{L_1(\mathbb{R}^n,q(x)\,dx)}.$$

Hence, $W(t)$ is infinitely differentiable ($t \in (0,\infty)$) on $L_1(\mathbb{R}^n, q(x)\,dx)$. Since $\|W'(t)f\|_{L_\infty(\mathbb{R}^n,\,dx)} \le e^{\frac{1}{4}} 2^{\frac{n}{2}} \frac{n}{t} \|f\|_{L_\infty(\mathbb{R}^n,\,dx)}$, the semigroup $W(t)$ is infinitely differentiable ($t \in (0,\infty)$) on $L_\infty(\mathbb{R}^n, dx)$ too. Applying once more the Riesz–Thorin theorem we establish that $W(t)$ is infinitely differentiable on the spaces $L_p(\mathbb{R}^n, q(x)\,dx)$ ($p < \infty$). In exactly the same way its differentiability on $L_\infty(\mathbb{R}^n, q(x))$ is obtained.

If c_t in (6.3) does not depend on t, then $\sup_{0<t\le 1} \|W(t)\| < \infty$ in the corresponding spaces. Since $C_0^\infty(\mathbb{R}^n) \subset \mathfrak{D}(A)$ (A is the generator of $W(t)$ in the corresponding space), then $\mathfrak{D}(A)$ is dense in $L_p(\mathbb{R}^n, q(x)\,dx)$, hence $W(t)$ is a semigroup of class C_0 if $c_t = \text{const}$. Moreover, in this case it is a holomorphic semigroup. If $c_t \ne \text{const}$, then the estimate $q(x) \le \exp(a|x|^2)$ implies, because of the uniqueness theorem for solutions of the Cauchy problem for partial differential equations (see Gelfand–Šilov [2, p. 52]), that in all spaces considered $\operatorname{Ker} W(t) = \{0\}$. Now we can employ the results of Section 3 if $p \in (1,\infty]$. For $p = 1$ it should be noted here that $L_1(\mathbb{R}^n, q(x)\,dx)$ is not dual to any Banach space. However, we can imbed it into the space $\operatorname{rca}(\mathbb{R}^n, q(x))$ of Baire measures μ on $\mathbb{R}^n$ for which $\int_{\mathbb{R}^n} q(x)\,d|\mu| < \infty$ ($|\mu|$ is the total variation of a measure μ; the norm in $\operatorname{rca}(\mathbb{R}^n, q(x))$ coincides with the latter integral) as the space of all absolutely continuous measures with the same property. This space is the dual of the space of continuous functions $f(x)$ on $\mathbb{R}^n$ with compact support, with norm given by $\sup_{x\in\mathbb{R}^n} \frac{f(x)}{q(x)}$. After this remark, on the basis of Theorem 3.3 and Corollary 3.3 we arrive at the following conclusion.

THEOREM 6.1. *Let $q(x)$ be a positive function satisfying condition (6.3) where c_t does not depend on t and let $u(t,x) : (0,\infty) \times \mathbb{R}^n \to \mathbb{R}^1$ be a solution in the half-space $(0,\infty)\times\mathbb{R}^n$ of the heat equation (6.1). Then*

a) *in order that a function $u(x) \in L_p(\mathbb{R}^n, q(x)\,\mathrm{d}x)$ $(p \in (1,\infty))$ such that*

$$\lim_{t\to 0} \int_{\mathbb{R}^n} |u(t,x) - u(x)|^p q(x)\,\mathrm{d}x = 0$$

exists, it is necessary and sufficient that

$$\sup_{t\in(0,1)} \int_{\mathbb{R}^n} |u(t,x)|^p q(x)\,\mathrm{d}x < \infty;$$

b) *in order that a function $u(x) \in L_\infty(\mathbb{R}^n, q(x))$ such that*

$$u(t,x) = \frac{1}{(4\pi t)^{\frac{n}{2}}} \int_{\mathbb{R}^n} \exp\Big(-\frac{|x-y|^2}{4t}\Big) u(y)\,\mathrm{d}y$$

exists, it is necessary and sufficient that

$$\sup_{t\in(0,1)} \sup_{x\in\mathbb{R}^n} \frac{|u(t,x)|}{q(x)} < \infty;$$

c) *there exists a Baire measure μ such that $\int_{\mathbb{R}^n} q(x)\,\mathrm{d}|\mu| < \infty$ and*

$$u(t,x) = \frac{1}{(4\pi t)^{\frac{n}{2}}} \int_{\mathbb{R}^n} \exp\Big(-\frac{|x-y|^2}{4t}\Big)\,\mathrm{d}\mu(y)$$

if and only if

$$\sup_{t\in(0,1)} \int_{\mathbb{R}^n} |u(t,x)| q(x)\,\mathrm{d}x < \infty.$$

If, in addition, we assume that $q(x) \le \exp(a|x|^2)$ $(\forall\, a > 0)$, then this theorem is also valid when c_t depends on t. However, in a) we must use the representation by the Gauss–Weierstrass integral instead of approach to zero and replace everywhere "necessary and sufficient" by "sufficient".

6.2. THE LAPLACE EQUATION IN A STARLIKE DOMAIN

Let G be a bounded starlike domain in $\mathbb{R}^n$ with infinitely smooth boundary ∂G. Thus, for a certain $x \in \mathbb{R}^n$ and all $y \in \partial G$ the interval $[x, y)$ lies in G. We may assume $x = 0$. In this case $\alpha G \subset G$ for all $\alpha \in [0,1]$.

It is well-known (see Miranda [1]) that for any function $f \in C[\partial G]$ there exists a $u \in C[\overline{G}]$ such that $\Delta u = 0$ in G and $u = f$ on ∂G. On $C[\partial G]$ we define operators

P_α ($\alpha \in [0,1]$) as follows: $(P_\alpha f)(x) = u(\alpha x)$. According to the maximum modulus principle,

$$\|P_\alpha f\|_{C[\partial G]} = \max_{x\in\partial G} |u(\alpha x)| \le \max_{x\in\partial G} |f(x)| = \|f\|_{C[\partial G]},$$

which shows that operators P_α are contractions. Their norm $\|P_\alpha\| = 1$ is reached on $f(x) = 1$. We prove that $P_{\alpha_1\alpha_2} = P_{\alpha_1}P_{\alpha_2}$ ($\forall\,\alpha_1, \alpha_2 \in [0,1]$). By virtue of the structure of the domain G and of the Laplace operator, $u(\alpha x)$ ($\forall\,\alpha \in [0,1]$) is a harmonic function in G if $\Delta u = 0$ in G. The relation required for P_α is deduced from the uniqueness of a solution of the Dirichlet problem. After the change $t = \ln\frac{1}{\alpha}$ we arrive at the semigroup $Q(t) = P_{e^{-t}} = P_\alpha$ of contractions on $C[\partial G]$. Since harmonic functions are infinitely differentiable in G, $\mathfrak{R}(Q(t)) \subset C^\infty[\partial G]$ as $t > 0$. Moreover, $\operatorname{Ker} Q(t) = \{0\}$. Thus, the theory of Section 3 can be applied to the semigroup $Q(t)$.

Let $\mathfrak{B}_{\{-\}}$ be the corresponding space constructed using the semigroup $Q(t)$. For the extension of $Q(t)$ to $\mathfrak{B}_{-t}$ we preserve the same notation. The space $\mathfrak{B}_{\{-\}}$ contains the set $\mathcal{D}'(\partial G)$ of all distributions on ∂G. Indeed, if D is any differential operator on $C[\partial G]$ with constant coefficients, then by virtue of the closed graph theorem $\|DQ(t)\|_{C[\partial G]} \le a(D,t)$. But D and $Q(t)$ commute, whence $\|Q(t)D\|_{C[\partial G]} \le a(D,t)$. This inequality implies the topologic imbedding $\mathcal{D}'(\partial G) \subset \mathfrak{B}_{\{-\}}$. Hence, $Q(t)$ acts not only on $C[\partial G]$ but also on $\mathcal{D}'(\partial G)$, and $Q(t)\mathcal{D}'(\partial G) \subset C^\infty[\partial G]$ ($t > 0$). In particular, $Q(t)$ is defined on $\mathrm{rca}(\partial G)$. Moreover, weak convergence in $\mathrm{rca}(\partial G)$ implies convergence in $\mathcal{D}'(\partial G)$. Therefore, if $\{f_n\}_{n=1}^\infty$ converges weakly to f in $\mathrm{rca}(\partial G)$, then $Q(t)f_n \to Q(t)f$ in $C[\partial G]$.

Assume that $\mu \in \mathrm{rca}(\partial G)$. Then $Q(t)\mu$ can be represented by

$$(Q(t)\mu)(x) = \int_{\partial G} k_t(x,y)\,d\mu(y), \quad k_t(x,y) = (Q(t)\delta_y)(x), \tag{6.4}$$

where δ_y is the Dirac δ-function concentrated at the point y. This is true because $Q(t)$ is defined on δ-functions, and linear combinations of δ-functions are weakly dense in $\mathrm{rca}(\partial G)$. The kernel $k_t(x,y)$ (the kernel of the solution of the Dirichlet problem) is continuous with respect to both variables x and y, and also $Q(t)1 = \int_{\partial G} k_t(x,y)\,dy = \int_{\partial G} k_t(x,y)\,dx = 1$. In addition, $(Q(t)f)(x) \ge 0$ if $f(x) \ge 0$. If this were false we would obtain a contradiction to the fact that a harmonic function cannot receive its maximum or minimum inside its domain. Now we are able to conclude that the kernel $k_t(x,y)$ is non-negative. The representation (6.4) implies the following important equalities:

$$\|Q(t)\|_{[L_\infty[\partial G]]} = \|Q(t)\|_{[C[\partial G]]} = 1; \quad \|Q(t)\|_{[\mathrm{rca}(\partial G)]} = \|Q(t)\|_{[L_1(\partial G)]}.$$

Further,

$$\|Q(t)f\|_{L_1(\partial G)} = \int_{\partial G} |(Q(t)f)(x)|\,dx \le \int_{\partial G}\Big(\int_{\partial G} k_t(x,y)|f(y)|\,dy\Big)\,dx$$
$$= \int_{\partial G}\Big(\int_{\partial G} k_t(x,y)\,dx\Big)|f(y)|\,dy = \int_{\partial G}|f(y)|\,dy = \|f\|_{L_1(\partial G)},$$

i.e.

$$\|Q(t)\|_{[L_1(\partial G)]} \leq 1.$$

We now consider the spaces $\mathfrak{B}$ interpolational for the pair $(L_1(\partial G), L_\infty(\partial G))$ in the sense of S. Krein, Petunin, Semyonov [1]. All Orlicz spaces (in particular, $L_p(\partial G)$), Marcinkiewicz spaces, etc. are among them. We also note that if $C[\partial G]$ is dense in such a $\mathfrak{B}$, then $Q(t)$ is a C_0-semigroup on it, since the domain of its generator is dense in $C[\partial G]$, hence in $\mathfrak{B}$. From the above-mentioned facts we arrive at the following statement.

THEOREM 6.2. *Let G be a starlike domain relative to zero with infinitely differentiable boundary ∂G. Assume also that u is a harmonic function in G and that $k_t(x, y)$ is the kernel of the solution of the Dirichlet problem. Then*

a) $\sup_{x \in G} |u(x)| < \infty$ *if and only if*

$$u(e^{-t}x) = \int_{\partial G} k_t(x, y) f(y)\, dy \quad (t > 0,\ x \in \partial G),$$

where $f \in L_\infty(\partial G)$;

b) $\sup_{\frac{1}{2} < \alpha < 1} \|u(\alpha x)\|_{L_1(\partial G)} < \infty$ *if and only if*

$$u(e^{-t}x) = \int_{\partial G} k_t(x, y)\, d\mu(y),$$

where μ is a finite measure;

c) *if $\mathfrak{B}$ is an interpolation space for the pair $(L_1(\partial G), L_\infty(\partial G))$, $C[\partial G]$ is dense in $\mathfrak{B}$ and $\mathfrak{B}$ is the dual of a Banach space (in particular, if $\mathfrak{B} = L_p$, $1 < p < \infty$), then the existence of a function $v \in \mathfrak{B}$ for which $\|u(t, x) - v(x)\|_{\mathfrak{B}} \to 0$ $(t \to 1)$ is equivalent to the validity of the estimate*

$$\sup_{\frac{1}{2} < t < 1} \|u(t, x)\|_{\mathfrak{B}} < \infty.$$

Bibliographical Comments

Chapter 1

Section 1. Good references on Banach spaces and operators on them are the books Dunford–Schwartz [1] and Naimark [1], which contain proofs of all cited facts.

Section 2. The theory of bounded operators on a Hilbert space in the extent necessary for our purposes is contained, for example, in the books Ahiezer–Glazman [1] and Naimark [1].

Section 3. The basic statements regarding vector-valued functions cited in Subsections 3.1–3.3, 3.5 are contained in the book Hille–Phillips [1] with complete proofs. The well-known facts concerning the space $L_2(\mathfrak{H}, (a, b), d\sigma)$ are given in more detail as far as this space appears throughout the whole book. Subsection 3.6 is written on the basis of Laptev's work [1].

Section 4. The theory of unbounded operators on a Hilbert space (see Subsections 4.1–4.6) is presented according to the books Ahiezer–Glazman [1] and Dunford–Schwartz [2]. The tensor product of Hilbert spaces and operators (Subsection 4.7) is presented on the basis of Berezansky's book [4].

Section 5. The operational calculus for various classes of operators are contained in the books Ahiezer–Glazman [1], Dunford–Schwartz [1, 2], and S. Krein [1]. More detailed information about the exponential function of a bounded operator can be found in the book Daletsky–M. Krein [1].

Section 6. The minimax principle for self-adjoint completely continuous operators goes back to Fisher [1] and Courant [1]. S-numbers were introduced by Schmidt [1] while studying integral equations with non-symmetric kernels. Theorem 6.2 was obtained by Allakhverdiev [1]. Inequality (6.16) for eigenvalues of the sum of self-adjoint operators was established by Weyl [1, 2]. The analogous inequalities (6.14) and (6.15) for s-numbers were proved by Ku Fan [1]. The assertion of lemma 6.2 is well-known and frequently used. Its brief proof is given for completeness of presentation. The asymptotic Theorems 6.3 and 6.4 were established by Mikhailets [2]. They were preceded by Corollary 6.5, which is due to Ku Fan [1]. For self-adjoint operators it was already known to Weyl. As has been shown in the work of Mikhailets [7] the dominant term of the asymptotics of the spectrum in Corollary 6.5 need not be of power form. More detailed consideration of the ideals $\mathfrak{S}_p$ is contained in the book by Schatten [1]. Theorem 6.5 concerning the completeness of the system of eigenvectors and adjoined vectors of a dissipative nuclear operator belongs to Lidsky [1] and Theorem 6.6. – to Sakhnovich [1] (if $p = 2$) and Matsaev

[1] (for arbitrary p).

Section 7. The elementary notions of the theory of linear topological spaces collected in Subsections 7.1, 7.2 are presented on the basis of the books Kantorovich–Akilov [1], Gelfand–Shilov [1], Yosida [1] and Reed–Simon [1], Liu Gui-Zhong [1]. The theory of the Schwartz spaces of test and generalized functions, as well as of Sobolev spaces, can be found in the monographs by Schwartz [1], Vladimirov [2], Sobolev [1]. The proof of the intermediate derivative theorem (Theorem 7.2) is contained in the book Lions–Magènes [1].

Chapter 2

Section 1. The presentation of the theory of positive and negative spaces follows Berezansky's books [4, 5] (see also the paper [3]), which contain the complete information on this topic. Lemma 1.1 belongs to Richter [1].

Section 2. The construction of riggings of a Hilbert space by Hilbert spaces or their inductive and projective limits using a family of functions of a positive self-adjoint operator (Subsections 2.1 and 2.4) is done as in the authors' paper [9].

Various classes of infinitely differentiable vectors of the differentiation operator on concrete function spaces (analytic functions, Gevrey classes, entire functions of exponential type, quasi-analytic classes, etc.) were investigated in classical mathematical analysis (see Lions–Magènes [2]). In the abstract case the analytic vectors of an operator were introduced in Harish-Chandra [1] and Cartier–Dixmier [1], and were comprehensively investigated by Nelson [1]. The question of the denseness of the set of such vectors in the initial space is one of the first problems arising here. This problem is elucidated in Gelfand [1], Beals [1], Nelson [1], Nussbaum [1], Khrushchev [1], Radyno [1], Knyazyuk [2]. Theorem 2.5 was proved by Nelson, 2.4 – by Nussbaum. More detailed information about the theory of analytic vectors can be obtained from Reed–Simon [2]. The abstract analogue of the Paley–Wiener theorem (Theorem 2.2) belongs to V. I. Gorbachuk [4]. Analytic domination of operators was considered by Nelson [1], detailed information on this problem can be found in the book Barut–Rączka [1]. An important role here is occupied by the study of monotonic operator-valued functions. In the case of a finite-dimensional space they were investigated in the paper of Löwner [1]; for bounded operators on an infinite dimensional space – in Bendat–Sherman [1]. A systematic treatment of monotonic operator-function theory is presented in Donaghue's book [1]. Lemma 2.3 for bounded operators A and B is contained in Berezansky's monograph [4]. Theorem 2.6 and ensuing corollaries belong to Kashpirovsky [3].

Section 3. A great deal of literature on the theory of semigroups (see, for example, Hille–Phillips [1]) is associated with the study of properties of the operator exponential function e^{-At}, where A is, generally speaking, an unbounded operator with dense domain in a Hilbert space $\mathfrak{H}$. Vector-functions of the form $y(t) = e^{-At} f$ ($t \in [a, b]$), where f runs through, unlike in previous works, various spaces of generalized elements wider than $\mathfrak{H}$, guaranteeing the infinite differentiability of $y(t)$

inside the interval (a, b), were investigated in the authors' papers [6, 9], which are the basis for this section.

Section 4. Equations of the kind (4.1) and more general ones were being studied by various authors (see, for example, the books by Lions–Magènes [1] and S. Krein [1]). However, in describing various classes of solutions of those equations additional conditions on the behaviour of a solution at the ends of the interval were being imposed; for instance, the solutions had to take their values at the points a and b from a certain subset of the initial space $\mathfrak{H}$ or had to belong to a certain L_p space. The purpose of this section is to describe all solutions of equation (4.1) which are smooth only inside (a, b) and classify them in dependence on their growth near the ends of the interval. The considerations are carried out in the scheme of the authors' papers [6, 9].

Section 5. The results of the authors' papers [6, 9, 10] form the basis of Subsections 5.1–5.3, 5.6 of this section. The theory of vector-valued distributions was constructed by Schwartz [2]. Generalized solutions of equation (5.1) were thoroughly studied in Lions' monograph [1]. Theorem 5.4 about the coincidence of a generalized solution of this equation with a classical one was established by M. Gorbachuk and Kashpirovsky [1], in a more special case – by V. I. Gorbachuk [1]. The main results of this section were generalized by Kashpirovsky [2, 3], Knyukh [1] and Fishman [1, 2] to a certain class of higher-order equations.

Section 6. The investigation of the boundary values of functions analytic in a disc and other domains starts probably from Fatou's theorem, stating that if an analytic function $f(z)$ in the unit disc is bounded inside the disc then it has boundary values almost everywhere as z approaches the boundary along non-tangential directions (see, for example, Privalov [1]). These boundary values form a function on the boundary called the boundary value of $f(z)$. By relaxing the uniform boundedness of $f(z)$ and imposing restrictions on its growth near the boundary many authors established the existence of the boundary value of $f(z)$ in function spaces. So, F. Riesz [1] showed that the condition $\int_0^{2\pi} |f(re^{i\varphi})|^p d\varphi \le \text{const} < \infty$ is necessary and sufficient for the existence of the boundary value of $f(z)$ in the space L_p $(p > 1)$. Since the theory of generalized functions appeared, the question arose whether it is possible to ascribe a generalized boundary value for any analytic function in the disc. The answer was given by Köthe [1]. He has established that for such a function there always exists a boundary value in the class of generalized functions which are linear continuous functionals on the linear topological space of functions analytic on the unit circle, i.e. in the class of hyperfunctions. The boundary value of $f(z)$ belongs to the class of Schwartz distributions if and only if the function $f(z)$ has power growth near the boundary (see Köthe [1]). For analytic functions of several variables on non-compact domains (in a half-space, for example) this fact was established by Vladimirov [1] and Tillman [1]. If the growth of $f(z)$ when approaching the boundary is exponential, then, as was shown by Komatsu [1, 2], its boundary value is an ultradistribution. Certainly, the existence of boundary values of analytic functions is equivalent to that of harmonic functions, i.e. of solutions of the Laplace equation. With the aim of further generalization of the above-mentioned results concerning boundary values of harmonic functions, the Laplace equation was replaced by an arbitrary elliptic (see Stein [1], Stein–Weiss [1], Hirschman–Widder [1], Lions–Magènes [1], Mazya [1], V. Mikhailov [1, 3], Gushchin–V. Mikhailov [1, 2], Roitberg [1–3], Ju. Mikhailov [1], Petrushko [1]) and parabolic (see Johnson [1], Zhitarashou [1, 2], Ivasishen [1], Petrushko [2, 3], Chabrowski [1], and others) partial differential equations. In this section the abstract results of the previous one are illustrated when A in equation (5.1) coincides with the operator $-\frac{d^2}{dx^2}$ defined on periodic functions. As a special case of the theorems on the boundary values of solutions of operator differential equation (5.1), all above-mentioned results are obtained for a function that is 2π-periodic with respect to one of the variables and harmonic in the upper half-plane. Successful means for reaching this purpose are formal trigonometric series, whose theory was developed in the authors' work [12]. Theorems 6.3, 6.4 are contained in the books by Schwartz [1], Antosik–Mikusiński–Sikorski [1] and in a paper of Walter [1].

Section 7. The purpose of this section is to illustrate the results of Section 5 for the special case where $A = -\frac{d^2}{dx^2}$ is considered on $L_2(-\infty, \infty)$. Theorem 7.2, belonging to Paley and Wiener [1], is used here. The presentation is being conducted in the style of the authors' work [9]. Theorem 7.1 and 7.5 contain results of F. Riesz [1] and Vladimirov [1], as well as those of Komatsu [1, 2] regarding analytic functions in a half-plane. Theorems 7.3, 7.4, 7.6 were established in V. Gorbachuk–M. Gorbachuk [9], V. Gorbachuk–Fyodorova [1]. The notion of quasi-analytic class of functions is classical (in this connection see, for example, the books by Mandelbroit [1] and Shilov [2]). Theorem 7.7 is known as the Carleman–Ostrowski theorem. Theorems 7.8 and 7.9 belong to M. Gorbachuk, Dudnikov and Fyodorova [1].

Section 8. The operational calculus of non-self-adjoint operators with spectrum on the real axis and whose resolvent has power growth near the spectrum was constructed for the classes of n-times continuously differentiable functions on the spectrum (n depends on the rate of growth) in Wolf [1], Berezansky [2], Volk [1], Fojas [1]. When the growth of the resolvent near the spectrum exceeds power growth it was developed in the case of a bounded operator for certain classes of infinitely differentiable functions by Dyn'kin [1] and Cioranescu [1]. The results of this section are presented according to V. Gorbachuk–Fyodorova [1], where the operational calculus is constructed, on the basis of the boundary value theory for harmonic functions in the upper half-plane, of unbounded operators with spectrum on the real axis and whose resolvent has growth near the spectrum regulated by an arbitrary function $\gamma(t)$. In this case the class of functions suitable for the operational calculus is determined by the function γ. It should be noted that Lyubich–Matsaev [1], devoted to separation of spectrum of non-self-adjoint operators, is close to this subject.

Section 9. The space of initial data of smooth solutions of the Cauchy problem for equation (9.1) is described in M. Gorbachuk–Dudnikov [1]. The existence of the boundary values of solutions for this equation in various classes of generalized functions (Theorems 9.2 and 9.3) was established in the authors' work [9]. The spaces of S type were introduced by Shilov [1]. In the same book the problem on the relation between the spaces S_α^β and $S_\alpha \cap S^\beta$ was set. It was solved by Kashpirovsky [1] (Theorem 9.4). Kashpirovsky [2, 3] considered also the question on representation of the functions from spaces of S type as the boundary values of solutions of equation (9.2) in the upper half-plane with definite growth when approaching the real axis. The results of subsection 9.4 belong to him.

Chapter 3

Section 1. The results presented in subsections 1.1, 1.2 (except probably Theorem 1.3) are well-known (see Naimark [2], S. Krein [1], Sz.-Nagy–Foias [1]). Theorem 1.3 belongs to Phillips [1] (see also Strauss [2]). Linear relations were studied first by Arens [1] who introduced the notions of symmetric and self-adjoint (Hermitian) relations and by Rofe–Beketov [2] who proved the assertion of Theorem 1.4 regarding Hermitian relations. The other assertions of Theorem 1.4 (and the notions in it) are due to M. Gorbachuk–Kochubei–Rybak [1, 2]. In these works a proof of Theorem 1.4 is given which differs from that in this book. We would like to note some of the later works on the theory of linear relations on a Hilbert space. They are: Coddington [1], Dijksma–de Snoo [1], Coddington–Dijksma [1], Kochubei [1, 3], Bruk [4], Langer–Textorius [1], Bennewitz [1], Brown [1], Rofe–Beketov [3]. Only some of the publications are listed here. A lot of papers pertaining, for example, to linear relations on spaces with indefinite metrics are not mentioned here.

The problem on the description of all self-adjoint extensions of a symmetric operator in terms of abstract boundary conditions was put forward for the first time in Calkin [1] (see also Dunford–Schwartz [2]) and solved by him and (in another way) by Shtraus [1] for operators with finite deficiency numbers. The idea of the method presented in subsection 1.4 goes back to the paper by Rofe–Beketov [2] where, however, operators of a special kind were considered. Theorems 1.5, 1.6 were proved by Kochubei [1] and, independently, by Bruk [3]. The notion of boundary value space appeared first in the work by Talyush [1], where (as in the work by Kochubei [1]) an additional assumption ($\Gamma_1 f = \Gamma_2 f = 0$ if $f \in \mathfrak{D}(A)$) was needed. Bruk [3] observed that this condition is a consequence of the other assumptions.

A number of applications of the abstract boundary condition method are cited and commented in sections 2–5 and in Chapter 4. We would also like to pay attention to the results obtained by Bruk [3, 5] (a description of generalized resolvents and related questions), Kochubei [3–6, 9] (non-classical problems; multi-point problems with integral boundary conditions, construction of the characteristic function; a description of extensions commuting with a given family of unitary operators defined on $C_0^\infty(\mathbb{R}^n \backslash S)$, $\operatorname{mes} S = 0$; generalizations to the case of J-symmetric operators where J is an involution), Lyance–Storozh [1] (conditions of self-adjointness of differential-boundary operators), Etkin [1] (problems with rationally entering spectral parameter in boundary conditions), Kholkin [1] (the oscillation theorems), V. I. Gorbachuk [3] (solvable extensions of elliptic operators), which were not included but are close in the themes.

Section 2. About the Friedrichs extension see, for example, the book by Dunford–Schwartz [2]. The results of subsection 2.2 belong to Kochubei [7]. Vishik [1] was the first who described proper solvable extensions (of operators more general than symmetric); his description can be derived from Theorem 2.2 if we take a positive boundary value space of the form (2.4). The modified presentation of Vishik's results is contained in the work by Grubb [1]. Vainerman [6] generalized Theorem 2.2 to non-symmetric operators. The other classes of extensions (being restrictions of A^*) – closed, normally solvable, Fredholm, sectorial, symmetric with non-equal deficiency numbers, and so on – were described by Vainerman [6], Mikhailets [6], Storozh [1, 2], Arlinsky [1], Derkach–Malamud–Tsekanovsky [1]. The book by Dezin [1] is devoted to constructing solvable extensions of differential operators. Derkach–Malamud [1] studied self-adjoint extensions of a symmetric operator with spectral gaps with preservation of the gaps or when a finite number of eigenvalues is added into each gap.

Section 3. Subsections 3.1–3.3 reproduce (in a somewhat more general form) Bruk's results [4]. The methods go back to earlier works by the authors [3–5], where a special situation was considered. The paper by M. Gorbachuk–Kutovoi [1] also belongs to this range. The work by Gehtman [1] was used as the initial point for obtaining Theorem 3.3. It is shown in it that if the spectrum of the Friedrichs extension $\widetilde{A}$ of a symmetric operator A with infinite deficiency numbers is discrete and $\lambda_n(\widetilde{A}) \sim dn^\delta$, then for any β, $0 < \beta < \delta$, there exists a self-adjoint extension A_β of A with discrete spectrum $\lambda_n(A_\beta) \sim bn^\beta$. Corollary 3.1 generalizes and makes more precise Kochubei's theorem [2]. In his work [8] Kochubei investigated the structure of the non-real part of the spectrum of maximal dissipative (accretive) extensions A_K when the operator K is in a certain sense close to unitary. Theorem 3.4 was proved by Kochubei [2] (in the case when the boundary value space is of the form (2.4) and $p = \infty$, by Vishik [1]). Kochubei [2] also obtained sufficient conditions for completeness of the system of eigenvectors and adjoined vectors of a completely solvable extension. A result close to Theorem 3.5 was established in Mikhailets's work [6], which contains also a number of theorems about the spectrum of self-adjoint extensions of a semi-bounded operator given by abstract boundary conditions. It should be stressed here that the basic role in the development of the theory of semi-bounded self-adjoint extensions of a semi-bounded symmetric operator belongs to M. Krein [1, 2] and Birman [1].

Section 4. The structure of the domains of the minimal and maximal operators, generated by the expression (4.1) with a bounded operator coefficient, was studied by Ziatdinov [1], Rofe-Beketov [1] and M. Gorbachuk [1] when $\dim \mathfrak{H} = \infty$. As to the case of $\dim \mathfrak{H} < \infty$, see Naimark's book [2]. Theorem 4.2 for self-adjoint extensions is due to Rofe-Beketov [2] in the more general situation when the order of an equation is arbitrary. When $\dim \mathfrak{H} < \infty$, a description of self-adjoint extensions in a somewhat different form was obtained by M. Krein [2]. Theorems 4.3 and 4.4 are well-known and their proofs are given for completeness of presentation.

Section 5. A detailed bibliography of works devoted to the study of the Cauchy problem for expression (5.1) and more general expressions can be found in the monographs by S. Krein [1], Lions [1] and Sova [1]. There are a lot of subsequent works, which we however do quote. The domains of the operators L_0 and L_0^* generated by the expression (5.1) with additional condition $y'(0) = By(0)$, are described by the authors [1, 2], Vainerman [1] and by Vainerman–M. Gorbachuk [1] without this condition. In the same work, Vainerman and M. Gorbachuk obtained a description of all maximal dissipative (maximal accumulative), including self-adjoint, extensions of the operator L_0, as well as of special classes of these extensions – with discrete spectrum and solvable, α-smooth, etc. – in terms of boundary conditions. A special class of boundary value problems for the expression (5.1) on a finite interval was studied by Shishatsky [1]. The reasoning used when considering example 2 (subsection 5.8) was already used by Berezansky [1, 4]. We would like to note that the Dirichlet problem for the string equation was investigated by a number of mathematicians; besides the works by Berezansky [1, 4], see in this connection the works by Sobolev [2, 3], Alexandryan [1, 2], Vakhaniya [1], Denchev [1], Virabyan [1] and the bibliography in these works. The questions considered in section 5 for the expresssion (5.1), were being studied also for more general expressions – with a variable coefficient $A(t)$ by Vainerman [2] and Bruk [2], with degeneration – by Vainerman [4]. Solvable and completely solvable (but not necessary dissipative or accumulative) extensions of the minimal operator were described in Vainerman [6].

Chapter 4

Section 1. The investigation of the domains of the minimal and maximal operators generated by the expression (1.1) on a finite interval (Theorems 1.1 and 1.3) was done by M. Gorbachuk [2]. In this work he also introduced the notion of regularized boundary values for vector-functions from the domain of the maximal operator, established Green's formula (1.12), and on its basis all self-adjoint extensions of the minimal operator were described (Theorem 1.4 in the case of unitary K). A description of maximal dissipative and accretive, φ–accretive and other extensions is contained in the works by M. Gorbachuk–Kochubei–Rybak [1, 2], Mikhailets [1], M. Gorbachuk–Mikhailets [1]. These results were transfered by M. Gorbachuk–Kochubei [1] and Orudzhev [1, 2] to the case of higher order two-term equations. The case of semi-axis was considered by Kutovoi [1] and Nguen Kuok Zan [1].

Section 2. Theorem 2.1, on the existence of extensions of the minimal operator with the spectrum filling the whole numerical axis when $A(t) \equiv A$, is due to M. Gorbachuk [2]. In the concrete situation, when the operator A is generated on the space $L_2(c,d)$ $(-\infty < c < d < \infty)$ by the expression $-\frac{d^2}{dx^2}$ on functions satisfying the condition $y(c) = y(d) = 0$, it was established by Ilyin and Filippov [1]. All self-adjoint extensions of the minimal operator with discrete spectrum were described by the authors [5]. A description of maximal dissipative and maximal accumulative extensions whose spectrum is discrete, is presented in the works by M. Gorbachuk–Kochubei–Rybak [1, 2] for second-order expressions, and M. Gorbachuk–Kochubei [1], Orudzhev [1, 2] for expressions of higher orders. Lemma 2.1 was established by V. Gorbachuk–M. Gorbachuk [4]. A description of various classes of maximal dissipative and accumulative extensions of the minimal operator (with resolvents from the ideal $\mathfrak{S}_p$, smooth up to the boundary, with finite quadratic functional, D-extensions) was obtained by M. Gorbachuk–Kochubei–Rybak [1, 2]; information regarding the self-adjoint case can be found in the earlier works of the authors [3, 4].

Section 3. The results included in this section were obtained by Mikhailets [2, 4, 5]. The asymptotic equality (3.7) was established by the authors [4] before. In connection with studying the spectrum of elliptic problems with strong degeneration, similar questions were considered in Vulis–Solomyak [1, 2] for a second-order operator equation with degeneration.

Section 4. Positive boundary value spaces of the minimal operator generated by the expression (1.1), being an analogue of regularized boundary values for second-order elliptic operators in a bounded domain introduced by Vishik [1], were considered by Mikhailets [1, 6]. Lemmas 4.1 and 4.2 are also due to him. The sufficient conditions in Theorems 4.1, 4.2 were established by the authors [3, 4], the necessary ones – by Mikhailets [6]. Corollaries 4.1–4.3 and Theorems 4.4, 4.5 are also due to the authors [4]. Theorem 4.3 and its generalization to higher-order equations are contained in V. I. Gorbachuk [2]. Many of the results in this section also admit generalizations to higher-order equations (see M. Gorbachuk–Kochubei [1], Orudzhev [1, 2]), as well as to other kinds of extensions (see M. Gorbachuk–Kochubei–Rybak [1, 2]).

Section 5. Boundary value problems of the kind of the third boundary value problem for equation (1.1) were considered by the authors [4]. In that work Theorems 5.1 and 5.2 were established. Theorems 5.3, 5.4 are due to Mikhailets [3]. The periodic boundary value problem was studied by M. Gorbachuk [3].

Section 6. The method of constructing Green's function for the Neumann problem presented in subsection 6.1 is due to Laptev [2], the domains of the operators L_0 and L_0^* were investigated by Vainerman [2, 3]. The description of all maximal dissipative (including self-adjoint) extensions of the minimal operator as well as the special classes of extensions – with discrete spectrum, with definite asymptotics of eigenvalues, α–smooth and others – in terms of boundary value problems was obtained in the same works. Bruk [1] observed that under additional condition 4) from subsection 6.1 on the operators $A(t)$ it is possible to construct the regularized boundary vectors by formulas (6.20), which are simpler than (6.19) but are not valid in the general case. The example proving the necessity of condition 4) for the choice of regularizations in the form (6.20) is also due to him. Note that Theorems 6.3–6.5 are formulated provided that the eigenvalue distribution function $N(\lambda; L_N)$

of the extension L_N has asymptotics of the kind $a\lambda^{\alpha}$. Mikhailets [5] deduced conditions on the coefficient $A(t)$ guaranteeing this property of $N(\lambda; L_N)$. Note also that the results of this section may be extended to the case where the operator coefficient $A(t)$ is degenerated at one end of the interval (see Vainerman [5]).

Section 7. the examples cited in this section illustrate the general statements of the previous ones and are taken from the works mentioned above. We emphasize that throughout this chapter we considered expression (1.1) only on a finite interval, though the spectral theory of operator differential equations has started, probably from the paper Kostyuchenko–Levitan [1], to pertain the case of the whole axis. We do not refer to subsequent works concerning an infinite interval. A sufficient, if not complete, bibliography of these works is contained in the authors' reviews [7, 8], Birman–Solomyak [1] and in the books by Levitan–Sargsyan [1] and Kostyuchenko–Sargsyan [1].

Chapter 5

Section 1. The principal statements of the theory of semigroups of bounded linear operators on Banach spaces cited in the section had been taken from monographs by Balakrishnan [1], Yosida [1], S. Krein [1], Pazy [2], Reed–Simon [2], Hille–Phillips [1] (see also the review by S. Krein–Khazan [1]) which contain detailed proofs. Theorem 1.5 is presented on the basis of the paper by Shaw [1] and the book by Hille–Phillips [1].

Section 2. The material of this section is expounded mainly following the preprint by Knyazyuk [2]. Absolutely concave functions are closely connected with the class S of functions introduced and studied by M. Krein (see, for example, M. Krein [3, 4] and the monograph by M. Krein–Nudelman [1]). Absolutely concave functions of weakly positive operators were defined by Hirsh [1] and investigated by Hirsh [1] and Pustylnik [1]. Hirsh proved, in particular, that if A generates a C_0-semigroup of contractions, then $\varphi(A)$ is the generator of a semigroup with the same properties. In the case of a Hilbert space this result belongs to von Neumann (see, for example, Riesz–Sz.-Nagy [1]). However, the explicit form of the semigroup generated by $\varphi(A)$ was not indicated in these works. It was found by Knyazyuk [2]. The principal role in obtaining it was played by Theorem 2.3, proved by him. Theorem 2.5 is also due to Knyazyuk [2]; in the case of a Hilbert space it is also contained in the paper of Kolmanovich–Malamud [1].

Section 3. The results of this section generalize the considerations of sections 3, 4 from Chapter 2 to the case of a Banach space. They are due to Knyazyuk [1, 2]. Theorem 3.6 is contained in V.M. Gorbachuk–Matsishin [1].

Section 4. When A is a bounded operator on a Banach space equation (4.1) was investigated in Glushko–Krein [1] and Glushko [1]. The results expounded in this section concern the case where an unbounded coefficient A generates an analytic semigroup and are contained in V.M. Gorbachuk–Matsishin [1], which discusses the weakly degenerated equation. In it also a representation of the general solution of the inversely parabolic equation (4.3) (Theorem 4.2) is given. Theorems 4.3–4.5 are due to V.M. Gorbachuk [2]. The case of a positive self-adjoint operator on a Hilbert space had been earlier investigated by M. Gorbachuk–Pivtorak [1].

Section 5. The existence of generalized Cesàro and Abel limits at infinity of solutions of equation (5.4) (Theorems 5.1–5.3) was established by V.M. Gorbachuk [1]. Theorem 5.4, contained in M. Gorbachuk–V. M. Gorbachuk [1], generalizes Pazy's result [1], which is formulated in Corollary 5.1. The assertion of this corollary for $p = 2$ and a Hilbert space was proved by Datko [1]. In the special case of a bounded A, Datko's was also proved by Yakubovich [1], and Pazy's result – by M. Krein [5]. Datko's complete result is also contained in Likhtarnikov [1]. Theorems 5.5 and 5.6 are presented following Vinnishin–M. Gorbachuk [1].

Section 6. Here we have presented the results of Knyazyuk [2] regarding the theory of boundary values of solutions of partial differential equations which are generalizations of well-known theorems about boundary values of functions harmonic in a disc in L_p-spaces (Riesz [1]), in Orlicz spaces (Zygmund [1]) and their n-dimensional versions (Stein [1], Stein–Weiss [1]), as well as boundary values of solutions of the heat equation in a half-space (Hirshman–Widder [1]). It should be noted that they are close to the L_p-theory of boundary values of solutions of general elliptic differential equations in bounded domains (Mazya [1], V. Mikhailov [1–3], Ju. Mikhailov [1], Gushchin–Mikhailov [1, 2], Petrushko [1], Roitberg [3]) and parabolic equations in a half-space and tubes, (Johnson [1], Zhitarashu [1, 2], Ivasishen [1], Petrushko [2, 3], Chabrovski [1, 2]), developed in the 1970's – 1980's. Throughout the whole chapter we dealt with the abstract first-order equation of the parabolic type. Some results concerning second-order equations are contained in Knyazyuk [3].

References

Chapter 1.

Ahiezer N. I., Glazman I. M.

1. *Theorie der linearen Operatoren im Hilbert-Raum* (transl. from the Russian). Akad. Verlag, 1954.

Alexandryan R. A.

1. On the Dirichlet problem for the string equation and on completeness of a system of functions in the disc (in Russian), *Dokl. Akad. Nauk SSSR* **73** (1950), No. 5, 869–872.
2. Spectral properties of operators generated by systems of differential equations of Sobolev type (in Russian), *Trudy Moskov. Mat. Obšč.* **9**, (1960), 455–505.

Allakhverdiev D. E.

1. On the rate of approximating completely continuous operators by finite-dimensional operators (in Russian), *Učen. Zap. Azerb. Univ.* **2** (1957), 27–35.

Antosik, P., Mikusiński J., Sikorski R.

1. *Theory of distributions: the sequential approach*, Elsevier, 1973.

Arens R.

1. Operational calculus of linear relations, *Pacif. J. Math.* **11** (1961), No. 1, 9–23.

Arlinsky, Ju. M.

1. Positive spaces of boundary values and sectorial extensions of a nonnegative symmetric operator, *Ukrain. Mat. Zh.* **40** (1988), No. 1, 8–14. (Transl.: *Ukr. Math. J.* **40** (1988), 5–10).

Balakrishnan A.

1. *Applied functional analysis*, Springer, 1976.

Bary N. K.

1. *A treatise on trigonometric series* (transl. from the Russian). Pergamon, 1964.

Barut A., Rączka R.

1. *Theory of group representation and applications,* P.W.N., 1977.

Beals R.

1. Semigroups and abstract Gevrey spaces, *J. Funct. Anal.* **10** (1972), No. 3, 300–308.

Bendat J., Sherman S.

1. Monotone and convex operator functions, *Trans. Amer. Math. Soc.* **79** (1955), No. 1, 58–71.

Bennewitz C.

1. Symmetric relations on a Hilbert space, *Lect. Notes Math.* **280** (1972), 212–218.

Berezansky Ju. M.

1. On the Dirichlet problem for the string equation (Russian), *Ukrain Mat. Zh.* **12** (1960), No. 4, 363–372.
2. Some applications of spaces with negative norm, *Studia Math. Ser. Spec.* **1** (1963), 25–96.
3. Spaces with negative norms, *Uspekhi Mat. Nauk* **18**, (1963), No. 1, 63–96. (Transl.: *Russian Math. Surveys* **18** (1963), 63–95).
4. Expansion in eigenfunctions of self-adjoint operators (transl. from the Russian). *Amer. Math. Soc.*, 1968.
5. Self-adjoint operators in spaces of functions of infinitely many variables (transl. from the Russian). *Amer. Math. Soc.*, 1986

Billingsley P.

1. *Convergence of probability measures.* Wiley, 1968.

Birman M. Sh.

1. On the theory of self-adjoint extensions of positive definite operators (in Russian), *Mat. Sbornik* **38** (1956), No. 4, 431–450.

Birman M. Sh., Solomyak M. Z.

1. Asymptotic behaviour of the spectrum of differential equations, Itogi Nauk. i. Tekhn. VINITI, *Ser. Mat. Anal.* **14** (1977), 5–59. (Transl.: *J. Soviet Math.* **12** (1979), 247–283)

Brown R. C.

1. Notes on generalized boundary value problems in Banach spaces. 1. Adjoint and extension theory, *Pacif. J. Math.* **85** (1979), No.2, 295–322.

Bruk V. M.

1. Dissipative extensions of elliptic differential operator (in Russian), *Funkts. Anal.* (Ulyanovsk) **3** (1974), 35–43.
2. On spectral theory of differential equations with unbounded operator coefficients (in Russian), *Funkts. Anal.* (Ulyanovsk) **5** (1975), 14–24.
3. On a class of boundary value problems with spectral parameter in the boundary condition, *Mat. Sbornik*, **100** (1976), No. 2, 210–216. (Transl.: *Math. USSR Sb.* **29** (1976), 186–192).
4. Extensions of symmetric relations, *Mat. Zametki* **22** (1977), No. 6, 825–834. (Transl.: *Math. Notes* **22** (1977), 953–965).

5. On extensions of symmetric operator depending on a parameter (in Russian), *Funkts. Anal.* (Ulyanovsk) **10** (1978), 32–40.

Calkin J. W.

1. Abstract symmetric boundary conditions, *Trans. Amer. Math. Soc.* **45** (1939), No. 3, 369–442.

Cartier P. et Dixmier J.

1. Vecteurs analytiques dans les répresentations de groups de Lie, *Amer. J. Math.* **80** (1958), No. 1, 131–145.

Chabrowski I.

1. Representation theorems and Fatou theorems in the sense of Petrovskii, *Colloq. Math.* **31** (1974), No. 2, 301–315.
2. Representation theorems for parabolic systems, *J. Austral. Math. Soc.* **A32** (1982), No. 2, 246–288.

Cioranescu J.

1. Operator-valued ultradistribution in spectral theory, *Math. Ann.* **223** (1976), No. 1, 1–12.

Coddington E. A.

1. Extension theory of formally normal and symmetric subspaces, *Mem. Amer. Math. Soc.* **134** (1973), 1–80.

Coddington E. A., Dijksma A.

1. Adjoint subspaces in Banach spaces, with applications to ordinary differential subspaces, *Ann. Mat. Pura Appl.* **118** (1978), No. 1, 1–118.

Coddington E. A., de Snoo H. S. V.

1. Positive self-adjoint extensions of positive symmetric subspaces, *Math. Z.*, **159** (1978), 203–214.

Courant R.

1. Über die Eigenwerte bei den Differentialgleichungen der Mathematischen Physik, *Math. Z.* **7** (1920), 1–57.

Daletskiĭ Ju. L., Kreĭn M. G.

1. Stability of solutions of differential equations in Banach space (transl. from the Russian). *Amer. Math. Soc.*, 1974.

Datko R.

1. Extending a theorem of A.M. Liapunov to Hilbert space, *J. Math. Anal. and Appl.* **32** (1970), 610–611.

Denchev P. T.

1. On spectrum of an operator, *Dokl. Akad. Nauk SSSR* **151** (1963), No. 2, 258–261

Derkach V. A., Malamud M. M.

1. On the Weyl function and Hermitian operators with gaps, *Dokl. Akad. Nauk SSSR* **293** (1987), No. 5, 1041–1046. (Transl.: *Soviet Math. Dokl.* **35** (1987) 393–398).

Derkach V. A., Malamud M. M., Tsekanovskii E. K.

1. Sectorial extensions of a positive operator, and the characteristic function, *Dokl. Akad. Nauk SSSR* **298** (1988), No. 3, 537–541. (Transl.: *Soviet Math. Dokl.* **37** (1988), 106–110).

Dezin A. A.

1. *Partial differential equations: an introduction to a general theory of linear boundary value problems* (transl. from the Russian), Springer, 1987

Dijksma A., de Snoo H. S. V.

1. Self-adjoint extensions of symmetric subspaces. *Pacif. J. Math.* **54** (1974), No. 1, 71–100.

Donoghue William F. jr.

1. *Monotone matrix functions and analytic continuations*. Springer, 1974.

Dunford N., Schwartz J. T.

1. Linear operators: General theory, *Interscience*, 1958.
2. Linear operators: Spectral theory, *Interscience*, 1963.

Dyn'kin E. M.

1. The operational calculus based on the Cauchy–Green formula and quasi-analiticity of classes $D(h)$ (in Russian), *Zap. Nauchu. Sem. Leningradsk. Otdel. Mat. Inst. Steklov* **19** (1970), No. 1, 221–227.

Etkin A. E.

1. On boundary-value problem with spectral parameter, rationally entering the boundary conditions (Russian), *Funkts. Anal.* (Ulyanovsk) **15** (1980), 192–197.

Evgrafov M. A.

1. *Analytic functions* (transl. from the Russian), Saunders, 1966.

Fisher E.

1. Über quadratische Formen mit reelen Koeffizienten. *Monatsh. Math. und Phys.* **16** (1905), 234–249.

Fishman I. P.

1. Representation of the general solution of a differential operator equation, *Ukrain. Mat. Zh.* **36** (1984), No. 6, 804–808. (Transl. *Ukr. Math. J.* **36** (1984), 113–116).
2. Boundary values of the solutions of operator differential equations, *Ukrain. Mat. Zh.* **37** (1985), No. 3, 388–393. (Transl.: *Ukr. Math. J.* **37** (1985),

310–314.

Foias C.

1. Une application des destributions vectorielles á la theorie spectrale, *Bull. Sci. Math.* **84** (1960), No. 2, 147–158.

Gehtman M. M.

1. On the problem of the spectrum of self-adjoint extensions of a symmetric semi-bounded operator, *Dokl. Akad. Nauk SSSR*, **186** (1969), No. 6, 1250–1252. (Transl.: *Soviet Math. Dokl.* **10**, No. 3 (1969), 737–739).

Gel'fand I. M.

1. On one-parameter groups of operators in a normed space (in Russian), *Dokl. Akad. Nauk SSSR* **25** (1939), No. 9, 713–718.

Gel'fand I. M., Shilov G. E.

1. *Generalized functions Vol. 2*, Acad. Press, 1968 (transl. from the Russian).
2. *Generalized functions Vol. 3*, Acad. Press, 1968 (transl. from the Russian).

Glazman I. M.

1. *Direct methods of qualitaty spectral analysis of singular differential operators* (in Russian). Fizmatgiz, Moscow, 1963.

Glushko V. P.

1. Linear degenerate differential equations (in Russian). *Manual for students math. department*, Voronezh. Univ., 1972.

Glushko V. P., Krein S. G.

1. Degenerate linear differential equations in Banach space, *Dokl. Akad. Nauk SSSR*, **181** (1968), No. 4, 784–787. (Transl.: *Soviet Math. Dokl.* **9** (1968), 919–922).

Gorbachuk M. L.

1. On spectral functions of a second order differential equation with operator coefficients (in Russian), *Ukrain. Mat. J.* **18** (1966), No. 2, 3–21.
2. Self-adjoint boundary problems for a second-order differential equation with unbounded operator coefficient. *Funkts. Anal. i Prilozh.* **5** (1971), No. 1, 10–21. (Transl.: *Funct. Anal. Appl.* **5** (1971), 9–18).
3. Some questions of the spectral theory of differential equations in the space of vector-functions (in Russian). *Avtoref. Dis. Dr. Fiz-Mat Nauk*, Kiev, 1972.

Gorbachuk M. L., Dudnikov P. I.

1. On initial data of the Cauchy problem for parabolic equations for which solutions are infinitely differentiable (in Russian), *Dokl. Akad. Nauk Ukrain.SSR*, Ser. **A** (1981), No. 4, 9–11.

Gorbachuk M. L., Dudnikov P. I., Fyodorova L. B.

1. On boundary values of analytic functions. In *"Mat. Anal. i Teor. Veroyatn."*

(in Russian). Naukova Dumka, Kiev, 1978, pp. 36–40.

Gorbachuk M. L., Gorbachuk V. M.

1. Behaviour at infinity of solutions of a first-order operator differential equation (in Russian), *Funkts. Metody v Prikladn. Mat. i Mat. Fiz.*, Tezisi Dokl. Vsesoyuzn. Shkoly Mol. Uč. Tashkent. Gos. Univ., Tashkent, 1988, p. 70.

Gorbachuk M. L., Kashpirovskiĭ A. I.

1. Weak solutions of differential equations in Hilbert space, *Ukrain. Mat. Zh.* **33** (1981), No. 4, 513–518. (Transl.: *Ukr. Math. J.* **33** (1981), 392–395).

Gorbachuk M. L., Kochubei A. N.

1. Self-adjoint boundary value problems for certain classes of operator differential equations of higher order, *Dokl. Akad. Nauk SSSR* **201** (1971), No. 5, 1029–1032. (Transl.: *Soviet Math. Dokl.* **12** (1971), 1759–1762).

Gorbachuk M. L., Kochubei A. N., Rybak M. A.

1. Dissipative extensions of differential operators in a space of vector-functions, *Dokl. Akad. Nauk SSSR* **205** (1972), No. 5, 1029–1032. (Transl.: *Soviet Math. Dokl.* **13** (1972), 1063–1067).
2. On some classes of extensions of differential operators in the space of vector-functions. In *"Primenen. Funktsionaln. Anal. k Zadach. Mat. Fiz."* (in Russian). Mat. Inst. Akad. Nauk Ukrain.SSR, Kiev, 1973, pp. 56–82.

Gorbachuk M. L., Kutovoi V. A.

1. Resolvent comparability of boundary problems for an operator Sturm–Liouville equation, *Funkts. Anal. i Prilozh.* **12** (1978), No. 1, 68–69. (Transl.: *Funct. Anal. Appl.* **12** (1978), 52–54).

Gorbachuk M. L., Mikhailets V. A.

1. Semibounded self-adjoint extensions of symmetric operators, *Dokl. Akad. Nauk SSSR* **226** (1976), No. 4, 765–768. (Transl.: *Soviet Math. Dokl.* **17** (1976), 185–187).

Gorbachuk M. L., Pivtorak N. I.

1. Solutions of evolution parabolic equations with degeneration, *Differentsialnye Uravneniya* **21** (1985), No. 8, 1317–1324. (Transl.: *Differential Eq.* **21** (1985), 892–896).

Gorbachuk V. I.

1. Boundary values of generalized solutions of a homogeneous Sturm–Liouville equation in a space of vector functions, *Mat. Zametki* **18** (1975), No. 2, 243–252. (Transl.: *Math. Notes* **18** (1975), 732–737).
2. Asymptotics of the eigenvalues of boundary–value problems for differential equations in a space of vector functions, *Ukrain. Mat. Zh.* **27** (1975), No. 5, 657–664. (Transl.: *Ukr. Math. J.* **27** (1975), 657–663).
3. On boundary value problems for elliptic differential equations (in Russian),

Dokl. Akad. Nauk Ukrain.SSR, Ser. **A** (1981) No. 1, 7–11.

4. Spaces of infinitely differentiable vectors of a nonnegative self-adjoint operator, *Ukrain. Mat. Zh.* **35** (1983), No. 5, 617–621. (Transl.: *Ukr. Math. J.* **35** (1983), 531–535).

Gorbachuk V. I., Fyodorova L. B.

1. Operator calculus for some classes of non-self-adjoint operators, *Ukrain. Mat. Zh.* **31** (1979), No. 2, 123–131. (Transl.: *Ukr. Math. J.* **31** (1979), 95–101).

Gorbachuk V. I., Gorbachuk M. L.

1. Expansion in eigenfunctions of a second-order differential equation with operator coefficients, *Dokl. Akad. Nauk SSSR* **184** (1969), No. 4, 774–777. (Transl.:*Soviet Math. Dokl.* **10** (1969), 158–162).
2. Problems of the spectral theory of the second-order linear differential equation with unbounded operator coefficients, *Ukrain. Mat. Zh.* **23** (1971), No. 1, 3–15. (Transl.: *Ukr. Math. J.* **23** (1971), 1–12).
3. Self-adjoint boundary value problems with discrete spectrum generated by the Sturm–Liouville equation with unbounded operator coefficient, *Funkts. Anal. i Prilozh.* **5** (1971), No. 4, 67–68. (Transl.: *Funct. Anal. Appl.* **5** (1971), 322–324).
4. Classes of boundary-value problems for the Sturm–Liouville equation with an operator potential, *Ukrain. Mat. Zh.* **24** (1972), No. 3, 291–304. (Transl.: *Ukr. Math. Zh.* **24** (1972), 241–251).
5. The spectrum of self-adjoint extensions of the minimal operator generated by a Sturm–Liouville equation with operator potential, *Ukrain. Mat. Zh.* **24** (1972), No. 6, 726–734. (Transl.: *Ukr. Math. J.* **24** (1972), 582–589).
6. On the boundary values of solutions of homogeneous differential equations, *Dokl. Akad. Nauk SSSR* **228** (1976), No. 5, 1021–1024. (Transl.: *Soviet Math. Dokl.* **17** (1976), 852–863).
7. Some questions of spectral theory of elliptic differential equations in the space of vector-functions on finite interval (in Russian), *Ukrain. Mat. Zh.* **28** (1976), No. 1, 12–26.
8. Some questions of the spectral theory of differential equations of elliptic type in the space of vector-functions (in Russian), *Ukrain. Mat. Zh.* **28** (1976), No. 3, 313–324. (Transl.: *Ukr. Math. J.* **28** (1976), 244–253).
9. Boundary values of solutions of some classes of differential equations, *Mat. Sbornik* **102** (1977), No. 1, 124–150. (Transl.: *Math. USSR Sb.* **31** (1977), 109–133).
10. Dirichlet problem for an operator Sturm–Liouville equation, *Mat. Zametki* **24** (1978), No. 6, 801–807. (Transl.: *Math. Notes* **24** (1978), 925–929).
11. On initial data of smooth solutions of some classes of parabolic equations (in Russian), *Uspekhi Mat. Nauk* **34** (1979), No. 4, p. 164.
12. Trigonometric series and generalized periodic functions, *Dokl. Akad. Nauk SSSR* **257** (1981), No. 4, 799–804. (Transl. *Soviet Math. Dokl.* **23** (1981),

342–346).

Gorbachuk V. M.

1. On asymptotic behaviour of solutions of an operator differential equation in Banach space (in Russian), *Uspekhi Mat. Nauk* **42** (1987), No. 4, p. 162.
2. Behaviour at infinity of the solution of a first-order operator differential equation in Banach space (in Russian), *Ukrain. Mat. J.* **40** (1988), No. 5, 629–632.

Gorbachuk V. M., Matsishin I. T.

1. On solutions of evolution equations with degeneration in a Banach space. In *"Spektraln. Teor. Differentsialno-Operator. Uravn."* (in Russian). Mat. Inst. Akad. Nauk Ukrain.SSR, Kiev, 1986, pp. 5–10.

Grubb G.

1. A characterization of the non-local boundary value problems associated with an elliptic operator, *Ann. Sci. Norm. Sup. Pisa* **22** (1968), 425–523.

Gushchin A. K., Mikhailov V. P.

1. On boundary values in L_p, $p > 1$, of solutions of elliptic equations, *Mat. Sbornik* **108** (1979), No. 1, 3–21. (Transl.: *Math. USSR Sb.* **36** (1980), 1–19).
2. On boundary values of solutions of elliptic equations. In: *Obobshchionnije funktsii i ikh primenenije v matematicheskoi fizike*, Mat. Inst. V.A. Steklova, Moscow, 1981, pp. 189–205.

Harish–Chandra

1. Representations of a semi-simple Lie group on a Banach space, 1, *Trans. Amer. Math. Soc.* **75** (1953), No. 2, 185–243.

Hille E., Phillips R. S.

1. Functional analysis and semi-groups, *Amer. Math. Soc.*, 1957.

Hirsh F.

1. Integrales de rèsolvantes et calcus symbolique, *Ann. Inst. Fourier* **22** (1972), No. 4, 239–264.

Hirshman I. I., Widder D. V.

1. *The convolution transform*. Princeton Univ. Press, 1955,

Hörmander L.

1. *The analysis of linear partial differential operators,* Springer, 1983–1985, Vol. 1–4.

Ilyin V. A., Filippov A. F.

1. The nature of the spectrum of a self-adjoint extension of the Laplace operator in a bounded region, *Dokl. Akad. Nauk SSSR* **191** (1970), No. 2, 267–269. (Transl.: *Soviet Math. Dokl.* **11** (1970), 339–342.

Ivasishen S. D.

1. On integral representations and Fatou property for solutions of parabolic sys-

tems (in Russian), *Uspekhi Mat. Nauk* **41** (1986), No. 4, 173–174.

Johnson R.

1. Representation theorems and Fatou theorems for second order linear parabolic partial differential equations, *Proc. London Math. Soc.* **23** (1971), No. 3, 325–347.

Kantorovich L. V., Akilov G. P.

1. *Funktionalanalysis in normierten Räumen* (transl. from the Russian), Akad. Verlag, 1964.

Kashpirovskiĭ A. I.

1. Equality of the spaces S_α^β and $S_\alpha \cap S^\beta$, *Funkts. Anal. i Prilozh.* **14** (1980), No. 2, p. 60. (Transl.: *Funct. Anal. Appl.* **14** (1980), p. 129).
2. Analytic representation of generalized functions of S-type (in Ukrainian), *Dokl. Akad. Nauk Ukrain.SSR.* Ser **A** (1980), No. 4, 12–14.
3. Boundary values of solutions of some classes of homogeneous differential equations in Hilbert space (in Russian): *Avtoref. Dis. Kand. Fiz.-Mat. Nauk. Mat. Inst. Akad. Nauk Ukrain.SSR*, Kiev, 1981.

Khinchin A. Ja.

1. *Kettenbrüche,* Teubner, 1956 (transl. from the Russian).

Kholkin A. M.

1. Oscillation theorems for Sturm–Liouville and Dirac systems on the real axis. In *"Issled. Teor. Oper. i Prilozhen."* (in Russian). Naukova Dumka, Kiev, 1979.

Khrushchev S. V.

1. Uniqueness theorems and essentially self-adjoint operators (in Russian), *Zap. Naučn. Sem.* LOMI **107** (1982), 169–178.

Knyazyuk A. V.

1. On the boundary values of solutions of differential equations in a Banach space (in Russian), *Dokl. Akad. Nauk Ukrain.SSR*, Ser. **A** (1984), No. 9, 12–14.
2. Boundary values of infinitely differentiable semigroups (in Russian). *Preprint No. 69.* Mat. Inst. Akad. Nauk Ukrain.SSR, Kiev, 1985.
3. The Dirichlet problem for second-order differential equations with operator coefficient (in Russian), *Ukrain. Mat. Zh.* **37** (1985), No. 2, 256–260.

Knyukh B. I.

1. Representation and boundary values of solutions of homogeneous second-order operator differential equation, *Ukrain. Mat. Zh.* **38** (1986), No. 1, 101–104. (Transl.: *Ukrain. Mat. J.* **38** (1988), 91–94).

Kochubei A. N.

1. Extensions of symmetric operators and symmetric binary relations, *Mat. Zametki* **17** (1975), No. 1, 41–48. (Transl.: *Math. Notes* **17** (1975), no. 1, 25–28).

2. The spectra of self-adjoint extensions of a symmetric operator, *Mat. Zametki* **19** (1976), No. 3, 429–434. (Transl.: *Math. Notes* **19** (1976), No. 3, 262–265).
3. The extension of a nondensely defined symmetric operator, *Sibirsk. Mat. Zh.* **18** (1977), No. 2, 314–320. (Transl.: *Sib. Math. J.* **18** (1977), No. 2, 225–229).
4. Symmetric operators and nonclassical spectral problems, *Mat. Zametki* **25** (1979), No. 3, 425–434. (Transl.: *Math. Notes* **25** (1979), No. 3, 224–228).
5. Symmetric operators commuting with a family of unitary operators, *Funkts. Anal. i Prilozh.* **13** (1979), No. 4, 77–78. (Transl.: *Funct. Anal. Appl.* **13**, No. 4 (1979), 300–301.
6. On extensions of J-symmetric operators (in Russian), *Teor. Funktsii, Funkts. Anal. i Prilozh.* **31** (1979), 74–80.
7. On extensions of a positive definite symmetric operator (in Russian), *Dokl. Akad. Nauk Ukrain.SSR*, Ser. **A** (1979), No. 3, 168–171.
8. On characteristic functions of symmetric operators and their extensions (in Russian), *Izv. Akad. Nauk Arm.SSR* **15** (1980), No. 3, 219–232.
9. Elliptic operators with boundary conditions on a subset of measure zero, *Funkts. Anal. i ego Pril.* **16** (1982), No. 2, 78–79. (Transl.: *Funct. Anal. Appl.* **16** (1982), No. 2, 137–139).

Kolmanovich V. Ju., Malamud M. M.

1. An operator analogue of the Schwartz–Löwner lemma (in Russian), *Teor. Funktsii, Funkts. Anal. i Prilozh.* **48** (1987), 71–74.

Komatsu H.

1. An introduction to the theory of hyperfunctions, *Lecture Notes Math.* **287** (1973), 3–41.
2. Ultradistributions and hyperfunctions, *Lecture Notes Math.* **287** (1973), 180–192.

Kostyuchenko A. G., Levitan B. M.

1. Asymptotic behaviour of the eigenvalues of the Sturm–Liouville operator problem, *Funkts. Anal. i Prilozh.* **1** (1967), No. 1, 86–96. (Transl.: *Funct. Anal. Appl.* **1**, No. 1 (1967), 75–84).

Kostyuchenko A. G., Sargsyan I. S.

1. *Distribution of eigenvalues* (in Russian). "Nauka", Moscow, 1979.

Köthe G.

1. Die Randwerte einer analytischen Funktionen, *Math. Z.* **57** (1952), 13–33.

Krein M. G.

1. On resolvents of a Hermitian operator with deficiency index (m, m) (in Russian), *Dokl. Akad. Nauk SSSR*, **52** (1946), No. 8, 657–660.
2. The theory of self-adjoint extensions of semi-bounded Hermitian operators and its applications, I–II (in Russian), *Mat. Sbornik* **20** (1947), No. 3, 431–495; **21** (1947) No. 3, 365–404.

3. Solution of the inverse Sturm–Liouville problem (in Russian) *Dokl. Akad. Nauk SSSR* **76** (1951), No. 1, 21–24.
4. On a generalization of Stieltjes' investigations (in Russian), *Dokl. Akad. Nauk SSSR* **87** (1952), No. 6, 881–884.
5. A Remark on a theorem in the paper of V.A. Yakubovich titled "A frequency theorem for the case when...., II", *Sibirsk Mat. Zh.* **18** (1977), No. 6, 1411–1413. (Transl.: *Sib. Math. J.* **18**, No. 6 (1977), 1001–1003).

Krein M. G., Nudelman A. A.

1. The Markov moment problem and extremal problems (transl. from the Russian). *Amer. Math. Soc.* 1971.

Krein S. G.

1. Linear differential equations in Banach space (transl. from the Russian), *Amer. Math. Soc.* (1971).

Krein S. G., Khazan M. I.

1. Differential equations in a Banach space, *Itogi Nauki i Tekhn. Mat. Anal.* (VINITI, Moscow) **21** (1983), 130–264. (Transl.: *J. Soviet Math.* **30** (1985), 2154–2239).

Krein S. G., Petunin Ju. I., Semyonov E. M.

1. Interpolation of linear operators (transl. from the Russian). *Amer. Math. Soc.* 1982.

Ku Fan

1. Maximum property and inequalities for the eigenvalues of completely continuous operators, *Proc. Nat. Acad. Sci. USA* **37** (1951), 760–766.

Kutovoi V. A.

1. Spectrum of Sturm–Liouville equation with unbounded operator coefficient, *Ukrain. Mat. Zh.* **28** (1976), No. 4, 473–482. (Transl.: *Ukr. Math. J.* **28** (1976), No. 4, 365–372).

Langer H., Textorius B.

1. On generalized resolvents and Q-functions of symmetric linear relations (subspaces) in Hilbert space. - *Pacif. J. Math.* **72** (1977), No. 1, 135–165.

Laptev G. I.

1. Eigenvalue problems for second-order differential equations in Banach and Hilbert spaces (in Russian), *Differentsialn. Uravn.*, **11** (1966), No. 9, 1151–1160.
2. Strongly elliptic second order equations in a Hilbert space (in Russian), *Lit. Mat. Sbornik*, **8** (1968), No. 1, 87–89.

Levitan B. M., Sargsyan I. S.

1. Introduction to spectral theory (transl. from the Russian). *Amer. Math. Soc.* 1975.

Lidsky V. B.
1. The completeness conditions of the system of root subspaces of nonself-adjoint operators with discrete spectrum (in Russian), *Trudy Moskov. Mat. Obšč.* **8** (1959), 84–220.

Likhtarnikov A. L.
1. Absolute stability criteria for nonlinear operator equations, *Izvestiya Akad. Nauk SSSR. Ser. Mat.* **41** (1977), No. 5, 1064–1083. (Transl.: *Math. USSR Izv.* **11** (1977), No. 5, 1011–1029).

Lions J.–L.
1. *Equations différentielles opérationelles et problèmes aux limites.* Springer, 1961.

Lions J.–L., Magènes E.
1. *Non-homogeneous boundary value problems and applications. Vol. 1.* Springer, 1972 (transl. from the French).
2. *Non-homogeneous boundary value problems and applications. Vol. 3.* Springer, 1972 (transl. from the French).

Liu Gui-Zhong
1. Evolution equations and scales of Banach spaces, Ph. D. Thesis, Eindhoven University of Technology, 1989. (ISBN 90-9002908-7).

Löwner K.
1. Über monotone Matrix Funktionen, *Math. Z.* **38** (1934), 177–216.

Lyance V. E., Storozh O. G.
1. The neutral contingency conditions of some closed operators in terms of abstract boundary operators (in Russian), *Dokl. Akad. Nauk Ukrain.SSR.* Ser **A** (1980), No. 6, 29–30.

Lyubich Ju. I., Matsaev V. I.
1. On operators with separable spectrum (in Russian), *Mat. Sbornik,* **56** (1962), No. 4, 433–468.

Mandelbroit S.
1. *Series de Fourier et classes quasi-analytiques des fonctions,* Gauthier–Villars, 1935.

Markushevich A. I.
1. *Theory of functions of a complex variable* (transl. from the Russian). Chelsea, 1977.

Matsaev V. I.
1. Volterra operators produced by perturbation of self-adjoint operators, *Dokl. Akad. Nauk SSSR* **139** (1961), No. 4, 810–814. (Transl.: *Soviet Math. Dokl.* **2** (1961), 1013–1016).

Mazya V. G.

1. On a degenerating problem with directional derivative, *Mat. Sbornik*, **87** (1972), No. 3, 417–454. (Transl.: *Math. USSR Sb.* **16** (1972), No. 3, 429–469).

Mikhailets V. A.

1. On solvable and sectorial boundary value problems for Sturm–Liouville operator equation, *Ukrain. Mat. Zh.* **26** (1974), No. 4, 447–455. (Transl.: *Ukr. Math. J.* **26** (1974), 370–377).
2. Spectral asymptotics of boundary value problems for a Sturm–Liouville operator equation, *Dokl. Akad. Nauk SSSR*, **228** (1976), No. 5, 1037–1040. (Transl.: *Soviet Math. Dokl.* **17** (1976), 871–875).
3. Asymptotic representation of the eigenvalues of the third boundary-value problem for the Sturm–Liouville operator equation, *Funkts. Anal. i Prilozh.* **10** (1976), No. 1, 83–84. (Transl.: *Funct. Anal. Appl.* **10** (1976), No. 1, 72–74).
4. Distribution of the eigenvalues of the Sturm–Liouville operator equation, *Izv. Akad. Nauk SSSR*, Ser. Mat. **41** (1977), No. 3, 607–619. (Transl.: *Math. USSR Izv.* **11** (1977), No. 3, 571–582).
5. Asymptotics of the eigenvalues of a Sturm–Liouville equation with variable operator coefficients, *Funkts. Anal. i Prilozh.* **11** (1977), No. 1, 71–72. (Transl.: *Funct. Anal. Appl.* **11** (1977), No. 1, 62–64).
6. Spectra of operators and boundary-value problems. In: *"Spektraln. Anal. Differentialn. Oper."* (in Russian), Mat. Inst. Akad. Nauk Ukrain.SSR, Kiev, 1980, pp. 106–131.
7. Self-adjoint extensions of operators with preassigned asymptotics of the spectrum (in Russian), *Dokl. Akad. Nauk Ukrain.SSR.* Ser. **A** (1980), No. 5, 337–339.

Mikhailov V. P.

1. On the boundary values of solutions of elliptic equations in domains with a smooth boundary, *Mat. Sbornik*, **101** (1976), No. 2, 163–188. (Transl.: *Math. USSR-Sb.* **30** (1976), No. 2, 143–166).
2. Dirichlet's problem for a second-order elliptic equation, *Differentsialn. Uravnen.*, **12** (1976), No. 10, 1877–1891. (Transl.: *Diff. Eq.* **12** (1976), No. 10, 1320–1330.
3. On boundary properties of solutions of elliptic equations (in Russian), *Mat. Zametki*, **27** (1980), No. 1, 137–145.

Mikhailov Ju. A.

1. Boundary values in L_p, $p > 1$, of solutions of second-order linear elliptic equations, *Differentisialn. Uravnen.*, **19** (1983), No. 2, 318–339. (Transl.: *Diff. Eq.* **19** (1983), No. 2, 243–258).

Miranda C.

1. *Partial differential equations of elliptic type*. Springer, 1970.

Naimark M. A.

1. *Normed rings* (transl. from the Russian). Reidel, 1984.
2. *Lineare Differential Operatoren* (transl. from the Russian). Akad. Verlag, 1960.

Nelson E.

1. Analytic vectors, *Ann. Math.* **79** (1959), 572–615.

Nguen Kuok Zan

1. On a boundary problem for the Laplace equation in the disk, *Ukrain. Mat. Zh.* **24** (1972), No. 6, 763–772. (Transl.: *Ukr. Math. J.* **24** (1972), No. 6, 613–620).

Nussbaum A.

1. Quasi-analytic vectors, *Ark. Mat.* **6** (1965), 179–191.

Orudzhev G. D.

1. Self-adjoint extensions of a fourth order differential expression with unbounded operator coefficients (in Russian), *Funkts. Anal., Teor. Funktsii i Primenen., Makhachkala*, **2** (1975), No. 2, 58–67.
2. Description of self-adjoint extensions of the high order operator differential expressions (in Russian), *Dokl. Akad. Nauk Az.SSR* (1976), No. 8, 9–12.

Paley R., Wiener N.

1. Fourier transforms in the complex domain. *Amer. Math. Soc.* 1934.

Pazy A.

1. On the applicability of Liapunov's theorem in Hilbert space, *SIAM J. Math. Anal.* **3** (1972), 291–294.
2. *Semigroups of linear operators and applications to partial differential equations.* Springer, 1983.

Petrushko I. M.

1. On boundary values in L_p, $p > 1$, of solutions of elliptic equations in domains with a Liapunov boundary, *Mat. Sbornik*, **120** (1983), No. 4, 569–588. (Transl.: *Math. USSR-Sb.* **48** (1984), No. 2, 565–585)
2. On the boundary values of solutions of parabolic equations, *Mat. Sbornik* **103** (1977), No. 3, 404–429. (Transl.: *Math. USSR-Sb.* **32** (1977), No. 3, 347–370).
3. On boundary and initial conditions in L_p, $p > 1$, of solutions of parabolic equations, *Mat. Sbornik*, **125** (1984), No. 4, 489–521. (Transl.: *Math. USSR-Sb.* **53** (1986), No. 2, 489–522).

Phillips R. S.

1. Dissipative operators and hyperbolic systems of partial differential equations. *Trans. Amer. Math. Soc.* **90** (1959), 193–254.

Privalov I. I.

1. *Boundary properties of single-valued analytic functions* (in Russian). Gostekhizdat, Moscow–Leningrad, 1950.

Pustylnik E. I.

1. On functions of a positive operator, *Mat. Sbornik*, **119** (1982), No. 1, 32–47. (Transl.: *Math. USSR-Sb.* **47** (1984), No. 1, 27–42).

Radyno Ya. V.

1. The space of vectors of exponential type (in Russian), *Dokl. Akad. Nauk BSSR* **27** (1983), No. 9, 791–793.

Reed M., Simon B.

1. *Methods of modern mathematical physics. 1. Functional analysis.* Acad. Press, 1972.
2. *Methods of modern mathematical physics. 2. Fourier Analysis. Self-adjointness.* Acad. Press, 1975.

Richter, P.

1. *Unitary representations of countable infinite dimensional Lie group.* Karl Marx Universität Leipzig, 1977, Mph 5.

Riesz F.

1. Über die Randwerte einer analytischen Funktion, *Math. Z.* **18** (1922), 87–95.

Riesz F., Sz.–Nagy B.

1. *Functional analysis,* F. Ungar, 1955 (transl. from the French).

Rofe–Beketov F. S.

1. The eigenfunction expansion of infinite systems of differential equations. In *"Funkts. Anal. i Prilozhen."* (in Russian). Trudy V. Vsesoyuzn. Konf. Funkts. Anal. i Primenen. Baku, 1961, pp. 230–237.
2. On self-adjoint extensions of differential operators in the space of vector-functions (in Russian), *Teor. Funktsii, Funkt. Anal. i Prilozh.* **8** (1969), 3–24.
3. Square-integrable solutions, self-adjoint extensions and spectrum of differential systems. In *"Differential Eq."* Proc. Int. Conf. Differential Eq., Uppsala, 1977, pp. 169–178.

Roitberg Ja. A.

1. On boundary values of generalized solutions of elliptic equations, *Mat. Sbornik,* **86** (1971), No. 2, 248–267. (Transl.: *Math. USSR-Sb.* **15** (1971), No. 2, 241–260).
2. On limiting values on surfaces, parallel to the boundary, of generalized solutions of elliptic equations., *Dokl. Akad. Nauk SSSR,* **238** (1978), No. 6, 1303–1306. (Transl.: *Soviet Math. Dokl.* **19** (1978), 229–233).
3. Existence of boundary values of generalized solutions of elliptic equations at boundary of region, *Sibirsk Mat. Zh.* **20** (1979), No. 2, 386–396. (Transl.: *Sib. Math. J.* **20** (1979), 276–284).

Rudin W.

1. *Functional analysis.* McGraw–Hill, 1973.

Sakhnovich L. A.
1. Investigation of a "triangle model" of non self-adjoint operators (in Russian), *Izv. Vuzov. Mat.* (1959), No. 4, 141–149.

Schatten R.
1. *Norm ideals in completely continuous operators.* Springer, 1960.

Schmidt E.
1. Auflösung der allgemeinen linearen Integralgleichung, *Math. Ann.* **64** (1907), 161–164.

Schwartz L.
1. *Théorie des distributions, I et II,* Hermann, 1950–1951.
2. Théorie des distributions à valeur vectorièlles, *Ann. Inst. Fourier,* **7** (1957), 1–141.

Shaw S.–Y.
1. Ergodic projections of continuous and discrete semigroups, *Proc. Amer. Math. Soc.* **78** (1980), No. 1, 69–76.

Shilov G. E.
1. On one quasi-analyticity problem (in Russian), *Dokl. Akad. Nauk SSSR,* **102** (1955), No. 5, 893–896.
2. *Mathematical analysis.* M.I.T., 1974 (transl. from the Russian).

Shishatsky S. P.
1. The correctness of boundary value problems for a second-order differential equation with self-adjoint negative operator in a Hilbert space, *Funkts. Anal. i Prilozh.* **2** (1968), No. 2, 81–86. (Transl.: *Funct. Anal. Appl.* **2** (1968), No. 2, 169–173).

Shtraus A. V.
1. Some problems of the extension theory of symmetric non-self-adjoint operators. In *"Trudy II Nauchn. Konf. Mat. Kafedr Ped. Inst. Povolzhia"* Vyp. 1 (in Russian). Kuibyshev Ped. Inst., Kuibyshev, 1962, pp. 121–124.
2. On the extensions and the characteristic function of a symmetric operator, *Izv. Akad. Nauk SSSR, Ser. Mat.* **32** (1968), No. 1, 186–207. (Transl.: *Math. USSR-Izv.* **2**, no. 1 (1968), 181–204).

Sobolev S. L.
1. Applications of functional analysis in mathematical physics. (transl. from the Russian), *Amer. Math. Soc.,* 1963.
2. *Partial differential equations of mathematical physics* (transl. from the Russian). Pergamon, 1964.
3. An example of correct boundary value problem for the string equation with data along the entire boundary. (in Russian), *Dokl. Akad. Nauk SSSR.* **109** (1956), No. 4, 707–709.

Sonin N. Ja.

1. On one definite integral, containing the numerical function $[x]$, *Izv. Varsh. Univ.*, 1885. (in Russian).

Sova M.

1. *Equations différentielles opérationnelles linéaires du second ordre à coefficients constants*. Academia, Praha, 1970.

Stein E. M.

1. *Singular integrals and differentiability properties of functions*. Princeton U.P., 1970.

Stein E. M., Weiss G.

1. *Introduction to Fourier analysis on Euclidean spaces*. Princeton U.P., 1971.

Storozh O. G.

1. Extensions of symmetric operators with distinct deficiency numbers, *Mat. Zametki*, **36** (1984), No. 5, 791–796. (Transl.: *Math. Notes* **36**, No. 5 (1984), 791–804.
2. A description of some classes of extensions of a non-negative operator (in Russian), *Dokl. Akad. Nauk Ukrain.SSR*, Ser. **A** (1987), No. 10, 15–17.

Sz.-Nagy B., Foias C.

1. *Harmonic analysis of operators on Hilbert space*. North–Holland, 1970 (transl. from the French).

Talyush M. O.

1. Typical structure of dissipative operators (in Russian), *Dokl. Akad. Nauk Ukrain.SSR*, Ser. **A** (1973), 993–996.

Tillman H. G.

1. Distributionen als Randverteilungen analytischer Funktionen, *Math. Z.* **76** (1961), 5–21.

Vainerman L. I.

1. On the existence of a distribution function of a second-order differential equation with operator coefficients (in Russian), *Dokl. Akad. Nauk Ukrain.SSR*. Ser. **A** (1972), No. 1, 3–5.
2. Self-adjoint boundary-value problems for strongly elliptic and hyperbolic equations of second order in Hilbert space, *Dokl. Akad. Nauk SSSR*, **218** (1974), No. 4, 345–348. (Transl.: *Soviet Math. Dokl.* **15** (1974), No. 5, 1391–1395).
3. Boundary-value problems for a second-order strongly elliptic equation in Hilbert space. In "*Spektraln. Teoria Oper. i Beckonechnomernyi Anal.*" (in Russian). Mat. Inst. Akad. Nauk Ukrain.SSR, Kiev, 1975, pp. 5–25.
4. A hyperbolic equation with degeneracy in Hilbert space, *Sibirsk Mat. Zh.* **18** (1977), No. 4, 736–746. (Transl.: *Sib. Math. J.* **18** (1977), No. 4, 520–526).
5. A degenerate elliptic equation with variable operator coefficients, *Ukrain. Mat.*

Zh. **31** (1979), No. 3, 247–255. (Transl.: *Ukr. Math. J.* **31** (1980), No. 3, 191–197).

6. Extensions of closed operators in Hilbert space, *Mat. Zametki*, **28** (1980), No. 6, 833–841. (Transl.: *Math. Notes* **28** (1980), No. 6, 871–875)..

Vainerman L. I., Gorbachuk M. L.

1. On boundary value problems for a second-order differential equation of hyperbolic type in a Hilbert space, *Dokl. Akad. Nauk SSSR*, **221** (1975), No. 4, 763–766. (Transl.: *Soviet Math. Dokl.* **16** (1975), No. 2, 401–405).

Vakhaniya N. N.

1. On one boundary value problem with data along the entire boundary for the hyperbolic system equivalent to the string equation (in Russian), *Dokl. Akad. Nauk SSSR*, **116** (1957), No. 6, 906–909.

Vinnishin Ja. F., Gorbachuk M. L.

1. Behaviour at infinity of solutions of operator differential equations, *Ukrain. Mat. Zh.* **35** (1983), No. 4, 489–493. (Transl.: *Ukr. Math. J.* **35** (1983), no. 4, 412–415).

Virabyan G. V.

1. The resolvent of an operator (in Russian), *Dokl. Akad. Nauk SSSR*, **151** (1963), No. 2, 258–261.

Vishik M. I.

1. On general boundary value problems for elliptic differential equations, *Trudy Moskov. Mat. Obšč.*, **1** (1952), 187–246. (Transl.: *Amer. Math. Soc. Transl.* (2) **24** (1963), 107–172).

Vladimirov V. S.

1. On constructing envelopes of holomorphy for domains of special type and their applications (in Russian), *Trudy Mat. Inta Steklov.*, **60** (1960), 101–144.
2. *Generalized functions in mathematical physics* (in Russian). "Nauka", Moscow, 1976.

Volk V. Ja.

1. Spectral resolution for a class of non-self-adjoint operators, *Dokl. Akad. Nauk SSSR*, **152** (1963), No. 2, 259–261. (Transl.: *Soviet Math. Dokl.* **4** (1963), No. 5, 1279–1281).

Vulis I. L., Solomyak M. Z.

1. Spectral asymptotics of degenerate elliptic operators, *Dokl. Akad. Nauk SSSR*, **207** (1972), No. 2, 262–265. (Transl.: *Soviet Maht. Dokl.* **13** (1976), No. 6, 1484–1488.
2. Spectral asymptotics of second-order degenerate ellliptic operators, *Izv. Akad. Nauk SSSR.* Ser. Mat. **38** (1974), No. 6, 1362–1392. (Transl.: *Soviet Math. Dokl.* **8** (1974), No. 6, 1343–1371).

Walter G.

1. Series de Fourier tipo de convergencia, *Bol. Ciencias Univ. Cat. Peru ANO II*, **40** (1969), No. 4, 40–53.

Weyl H.

1. Über die asymptotische Verteilung der Eigenwerte, *Gött. Nachr.* (1911), 110–117.
2. Das asymptotische Verteilungsgesetz der Eigenschwingungen eines beliebig gestalteten elastischen Körperns, *Rend. Circ. Mat. Palermo*, **39** (1915), 1–49.

Wolf F.

1. Operators in Banach space which admit a generalized spectral decomposition, *Nederl. Akad. Wetensch.* Proc. Ser A, **60** (1957), 302–311.

Yakubovich V. A.

1. A frequency theorem for the case in which the state and control spaces are Hilbert spaces, with an application to some problems of synthesis of optimal controls, II, *Sibirsk Mat. Zh.* **16** (1975), No. 5, 1081–1102. (Transl.: *Sib. Math. J.* **16** (1975), No. 5, 828–845).

Yosida K.

1. *Functional analysis,* Springer 1965.

Zhitarashu N. V.

1. On correct solvability in generalized functions of a class of nonlocal parabolic boundary value problems and conjunction problems (in Russian), *Izv. Akad. Nauk Mold.SSR, Ser. Fiz-Tekhn. Mat. Nauk* (1986), No. 3, 12–16.
2. On correct solvability of general model parabolic boundary value problems in spaces $H^s(\Omega)$ $(-\infty < s < \infty)$ (in Russian), *Izv. Akad. Nauk SSSR,* Ser. Mat. **51** (1987), No. 5, 962–993.

Ziatdinov F. Z.

1. On linear second-order differential operators in a Hilbert space of vector-functions with values in an abstract separable Hilbert space (in Russian), *Izv. Vuzov. Mat.* (1960), No. 4, 89–100.

Zygmund A.

1. *Trigonometrical series,* 1, Cambridge UP, 1959.

Subject Index

9 780792 303817